# STRUCTURES OF OPHIOLITES AND DYNAMICS OF OCEANIC LITHOSPHERE

**PETROLOGY AND STRUCTURAL GEOLOGY**

Volume 4

---

*Series Editor:*

A. NICOLAS

*Department of Earth Sciences,*
*University of Montpellier,*
*France*

*The titles published in this series are listed at the end of this volume.*

# STRUCTURES OF OPHIOLITES AND DYNAMICS OF OCEANIC LITHOSPHERE

*by*

A. NICOLAS

*Department of Earth Sciences,*
*University of Montpellier, France*

KLUWER ACADEMIC PUBLISHERS

DORDRECHT / BOSTON / LONDON

Library of Congress Cataloging in Publication Data

Nicolas, A. (Adolphe), 1936-
Structures of ophiolites and dynamics of oceanic lithosphere / A. Nicolas.
p. cm. -- (Petrology and structural geology)
Includes index.
ISBN 0-7923-0255-9
1. Ophiolites. 2. Submarine geology. I. Title. II. Series.
QE462.06N53 1989
552'.3--dc20 89-32244

ISBN 0-7923-0255-9

Published by Kluwer Academic Publishers,
P.O. Box 17, 3300 AA Dordrecht, The Netherlands.

Kluwer Academic Publishers incorporates
the publishing programmes of
D. Reidel, Martinus Nijhoff, Dr W. Junk and MTP Press.

Sold and distributed in the U.S.A. and Canada
by Kluwer Academic Publishers,
101 Philip Drive, Norwell, MA 02061, U.S.A.

In all other countries, sold and distributed
by Kluwer Academic Publishers Group,
P.O. Box 322, 3300 AH Dordrecht, The Netherlands.

*printed on acid free paper*



Printed in The Netherlands

# ACKNOWLEDGMENTS

This book of 'comparative ophiolitology' is based on a great number of structural observations which have been made in peridotite massifs and ophiolites by members and students of my group over the last 20 years. Several ideas developed here were first formulated by them. I wish to thank you all, Françoise Boudier, Jean-Luc Bouchez, Jean-Claude Mercier, Anne-Marie Boullier, Michel Darot, Yves Gueguen, Philippe Coisy, Marie Jackson, Alain Prinzhofer, Daniel Cassard, Jean-François Violette, Dominique Sécher, Jacques Girardeau, Maxime Misseri, François Cordellier, Mathilde Cannat, Georges Ceuleneer and Keith Benn.

Colleagues who participated directly in the preparation of this book were F. Boudier, C. Dupuy and M. Rabinowicz, who wrote or rewrote a few sections, and Y. Bottinga, P.J. Fox, K. Benn, P. Nehlig, H.G. Avé Lallemant, D. Mainprice, A. Prinzhofer, G. Ceuleneer, M. Leblanc, C.J. MacLeod, G.Suhr, R.S. Coe, C. Mével, J.F. Karson, J. Girardeau, M. Cannat, R.G. Coleman, J.C. Bodinier, and T. Juteau who reviewed parts of the manuscript. Thanks also go to all those who helped with the final draft of the manuscript. Finally, the book was prepared camera-ready thanks to the careful collaboration of M.C. Bréhier, A. Cossard, R. Bonnet, B. Allard and F. Pialoux with a special mention for E. Ball who composed and drew the illustrations and S. Fournier who coordinated the text.

Thanks to all of you,
Montpellier,
December 1st, 1988,
A. NICOLAS

# TABLE OF CONTENTS


*PART I - INTRODUCTION AND ANALYTICAL METHODS*

***Chapter 1. Introduction*** 3

1.1. Historical development of the ophiolite concept 3

1.2. Interest of ophiolite studies 6
1.2.1. Ophiolites as key for the study of oceanic lithosphere and asthenosphere 6
1.2.2. Ophiolites as markers of past plate tectonics 7

1.3. Scope and structure of the book 8

***Chapter 2. Analytical methods in ophiolites*** 9

2.1. Introduction 9

2.2. The oceanic reference frame 9
2.2.1. The ridge referential 9
2.2.2. Ridge side of origin of a given ophiolite 10

2.3. Structural studies in the hypovolcanic and volcanic sequences 12

2.4. Structural studies in the plutonic sequence 13
2.4.1. Principal structures 13
2.4.2. Viscous/plastic deformation 13
2.4.3. Importance of viscous flow 20

2.5. Structural studies in the ultramafic section 20
2.5.1. Homogeneity of mantle structures 20
2.5.2. Principal structures 20
2.5.3. Melt products : evidence for segregation/impregnation 23
2.5.4. Microstructures in peridotites and kinematic analysis 24
2.5.5. Microstructural imprint of asthenospheric and lithospheric flow 27
2.5.6. Serpentinization and low temperature deformations 29

2.6. Expected asthenospheric flow patterns 29

***PART II - TYPICAL OPHIOLITE COMPLEXES***

***Introduction*** 35

***Chapter 3. Oman ophiolite : the harzburgite ophiolite type*** 37

3.1. Introduction 37

3.2. Geological setting 40
3.2.1. Geodynamic setting 40
3.2.2. History of the Hawasina basin 40

3.3. General description of the ophiolite 46
3.3.1. Introduction 46
3.3.2. Mafic section 49
3.3.3. Ultramafic section 61
3.3.4. Metamorphic aureoles 67
3.3.5. High pressure metamorphism 68

3.4. Structure of the Oman ophiolite 68
3.4.1. Introduction-main structural events 68
3.4.2. Structures related to accretion at the spreading center 70
3.4.3. Structures related to oceanic thrusting and obduction 78

3.5. General interpretation of the Oman ophiolite 85
3.5.1. Introduction 85
3.5.2. Spreading rate estimation 85
3.5.3. Paleo-environment of origin and obduction history 85

***Chapter 4. Xigaze and Trinity ophiolites-Plagioclase lherzolite massifs : the lherzolite ophiolite type*** 91

4.1. Introduction 91

4.2. Xigaze ophiolite 91
4.2.1. Introduction 91
4.2.2. Geological setting 91
4.2.3. Description 94
4.2.4. Structural analysis 98
4.2.5. Geochemistry 100
4.2.6. Discussion 102

4.3. Trinity ophiolite 105
4.3.1. Introduction 105
4.3.2. Geological setting 105
4.3.3. Description 106
4.3.4. Structural analysis 111
4.3.5. Melt extraction and melt reaction 111
4.3.6. Petrology and geochemistry 112
4.3.7. Discussion 113

4.4. The western Alps ophiolites 115

4.5. The spinel-plagioclase lherzolite massifs 120
4.5.1. Petrological zonation 120
4.5.2. Structural zonation 120
4.5.3. Structure and geodynamic environment 126
4.5.4. Contact metamorphism and nature of metamorphosed formations 126

***Chapter 5. Bogota Peninsula and N.E. districts of New Caledonia - Wadi Tayin in Oman - Coastal Complex of Newfoundland : possible origin in transform faults*** 127

5.1. Introduction 127

5.2. Bogota Peninsula and N.E. ophiolitic districts of New-Caledonia 127
5.2.1. Introduction 127
5.2.2. Geological setting 129
5.2.3. Description of the Bogota Peninsula shear zone 132
5.2.4. Description of the Tiebaghi-Poum-Belep shear zone 135
5.2.5. Discussion 140

5.3. Coastal Complex of Newfoundland 142
5.3.1. Introduction 142
5.3.2. Geological setting 145
5.3.3. Description 145
5.3.4. Petrology and geochemistry 148
5.3.5. Interpretation 148

5.4. Wadi Tayin massif in Oman 153
5.4.1. Introduction 153
5.4.2. Structural description 153
5.4.3. Discussion 153

5.5. Conclusion 155
5.5.1. The diversity of ophiolitic transforms 155
5.5.2. Dike orientation in transform zones 157

***Chapter 6. Canyon Mountain ophiolite : possible origin in an island arc*** 159

6.1. Introduction 159

6.2. Geological setting 159

6.3. Description 161

6.4. Structural analysis 163

6.5. Petrology and geochemistry 164

6.6. Discussion 166
6.6.1. Specific characteristics of the Canyon Mountain ophiolite 166
6.6.2. Structural models 166
6.6.3. Geodynamic environment of origin 167

***PART III - ACTIVITY OF OCEANIC SPREADING CENTERS AND THE ORIGIN OF OPHIOLITES***

***Introduction*** 169

***Chapter 7. Melt generation and extraction in mantle diapirs*** 169

7.1. Introduction 171

7.2. Melt extraction from the asthenosphere 171
7.2.1. Conditions of adiabatic melting 171
7.2.2. Asthenospheric path and the meeting with lithospheric conditions 173
7.2.3. Depth of first melting 175
7.2.4. Maximum depth of melt extraction 177

7.3. Physical mechanisms of melt extraction 177
7.3.1. Fraction of stable melt in a peridotite 177
7.3.2. Melt extraction 178

7.4. A model of melt extraction by hydrofracturing 180
7.4.1. The model 180
7.4.2. Melt velocity within dikes, episodicity and duration of episodes of melt extraction 181
7.4.3. Geochemical implications 182

7.5. Melt extraction by solid compaction and melt percolation in transition zones of ophiolites 183

7.6. Focusing of melt extraction below oceanic ridges 184

***Chapter 8. The various ophiolites and their oceanic environments of origin*** 187

8.1. Introduction 187
8.2. Harzburgite and lherzolite types of ophiolites - Role of spreading rate 188
8.2.1. Distinctive characteristics 188
8.2.2. Harzburgite and lherzolite types of ophiolites and mantle partial melting 193
8.2.3. Harzburgite and lherzolite types of ophiolites and oceanic environments 193
8.3. Island-arc, back-arc or mid-ocean ophiolites 199
8.3.1. Geochemical characteristics 199
8.3.2. Other criteria 200

*Chapter 9 . Mantle flow, lithospheric accretion and segmentation of oceanic ridges* 203

9.1. Introduction 203

9.2. Mantle flow in the Oman ophiolite 205
9.2.1. Introduction 205
9.2.2. Homogeneous mantle flow away from the ridge-Relation with seismic anisotropy 207
9.2.3. Channeling of mantle flow along the ridge axis 207
9.2.4. Mantle flow in transform faults 209
9.2.5. Mantle flow in diapirs 209
9.2.6. Mantle flow patterns beneath the Oman paleo-ridge 209

9.3. Mantle flow in the Trinity ophiolite and lherzolite massifs 210

9.4. Mantle diapirism and ridge segmentation 211
9.4.1. Introduction 211
9.4.2. Models of mantle diapirs 213
9.4.3. Return flow and thickness of the buoyant layer 215
9.4.4. Spacing of mantle diapirs and ridge segmentation 215
9.4.5. Stability of mantle diapirs 220

*Chapter 10. Magmatic processes in the uppermost mantle at oceanic spreading centers* 223

10.1. Introduction 223

10.2. Principal characteristics of transition zones 223

10.3. Origin of the wehrlitic intrusions 224

10.4. Origin of dunites 225
10.4.1. Introduction 225
10.4.2. Field occurrences 226
10.4.3. Residual/magmatic origin 227
10.4.4. Mechanism of formation of residual dunites 233
10.4.5. Geochemical reequilibration 235
10.4.6. Conclusion as to the origin of dunites 236

10.5. Structure and origin of the chromite deposits 237
10.5.1. Introduction 237
10.5.2. Setting of chromite deposits 237
10.5.3. Structure of chromite deposits 238
10.5.4. Composition of chromite deposits 247
10.5.5. Origin of chromite deposits 251

*Chapter 11 - Generation of oceanic crust* 253

11.1. Introduction 253

11.2. Lithology of ophiolites and seismic structure of the oceanic crust 254

11.3. Serpentinite sea-floor in slow spreading environments and LOT 258
11.3.1.Abyssal and ophiolitic peridotites 258
11.3.2. Serpentinized peridotites as sea-floor 258
11.3.3. Nature of the Moho 260

11.4. The plutonic section and the problem of magma chambers 261
11.4.1. Introduction 261
11.4.2. Origin of the layering in the plutonic gabbro sequence 262
11.4.3. Magma chamber models 263
11.4.4. Conclusions about magma chamber models 268
11.4.5. Plating of gabbros and diking at the roof of magma chambers 270
11.4.6. Initiation of a new magma chamber 272

11.5. Sheeted dikes and volcanic units 274
11.5.1. Introduction 274
11.5.2. Generation at rifts and ridges 274
11.5.3. Structural evolution of the volcanic-hypovolcanic units 277

11.6. Crustal discontinuities in lherzolite type of ophiolite and episodic oceanic spreading 279
11.6.1. Variable basalt delivery along ridge-strike 279
11.6.2. Episodic basalt delivery in time 280

11.7. Early metamorphism in ophiolites and hydrothermal activity at oceanic ridges 281
11.7.1. Introduction 281
11.7.2. Metamorphic-zonation in ophiolites 282
11.7.3. Relationship with the sequence of hydrothermal alteration in oceanic crust 284

***PART IV - EMPLACEMENT OF OPHIOLITES TROUGH SPACE AND TIME***

***Chapter 12 - Ophiolites emplacement*** 289

12.1. Introduction 289

12.2. Ophiolite belts 292
12.2.1. Passive margins of continents 292
12.2.2. Active margins of continents 292
12.2.3. Collision belts 294

12.3. Emplacement-related features in ophiolites 294
12.3.1. Introduction 294
12.3.2. Ophiolite nappes and high temperature aureoles 294
12.3.3. Ophiolitic mélanges and high pressure metamorphism 300

12.4. Mechanisms of ophiolite emplacement 300
12.4.1. Introduction 300

12.4.2. Thrusting on passive continental margins 303
12.4.3. Upheaval in the accretionary prism of active margins 308

12.5. Summary and concluding remarks 310

***Chapter 13 - Ophiolite belts through time*** 313

13.1. Introduction : a reappraisal of ophiolites and their oceanic environments 313

13.2. Ophiolites generation and emplacement through time 313

13.3. Ophiolites as witness of pangean cycles 316

***Bibliography*** 321

***Index*** 359

# PART I

# INTRODUCTION AND ANALYTICAL METHODS

# Chapter 1

# INTRODUCTION

## *1.1. HISTORICAL DEVELOPMENT OF THE OPHIOLITE CONCEPT.*

Ophiolite, Greek for 'the snake stone', appears to have received its first written definition by Brongniart (1813) as a serpentine matrix containing various minerals. Later in 1821 and 1827, Brongniart determined that volcanic and gabbroic rocks were also present, associated with cherts, and he ascribed an igneous origin to the ophiolite. Amstutz (1980) gives an excellent exegesis of these early contributions and traces the further use of the term and concept of ophiolite. This concept had been forged in the western Alps and Apennines where, thanks to talented Italian geologists, in particular A. Sismonda, B. Gastaldi, V. Novarese and S. Franchi, the study on metamorphic ophiolites (the 'pietre verdi') has rapidly progressed.

At the turn of the century the association of radiolarite, diabase, gabbro (euphotide), and serpentinite-peridotite was clearly identified, even through their metamorphic transformations. In 1902, Franchi developed the hypothesis introduced earlier by Lotti (1886), of a submarine outflow to explain the 'pietre verdi' association, on the basis of the attribution of the variolites and metamorphic prasinites to an hypabyssal volcanism, also responsible for the formation of radiolarites. Thus, before the popular work of Steinmann in 1927, the various components constituting an ophiolite had been identified and its hypabyssal origin proposed. As recalled by Amstutz (1980), the so-called 'Steinmann trinity', which consists of the association of radiolarites, diabases and serpentinites, was more completely and better defined in these earlier works.

The subsequent studies on ophiolites, mainly conducted in the Mediterranean basin, were marked by the conflict between the tenants of a purely magmatic origin of the ultramafic section and those of an intrusive origin. As recalled by Coleman (1977), this conflict was reflected in America by the controversy about the genesis of peridotite massifs : Bowen (1927), drawing on his experimental work and on field reports on stratiform complexes, favored a crystal settling interpretation while Benson (1926) inspired by field work in peridotite massifs inserted in mountain belts (the 'Alpine peridotites'), proposed the interpretation of a plutonic intrusion.

The purely magmatic model for ophiolites was introduced by Routhier (1946, 1953) and Dubertret (1953) and further developed by Brunn (1956, 1960). A vast pouch of mafic magma was supposed to be extruded on the sea floor, presumably along deep faults (Kundig, 1956). Below a skin of chilled volcanics the gabbro-peridotite segregation was produced by crystal settling. As clearly presented by Vuagnat (1963) in his review of the various interpretations of ophiolites, just before the emergence of plate tectonics, it is impossible in the pouch model to balance the smaller mafic section with the dominant section of peridotites if both are supposed to be formed by differentiation of the basaltic melt. An alternative was to suppose that the parent magma was ultramafic (Hess, 1938; Bailey and McCalhen, 1953 ; Rittmann, 1960). Vuagnat evokes and discusses critically this interpretation and a few others, and finally gives his preference for the 'subcrustal' model. Best expressed by De Roever (1957), this model is also the closest to modern views. It suggests a mantle origin for peridotites which are tectonically intruded in the solid state through oceanic crust. The consanguinity of mafic and ultramafic formations is explained by the former being generated by partial melting due to decompression during ascent of the peridotites. This seems to be the first clear ascription of the ophiolite peridotite section to the mantle underlying the oceanic crust. Hess (1960) also proposed a similar origin for peridotites in Puerto Rico.

As recalled by Moores (1982), in 1960 two camps existed. The European camp, mainly represented by the French workers in the Mediterranean basin, had more or less adopted the 'pouch' model ; following Hess's opinion (1955) that the ophiolite concept unnecessarily confused the issue, the American camp was referring to 'alpine peridotites' and 'peridotite -gabbro' complexes, denying any connection with the associated volcanics as examplified by Thayer (1963). However, in this very paper, Thayer recognizes the affinity of the Canyon Mountain Complex of Oregon with Mediterranean complexes, in particular the Troodos.

The modern attitude of equating ophiolites with oceanic floor, perhaps too dogmatically, immediately followed the surge of the new concept of plate tectonics. It had been already proposed implicitly by De Roever (1957) and quite explicitly by Brunn (1959), who pointed to the remarkable analogy between ophiolites and the Mid-Atlantic ridge. By 1970, the two camps sitting on each side of this ridge had largely accepted this new interpretation of ophiolites (Hess, 1965 ; Gass, 1967, 1968 ; Moores, 1969 ; Peters, 1969 ; Reinhardt, 1969 ; Dercourt, 1970 ; Dewey and Bird, 1970, 1971 ; Moores and Vine, 1971 ; Bezzi and Piccardo, 1971 ; Coleman, 1971).

Interestingly, a large part of the community of marine geologists and geophysicists was reluctant to accept the ophiolite-oceanic floor analogy for reasons recalled by Moores (1982). These deal with differences in composition between dredged specimens and ophiolites and with the thickness of the mafic section of ophiolites found to be insufficient compared to the 6 km of ocean crust (Coleman, 1971). The magma chamber issue also separated the ophiolite community from that of marine geophysicists. A large magma chamber seemed necessary to account for generally well developed layered gabbros in some ophiolites (Greenbaum, 1972 ; Parrot and Ricou, 1976 ; Pallister and Hopson, 1981) whereas at first, no evidence of it was found below oceanic ridges. The ophiolite analogy became less suspicious for marine geophysicists when evidence for magma chambers, admittedly smaller than expected, was reported along portions of fast spreading ridges (chapter 11). Better knowledge of both the oceanic crust and of ophiolites, for instance the discovery of ophiolites with small and discontinuous magma chambers (chap. 4) and that of the complexity and variety of oceanic lithosphere (transform faults, back arc or fore arc basins, ...) which extend the range for possible comparisons with ophiolites, have reinforced the association of ophiolites with oceanic lithosphere and altogether rendered it richer.

However, there is in the comparison of ophiolites with oceanic lithosphere an instructive feedback effect which enlightens the problem of scientific amplification of certain concepts when they are studied by distinct communities. The seismic layering of oceanic crust was suggested by Hess (1962) to be a result of a serpentinized mantle beneath a carapace of basalts. In spite of the reluctance mentioned above, the ophiolite concept was penetrating the marine geophysicists community which progressively adopted the ophiolitic model for the oceanic crust, layer 2 being equated with volcanics and hypovolcanics and layer 3 with plutonics (Fox et al., 1973 ; Moores and Jackson, 1974 ; Cann, 1974). The ophiolite community, ignoring its own influence on the opinion of the other community, was thus reinforced in its conclusion that ophiolites could be equated with oceanic crust. As a result, very little attention was paid to the common dredging and drilling of peridotites and serpentinites specimens from the oceanic floor and on the other hand, to particular relations between peridotites and basalts or sediments in the ophiolitic environments (see § 4.4 and 11.3). It is now apparent that situations exist in the oceans where the ophiolite dogma does not apply and that this question needs further examination.

Acceptance of the oceanic lithosphere as the source of ophiolites was greatly helped in the sixties and early seventies by the evolution of ideas on the nature and origin of their ultramafic component. A first step was accomplished in 1960 thanks to T.P. Thayer pointing to critical differences between 'alpine-type' peridotites and those associated with stratiform complexes. Using petrofabric analysis, Andreatta (1934), Ernst (1935) and Turner (1942), had been able to recognize the effects of solid state deformation in various peridotites. Den Tex (1969) reintroduced this powerful tool and showed the tectonic nature of the structures in the 'alpine-type' or ophiolitic peridotites. A similar conclusion had been attained by Ragan (1963, 1967) for the Twin Sisters peridotite body. The reliability of structural and petrofabric studies was considerably increased by the first experimental results on deformation of olivine (Raleigh, 1968) and olivine aggregates (Carter and Avé Lallemant, 1970 ; Avé Lallemant and Carter, 1970 ; Nicolas et al., 1973). This opened the way to the kinematic analysis of plastic flow in peridotites (Chapter 2). Simultaneously, petrological studies on peridotite massifs (Green, 1964) and experimental data on phase equilibrium in peridotites (O'Hara, 1967) contributed results indicating a mantle origin of the various peridotite groups. Jackson and Thayer (1972) introduced a division of the 'alpine-type' peridotites group, whose tectonic-metamorphic fabric was by then widely accepted, into the lherzolite and the harzburgite subtypes. The harzburgite subtype, closely associated with ophiolites, was thought to represent the uppermost oceanic mantle and the less depleted lherzolite subtype, either the subcontinental mantle or the deeper oceanic mantle where partial melting is less severe (Nicolas and Jackson, 1972). American and European geologists meeting to consider ophiolites of the western United States (Anonymous, 1972), adopted a common definition of ophiolite, the 'Ophiolite-Manifesto', now largely accepted which states as follows : 'Ophiolite refers to a distinctive assemblage of mafic to ultramafic rocks. It should not be used as a rock name or as a petrologic unit in mapping. In a completely developed ophiolite, the rock types occur in the following sequence, starting from the bottom and working up :

- Ultramafic complex, consisting of variable proportions of harzburgite, lherzolite and dunite, usually with a metamorphic tectonic fabric (more or less serpentinized) ;
- Gabbroic complex, ordinarily with cumulus textures commonly containing cumulus peridotites and pyroxenites and usually less deformed than the ultramafic complex ;
- Mafic sheeted dike complex ;
- Mafic volcanic complex, commonly pillowed.
- Associated rock types include (1) an overlying sedimentary section typically including ribbon cherts, thin shale interbeds, and minor limestones ; (2) podiform bodies of chromite generally associated with dunite ; and (3) sodic felsic intrusive and extrusive rocks.

Faulted contacts between mappable units are common. Whole sections may be missing. An ophiolite may be incomplete, dismembered, or metamorphosed. Although ophiolite generally is interpreted to be oceanic crust and upper mantle, the use of the term should be independent of its supposed origin'.

A new major debate on ophiolites was prompted in 1973 by taking account of geochemical data. On the basis of major and trace elements distribution mainly in basalts, Miyashiro claimed that the Troodos ophiolite had been formed in an island arc environment and not along a mid-oceanic ridge. This interpretation was criticized both on the ground of the significance of major elements analysis and because of the contradiction between the expected absence of spreading in an island arc environment and that deduced for the dike swarm extension in the Troodos ophiolite. It was, however, a benchmark publication. At the same time, the analogy in the trace elements signature of ophiolitic and oceanic assemblages was emphasized by Allègre et al. (1973), and further supported by isotopic data, mainly the $^{143}Nd/^{144}Nd$ ratio which is insensitive to sea-water alteration ; these data were obtained from the ultramafic and mafic plutonic sections of ophiolites (Jacobsen and Wasserburg, 1979 ; McCulloch et al., 1980). New diagrams based on minor and trace elements mainly in the upper extrusives of ophiolites (Pearce and Cann, 1973) pointed to a departure in many ophiolites from mid-oceanic ridge basalt compositions, another possible candidate being a marginal basin ridge. As discussed in the next section, this problem has yet to be solved and is the focus of ongoing research.

## *1.2. INTEREST OF OPHIOLITE STUDIES*

Since the realization that ophiolites represent fragments of oceanic lithosphere, interest in their study has greatly increased. Two complementary investigative strategies contribute to our understanding. Thanks to the fact that ophiolite sections are representative of formations corresponding to deep parts of oceanic crust and upper mantle, which are normally inaccessible, one can use ophiolites to obtain critical information about these inaccessible oceanic levels. On the other hand, the understanding of ophiolites is greatly assisted by progress in oceanic lithosphere studies.

Apart from this thematic interest, ophiolite studies can also help to understand regional history. Assuming that the ophiolite-oceanic lithosphere relationship is established, one can use characteristics within ophiolites to reconstruct past tectonic environments.

### *1.2.1. Ophiolites as key for the study of oceanic lithosphere and asthenosphere.*

The in-situ study of oceanic lithosphere is limited by the tools available. Its general structure and activity at constructing, consuming and transform plate margins are deduced from indirect geophysical soundings, including acoustic imaging, seismology, magnetism, magneto-tellurics, gravimetry and heat flow measurements. In-situ specimens are obtained through dredging, drilling programs and sampling by deep sea submarines. Specimens thought to be representative of deep crust and mantle, including serpentinized peridotites and amphibolites, are routinely recovered from transform faults, but their relative position in an oceanic lithosphere sequence is not known. In 'normal' crust, the deepest drilling is at present the DSDP hole 504B, which provided more than 1.5 km of cored specimens and geophysical logging in pillows basalts and sheeted dikes of layer 2 (Becker et al., 1988). Drilling during the recent leg 118 on the flank of the South Indian Ridge has penetrated 500 m of gabbros and flasergabbros (leg 118 Shipbord sc. party, 1988).

As a result of these limitations, the best information on oceanic lithosphere is the shape, relief, structure and segmentation geometry of the crust created along the global system of ridges and the seismically determined velocity structure of the crust and upper mantle. The petrology and geochemistry of the basalts capping the oceanic crust and the hydrothermal circulation at ridges are also fairly well known. As will be explained in part III of this book, detailed studies along these trends are bringing important results and deductions such as the recognition of magma chambers below fast spreading ridges, of asthenosphere upwelling as deduced from ridge segmentation, and associated flow directions deduced from seismic anisotropy. However, we still have no direct knowledge of the deep parts of oceanic crust and of the top of the underlying mantle in the oceanic lithosphere. This is where study of ophiolites can be of great help, as complete and tectonically undismembered complexes offer continuous sections from the sedimentary cover overlying basalts down to around 10 km into the mantle section, below the mafic crust. For example, the postulated existence of magma chambers in ophiolites required to explain the layered gabbros has fostered the search for such structures below spreading centers, leading finally to their recognition below the East Pacific Rise.

A current debate in the ophiolite community is the search for specific oceanic environments of origin, a search relying mainly on the geochemistry of lavas. However, integrating all the available informations on ophiolites, including those on the ultramafic sections which have been somewhat neglected, one discovers a surprisingly large variety of ophiolites. Such a variety almost certainly reflects several distinct sites of origin in the oceans (i.e. mid-oceanic ridge, back-arc basin) and also other controlling parameters, of which the most important could be the spreading rate.

Thus, the diversity of oceanic situations is increasingly matched by a diversity in ophiolites, which presumably in the future will appear equally as rich. Bringing together the two subjects has therefore great potentials which so far have not been explored in a systematic way. This is the main object of the present book. One of the major difficulties in this enterprise derives from the different nature of the information obtained in ophiolites and in oceanic lithosphere. As already mentioned, the information obtained in the oceans concerns essentially the geochemistry of lavas and the large-scale geophysical structure. Dealing with the first point, the comparison with ophiolites is commonly obscured by the facts that in these, the volcanics may have been eroded or tectonically separated from the other sections and that, due to a possibly complex history, they may be altered and/or mixed with, or overlain by, the products of independent volcanic events (seamounts, island-arc volcanism...). Dealing with the structure, comparison between the oceanic lithosphere and ophiolites is made difficult by the differing scales of observation. The structures described in ophiolites and even the size of many ophiolite massifs are commonly below the scale of resolution of the geophysical methods used in marine exploration.

On the other hand, the recognition of magma chambers below ridges has required the use of fine scale seismological techniques and it approaches the limits of detection by such techniques. It should also be recalled that ophiolite sections do not sample deeper than around 15 km into the lithosphere.

This analysis points to the paramount interest of studying ophiolite complexes which are as little dismembered as possible and which extend over areas large enough to be able to make a comparison with the oceanic geophysical structures. Such complexes are unfortunately rare at the Earth surface ; this is why the Oman ophiolite, which is one such rare example, and is certainly the best studied so far, will be addressed with a special attention.

### *1.2.2. Ophiolites as markers of past plate tectonics.*

Assuming that the correlations between oceanic lithospheres and ophiolites in terms of structure and nature are well established and that the signs of lithospheric activity at constructive, consuming and transform plate margins are identified in ophiolites, it becomes possible by adequate studies in a given ophiolite to trace back the opening history of the ocean of origin, the aging and oceanic events which affected the corresponding oceanic lithosphere, and finally the closure and collision history, which are responsible for the ophiolite emplacement onto a continent. This paleogeographic evolution can be reset in its geographic framework if paleomagnetic studies are successfully associated to the geologic ones.

Dealing with the oceanic spreading stage, it will be shown that it becomes possible to determine the age of spreading, the orientation of the accreting ridge, the flank of the ridge from which the considered ophiolite is derived, presuming that it is formed at a ridge, and information about spreading rates and the nature of the oceanic environment of origin. Subsequently, during oceanic aging, the ophiolite may be modified by hydrothermal alteration or volcanism (e.g. seamounts, island arcs) whose identification would be valuable in tracing back the regional history.

Finally, the plate convergence episode, often culminating in continental collision, may be recorded in ophiolites. For instance, a common process of ophiolite emplacement onto continents begins by an intra-oceanic lithospheric thrusting related to oceanic convergence. The timing, presumed temperature, pressure conditions and kinematics of this thrusting event are registered in basal parts of the ophiolite and in its metamorphic aureole. In the study of a past subduction-collision event these pieces of information bring new and important constraints.

## *1.3. SCOPE AND STRUCTURE OF THE BOOK*

From the preceding section, it should be clear that the scope of this book is to establish a better comparison between ophiolites and the various oceanic environments, in order 1) to improve our understanding of the creation and evolution of oceanic lithosphere and 2) to be able to use ophiolites, in return, as markers of past plate tectonics history.

In essence, the approach in this book is structural. It is largely based on the structural mapping achieved in the author's group in some 15 ophiolite massifs over the last twenty years. This mapping has been mainly carried out in the ultramafic sections of the considered ophiolites. Although in ophiolites the ultramafic section is usually dominant in volume, most other studies have concentrated on the mafic section. Moreover, these studies have been mainly petrological and geochemical with a few remarkable exceptions.

This new approach of the problem of ophiolites, the methods of which are described in the next chapter, was aimed at retrieving in each ophiolite massif the overall structure and kinematic functioning, first at the oceanic spreading center of origin and, next, during emplacement onland. The chapters which follow the chapter on methodology include descriptions of a few selected ophiolite complexes for which the structural information is most complete. The choice of a limited number of ophiolites was determined by the desire to show their remarkable variety while nevertheless limiting this dominantly descriptive part to a reasonable length.

The contrasted typology of ophiolites which emerges from this review is related in the following part of the book to what seems to be, in oceanic spreading centers activity, the most important physical parameter : the spreading rate. The structural and kinematic picture of the functioning of oceanic ridges deduced from ophiolites is thus confronted with geological and geophysical data pertaining to fast and slow spreading environments.

The last part of the book deals with the subsequent history of an oceanic lithosphere bound to become an ophiolite by emplacement onto a continent.

It is believed that the most urgent problem to be solved in ophiolites as well as in oceanic lithosphere is that of obtaining a structural framework and some insight into the physical functioning of these systems. This should be based on systematic structural measurements and not on preconceived models, like the model of the great stratiform complexes evoked each time that a layered structure is observed. Physically, the static and cold-floored magma chambers of stratiform complexes have little in common with the moving and hot-floored chambers of ophiolites and oceanic ridges. At this early stage, petrological and geochemical data at hand are not discriminant enough to overrule structural data. It is hoped that the rapidly increasing amount of sophisticated geochemical data will improve and transcend the framework proposed in this book.

# Chapter 2

# ANALYTICAL METHODS IN OPHIOLITES

## *2.1. INTRODUCTION*

During the last several decades, an initial objective for geologists who accepted the ophiolite concept has been to identify a suite of rocks as an ophiolite. This objective has been realized by mapping of the main units and by petrological samplings to show that these units were parts of an ophiolite suite.

Detailed mapping and more systematic petrological and geochemical studies were fostered in the 1970's when it was realized that ophiolites could be derived from various oceanic environments and that they could reflect this variety. The present situation is still far from satisfactory. For example in most massifs the ultramafic section, which usually has the largest extension in the field, is only delineated and the crustal unit known only through cross-sections ; mapping at the scale appropriate to reveal the size and the shape of magma chambers is only now beginning. Petrological and geochemical sampling often reflects this lack of precise field knowledge. Consequently, one must stress the importance of a systematic detailed mapping in the ophiolite complexes which, if they have not been badly dismembered, deserve such studies.

In this chapter, the typical structures in each unit of an ophiolite will be considered and we will discuss how they contribute to defining the framework of origin of the ophiolite in its oceanic environment. In this context, the structural and kinematic data obtained in the ultramafic section of the ophiolite is a major contributor to the understanding of the origin and history of the ophiolite under consideration. The methods and techniques of structural and kinematic analysis in mantle peridotites must therefore be presented here.

## *2.2. THE OCEANIC REFERENCE FRAME*

In ophiolite complexes where the internal continuity between the various units has been preserved or can be restored, it is possible to reorient all the structural features into their presumed orientation at the oceanic site where the ophiolite originated. For a given ophiolite, it may be also important to identify from which side of the ridge it is derived.

### *2.2.1. The ridge referential*

The reference frame attached to an oceanic spreading center is defined by the horizontal plane and the ridge trend. The seismic layering of the oceanic lithosphere is generally horizontal and the Moho can be accepted as a horizontal surface, at least at the scale of resolution of seismic data and for medium to fast spreading ridges (§ 2.6). The ridge is a tensional system, and thus its trend can be recognized by considering the average orientation of extensional structures (§.2.3).

In ophiolites, the paleohorizontal is accordingly defined as the boundary between the base of the mafic unit, generally composed of layered gabbros, and the ultramafic unit, generally composed of tectonic peridotites. This boundary, which is commonly sharp, should correspond with the seismological Moho in the oceanic lithosphere and will be considered as such in this book. The ridge trend is taken as being parallel to the diabase dike swarm once the Moho has been rotated to the horizontal (§ 2.4). If part of an ophiolite is supposed to represent a transform fault (chapter 5), after rotation of the Moho to the horizontal the sheared domains in the mantle and/or crust sections must evidently be steeply dipping and at a high angle to the dike swarm azimuth outside this domain. Once established from the ophiolite structure, the oceanic frame can be used for paleogeographic reconstructions, but only if one has paleomagnetic data to account for possible rotations with respect to geographical coordinates. The procedure is summarized in figure 2.1. In an ophiolite where numerous structural data have been measured, it is necessary to operate the rotations on average values for each set of data. This is achieved by computing the best axis of point concentrations and the best pole of girdle concentrations for field measurements. This treatment can include confidence cones. These technical procedures are detailed in Nicolas and Poirier (1976, chapter 8).

Many difficulties complicate the measurement of the paleomoho as defined above :

- the peridotite-gabbro limit is often a serpentine zone of low-temperature shearing or thrusting because of the contrasted rheology between the formations above and below this limit ;
- in a few massifs like Bay-of-Islands the high-temperature plastic deformation has been imprinted in the lowermost gabbros (Casey and Karson, 1981 ; Girardeau and Nicolas, 1981). In this case the paleomoho may have been rotated by tectonic transposition;
- the postulate that this surface was horizontal in the ocean of origin may not be true at the scale of observation in ophiolites, which is much reduced compared to the resolution of seismic data.

On the other hand, in the Oman ophiolite where the outcrop conditions are exceptionally good it is *always* observed, except when faulted or deformed, that this peridotite-gabbro boundary is parallel to the layering plane in the basal layered gabbros. If this observation can be extended to the other similar ophiolites, the measurement of this layering plane would provide the paleo horizontal reference.

Although the attitude of the magmatic layering in the lowest layered gabbros gives a horizontal referential, such a relationship is probably not true for the highest gabbros. From systematic measurements made in Oman and Bay of Islands, it is concluded that the layering dip increases upsection and can become vertical (§ 3.3.2 ; Casey and Karson, 1981).

### *2.2.2. Ridge side of origin of a given ophiolite*

Several criteria have been used in ophiolite studies to determine on which flank of the spreading center the ophiolite has been created. They include the facing direction of chilled margins in the sheeted diabase dike unit, the analysis of sedimentary figures in layered gabbros, the direction of dip of layering in these gabbros and that of the constructed flow plane in the underlying peridotites, the shear sense in peridotites and the overall geometry of a large ophiolite system including in particular transform faults.

Starting from the observation that commonly diabase dikes of the sheeted dike unit intrude one into the other, if one assumes that the intrusion occurs recurrently along the same weaker zone, the dikes drifting in one sense should have chilled margins facing away from the spreading axis (Kidd and Cann, 1974 ; Kidd, 1977).

Statistical observations on the facing direction of the chilled margins confirm that dike intrusion occurs along zones only a few tens of meters wide (Kidd, 1977 ; Pallister, 1981; Rosencrantz, 1983). The bias in statistics on chilled margins is small. This may be due to the fact that feeding zones can jump, and thus they do not necessarily coincide

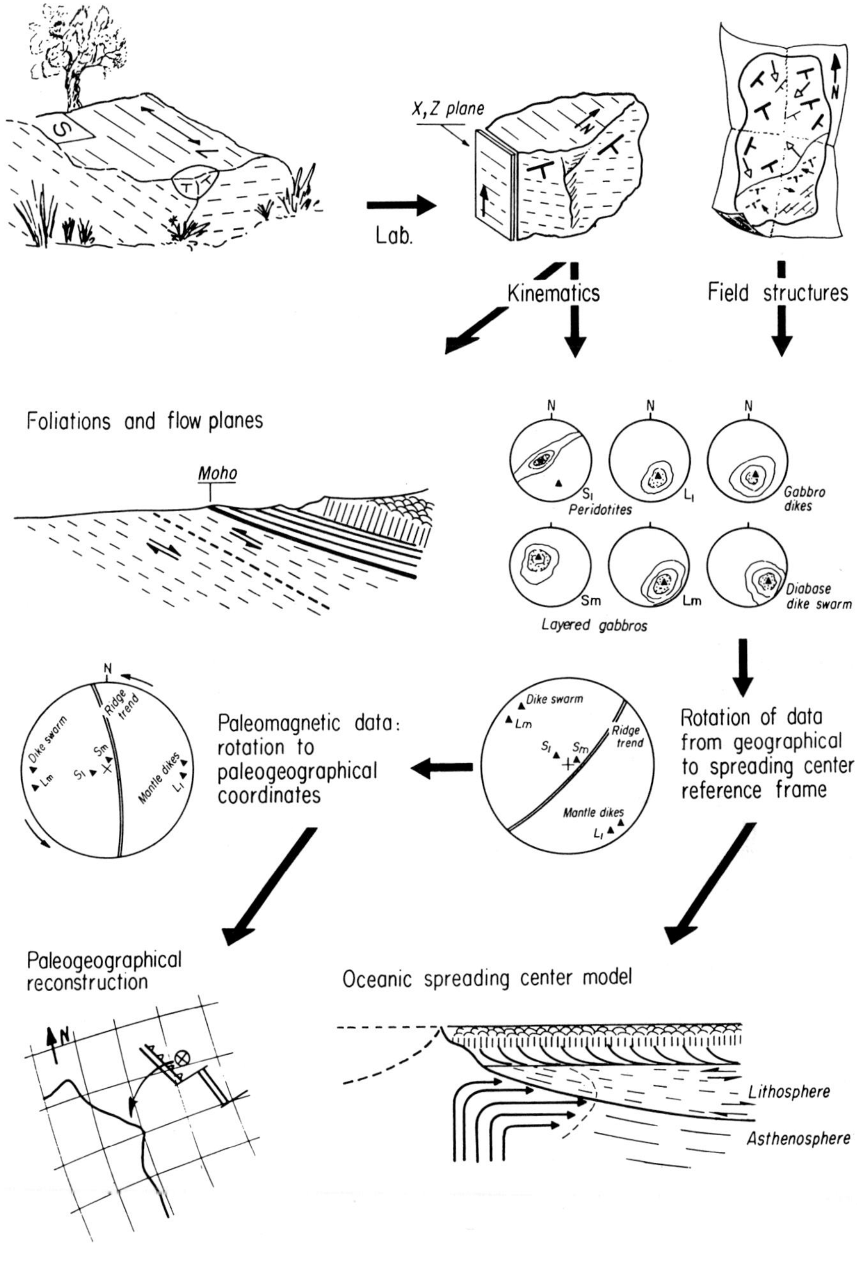


Fig. 2.1. General procedure for the structural analysis of ophiolite complexes.

strictly with the ridge axis. If this is true, then a large number of measurements are required to test the side of origin for the ophiolite in question.

Overturning sense of slumps, sense of movement on normal faults, cross bedding structures in layered gabbros of the plutonic sequence can be used as criteria to indicate the slope of the magma chamber floor, which is assumed to dip toward the chamber axis. However, Casey and Karson's (1981) observations in Bay of Islands have revealed contradictory relationships within small areas and in Oman ophiolites it has been shown that magmatic sedimentation structures are rare and can be readily mistaken for magmatic flow structures (§ 2.4.3.). The sense of motion deduced from these markers in Oman relate them to the shear sense of magmatic flow.

The direction of the upward rotation of the layering and magmatic foliation in the plutonic gabbro sequence can also be considered. Most authors envisaging this rotation have proposed that the layering in the upper gabbros dips toward the chamber axis (Cann, 1974 ; Dewey and Kidd, 1977 ; Casey and Karson, 1981 ; Pallister and Hopson, 1981 ; Smewing, 1981 ; Nicolas and Violette, 1982) ; on the contrary Nicolas et al. (1988) propose that the dip is away from the chamber axis (fig. 11.8). These opposite conclusions can result from two causes. First, measuring the rotation of the layering upsection may be problematic because one cannot exclude that a measured rotation has not been induced by a subsequent tectonic event : in a flat-lying massif the lowest and highest layered gabbros, which are vertically a few kilometers apart, will be separated by horizontal distances so large that it is difficult to exclude tectonic rotations; in a tilted massif, their horizontal distance is reduced in proportion to the tilting, but simultaneously tectonic rotations become probable. Second, these interpretations rely on independent criteria used to locate the ridge axis. The first group of authors used the facing direction of chilled margins in the diabase dikes of the dike swarm ; Casey and Karson (1981) also considered the overall geometry of the Bay of Islands Complex, including the Coastal Complex transform (fig.5.19). In Oman, Nicolas et al. (1988) derive the opposite conclusion on the basis of the analysis of shear flow in the underlying mantle (see below). It seems wise to conclude that it is premature to try to derive the side of origin of a given ophiolite with respect to the ridge axis from sense of the upsection rotation of the layering in the plutonic section.

The last criteria deal with the direction of dip of the constructed flow plane in the tectonic peridotites with respect to the Moho and with the sense of shear in these formations. The flow planes in the asthenospheric mantle flowing away from a ridge axis are expected to be tangential to the overlying lithosphere surface to which they are progressively incorporated on cooling (§ 2.6). Thus, the side of the ridge can be deduced from the dip of the frozen flow planes (fig. 2.10). More speculative is the idea of using the shear sense of the flowing asthenosphere because it depends on models of mantle flow pattern below ridges (§ 9.2). In both cases, it is necessary to consider the peridotite structures at a depth greater than 500 m below the Moho because above, a shear sense inversion is usually found (fig. 2.2).

## *2.3. STRUCTURAL STUDIES IN THE HYPOVOLCANIC AND VOLCANIC SEQUENCES*

The most important structural measurement in the diabase dike swarm is its average trend supposed to coincide with the ridge azimuth (Gudmundsson, 1983 ; Helgason and Zentilli, 1985 ; Karson, 1987 ; Auzende et al., in press). Otherwise, models have been developed predicting the rotation of the volcanic flows and the dike swarm attitudes (§ 11.5.3). Such rotations result either from progressive isostatic subsidence at distance from the axis in response to the volcanic discharge close to the axis or from block tilting along listric faults. These models result in opposite dips (fig. 11. 15). Thus, systematic dip measurements of volcanic flow planes and dikes could permit predicting the side of origin with respect to the ridge provided the process responsible for rotation is identified. This is rendered hazardous by subsequent tectonic rotations which are difficult to estimate.

Finally, mapping the magmatic flow direction in diabase dikes, using the anisotropy of magmatic susceptibility, may provide a means of locating the magmatic feeding centers along the paleo-ridge of origin (§ 11.5.2).

## 2.4. STRUCTURAL STUDIES IN THE PLUTONIC SEQUENCE

### 2.4.1. Principal structures

The plutonic part of the mafic section is structurally composed of layered, foliated and isotropic gabbroic rocks. Coherent structures are measured only in the two former types where they should be systematically mapped. Two distinct types of layering have been described in the plutonic section of ophiolitic sequences. The most conspicuous character defining this layering is the modal composition. Variations in grain size, texture and mineral chemistry of individual layers have been clearly reported in ophiolites. For this reason we will retain from Irvine's (1982) terminology of layered intrusions, the terms 'isomodal' and 'modally graded' layering. These terms are equivalent respectively to the 'uniform' and the 'stratified' layerings used by Casey and Karson (1981) in their study of Bay of Islands ophiolites. Isomodal layering is characterized by uniform proportions of minerals (plate 2.1b) but does not consider possible variations in other properties. Modally graded layering is characterized by a progressive change in mineral proportions ; the commonly observed graded bedding belongs to the category of modally graded layering (plate 2.1a). In layered gabbros of Bay of Islands, Oman and Cyprus, the isomodal layering is the dominant type (> 80% in Bay of Islands) and commonly layered sequences consist exclusively of successive isomodal layers. In Oman and in Cyprus the modally graded layering is mainly observed in basal gabbros. The magmatic foliation plane, also called the 'lamination plane', is defined by the preferred orientation of tabular undeformed minerals (plate 2.1e). It is commonly accompanied by the development of a mineral lineation (plate 2.1f) often parallel to the axis of magmatic folds, and making various angles with normal faults and magmatic shear zones (plate 2.1h). Magmatic foliation can appear in both gabbros which display a compositional layering and those which do not ; in the former case, the foliation and the layering planes are generally parallel or at a small angle. The angle between these two planes may indicate the sense of magmatic flow (Benn and Allard, 1989).

### 2.4.2. Viscous/plastic deformation

As shown by the study of the transition zone in Oman (§ 3.3.3), there is a sharp transition between the gabbro lenses from this zone which have been deformed in the solid state by plastic flow, and the overlying gabbros from the plutonic sections which have been deformed in the magmatic state by viscous flow. Distinguishing in a gabbro between these two modes of flow is possible by considering the substructures and the fabrics developed in minerals. The high temperature plastic deformation considered here induces in olivine a typical substructure (§ 2.5.4) which is generally absent in magmatic olivine, where a tight substructure typical of lower temperatures deformation may however be induced locally by a subsequent mild deformation. In plagioclase, plastic deformation results in mechanical twins which are thin, sinuous and pinching at

***Plate 2.1.***

***Plate 2.2.***

***Plate 2.3.***

***Plate 2.1. Magmatic structures in Oman layered gabbros***

a- Modally graded layering on the scale of 15 cm marked by an olivine/plagioclase ratio diminishing upward (photograph G. Ceuleneer)
b- Isomodal or uniform layering marked by wehrlite black layers alternating with gabbro light layers. Same outcrop as 21c. Mark is 10 cm long.
c- Modally graded layers at the base of the outcrop and isomodal layers above. The lenticular shape and the sharp boundaries of the wehrlite layers suggest that they were injected as sills.
d- Isomodal layering defined by anorthosite lenses in an otherwise poorly layered gabbro. These anorthosite lenses observed at any level within modally layered gabbros are regarded as flow-induced mineral segregations ; lens cap is 5 cm in diameter.
e- Magmatic foliation in poorly layered gabbros
f- Magmatic lineation within foliation plane, marked by mineral aggregates. Mark is 10 cm long.
g- Magmatic folds in a layered gabbro. The magmatic foliation is axial plane of these folds and the mineral lineation is parallel to their axis. They are analogous to sheath folds described in metamorphic rocks.
h- Magmatic sinistral shear zones in a layered gabbro. Mark is 10 cm long.

***Plate 2.2.***

***Magmatic structures in Oman layered gabbros***

a- Conjugate magmatic shear zones in a layered gabbro. Mark is 10 cm long.
b- Magmatic flow structures in a layered gabbro : isoclinal sheath folds near lower right corner, sinistral shear zones near upper left corner. The thinning and streaky aspect of the layering is ascribed to very large magmatic flow. Mark in the center is 10 cm long.
c- Boudinaged wehrlitic layers (black and weathered out) in a magmatically foliated gabbro.
d- Magmatic dispersal of anorthositic gabbro lenses within a wehrlitic gabbro.

***Partial melting and dike intrusions in peridotites***

e- Incipient melting in Lanzo plagioclase lherzolites, producing gabbroic lenses with depleted margins. The lenses are oblique to the foliation (EW on photograph) and parallel to the flow plane identified by fabric analysis.
f- Indigeneous gabbro dikelet in Lanzo plagioclase lherzolites. Note the irregular aspect of the dike and its dunitic margins (smoother relief with respect to the more rugged surrounding lherzolites).
g and h- Intrusive gabbro dikes with clear-cut walls and no contact reactions in Oman harzburgites. g- Tension fracturing (comb structure normal to dike walls). h- Brecciation of a cooling harzburgite ascribed to melt overpressure (~100 m below Moho, filling with microgabbro). For scale, tape recorder is 15 cm long.

***Plate 2.3. Plastic deformation structures in peridotites***

a- Ariegite compositional layering with oblique foliation (parallel to felter pen), in the hinge area of the km-sized fold of the Lanzo plagioclase herzolite massif (fig. 4.23).
b- Orthopyroxenite and dunite compositional layering in Antalya harzburgites (Turkey) ; foliation parallel to layering.
c- Boudinaged pyroxenite and gabbro layers in a mylonitic harzburgite from Oman. Marker is 10 cm long.
d- Trace of foliation in an exposure normal to foliation and parallel to lineation ; plagioclase lherzolites from Liguria (Italy).
e- Aggregate lineation in a high-T facies of Lanzo plagioclase lherzolites (photograph F. Boudier).
f- Lamellar enstatite lineation in a mylonitic garnet lherzolite from the NE margin of the Sierra Bermeja

massif. Slip-induced elongation in such pyroxenes can attain a 100/1 ratio.
g- Rounded hinge of an isoclinal fold in the websterite layering of Lanzo plagioclase lherzolites. The aggregate mineral lineation is parallel to the fold hinge.
h- Isoclinal folds on a meter scale in the hinge zone of the km-sized fold of Lanzo plagioclase lherzolite massif (fig. 4.23).

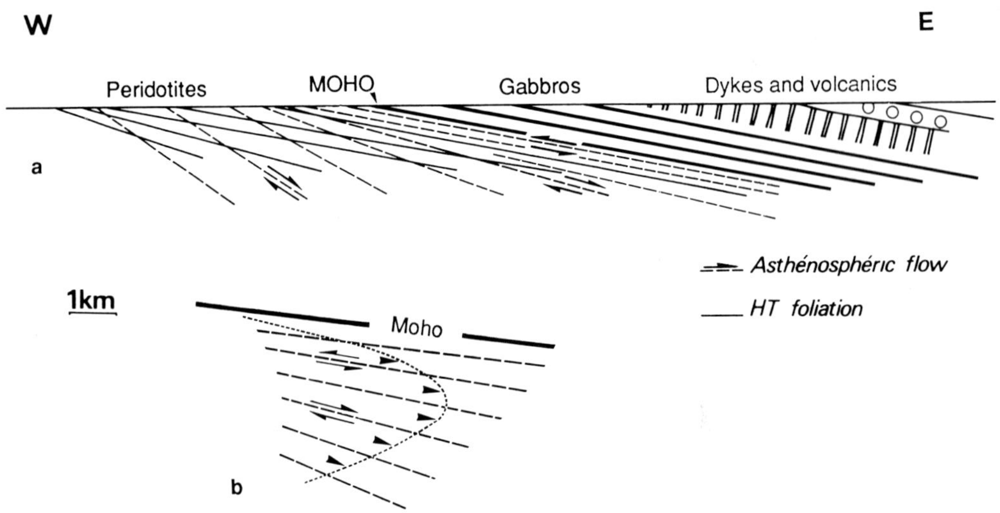


Fig. 2.2. a) Measured foliations and constructed flow planes attitudes in the peridotite section of the Hilti massif in Oman. The spacing of the flow planes reflects the shear strain (Ceuleneer et al., 1988). b) Scheme showing the flow inversion just below the Moho ; dotted line : velocity gradient.

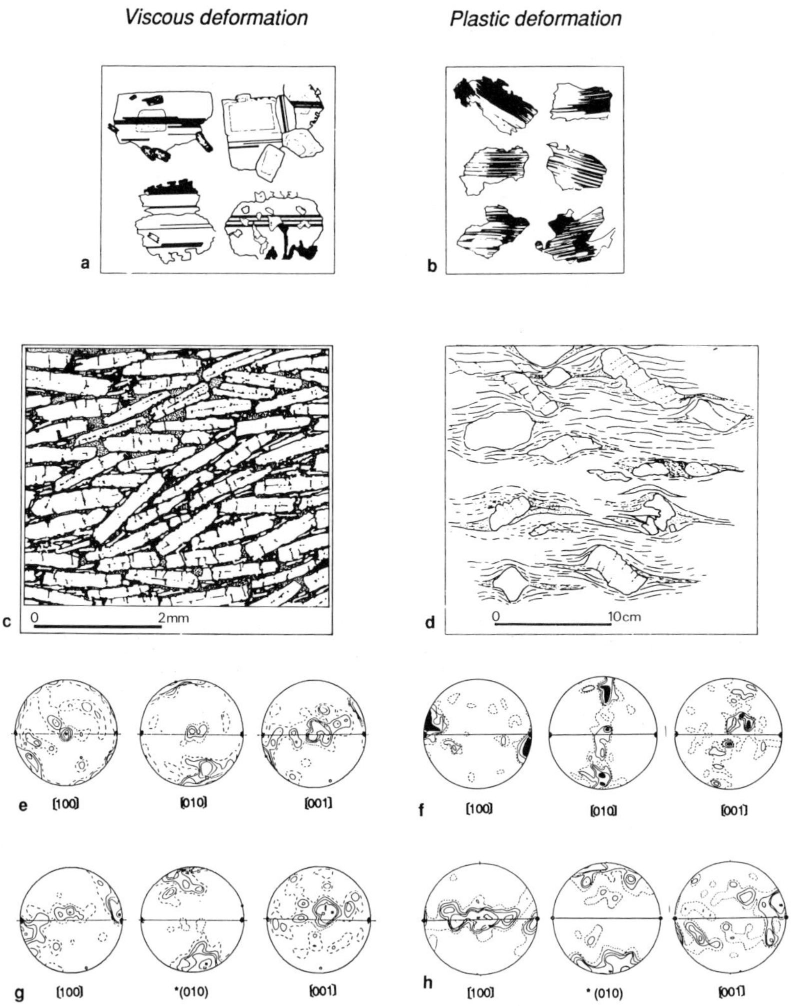


Fig. 2.3. Criteria used to distinguish solid state (plastic) deformation and magmatic (viscous) deformation. a) Few and rectilinear magmatic growth twins in plagioclase. b) Numerous, narrow and curved deformation twins in plagioclase. Note the tapering at crystal boundaries. c) Magmatic foliation, the plagioclase phenocrysts are euhedral and undeformed. d) Plastic foliation, the plagioclase porphyroclasts are augen-shaped and internally deformed. e) and f) Olivine fabrics, related to a vertical E-W foliation (straight line) and to an E-W lineation (dots), respectively in viscous and plastic deformation. g) and h) Plagioclase fabrics (*(010) is pole of (010) plane), same referential, respectively in viscous and plastic deformation. Note in these fabrics the stronger plastic maxima and the near coincidence of slip directions ([100] in olivine and [001] in plagioclase) with the lineation. (100 crystal measurements ; equal area projection in lower hemisphere ; contours 1,2,4,6 %). (a, b, c, d, after Nicolas, 1987 ; e, g, after Benn and Allard (1988); f, after Nicolas, 1986b ; h, after Ji and Mainprice, 1988).

the crystal boundary in contrast with magmatic growth twins which are wider and straight throughout the crystal (fig. 2.3). The fabrics also contrast (fig. 2.3).

### *2.4.3. Importance of viscous flow*

Structural and petrofabric studies conducted in layered and foliated gabbros of Oman and Cyprus (Benn et al., 1988 ; Nicolas et al., 1988a ; Benn and Allard, 1989) point to the importance of large viscous flow in the shaping of these rocks. The field evidence for this large flow is in the stretching of layers resulting in boudinage (plate 2.2c) and extreme dispersion of layers (plate 2.2d), in the development of magmatic shear bands and of isoclinal folds belonging to the category of sheath folds (plate 2.1g). In these folds the axis has been rotated into parallelism with the mineral lineation as a result of very large strain (Cobbold and Quinquis, 1978). Such folds should not be mistaken for magmatic slumps in which the mineral lineation should be only exceptionally parallel to the fold axis. In the same way, the magmatic flow rotates a layer, whatever its initial orientation into near-parallelism with the flow direction. This process is well known in plastic deformation as 'tectonic transposition' (Nicolas, 1987). The strong mineral shape fabrics obtained in these rocks (figs. 2.3e and g) are also incompatible with static settling or growth of the crystals which produce at best a weak fabric. Although this is not quantified, the fabrics obtained require a large flow.

## *2.5. STRUCTURAL STUDIES IN THE ULTRAMAFIC SECTION*

### *2.5.1. Homogeneity of mantle structures*

In the ophiolite massifs where peridotites are on average little or moderately serpentinized, say with less than 40-50 % serpentine, the ultramafic section tends to behave like a homogeneous block. Although this section is locally split by serpentinite shear zones, its internal structures display commonly over large areas a homogeneous or a progressively changing pattern. Domains of incoherent structures are indicated by a denser network of serpentine shear zones and breccias. This conclusion is important because it implies that a coherent mantle structure can be worked out in many ophiolites. This has been sometimes questioned for instance in northern Oman and in Xigaze (Tibet), on the basis of structural traverses along one or two major valleys which happened to be following large serpentinite bands. In both cases, a more complete mapping has demonstrated the overall coherence of the structures (figs. 3.8 and 4.6). A striking illustration of this coherence is given by the case of the Massif du Sud in New Caledonia, where the peridotite structures are remarkably homogeneous over 6000 $km^2$ (fig. 5.2), although the nappe is now at most 3 km thick. Such a behavior is ascribed to the fact that fresh peridotites below around 700°C are specially rigid, behaving elastically (Watts et al., 1980 ; Calmant, 1987) and yielding only along serpentinized fractures, later preferentially followed by valleys and by geologists.

### *2.5.2. Principal structures*

Mapping the ultramafic section of ophiolites is a dull task because the petrological differences can hardly be detected in the field and the structures are usually not conspicuous. The structure most easily recognized is *compositional layering,* which over the peridotite background is composed of parallel mineralogical segregations being either well-defined (plate 2.3a) or more diffuse (plate 2.3b, c). The layer thickness is variable, usually in the 1-5 cm range. Layering in lherzolites is dominantly formed by websterites and ariegites (Lensch, 1976), also called 'Cr-diopside' and 'Al-augite' pyroxenites by Wilshire and Shervais (1975). The websterites usually do not exceed 10 cm in thickness and the ariegites 100 cm. In harzburgites, the layering is more commonly formed by orthopyroxenites and dunites with rare chromitite layers. In contrast to dikes, layers are strictly parallel to one another.

The tectonic structure in mantle peridotites is characterized by a *foliation* plane which is the plane of mineral flattening (X,Y plane of the deformation ellipsoid, Nicolas and Poirier, 1976) (plate 2.3d) and by a mineral or mineral aggregate *lineation* which is usually parallel to X, the longest axis of the deformation ellipsoid (plate 2.3e, f). The foliation and the lineation attitudes are defined by the shape of pyroxenes and spinel (or feldspar when present) ; in difficult cases, they are determined in the laboratory on oriented specimens after repeated operations of bleaching by diluted HCl and saw-sectioning. All microstructure observations are carried in the X, Z plane. Procedures are described in detail by Nicolas and Poirier (1976).

In this volume and in recent publications (Nicolas et al., 1988), foliations and lineations are represented in maps by their trajectories for practical reasons. Figure 5.19 is an illustration of how trajectories relate to individual measurements. In areas of flat-lying foliations, the foliation trajectory map is a poor representation and, in contrast, the lineation map is well suited (compare for instance figs. 5.2 a and b) ; the opposite situation prevails when foliations are steep, in which case a lineation map becomes of little use (fig. 4.21).

The compositional layering is usually parallel to the foliation except in areas of folding (plate 2.3a). There, the foliation is parallel to the axial plane of the folds and the mineral lineation, to the fold axis, thus having an orientation close to the plastic flow direction (Nicolas and Boudier, 1975) (plate 2.3g). In fold hinges, the thickness of the layering can be increased by several orders of magnitude (plate 2.3 g and h).

Other compositional differentiates in peridotites are dikes and veins which can grade into irregular bodies. Boudier and Nicolas (1972, 1977), and Nicolas and Jackson (1982) have distinguished between 'in situ' or 'indigenous' and 'intrusive' dikes. These terms are somewhat ambiguous because they imply that the first category is *entirely* formed by local melting which is true only in a special case (see below).

*Indigenous dikes, dunite veins and bodies* - The indigenous dikes are pyroxenites and gabbros that display an irregular contact with their walls. The contact zones are composed of symmetrical screens of depleted dunite on each side of the dikes (plates 2.3f and 4.1c, d). Discordant dunite veins and bands, which in harzburgite massifs are entirely sterile or contain only relics of mafic dikes are related to these indigenous dikes (§10.4.2). Incorporating a contribution of melt from the surrounding peridotites, *these dikes were injected into a melting peridotite.*

*Intrusive dikes* - They are composed of pyroxenites and gabbros with sharp contacts, non-depleted walls, and internal magmatic structures (plate 2.2g, h). This indicates that in contrast with indigenous dikes *the magma was injected when the peridotite was well below its solidus* and could not significantly react with it. Texturally, intrusive gabbro dikes grade into finer grained diabase dikes when the temperature of the peridotite wall attains ~450°C (§ 11.4.4).

The thermal sequence from indigenous to intrusive dikes is confirmed by a deformational history showing that the indigenous dikes and dunite veins are commonly foliated and folded whereas the intrusive dikes are less or not deformed and transect the former dikes and veins.

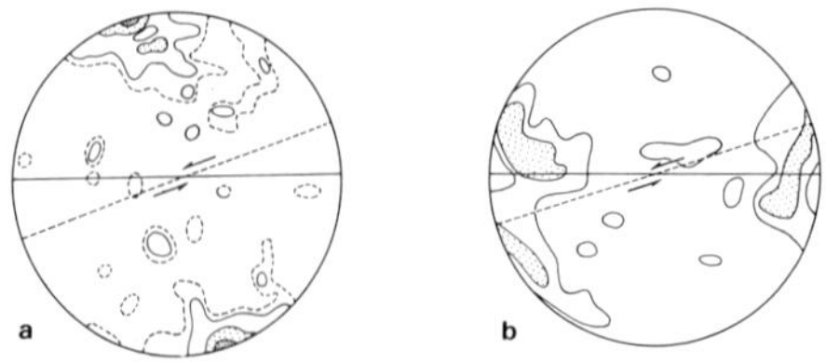


Fig. 2.4. Relation between gabbro dike and lens orientation and structural reference system in Lanzo lherzolite massif. a) Feldspathic lenses and veinlets, 81 measurements ; contours at approximately 1, 2, 4, 8 per cent. b) Gabbro dikes, 189 measurements ; contours at approximately 1, 2 per cent. Lower hemisphere projection, equal area net ; structural reference system : foliation vertical E-W, mineral lineation horizontal EW. Dashed line : trace of the shear plane. Density contours of poles to dikes, per 0.45 % area (Boudier and Nicolas, 1972).

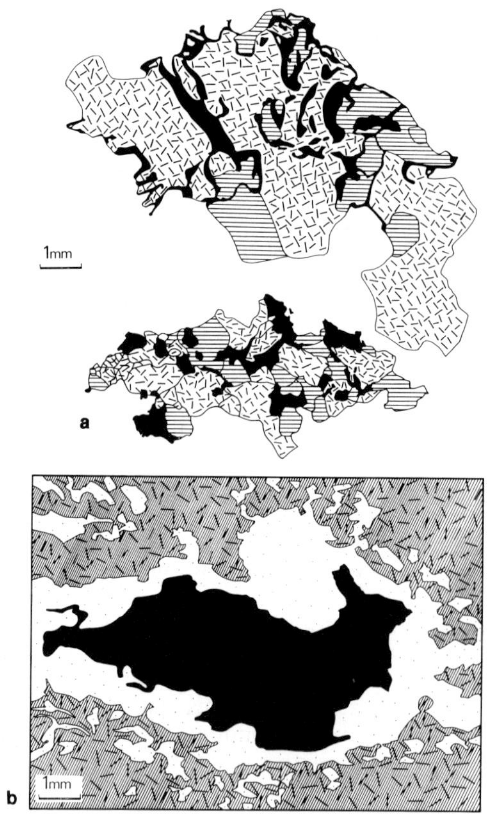


Fig. 2.5. Aluminous minerals associations in lherzolites thought to derive from melt reactions. a) Orthopyroxene-clinopyroxene-spinel clusters surrounded by olivine. b) Plagioclase corona around spinel. Black decoration : spinel ; dashes : orthopyroxene ; hatches : clinopyroxene ; mixed decoration when the two pyroxenes are not distinguished ; dots : plagioclase (Nicolas, 1986a, b).

Typical dike orientations in peridotite massifs, related to the foliation-lineation framework, have been studied by Jackson (1979), and Nicolas and Jackson (1982). In massifs which have suffered a large plastic flow after the dike injection, these dikes are now tectonically transposed parallel to the foliation and are assimilated to the ubiquitous compositional layering. Original dike orientations are best studied in plagioclase lherzolite massifs where the gabbroic melts are easily identified and where no large subsequent deformation intervened before the cooling. The sequence of progressive melting and dike formation has been studied in the Lanzo massif, where Boudier and Nicolas (1972, 1977) show that the first melt segregations forming 10-20 cm long lenses, described in the next section, are oriented parallel to the plastic flow plane (fig. 2.4a). They feed tension dikes of gabbro which are oriented at high angles to the mineral lineation (fig. 2.4b) and normal to the expected $\sigma_3$ principal stress direction.

### 2.5.3. *Melt products : evidence for segregation/impregnation*

During the last 15 years, the presence of melt products within the peridotites of the ultramafic section of ophiolites has been documented. The study of these melt products is very important when considering the composition of the peridotites, or their physical conditions during deformation (see § 2.5.5). In particular it is critical to be able 1) to identify in a peridotite the presence of melt products now cristallized in assemblages where usually the most abundant minerals are diopside and plagioclase, and 2) to distinguish whether these melt products are due to 'in situ' melting of the peridotite, or result from impregnation by transported or migrating melt. This subject has been recently discussed (Nicolas, 1986a et b).

The presence of melt products is obvious when one is dealing with dikes corresponding to sharp chemical and structural discontinuities like those considered above. The diagnosis becomes more difficult when the suspected molten phase is more intimately distributed throughout the peridotite. It is, however, particularly important to properly analyse this situation and to distinguish melt creation from melt introduction because, in dealing with the melt extraction process, it is at this scale that it takes its source.

*Partial melting evidence* - Partial melting has been documented in plagioclase lherzolite massifs both on structural (Boudier and Nicolas, 1972, 1977 ; Menzies, 1973 ; Le Sueur and Boudier, 1986) and geochemical (Menzies, 1976 ; Bodinier et al., 1988) grounds. In the Lanzo (western Alps) and Trinity (California, §.4.3) massifs, a complete sequence can be traced from the thin section to the massif scale, using the following *criteria for melt formation* :

i) Presence of interstitial minerals, mainly plagioclase, with concave interface with respect to olivine. Plagioclase also forms coronas around spinels and clinopyroxenes (fig.2.5b). Although these coronas are often considered to be subsolidus reaction products, they are perhaps better explained as being produced in a molten state because of their occurrence in areas where melt segregates (see next point).

ii) Continuity between these diffuse plagioclase-diopside segregations, clots and aligned lenses of these minerals (plate 2.2e), and finally, dikes of the indigenous type at the scale of the outcrop, the best scale for critical observations.

iii) Presence of halos that are depleted in clinopyroxene and/or plagioclase in the peridotite around clots, lenses and indigenous dikes. In the case of clots and small lenses, there is an overall compensation in the clinopyroxene + plagioclase fraction between the enriched and the adjacent depleted domains. This strongly suggests a local mineral segregation and thus a local origin for the melt represented by these two minerals.

iv) Regular distribution of melt products at the scale of the massif. This is best shown in the Lanzo massif (Boudier and Nicolas, 1977) where the mineral segregations, plagioclase lenses, gabbro dikes, and dunite veins and bodies are distributed over an area 2-3 km wide, but progressively disappear eastward giving way to a homogeneous lherzolite.

*Magmatic impregnation evidence* - Feldspar and clinopyroxene impregnation of harzburgites and dunites by a percolating magma is now a well documented process (Dick, 1977 ; Sinton, 1977 ; George, 1978 ; Savelyev and Savelyeva, 1979 ; Violette, 1980 ; Nicolas et al., 1980 ; Boudier and Coleman, 1981 ; Nicolas and Prinzhofer, 1983 ; Evans, 1985) (plates 3.2g, h and 3.3a, b, c, d). The impregnation is ascribed to mafic dikes being unable to further fracture the peridotites and dispersing their melt into them. In ophiolites, such impregnation features are restricted to the transition zone, a domain in which dikes propagating by hydrofracturing cannot easily pass (§ 7.5). Plagioclase lherzolites reconstituted in this way are difficult to distinguish from the pristine mantle ones (Nicolas and Dupuy, 1984). Impregnation of dunites also creates wehrlites and troctolites (fig. 2.6). In these wehrlites and troctolites, a structure which mimics an ultramafic cumulate can be created by the corrosion of residual olivine which produces euhedral facets (Donaldson, 1985 ; Nicolas, 1985), the poikilitic crystallization of diopside, and the growth of chromite grains into euhedral crystals (fig.2.6). *Fabric studies of olivine are necessary to distinguish between true cumulates and impregnated mantle rocks* . In the former case, olivine has a very weak shape-controlled fabric and the diopside oikocrysts are undeformed unless plastic deformation has been superimposed, in which case both olivine and diopside are deformed. In the latter case, olivine commonly has a strong lattice preferred orientation due to a plastic deformation, in contrast with the absence of deformation in the diopside oikocrysts. It is thus demonstrated that melt was injected into an already deformed peridotite (Nicolas and Prinzhofer, 1983).

*The following criteria for melt impregnation are proposed :*

i) At the thin section scale, interstitial diopside and/or plagioclase tend to develop a poikilitic habit by olivine corrosion (fig.2.6). The interstitial diopside may present the simple (100) growth twin (Nicolas and Poirier, 1976) ; this is never observed in the 'mantle' diopside which belongs to opx-cpx-sp clusters (fig.2.5a).

ii) At the scale of the massif, as typically observed in the harzburgites and dunites of ophiolites within the first kilometer below the mafic layered gabbros, melt products have a heterogeneous, local and discontinuous distribution. The dikes and associated clinopyroxene-plagioclase diffuse enrichment zones transect the harzburgite-dunite contacts, developing *local and irregular patches* of lherzolites, wehrlites or troctolites.

### *2.5.4. Microstructures in peridotites and kinematic analysis*

The typical sequence of microstructures and fabrics developed with increasing strain in mantle peridotites has been described by Mercier and Nicolas (1975). The protogranular or coarse-equant microstructure, which is found in peridotite xenoliths of basalts and kimberlites, and which reflects the absence of plastic strain, is unknown

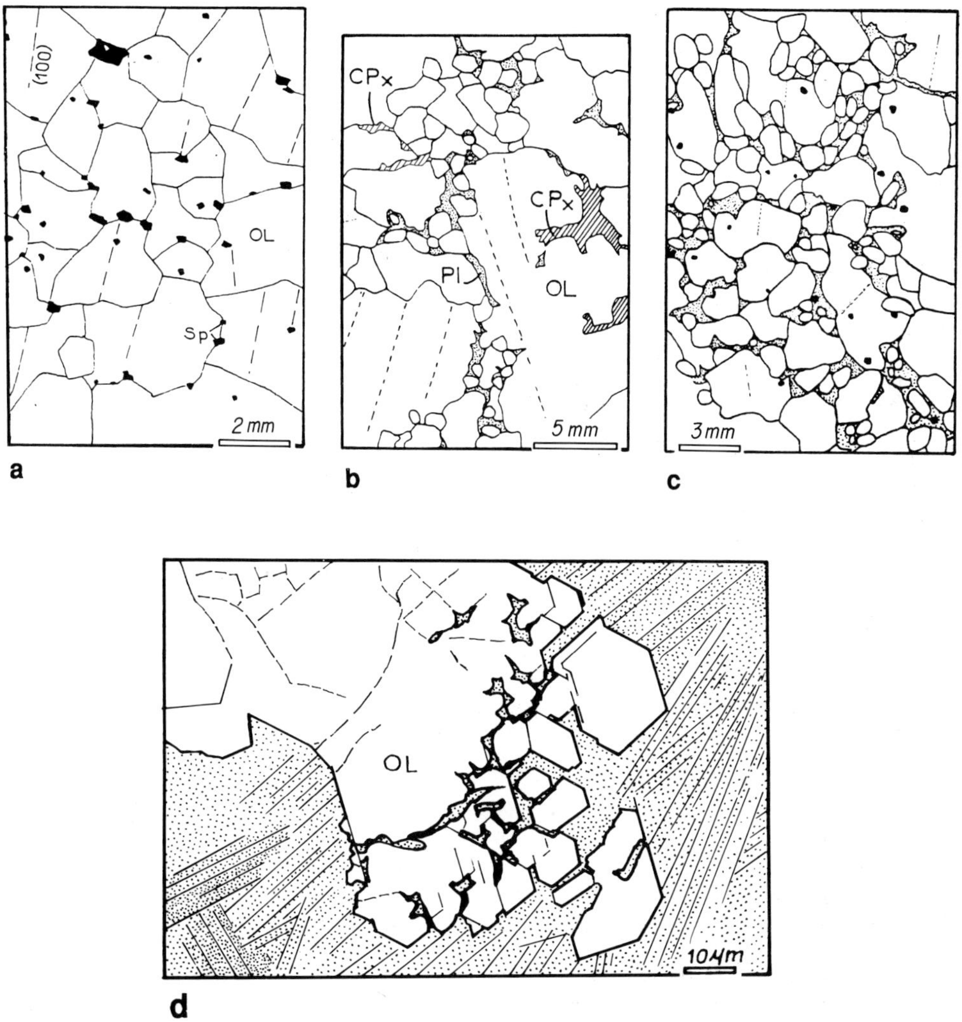


Fig. 2.6. Melt-impregnated dunites. a) Dunite with a strong lattice fabric (parallel orientation of the (100) dislocation walls), thought to have recrystallized in the presence of a melt. Melt-enhanced diffusion would be responsible for chromite recrystallization in near-euhedral grains and for their inclusion in olivine, due to grain boundary migration of olivine. b) and c) Plagioclase (dotted areas) and clinopyroxene (hatched areas) impregnation increasing from b) to c), starting from a dunite of the a) type. d) Idiomorphic olivine crystals due to corrosion by melt, in a peridotite partial melting experiment. (a, b, c : Violette, 1980 ; d : Nicolas and Prinzhofer, 1983).

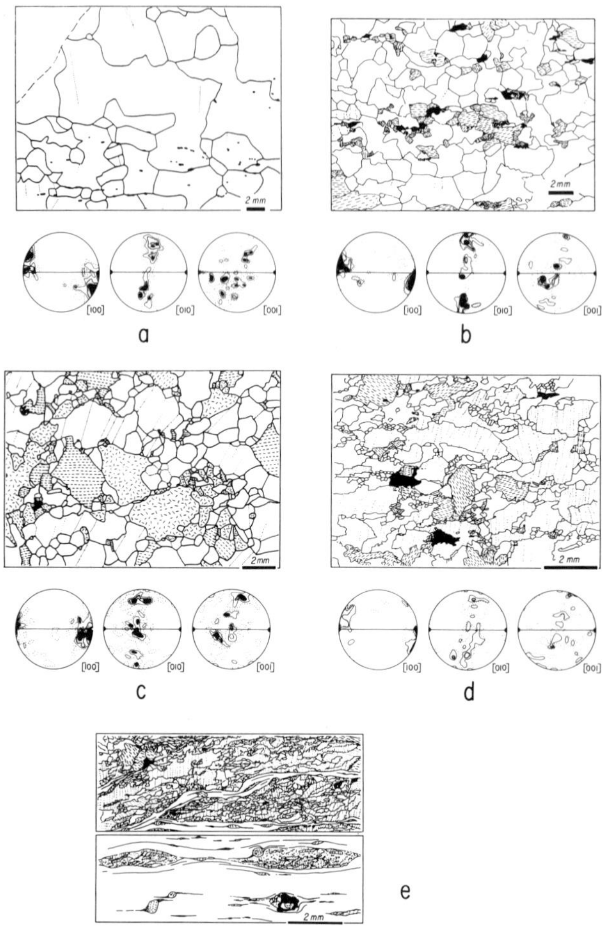


Fig. 2.7. Sequence of microstructures (with increasing magnifications) and corresponding lattice fabrics in peridotites from ophiolites. The drawings and fabrics illustrate a dextral shear regime. a) and b) Respectively dunite and harzburgite from the transition zone with the overlying crustal section, affected by an important grain boundary migration of olivine and, in the case of harzburgite, by orthopyroxene recrystallization ; note the remarkably strong fabrics explained by very large strain in hypersolidus conditions (recovery creep with a dominant activation of one slip system). c) Typical high-T, low-stress porphyroclastic microstructure and fabrics of the asthenospheric deformation in the harzburgite sequence. d) Typical low-T, high stress microstructure and fabrics of the lithospheric deformation at the base of the harzburgite sequence. e) Mylonitic and mylonitic-fluidal microstructures in the thrust plane at the base of a harzburgite sequence. Decoration : olivine, blank except for the trace of (100) dislocation walls ; orthopyroxene, dashes ; spinel, black. Stereonets : equal area projection ; 100 olivine measurements ; contours : 1,2,3,4,5 % per 0.45 % net area ; line : foliation trace ; dot : mineral lineation (Nicolas, 1986b).

in ophiolitic peridotites. The dominant microstructure is porphyroclastic (fig.2.7.c,d) with, in the olivine porphyroclasts, optically visible subgrains with a (100) tilt walls spacing of 0.2 mm and a neoblast size of 0.5 mm. As discussed in the next section these subgrains and neoblast dimensions decrease downsection in relation with the development of a higher stress-lower temperature deformation which culminates in mylonitic microstructures (fig.2.7.e). Coarse-mosaic microstructures (fig. 2.7.a,b), also discussed in the next section, are restricted to the dunites and harzburgites of the upper-most peridotite section including the transition zone.

The deformation proceeds by dislocation slip and climb with variable degrees of grain boundary migration and syntectonic recrystallization (Nicolas and Poirier, 1976). It is possible to know the orientations of the flow plane and the flow line by petrofabric analysis and, relating those to the foliation and mineral lineation, to deduce the flow regime (Nicolas et al., 1971 ; Nicolas and Poirier, 1976).This is illustrated by figure 2.8, which shows the textural evolution of a peridotite with increasing strain and the obtained lattice fabrics in the regime of simple shear. In ophiolitic peridotites, as in other mantle peridotites, large homogeneous deformation in a regime approaching simple shear is the most common natural situation, emphasizing the interest of this example. It can be seen in figure 2.8 that an imaginary circle delineated in an undeformed peridotite is transformed with increasing strain into an ellipse. This is the finite strain ellipse ($X \geq Z$) (an ellipsoid in 3-D, with $X \geq Y \geq Z$ ; here Y is an invariant axis). The X,Y,Z directions are materialized in plastically and homogeneously deformed rocks by the foliation (X, Y plane of mineral flattening) and the stretching lineation (X axis), commonly a mineral lineation. On the other hand, the slip planes and slip lines of the actively deforming minerals, olivine and orthopyroxene, become progressively oriented during flow, parallel to the plane and the line of simple shear, respectively. Thus the flow structure is recorded in the lattice preferred orientation of olivine and enstatite. From the obliquity between shape (foliation and lineation) and lattice (slip planes and slip lines) fabrics, the regime can be deduced and from the sense of rotation bringing the two fabrics in coincidence, the sense of shear can be deduced (dextral in figure 2.8). Theoretically, one can also deduce the shear strain $\gamma$. This strain depends on the $\alpha$ angle between foliation and flow plane through the formula : $\gamma = 2 \cot 2\alpha$. Finally, approximate estimates of stress can be derived from the dislocation substructure and neoblast size (Goetze, 1975 ; Mercier et al., 1977 ; Nicolas, 1978 ; Karato et al., 1980 ; Ross et al., 1980 ; Karato, 1984 ; Zeuch and Green, 1984).

### *2.5.5. Microstructural imprint of asthenospheric/lithospheric flow.*

Distinguishing which microstructures correspond to an asthenospheric flow and which correspond to a lithospheric flow is a problem of the first order. It is feasable provided that, 1) one accepts that a signature of asthenospheric deformation is the presence of basaltic melt within the deforming peridotite and, 2) one is able to identify this situation in the now frozen peridotite structure. In the case of harzburgitic ophiolite massifs, this criterion is complemented by independent geological considerations showing that the peridotite section is affected by two distinct plastic deformation episodes, which are respectively related to flow below the accreting center of origin and to flow during a subsequent oceanic thrusting (Nicolas et al., 1980). The microstructures associated with these two episodes of plastic deformation can be regarded as typical of asthenospheric and lithospheric conditions, respectively.

The microstructures of the asthenospheric deformation observed in dunites and depleted harzburgites at the top of the ultramafic section, just below the mafic crustal

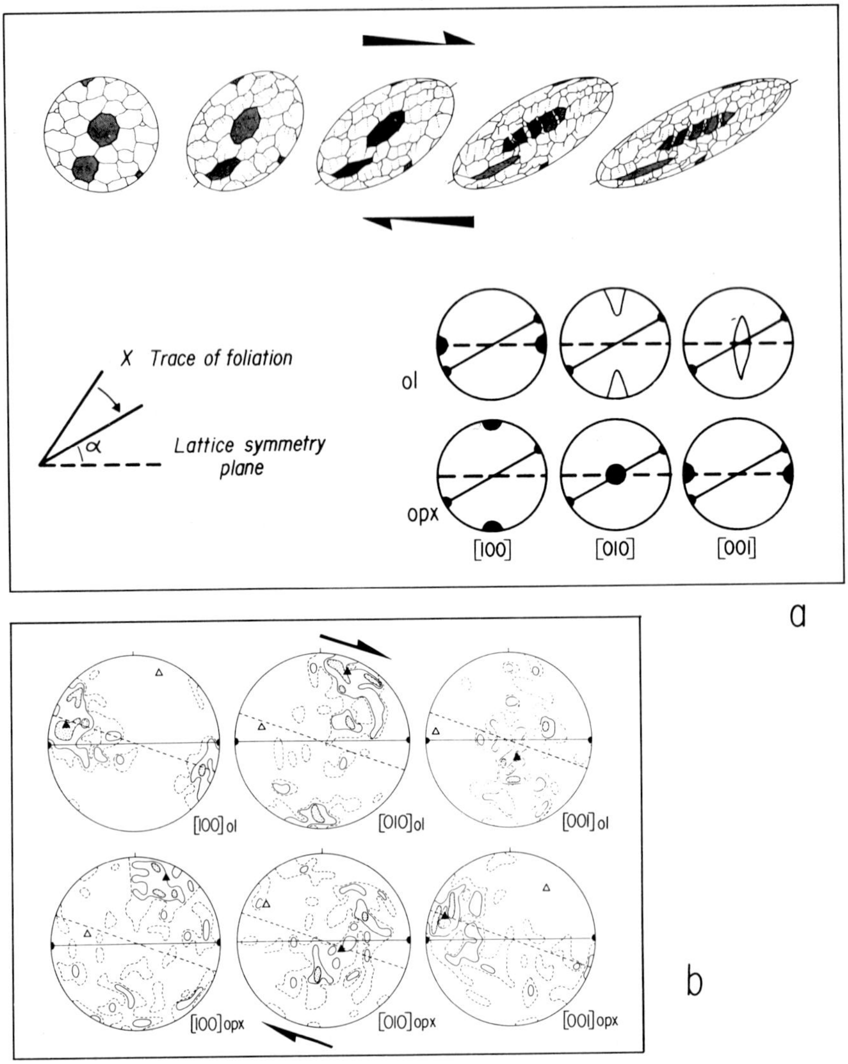


Fig. 2.8. Example of kinematic analysis in a peridotite. a) Theoretical sketches (keys as in figure 2.7). In a progressive deformation by simple dextral shear (shear plane E-W perpendicular to the figure plane), the foliation X is rotated and lengthened. The stereograms corresponding to the final stage show that the orientation of slip systems coincide with that of the shear plane (dashed line) and is oblique, in a sense reflecting the shear sense, with respect to the finite deformation axes (straigt line : trace of foliation ; small dots : trace of lineation X). b) Illustration in the case of a natural peridotite, dextral shear. Equal area projection, lower hemisphere ; contours : 1,2,4,8 %. Open triangle, best computed axis ; solid triangle, pole of best computed plane. 100 measurements for olivine and pyroxene (Nicolas, 1987).

section, are coarse-porphyroclastic (fig. 2.7c) and coarse-mosaic with equant to tabular olivine neoblasts 0.5 mm across (fig. 2.7b). The remarkably strong lattice fabrics (fig. 2.7a and 2.7b) reflect very large strains, achieved at hypersolidus temperatures. Such high temperatures explain the recovered nature of the structure and account for a very active grain boundary migration, probably also favored by the impregnation by melt during the flow of the rock (fig. 2.6.).

The microstructures of the lithospheric deformation imprinted in the lowest section of the harzburgites massifs grade upward, from mylonitic in or adjacent to the thrust plane (fig. 2.7e), to high stress porphyroclastic (fig.2.7d) a few hundred meters above. Equilibration temperatures of 900° - 950°C and 850° - 900°C are recorded respectively in these peridotites and in the underlying granulites (§ 12.3.2). By analogy with similar microstructures which have developed into peridotites xenoliths in basalts where temperature estimates directly relate to deformation (no recovery), 800 to 900°C can be assigned to the lowest temperature of deformation in the mylonitic peridotites just above the contact with the metamorphic aureole (Cabanes and Briqueu, 1986).

In conclusion, the high temperature structures recorded in peridotites would correspond to asthenospheric flow in the range of 1250° - 1200°C (hypersolidus to solidus conditions) and low temperature structures, to lithospheric flow between 1000°C and 800°C.

### *2.5.6. Serpentinization and low temperature deformations*

Peridotite massifs are usually serpentinized by the mesh-structured lizardite which results from a low-temperature and static alteration taking place as well in surface conditions (Barnes et al., 1978). This is suggested by the fact that cored specimens in ultramafic massifs are often less serpentinized than surface specimens. Less common but highly significant is the occurrence of antigorite replacing olivine, often associated with tremolite and chlorite replacing, respectively, pyroxenes and spinels. These minerals are often strongly oriented, defining a foliation. Associated gabbroic facies are rodingitized by a Ca-rich metasomatism with replacement of primary minerals by zoisite or epidote, diopside, grossularite, vesusianite,... Such dynamic transformations correspond generally to greenschist facies conditions (150° - 450°) and are now largely ascribed to the hydrothermal circulation taking place at oceanic spreading centers. In fact, this hydrothermal alteration can be initiated at higher temperatures, probably soon after the peridotite accretion below the ridge. Kimball et al. (1985) have documented, on ultramafics dredged in the Islas Orcados Fracture Zone, a sequence of mineral reactions starting around 900°C and illustrated by the reactions of figure 2.9. The related deformation may be associated to shear motion on listric faults and shear zones (Norrell et al., in press). When present, this antigorite serpentinization always predates the lizardite one. The large asbestos deposit of Thetford Mines (Canada) formed in a different way, when felsic intrusions penetrated a cold and serpentinized peridotite at high temperature (Clague et al., 1985).

## *2.6. EXPECTED ASTHENOSPHERIC FLOW PATTERNS*

Ophiolite complexes are fragments of oceanic lithosphere which have been created at a spreading center and frozen during sea floor spreading. Because they sample no deeper than the first 10-15 km in the lithosphere, ophiolites represent an asthenosphere frozen within the first few millions of years after spreading. At these shallow depths, that is in the vicinity of the lithospheric front, the slip lines in the asthenosphere must be parallel to this front because, in contrast to particle paths, slip lines cannot penetrate into the lithosphere (fig. 2.10). In hydrodynamics, a particle path is referred to as 'a stream line'

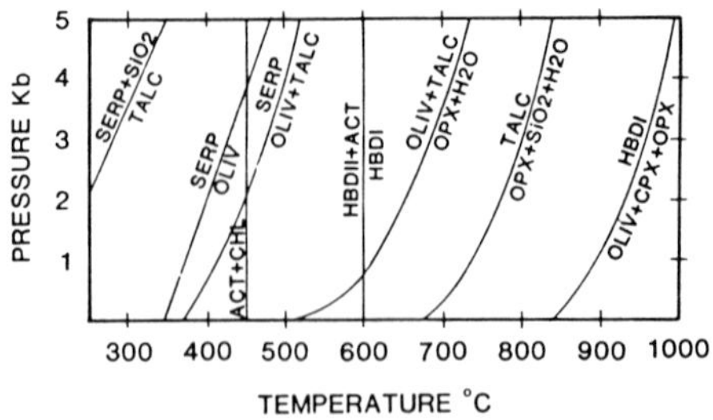


Fig. 2.9. Pressure-temperature diagram showing experimental curves for various reactions induced in a peridotite by hydrothermal alteration at decreasing temperatures (Kimball et al., 1985).

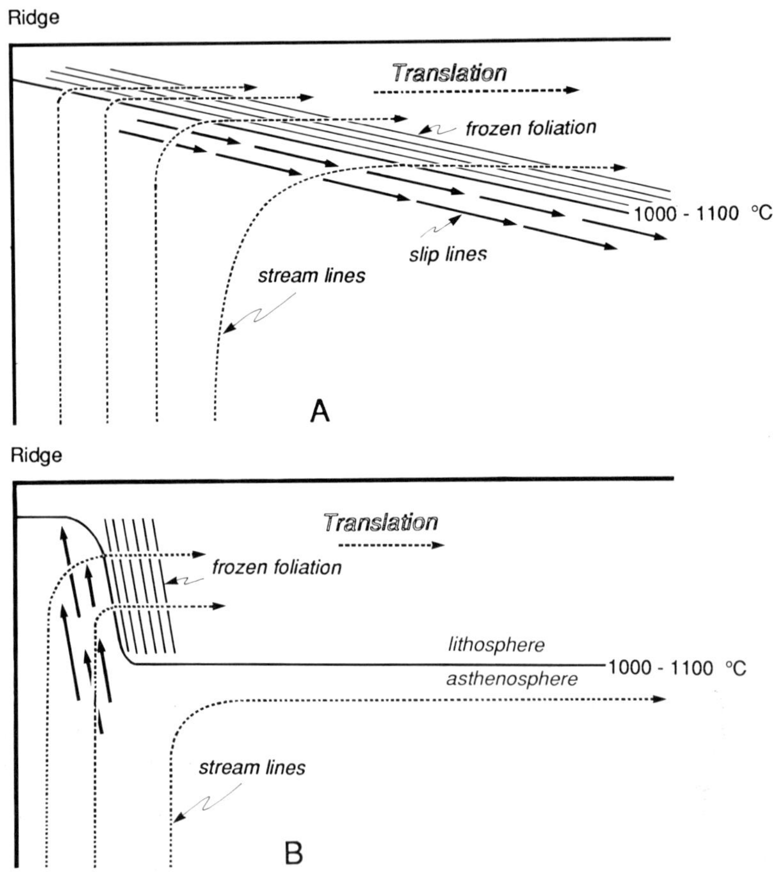


Fig. 2.10. Theoretical models of asthenospheric flow and lithospheric accretion in the thickening (a) and dike intrusion (b) models of a ridge as defined in § 9.1. Three flow-related entities are presented here : stream lines (aligned dashes) which describe the trajectories of solid particles, slip lines (thick arrows) which describe the active shear directions in the asthenosphere, and foliation traces (parallel thin lines) which are oriented at a small angle to the frozen shear surfaces (inside the lithosphere) (stream lines contours are from Phipps Morgan et al., 1987).

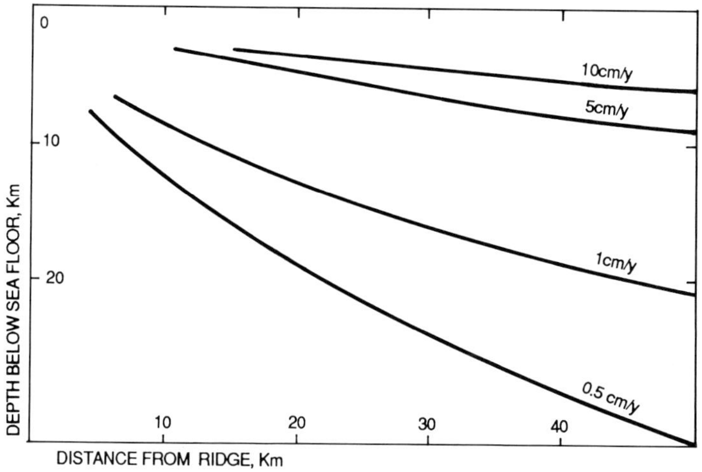


Fig. 2.11. Profiles of the 1100°C isotherm, taken as the boundary between lithosphere and asthenosphere, for different spreading rates. In spite of their inaccuracy for the young ages considered here, these profiles illustrate how this boundary changes in slope with spreading rate. In ophiolites, the dip of mantle foliation will be related to the slope of these isotherms and thus to spreading rate (based on data from Parker and Oldenburg (1972) for older ages and from Morton and Sleep (1985) for younger ages).

or a 'flow line'. In this book, we will use the term 'flow' to designate the displacement field in the asthenosphere. The slip lines in the asthenosphere are progressively slowed down within a boundary layer ; they are eventually frozen and accreted to the lithosphere tangentially at this boundary layer. The foliations and lineations observed in the lithosphere slab sampled by ophiolites thus record a frozen shear flow field whose plane was parallel to the asthenosphere-lithosphere boundary. In the present analysis this boundary corresponds to a layer through which the creep rate decreases rapidly and the lithosphere is regarded as a kinematic entity. Although this rate can also depend on stress or viscosity variations due to local causes, like the presence of partial melt into the peridotite (Phipps Morgan et al., 1987), it is primarily dependent on temperature. Hence, we will equate here the asthenosphere-lithosphere boundary with an isotherm ; in other words, we will consider the lithosphere as primarily a thermal entity. The boundary isotherm should be around 1000° - 1100° (§ 2.5.5).

*The foliations and lineations representing this frozen flow are also subparallel to the lithosphere surface*, accepting here for the sake of simplicity that in shear flow conditions, and for the large strain experienced by the corresponding peridotites, the flow plane and the resulting foliation are sub-parallel (Nicolas and Poirier, 1976). A common situation in ophiolites is that of high temperature foliations regionally parallel to the Moho. In this situation, because it theoretically coincides with an isothermal surface, the shear flow plane attitude deduced from these foliations should give two pieces of information:

- it should dip away from the ridge, thus indicating the side of the origin of ophiolite (§2.2.2) (fig. 2.10).
- its dip should depend on the spreading rate, following models of thermal structure below ridges (Parker and Oldenburg, 1972 ; Bottinga and Allègre, 1978) (fig. 2.11). This prediction seems to be verified in ophiolites (chapter 9).

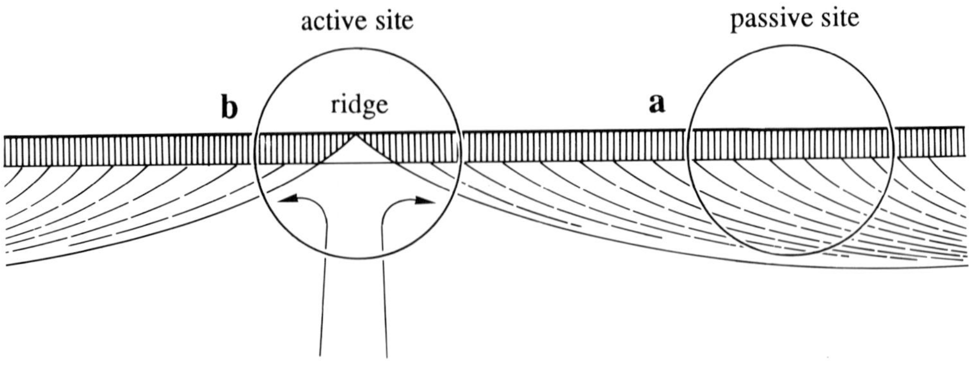


Fig. 2.12. Sketch illustrating a) the expected mode of ophiolite sampling of an oceanic lithosphere drifting steadily from a ridge and b) the sampling of an active structure, here the ridge itself.

Following this analysis, the observation of regionally steep foliations points to asthenosphere flowing along a steep lithospheric boundary, a situation expected in transform domains and possibly below a new ridge propagating into an older lithosphere. Steep structures related to a mantle diapir correspond to a particular situation examined below.

Another consequence of this analysis, investigated by Ceuleneer et al. (1988) is that the deepest parts of the mantle section of an ophiolite are also those which accreted farthest from the ridge axis, a point we must keep in mind when reconstructing asthenospheric flow patterns after the structures recorded by the peridotites. The distance of a given accretion zone from the ridge as a function of depth cannot be determined accurately : firstly, the actual thermal structure of the mantle near the ridge is poorly known due to the intense hydrothermal circulation taking place there (e.g. Davies and Lister, 1977) ; secondly, in a steady state expansion regime, the asthenospheric structures are not suddenly chilled when crossing the asthenosphere-lithosphere boundary ; finally, the actual spreading rate of the paleo-ridge where the ophiolite formed is largely unknown. However, we can try to fix plausible boundaries. The thermal structure of a spreading center down to a depth of about 4 km below the Moho has been computed by Morton and Sleep (1985) who show that the uppermost kilometer of the mantle section is accreted at a distance from the ridge between 0.25 and 0.50 Ma X half-spreading rate, assuming that a drop in temperature of about 200°C to 300°C is needed to prevent the peridotites from flowing under the expected deviatoric stress conditions (<1 MPa). As an example, in Oman, where the maximum thickness of the mantle section is around 10 km, the flow at that depth is frozen at about 2 Ma, the time lapse necessary for the 1000°C isotherm to reach 9 km below the Moho (Parson and Sclater, 1977). For medium to fast spreading rates this corresponds to distances of 100 km or more from the ridge.

This general analysis applies to the usual situation where the considered ophiolite is derived from a passively drifting lithosphere with already frozen structures (fig. 2.1a). Another situation may arise in which the ophiolite has sampled an active oceanic domain, such as a complete spreading center (fig. 2.12b), an overlapping spreading center, a propagating rift or possibly an off-axis volcano. In a few harzburgite massifs where steep foliations and lineations have been locally measured, mapping of these structures has revealed the contours of mantle diapirs from which the asthenosphere flow diverges in every direction (chapter 9). The flow lineations tend to be oriented normal to the ridge trend as defined by the diabase dike swarm only at a distance greater than 50 km from the diapiric structure. For these reasons, we have abandoned the former idea (Juteau et al., 1977 ; Girardeau and Nicolas, 1981) that the azimuth of lineations related to high-T flow in peridotites should be normal to the ridge azimuth and could be used to retrieve this azimuth. This expectation was based on the data of seismic anisotropy in the oceanic upper mantle and their interpretation in terms of plastic flow (review in Nicolas and Christensen, 1987), from which it was concluded that the azimuth of asthenospheric flow was at a high angle to the ridge trend. We believe now that this reorientation is not yet achieved in ophiolites which sample the flow active *very close to the ridge*. It is progressively achieved however at a greater distance to the ridge (corresponding to a domain of active flow deeper than the section observed in ophiolites) when the channeling effect of transform faults becomes important (fig. 9.1).

# PART II

# TYPICAL OPHIOLITE COMPLEXES

## *INTRODUCTION*

Ophiolites were first perceived as a formation repeatedly encountered at the Earth surface, and composed of the same elements, organised in a constant manner. The uniqueness of ophiolites is a concept which began to be broken down by geochemical studies showing that the nature of the lavas could be different from one complex to another. This led to the idea that ophiolites could be derived from different oceanic environments (Miyashiro, 1973 ; Pearce and Cann, 1973 ; Beccaluva et al., 1979). It was also found that a diabase sill swarm could replace the more common dike swarm (Hopson and Frano, 1977 ; Girardeau and Mercier, 1985), and that the layered gabbro section could be variously developed and locally absent (Nicolas et al., 1981 ; Le Sueur et al., 1984). Major differences also stemmed from structural studies showing that the internally layered structure in the crustal formations could be altered and those formations sheared, evoking a transform fault origin (Karson, 1984). Similar shearing could also affect the peridotite formations, leading to the same genetic conclusions (Prinzhofer and Nicolas, 1980 ; Reuber, 1985). The lherzolitic rather than harzburgitic nature of some peridotites was also noted and interpreted in terms of incipient rifting (Menzies, 1976 ; Boudier and Nicolas, 1985) and/or transform environments (Abbate et al., 1980 ; Nicolas and Dupuy, 1984). Systematic studies in the ultramafic section, mainly structural in nature, confirmed the diversity of ophiolites and contributed to introduce some rationale into this diversity (chapter 8).

In the present state of ophiolite studies, diversity appears as a prominent feature of ophiolites, a fact which is obviously related to the diversity of possible oceanic environments of origin. Therefore, the ophiolite complexes which are described in this part have been selected to illustrate the principal ophiolite types which are presently known. These complexes are among those which are the least dismembered and for which extensive descriptions have been found in the literature.

This selection of a few complexes out of the nearly forty for which descriptions are available does not reflect their relative abundance. The Oman ophiolite, taken here as an archetype, sharing many similar features with classical complexes such as Bay-of-Islands or Cyprus, in particular a harzburgitic ultramafic section, corresponds to the most common type and will be extensively described. After the Oman case, the Xigaze and the Trinity complexes will be described under the same heading (chapter 4) because they represent a distinct ophiolitic trend characterised by a thin mafic section and a lherzolitic ultramafic section. The case of ophiolites displaying features evoking oceanic fracture zones will be considered in chapter 5 and finally chapter 6 will discuss Canyon Mountain, an ophiolite with unusual structural and geochemical signature, evoking an island arc environment.

Although differences of opinion persist as to the geological history of some of these ophiolites, including the modes of formation and of emplacement, it was felt necessary to include this controversial section in the following chapters, otherwise devoted to descriptions, in order to delineate the currently outstanding problems.

# Chapter 3

# OMAN OPHIOLITE :

# THE HARZBURGITE OPHIOLITE TYPE

## *3.1. INTRODUCTION*

Although the first detailed report on the Oman ophiolite is recent (Reinhardt, 1969), it is now one of the best studied complexes and is often considered to be the best example of ophiolites in the world. It is indeed the largest, with a crescent-shaped extension over 500 km in length and 50-100 km in width (fig.3.1 and 3.2). The Papua-New Guinea ophiolite, which has a general setting and shape comparable to the Oman ophiolite, is not substantially smaller (400 km x 20-50 km) but, due to the heavy vegetal cover and to difficulties of access its study is far less advanced (Davies, 1971, 1980). On the other hand, in Oman the exposures are beautiful and the access easy.

As we shall see, the Oman ophiolite nappe, also called the Sumail nappe (although we wish to restrict this name to a massif crossed by the Wadi Sumail (fig.3.8)), has been obducted on the Arabic platform without any subsequent collision. For this reason the section is complete from the metamorphic sole resting on sedimentary nappes up to the volcanics and their deep-sea sedimentary cover. The internal structure in many places has escaped any obduction-related deformation and still represents the structure of a spreading center at the moment of its initial detachment.

Finally as illustrated by figure 3.1, the scale of the Oman ophiolite is superior to that of a representative specimen of oceanic ridge systems where a characteristic longitudinal dimension seems to be 50-200 km, a length corresponding to the spacing between transform segments or between overlapping centers. This is important because, thanks to integrated studies throughout the belt, one can discard the danger of dealing with some local and specific oceanic situation. On the contrary, with this ophiolite, one can test the homogeneity or variability of oceanic lithosphere at the scale of seismic soundings.

Before Reinhardt's publication, the first works dealing with the Oman ophiolites were those of Lees (1928) and Morton (1959). Lees recognized the allochtonous character of the ophiolites and of the underlying sediments which constitute the Hawasina nappes, whereas Morton and his followers regarded the ophiolites as autochthonous. In the late sixties, Glennie's group mapped the Oman mountains at the 1/500 000 scale, benefiting from the powerful logistics of oil companies. Their excellent map is the support of a large crop of new results (Glennie et al., 1973, 1974), in particular on the allochthonous character of the Hawasina and ophiolite (Sumail) nappes and on the geology of the ophiolites. As seen in chapter 1, Reinhardt (1969) is among the first authors to describe them as oceanic lithosphere formed at a ridge and obducted on a passive margin during a compressive event. The amphibolites of the metamorphic sole were then related to the obduction and dated (Allemann and Peters, 1972).

During the seventies, R.G. Coleman's American group worked in the south-east Wadi Tayin area and I.G. Gass' British group, in northern Oman. They were followed in the early eighties by two French groups (Nantes-Strasbourg Universities and Bureau de recherches Géologiques et Minières). The findings of these groups are partly published in special volumes ('Oman ophiolite', J. Geophys. Res., Vol. 86, 1981 ; 'The ophiolite of Northern Oman', Geol. Soc. mem., 11, 1986 ; 'The ophiolite of Oman', Tectonophysics, special issue, 1988), and are summarized below.

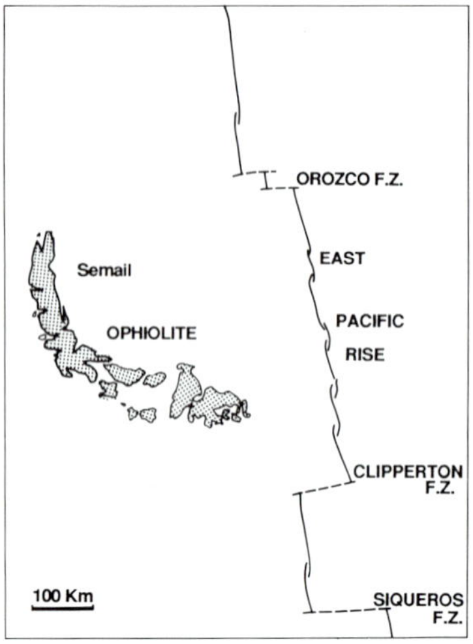


Fig. 3.1. Comparative dimensions of the Oman ophiolite and of the fast spreading East Pacific Rise with its typical segmentation (Ceuleneer, 1986).

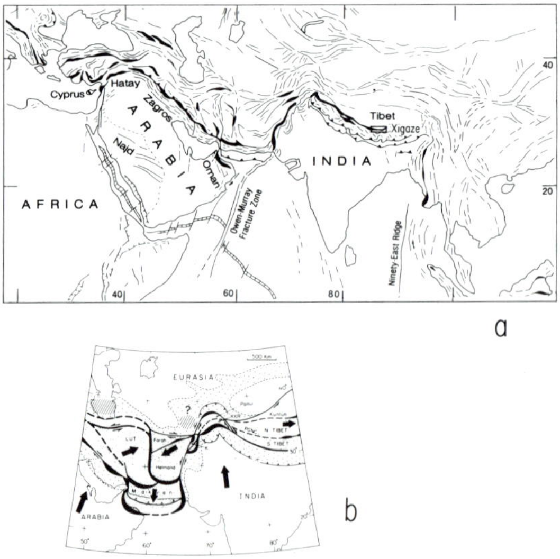


Fig. 3.2. The Asian section of the alpine belt with ophiolites and colored mélanges (black decoration) underlining the main sutures. a) Descriptive map after Coleman (1981) and Gansser (1966). b) Interpretive map after Tapponnier et al. (1981). The arrows indicate the approximate motions of intervening blocks ; dotted areas : zones of alpine deformation ; shaded areas : smaller blocks between the three major continental masses ; hatched areas : residual oceanic crust.

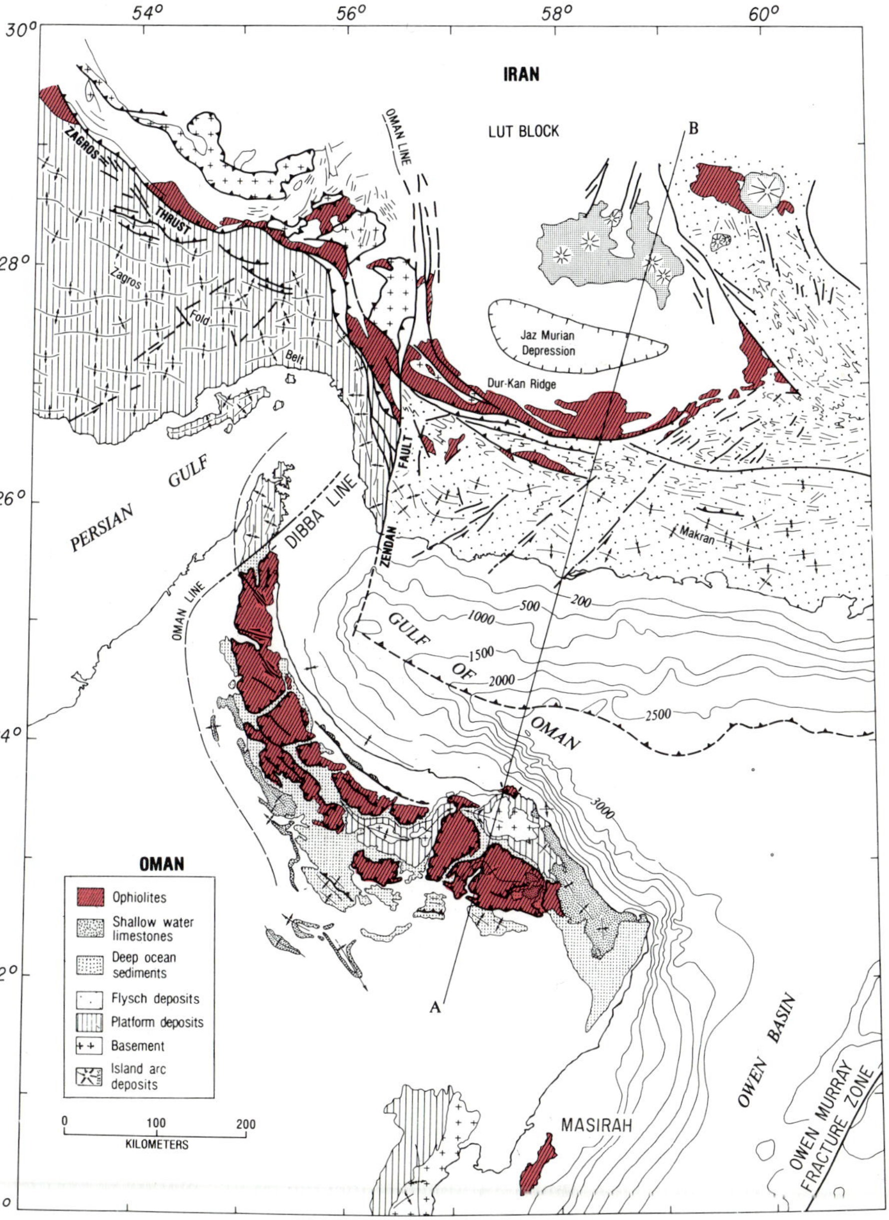


Fig. 3.3. Tectonic map of Oman and southeastern Iran. Ophiolites in color ; thrust fault in the central part of the Gulf of Oman : trace of present-day Makran subduction. Cross section A-B presented in figure 3.4 (Coleman, 1981).

## *3.2. GEOLOGICAL SETTING*

Among the ophiolite complexes dealt with in this book, two belong to the alpine belt : the Xigaze (Tibet) and the Oman ophiolites (fig. 3.2). The suture zones of this collision belt from the western Alps to the Tibet are underlined by ophiolites and mélanges. All of these ophiolites are thus in a collision environment (fig. 3.2b) with the exception of the Oman and possibly of the Troodos ophiolites. The following account on the Oman ophiolite geologic setting has borrowed much from Coleman's (1981) and Ceuleneer's (1986) reviews, where complementary information can be found.

### *3.2.1. Geodynamic setting*

The Oman ophiolite nappe represents the eastern extremity of the 'peri-arabic ophiolitic crescent' of Ricou (1971) which extends over 2000 km from Turkey and Cyprus to Oman, via the Zagros ophiolites and mélanges, and surrounds the northern promontory of the arabic shield (fig.3.2). The continental collision in the Zagros is transformed along the Dibba line and Zendam fault of northern Oman and southern Iran into an oceanic subduction beneath the Makran accretion prism (figs. 3.3 and 3.4). The Zendam fault extends through the Zagros and has been active from at least the Cretaceous to the present (White and Ross, 1979 ; Ross et al., 1987). Thus the Oman ophiolite obduction onto the arabic passive margin has not degenerated into a continental collision. At the geological scale of time, this is however a precarious situation as the Makran subduction is now swallowing the remnants of the part of the Indian ocean facing Oman at the rate of about 5cm/year. This small basin is limited Eastward by the Owen-Murray fracture zone, East of which the continental collision situation, here between India and Asia, is restored.

Below a sedimentary cover postdating the obduction, the eastern limit of the ophiolite could extend to the eastern margin of the Arabic shield, but possibly no further East than the Ras Jibsh-Masirah-Kuria Muria alignment of ophiolitic islands. This NE-SW alignment, 500 km long, would stake out an old fracture zone (Moseley and Abbotts, 1979) which may itself correspond to the former transform limit of the Oman basin. The largest of these islands, Masirah, displays a lower Cretaceous ophiolite locally deformed in a NNE-SSW trending mélange related to the transform activity. During the Eocene, transform activity shifted 300 km eastward on the Owen-Murray fracture zone.

During the Cretaceous, that is during the creation of the Oman ophiolite, the Dibba line and the Masirah fault may have been two major fracture zones limiting the Oman ophiolite ridge. Their 600 km spacing is compatible to that measured in present-day oceans between major fracture zones, like those of the eastern Pacific. However, along the Dibba line the ophiolites are thrust over the autochtonous formations of Northern Oman and there are no remaining signs of a presumed former transform activity.

Between these faults, the Hawasina and ophiolite nappes have been thrust on land in a SW direction over a distance of 150-200 km deduced from the present day SW limit of these nappes (fig.3.3). This implicitly suggests that the Arabic margin is parallel to the nappes front ; this is probably not the case (Glennie et al., 1974) as shown by a brief presentation of the history of the Hawasina basin where the ophiolites originated.

### *3.2.2. History of the Hawasina basin*

The Hawasina basin belongs to the Neo-Tethys which opened during the Lower

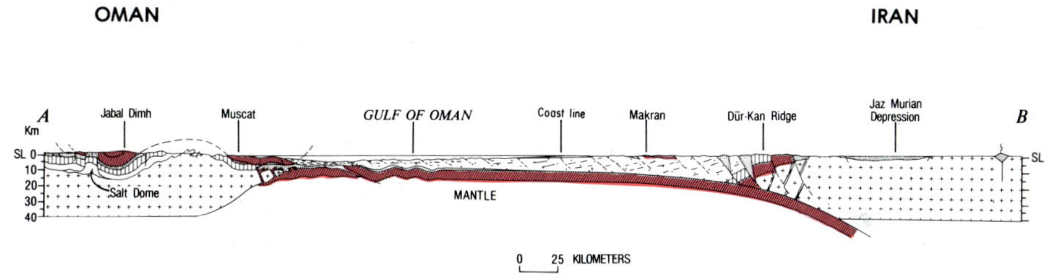


Fig. 3.4. Stylized cross section from Oman to Iran. For location and caption, see figure 3.3 (Coleman, 1981).

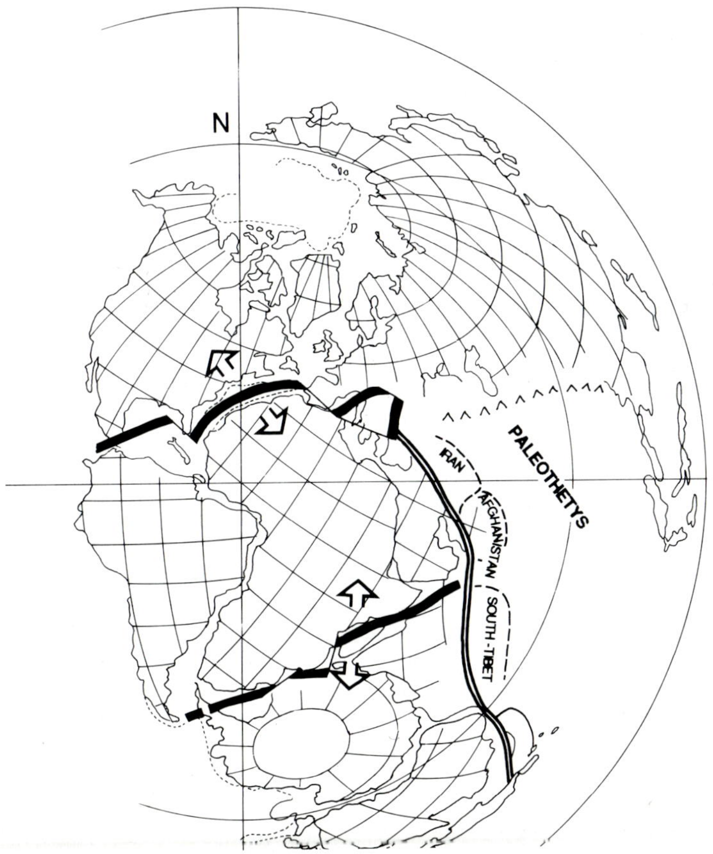


Fig. 3.5. Late-Paleozoic to Early Mesozoic riftings : the heavy black lines correspond to the Atlantic-Indian rifts between large continental masses, and the double line to the Neo-Tethys rift, along which much smaller blocks are being detached. Arrows indicate the Jurassic motions (modified from Patriat et al., 1982).

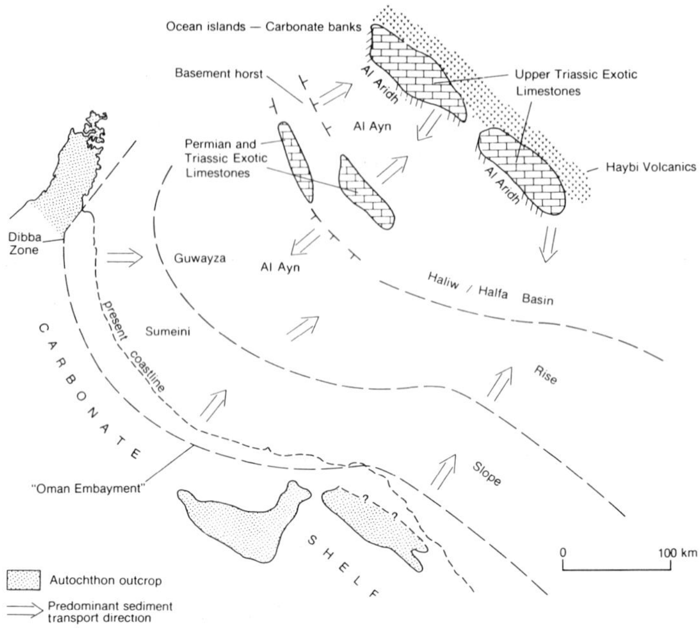


Fig. 3.6. Schematic paleogeographic reconstruction of the Oman continental margin (Lippard et al., 1986).

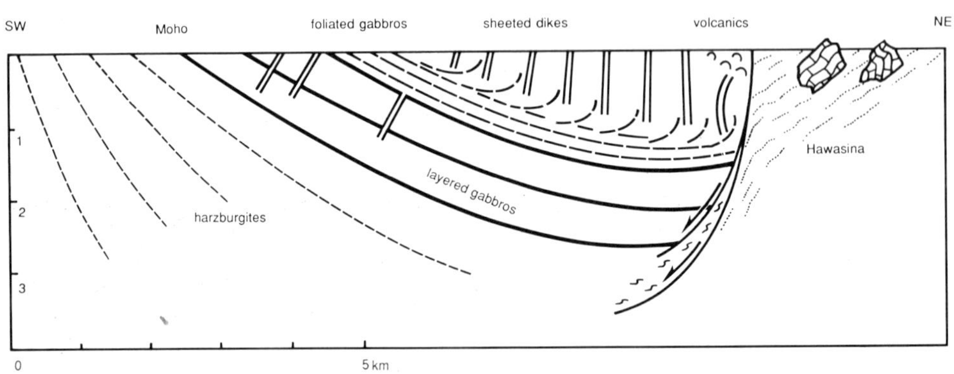


Fig. 3.7. Simplified cross section based on systematic structural measurements in the foothills of the Haylayn massif, showing the rotation of the ophiolite nappe close to the tectonic contact with the Hawasina inliers of the Batinah plain. This contact is interpreted as a major normal fault.

Mesozoic by fragmentation and rifting of the Gondwana continent (fig.3.5). It is this opening and the coeval subduction along the southern margin of Eurasia which progressively closed the Paleo-Tethys and welded to Eurasia a number of blocks detached from Gondwana. According to Patriat et al. (1982) the opening of the Neo-Tethys, which participates with the Atlantic and Indian ocean rifting to the dismembering of Pangea, presents an important difference with these latter riftings. They affect very large continental masses, whereas the Neo-Tethys detaches small blocks in front of a major subduction zone. This could explain why the average relative drift rates are so different : 1-2 cm/yr between Africa, North America, Eurasia and Madagascar-India, versus an average of 8cm/yr for both the South Tibet drift during Cretaceous, and for the more recent India drift. This suggests that the spreading rate for the Oman ophiolite was also large.

In this general scheme, the rifting which opened the Hawasina basin was initiated during Permo-Triasic time (Glennie et al., 1974 ; Graham, 1980). The former authors have estimated the width of the Hawasina basin between 400 km and 1200 km, on the basis of a palinspastic nappes reconstruction. This basin separated the northern margin of the Arabic shield from a block, possibly the Central Afghanistan block (Ceuleneer, 1986) (fig.3.2), now welded to Eurasia. The initial rift direction and thus the northern limit of the Arabic passive margin below the nappes are probably more E-W than SE-NW, which is the mature ridge trend in present day orientation (§ 3.5.3) (Glennie et al.,1974). This is deduced from the fact that only at the latitude of Mascate do the nappes ride over the autochtonous calcareous formations of the Arabic platform, cropping out in the Saih Hatat and the Djebel Akhdar. In the North, they are thrust over and against the Sumeini formations which represent a shelf- edge sedimentation taking place from the rifting initiation to Mid- Cretaceous (fig. 3.6.).

North of the passive margin, the Hawasina sedimentation proceeds from Permian (around 260 Ma) to Cenomanian (around 97 Ma). Permian and Triassic reef limestones, the latter associated with the Haybi volcanics which are alkali basalts dated at around 218 Ma (Lippard and Rex, 1982), have been interpreted by Glennie et al. (1974) as derived from seamounts inside the Hawasina basin and by Graham (1980) and Searle and Graham (1982) as derived from intrusive horsts in the border of the continent and related to the rifting. These formations now included within the mélange nappe form the 'exotics' at the base of the ophiolite nappe (fig.3.7). The Hawasina formations record a progressive subsidence probably reflecting the post-Triassic thermal relaxation following the episode of rifting and the transition to an oceanic situation. The increased subsidence during the Malm (150 Ma) (Murris, 1980) could correspond to this transition (Ceuleneer, 1986). Alkali basalt sills in the Hawasina pelagic sediments have ages within a spread from 162 to 92 Ma (Lippard and Rex, 1982).

Thus, the Mesozoic history of the Oman margin follows models for present Atlantic-type margins (Graham, 1980) until the ophiolite obduction occurred. The first compressive motion is dated in the metamorphic aureoles at 101-93 Ma and the final obduction on to the Oman margin at 75 Ma (table 3.I). These events are reflected in the sedimentary record of the Arabic platform (Ceuleneer, 1986). The sedimentation ceases in the Hawasina basin at around 97 Ma (Glennie et al., 1974). The calcareous platform sedimentation which had been monotonous since the Lias begins to be increasingly perturbed during the Albian (110 Ma), with a succession of transgression-regression cycles until the Coniacian (88-80 Ma) major uplift (Murris, 1980), also responsible for 4000 m of erosion in the shelf limestones of the margin and for the deposition of the Muti conglomerates (Glennie et al., 1974). The Hawasina and ophiolite nappes progression is marked by a fore-nappe trench which is filled by ophiolitic conglomerates (80-70 Ma) (Muris, 1980).

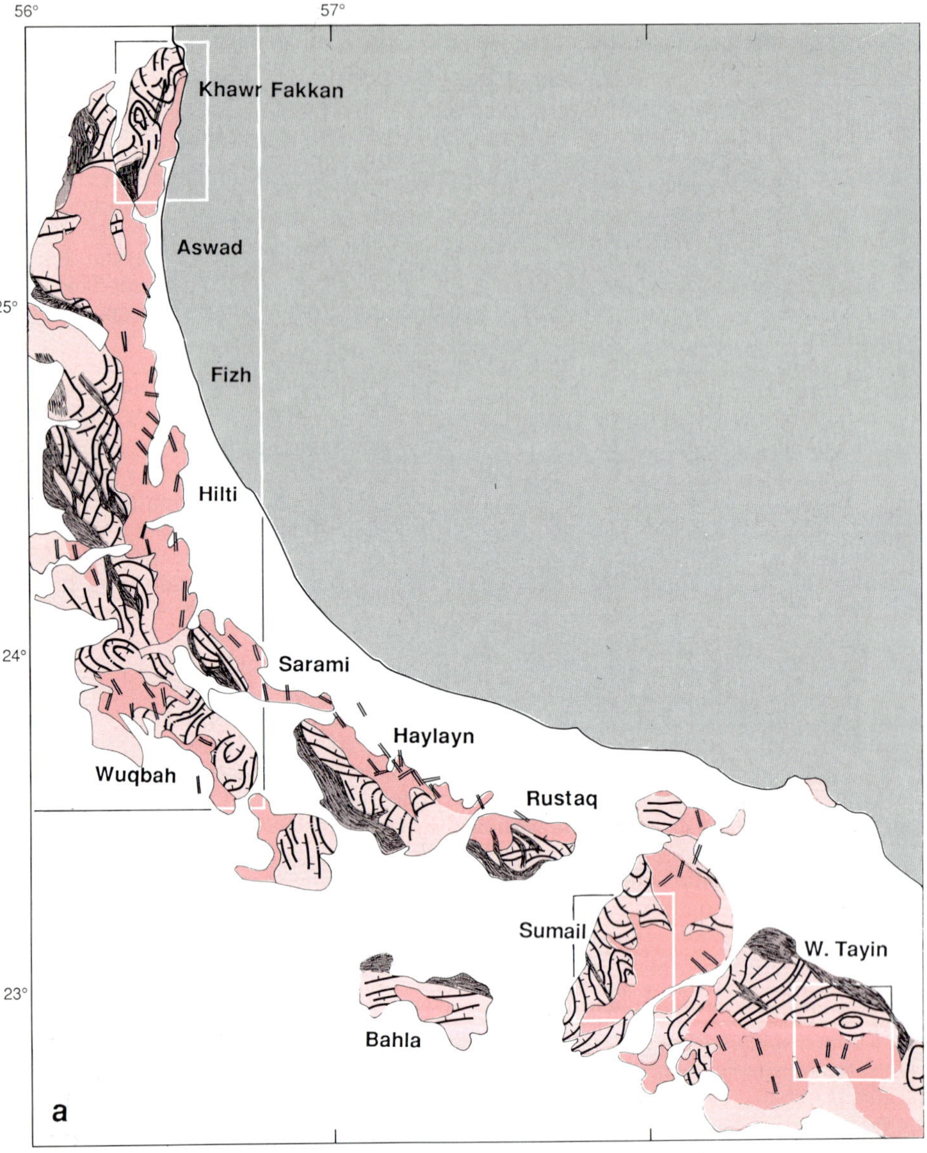


Fig. 3.8. General structural maps of the Oman ophiolites giving the location of the massifs in a) and of the principal wadis in b). a) Map of the trajectories of the foliation planes in the peridotites, showing also the strike of the diabase dike swarm (double lines) ; bold lines : high temperature foliation with indication of dip ; thin lines : low temperature foliation. b) Map of the mineral lineation trajectories ; bold lines : lineations in peridotites, plunging when terminated by an arrow head ; dashed lines : magmatic lineations in gabbros ; arrows : sense of shear flow, single arrow indicating the sense of motion of the upper compartment for flat-lying foliations and double arrow, the sense of shear in the case of steep foliations. Lighter tone : ultramafic section ; deeper tone : mafic section ; superimposed grey : low temperature deformation corridors (after Nicolas et al., 1988a).

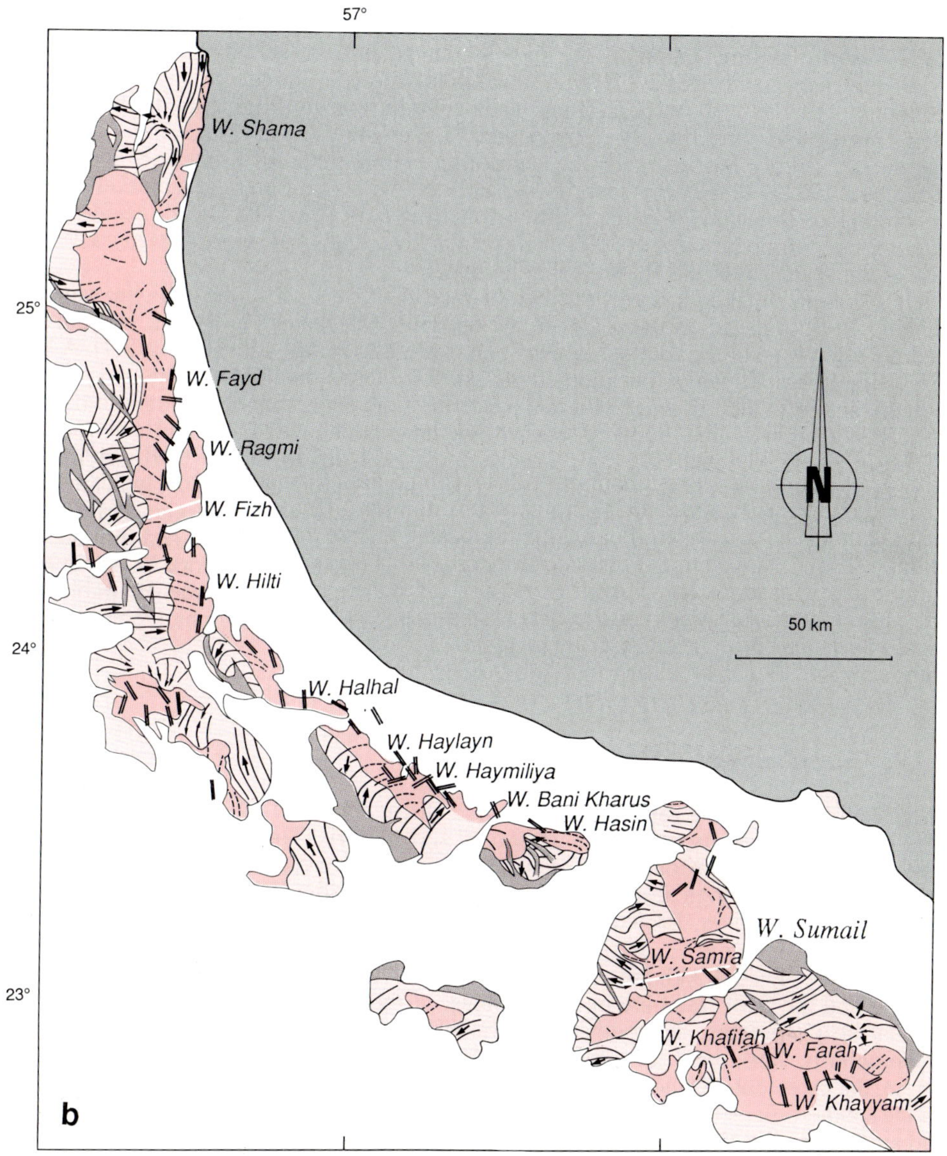
57°
25°
24°
23°
W. Shama
W. Fayd
W. Ragmi
W. Fizh
W. Hilti
W. Halhal
W. Haylayn
W. Haymiliya
W. Bani Kharus
W. Hasin
W. Sumail
W. Samra
W. Khafifah
W. Farah
W. Khayyam
N
50 km
b

Soon after the end of the obduction, during the Upper Maestrichtian (75 Ma), a shallow marine sedimentation on the autochthonous and allotochthonous formations of Oman indicates that the passive margin situation is restored. The Maestrichtian calcareous sediments rest as well on peridotites and contain lateritic fragments of peridotites. Thus, during and soon after its emplacement, the crustal section of the ophiolite nappe was locally deeply eroded and the peridotites submitted to a lateritic alteration (Coleman, 1981). Finally the margin was uplifted with a differential upheaval of the autochthonous limestone massifs of the surrounding Saih Hatat and Djebel Akhdar. The latter massif culminates now at 3000 m above sea level, that is at some 8000 m above the base of the Wadi Tayin and Rustaq ophiolite massifs.

In the present situation it is very difficult to trace the extension of the ophiolite nappe to the NE. In central and northern Oman the ophiolite disappears to the NE beneath the piedmont deposits of the Batinah coastal plain with an average 15° dip (fig. 3.7). A line of Exotics inliers, located a few kilometers East of the last ophiolitic outcrops and parallel to them, suggests the existence of a major normal fault separating the ophiolitic nappes from their eastern source. In many sections, we have documented a progressive rotation Eastward in the sheeted dike unit from a Westward dip to the vertical, within the easternmost kilometers of the ophiolite outcrops ; this rotation, around 20°, indicates that the nappe is similarly tilted Westward (fig. 3.7). It indicates that the Exotics derive from below and not from above the ophiolitic nappes (Lippard et al., 1986) and militates in favor of a major normal fault. The reconnaissance character of the work in gravimetry (Manghnani and Coleman, 1981 ; Shelton, 1984) and the absence of published off-shore seismic data make it impossible to define the relation of the ophiolite nappe with the oceanic lithosphere of the Gulf of Oman.

## *3.3. GENERAL DESCRIPTION OF THE OPHIOLITE*

### *3.3.1. Introduction*

This section presents the general features of the Oman ophiolite, while following sections deal more specifically with their structure and history. The ophiolite represents a fundamentally homogeneous piece of oceanic lithosphere over their 500 km extension. Structural and petrological differences inside the ophiolite belt deal mainly with the secondary magmatism (Browning, 1982 ; Alabaster et al., 1982) and the large scale shear zones (Boudier et al., 1988) which are both essentially developed in the massifs North of the Djebel Akhdar. As mentioned in the preceding section, the Djebel Akhdar and the Saih Hatat are uplifted domes of the continental shelf. The Djebel Akhdar also separates the southeastern Sumail and Wadi Tayin massifs where the paleomoho is horizontal or tilted to the South from the central and northern massifs where the paleomoho is horizontal or tilted to the North-East. The small Bahla massif on the southern flank of the Djebel Akhdar has, with respect to this Djebel, a setting and attitude similar to that of the Wadi Tayin massif with respect to the Saih Hatat (fig. 3.8).

The Oman ophiolite massifs can be thus separated into the group of southeastern massifs (Bahla, Sumail, Wadi Tayin) which have been thrust over the calcareous formations of the continental shelf locally uplifted as the Djebel Akhdar and Saih Hatat domes and the central and northern massifs (Rustaq, Haylayn, Sarami, Wuqbah, Hilti, Fizh, Aswad and Khawr Fakkan ; the two latter northern massifs being located in the Arab Emirates), thrust against the shelf-slope formations as proposed in § 3.2.2.

Because of this fundamental homogeneity, it is possible to organize the general description of the Oman ophiolite around a synthetic log (fig. 3.9). Singularities will be

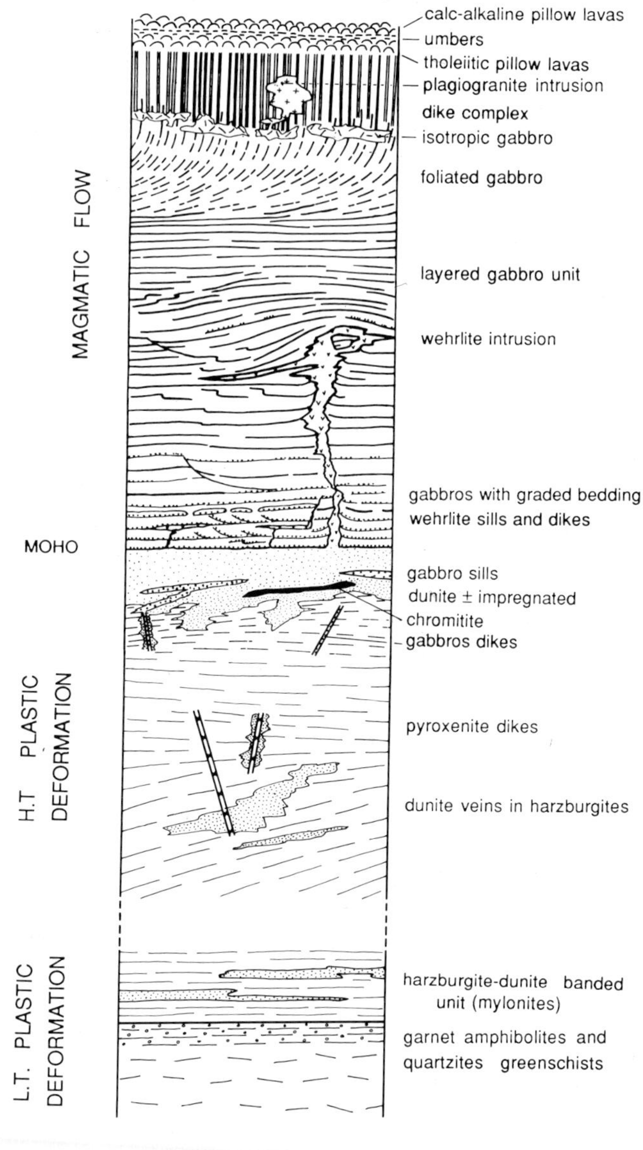


Fig. 3.9. Synthetic log of the Oman ophiolite. Structural orientations are respected, but not the relative thickness of units (modified from Nicolas et al., 1988b).

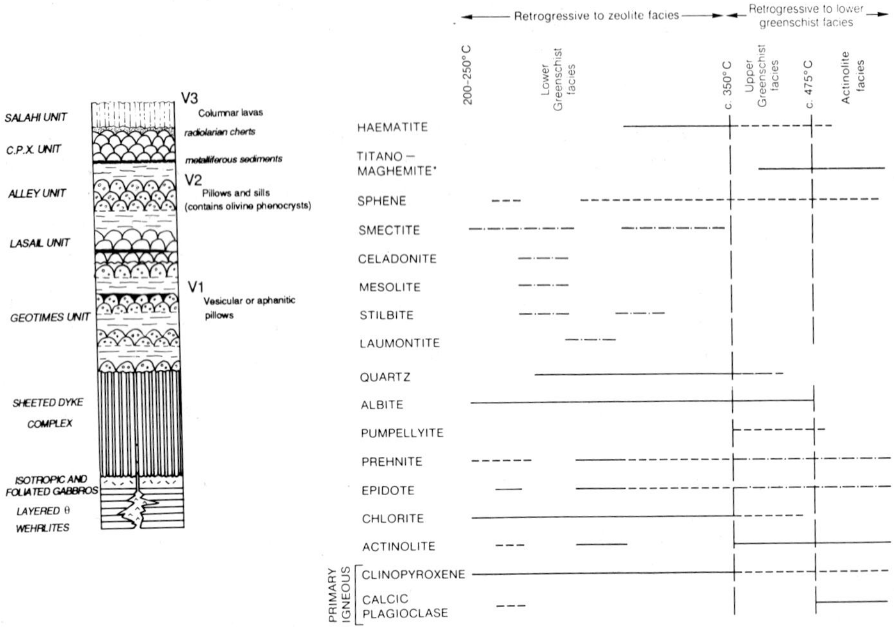


Fig. 3.10. Synthetic log of the ophiolite volcano-sedimentary cover and of the oceanic-floor metamorphism (Lippard et al., 1986 and F. Boudier, unpublished).

mentioned in the course of the description. In this description, it has been decided to skip over the detailed petrology and geochemistry of the mafic units and to insist on the structural aspects. This is in the spirit of this book, and is justified by the fact that most of the work carried out in Oman has been concerned with these aspects. The large pool of knowledge gained from them is summarized in the volumes referred to above (§ 3.1).

### *3.3.2. Mafic section*

*Volcanics and associated sediments* - The first formations belonging intrinsically to the ophiolites are the V1 volcanic flows and pillow lavas of the 'Geotimes unit' (Pearce et al., 1981 ; Alabaster et al., 1982 ; Ernewein et al., 1988) (plate 3.1a) which grade downward into the sheeted dike unit (fig.3.10). These V1 volcanics are locally interstratified with by a few meters of Cenomanian umbers, also called 'metalliferous sediments' ; they have been described by Fleet and Robertson (1980), Robertson and Fleet (1986) and Karpoff et al. (1988) and dated by Tippit et al. (1981). These sediments are also locally interstratified with, or covered by the pillow lavas and flows of a V2 volcanic series. All of these formations are covered by radiolarian mudstones or micritic limestones. Eventually, V3 volcanics cap these pelagic formations. This geological partitioning of the volcanic activity into three episodes is justified on petrological and geochemical grounds by Ernewein et al. (1988), who discuss a more detailed partitioning proposed earlier by Alabaster et al. (1980) (see fig. 3.10 where V3 is equated with the 'Salahi episode' and V2, with the 'Lasail', 'Alley' and 'clinopyroxene-phyric' episodes). The V1 'Geotimes' volcanics, the metalliferous sediments and associated V2 volcanics are interpreted as recording a spreading center activity, with the metalliferous sediments in particular being produced by high temperature hydrothermal activity at mounds, like those of the East Pacific Rise (Karpoff et al., 1988). The discovery of fossil worms in these metalliferous sediments reenforces this interpretation (Haymon et al., 1984). The V3 volcanism would be intraplate, produced some 15 to 20 Ma after crustal accretion, relying on datations made in associated radiolarian (table 3.I). The question of the environment of origin of these lavas and sediments is discussed in § 3.5.3.

*Sheeted dikes* - Below the 'Geotimes' volcanics, the contact with the underlying diabase dike complex is sharp, observed within a few tens of meters to 100 m (Lippard et al., 1986). These dikes, on average 0.8-1 m across, are intrusive one into the other, with one or two chilled margins at their contacts (plate 3.1b). Statistical measurements made by Pallister (1981), Lippard et al. (1986) and our group to detect a preferred facing direction for the chilled margin of the dikes with a single chilled margin have not been very conclusive. The dikes are dominantly tholeiitic in nature, locally highly hydrothermally altered in the greenschist facies (see below). A few picrite dikes are also observed.

*Isotropic and magmatically foliated gabbros* - The isotropic gabbros, diorites and associated plutonic rocks, should correspond to the high level gabbros and intrusives of the Open University (OU) maps and to the high level gabbros of the Bureau de Recherches Géologiques et Minières (BRGM) maps. However, because the limit between the isotropic gabbros and the gabbros affected by a magmatic foliation (the planar-laminated gabbros of the preceding authors) is difficult to trace upsection, and because the foliated gabbros become progressively layered downsection, correspondence between the various data sets are uncertain. Our foliated gabbro unit coincides roughly with the 'cumulate planar-laminated gabbro' unit of the BRGM. Together with the layered gabbros, it is incorporated to the 'cumulate gabbro' unit of the UO map and, on the other

***Plate 3.1.***

***Plate 3.2.***

***Plate 3.3.***

***Plate 3.1. Mafic section in the Oman ophiolite***

a- Pillow lavas from the Geotimes basalts (Wadi Jizzi).
b- Sheeted dike complex in the Hilti area
c- Plagiogranite intrusion in the lower layered gabbro section of Wadi Andam
d- Magmatic breccia with diabase dikes fragmented and partly assimilated by plagiogranite -diorite melt in Wadi Haymiliyah
e- Basal layered gabbros in Wadi Andam. In this dominantly modally graded sequence, the black wehrlitic lens in the lower central part of the photograph is interpreted as a sill.
f- Wehrlitic dikelets (black) cutting the basal layered gabbros in Wadi Hilti. These gabbros are dominantly modally graded, note however the anothosite layer, defining an isomodal layer.
g- Wehrlite dike (black), feeding a wehrlite sill (top of photograph) in the basal layered gabbros of Wadi Haylayn.
h- Wehrlite (black) intruding and folding the basal layered gabbros in Wadi Sumail.

***Plate 3.2. Transition zone in the Oman ophiolite***

a- The Moho in Wadi Andam with horizontal gabbros capping the hill, dunite and harzburgite below
b- Gabbro sills and impregnations in transition zone dunites, parallel to layered gabbros of the crustal section located 50 m above. These facies are strongly plastically deformed at high temperature (Wadi Khafifah). The remarkable parallelism of all layers is ascribed to tectonic transposition. Hammer in a circle for scale.
c- Gabbro sills in the dunites of the transition zone which are parallel to the layered gabbros cropping out 50 m above. The sill character is demonstrated by the rooting visible in the lower righ corner (Maqsad area).
d- High-T plastic deformation in impregnated dunites of the transition zone of Wadi Bani Kharus. The plastic and not magmatic character of the deformation can be demonstrated in the field when diopside augen are visible, as here in the center right part of the photograph. The large plastic flow is responsible for the tectonic transposition of these impregnations, to become parallel to the Moho (which is 20 m above).
e- The largest dunite body of Oman (13 km long and 2 km wide) in the Batin area. This flat body, between the overlying mafic section and the underlying harzburgites, constitutes the smoother and lighter colored foreground contrasting with the more rugged and darker background composed of harzburgites (view toward NE throughout the width of the dunite body).
f- Network of residual dunite veins and bodies (light color) within the harzburgite (darker color) section of Wadi Tayin.
g- Diffuse plagioclase-rich impregnation grading into a dikelet in dunites from the transition zone of Wadi Tayin.
h- Various stages of high-T plastic deformation (vertical foliation) in plagioclase-rich impregnations within dunites from the transition zone of Wadi Tayin.

***Plate 3.3.***

***Dispersal of dunites in the transition zone of Oman***

a- Diffuse gabbro dike, 1 m thick, grading into plagioclase-rich impregnations in the dunites from the Maqsad diapiric area.
b,c,d- Network of gabbroic irregular sills and dikes in the dunites of the transition zone of Wadi Bani Umar al Gharbi. Due to increasing melt/solid ratio, dunite fragments of various size (c,d) can be dispersed in a gabbroic matrix.

***Harzburgite section and basal aureoles in Oman***

e- Typical view of fresh harzburgite in the middle of the ultramafic section with parallel and moderately dipping layering and foliation (Wadi Hayl)
f- Dunite banding (light color) in the harzburgites (dark) a few hundred meters above the metamorphic aureole. The large low-T strain is responsible for tectonic transposition in parallel streaks of dunite veins which above would probably look like those of plate 3-2 f (Balah massif).
g- 'Green Pool' metamorphic aureole in Wadi Tayin. In the background, steep low-T foliation and layering in harzburgite. Top of the hill in the foreground, amphibolites, base of the hill, greenschists
h- Partial melting in the metamorphic aureole, developed within 50 m from the peridotites at the expense of phyllites (Masafi area, Emirates). This biotite migmatite is cut by a granitic dike which, although intersecting the foliation of the gneiss, is itself foliated. This demonstrates that the melt was produced during the deformation, being induced by the peridotite overthrusting.

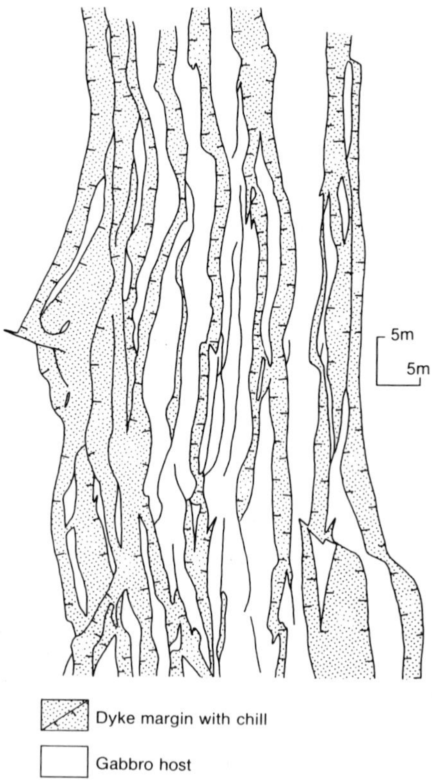


Fig. 3.11. Field sketch of the zone of rooting of the diabase sheeted dikes into the high level gabbros (Rothery, 1982).

hand, with the isotropic gabbro, to the 'massive gabbro' unit of the American Group map in Wadi Tayin.

At the base of the sheeted dike unit, amphibole-clinopyroxene gabbro screens become progressively more abundant between the diabase dikes. The gabbros are either poorly foliated or recrystallized into isotropic gabbros. The transition between the sheeted dike unit and the highest gabbros has been studied in detail in the Wuqbah massif by Rothery (1982)(fig. 3.11). The transition operates there through a zone of increasing diking which is from a few meters to a few tens of meters thick, as already reported by Pallister (1981) in Wadi Tayin. The transition is thus rather sharp, although the underlying gabbros can be locally invaded by swarms of diabase dikes. The diabase-gabbro transition zone coincides more or less with the horizon of extensive hydrous recrystallization of foliated and layered gabbros into isotropic gabbros. The isotropic gabbros display a large grain size variation probably related to water circulation, with development of secondary green amphiboles in more dioritic facies and wet anatexis responsible for injection of plagiogranite melts. The isotropic gabbros and associated plagiogranites show mutual intrusion relations with the diabase dikes, although they predominantly intrude the sheeted dike unit (Pallister, 1981 ; Smewing, 1981). In Haylayn massif, breccias of layered-foliated gabbros enclosed in an hydrous dioritic matrix mark the sheeted dike-gabbro transition. Sills and plugs of plagiogranite, several tens of meters across and commonly containing blocks of diabase, also intrude this level.

The upper gabbros have an homogeneous grain size. Compositionally, they range from dry clinopyroxene gabbros, locally noritic (presence of orthopyroxene) to amphibole gabbros. The norites, also marked by an enrichment in iron, represent for Juteau et al. (1988) a closed-system evolution of the magma chamber and could thus be used as indicators of a dying ridge system. The common black amphibole is foliated and lineated, whereas a prismatic bright green amphibole is often more disordered ; the black amphibole crystallized during magmatic flow and the green one, later. Close to the transition to the sheeted dike unit, the layering, dominant in the underlying gabbros, becomes fantomatic and the foliation itself tends to disappear. Downsection, foliation and layering are always associated. They are parallel or depart by less than 20° (§ 2.4.1.). Because of their progressive transition to layered gabbros, it is difficult to ascribe a given thickness to the isotropic and foliated gabbros unit. Locally they can practically disappear like in sections of the Wuqbah massif (Rothery, 1983) and in Wadi Haylayn where they are replaced by a gabbroic breccia.

An important structural feature of this gabbro unit, already noted by Browning (1982) and Rothery (1983), is the rotation of the magmatic foliation from the flat-lying attitude measured in the lower layered foliated gabbros, to an attitude parallel to that of the diabase sheeted dikes (figs. 3.9. and 3.12). The overall rotation of the foliation occurs within the last 500-1000 m below the sheeted dike unit, but it is only when the first diabase dikes appear in the section that the foliation swings into parallelism with their orientation. Accordingly, the mineral lineation steepens rapidly ; this final accordance is often difficult to trace because the foliation tends to vanish and to be destroyed by the hydrous recrystallization.

*Layered gabbros* - This unit is characterized by its ubiquitous layering, always associated with a magmatic foliation and a lineation, induced by magmatic flow (§ 2.4.1 ; Nicolas et al., 1988b). It can be equated with the 'cumulate layered gabbro' unit in the BRGM maps and the 'layered gabbro' unit in the American Group Map. It must be emphasized that all sections through this unit in Oman ophiolites show this magmatic deformation, more or

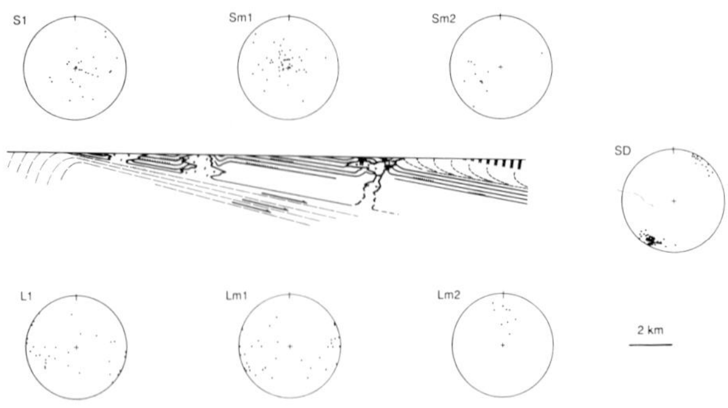


Fig. 3.12. Samra cross section through the crustal sequence, location in fig. 3.8b. Dashed thin lines : S1, high-T foliations in the margin of the Maqsad diapir ; small dots : dunites ; L1 : corresponding mineral lineations ; bold solid lines : Sm1 layering in gabbros, modally graded when dotted ; Lm1 : corresponding mineral lineations ; bold broken lines : Sm2 magmatic foliations in gabbros ; Lm2 : corresponding mineral lineations ; double lines : SD sheeted dike complex ; V : wehrlite intrusions.

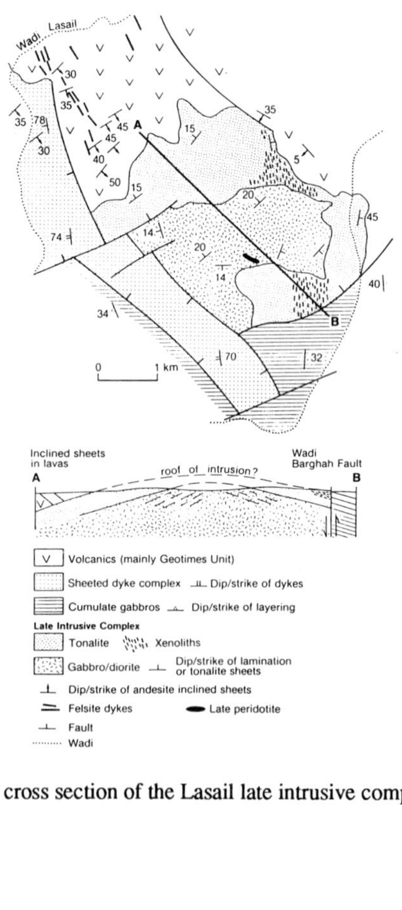


Fig. 3.13. Map and simplified cross section of the Lasail late intrusive complex (Lippard et al., 1986).

less severely imprinted in the gabbros. The layering is marked by various proportions of the three main rock-forming minerals : plagioclase, clinopyroxene and olivine. It is dominantly an isomodal layering in the middle to upper section, and a modally graded layering (graded bedding) in the lower section, following the definitions of § 2.4.1. The graded bedding evokes a sequential accumulation of crystals with olivine dominant at the base of the sequence and plagioclase at the top (plate 2.1a). Browning (1982) presented a detailed petrological study of a typical 37 cm thick sequence. A few tens of individual sequences can be followed in good outcrops. The isomodal layering is marked in the field by wehrlite (plates 2.1b and 3.1e) and anorthosite (plates 2.1d and 3.1f) lenses. Associated or not, they define a contrasted layering which is more conspicuous in the lower and middle parts of this gabbro unit. No rule has been found explaining their occurrence at a given level or within a given association ; their occurrence seems entirely random. The wehrlite lenses can be locally traced into wehrlite dikes (plate 3.1g), and consequently they are interpreted as sills or more irregular intrusions, tectonically transposed by the large magmatic flow into lenses (Nicolas et al., 1988b). The anorthosite lenses, usually only a few centimeters thick, can be inserted at any level within a sequence of graded bedding. They would represent an injected liquid and not the product of a magmatic accumulation. It is suggested that they derive from a melt segregated by the Bagnold effect, that is by the dispersive pressure induced by velocity gradients within a flowing magma.

As mentioned above, the transition between the layered and the foliated gabbros units is very gradual. It corresponds to a progressive disappearance of the layered character, at least partly due to the hydrous crystallization of amphiboles in a flowing medium. It contrasts with the lower contact with the mantle peridotites which usually takes place within a few meters, often less (see next section). The layering in the lower gabbros is always parallel to the plane of contact with the peridotites and to the foliation in the uppermost peridotites, provided this contact is not faulted. This is demonstrated by systematic structural measurements (Nicolas et al., 1988b). We have equated the gabbro-peridotite contact with the Moho (§.2.2.1), and considered that it was horizontal in the paleo-ridge reference frame. In most sections, the layering seems to remain flat-lying through most of the layered gabbro unit. Oblique intersection with diabase dike swarms in the upper layered gabbros (Nicolas et al., 1988b) suggests, however, that the upsection layering may progressively steepen. Estimating the layering attitude throughout this unit is a difficult task because the layering is commonly distorted by open folds. In a few sections, these folds can be related to the detachment and obduction (§ 3.4.3) ; however, it is now realized that most folds have been induced by wehrlite intrusions (see below). In Wadi Tayin, regionally steep attitudes may also be due to listric faulting in the ridge vicinity (§ 3.4.2.).

*Magmatic intrusions and recrystallizations* - As mentioned above, the magmatic intrusions tend to be more abundant in central Oman, where their petrology and geochemistry has been studied in detail (Lippard et al., 1986; Juteau et al., 1988a and b), although Pallister and Hopson (1981) have also described them in the southeastern Wadi Tayin massif.

The magmatic intrusions can be divided into the wide plagiogranitic group, ranging from gabbros and diorites to plagiogranites with their volcanic equivalents, and the wehrlitic group with dominantly wehrlites, troctolites, troctolitic gabbros and subordinate dunites ; picritic dikes are related to this wehrlitic group (Juteau et al., 1988b).

The intrusions of the *plagiogranitic group* range from diffuse segregations and dikes invading the upper gabbros and the sheeted dikes units, to kilometer-sized intrusions into these units and the lavas (fig. 3.13) (plate 3.1c). An interesting and common facies is composed by a breccia of doleritic fragments into a plagiogranite matrix (plate 3.1d). It

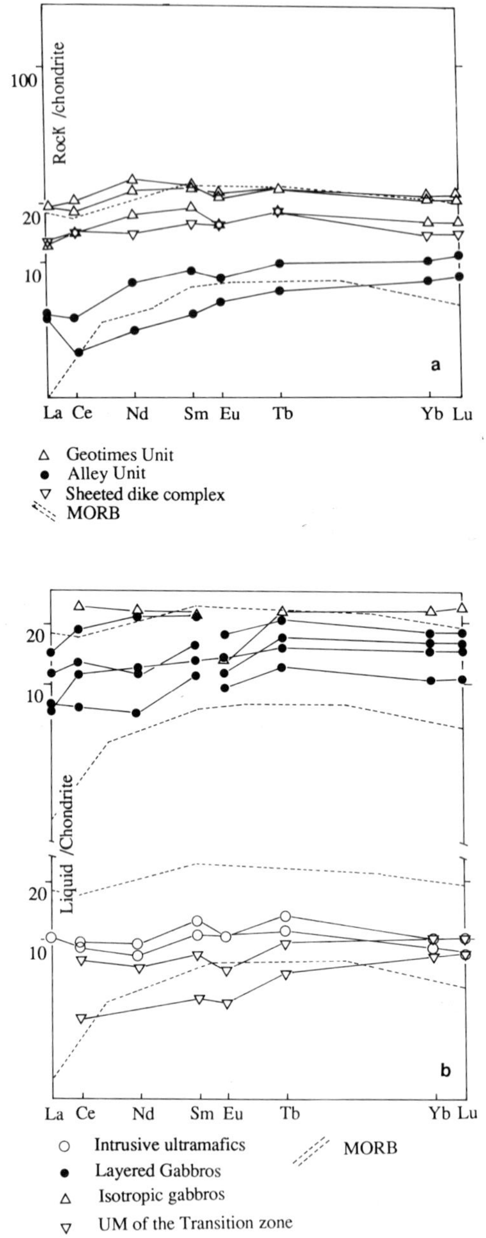


Fig. 3.14. Trace elements patterns normalized to chondrite values a) whole rock and b) separated clinopyroxenes. a) $V_1$ and $V_2$ volcanics and sheeted dike unit. b) Gabbros and wehrlites. Dashed contours : MORB range (a) modified from Alabaster et al., 1982 ; b) Lachize, Rapport Dea Montpellier, 1988).

constitutes dikes and small intrusive massifs. Mutual brecciation of the two components, lobate contours of the inclusions and common magmatic deformation of the inclusions and their matrix in the feeder dikes of larger intrusions suggest that they can represent a mixture of two magmas.

Some plagiogranite dikes are rooted into the isotropic gabbros and diorites formed by the hydrous recrystallization of the foliated layered gabbros. Hence, we ascribe them to the wet anatexis of these gabbros, a process documented elsewhere by Pedersen and Malpas (1984). This is not at all exclusive of another mode of origin by crystallization, in hydrous conditions, of the residual liquid of the magma chamber, the conclusion attained by Lippard et al. (1986) on the basis of geochemical analyses in the plagiogranites of central Oman. To these two possible origins, Boudier et al. (1988) propose to add a third one, which is the hydrous melting of the granulite-amphibolite metamorphic aureole below the peridotite nappe (§ 3.3.4.).

Andesite to dacite dikes, believed to belong to this plagiogranitic group, locally constitute swarms within the sheeted dikes unit, like in the Zabin and Rustaq areas, where they adopt a more westerly strike than the NW-SE diabase dikes.

The importance of the *wehrlitic intrusions* in the crustal section had been underestimated so far, as emphasized by Juteau et al. (1988b) and Benn et al. (1988), who also describe them in detail. In many massifs they constitute up to 30% of the volume, as well illustrated by Reuber's (1988) detailed maps of the crustal section of northern Oman. These intrusions are dominantly composed of olivine, diopside, some plagioclase and locally hornblende. They are observed at every level within the crustal sequence but are far more abundant in the plutonic section. In the basal gabbros, they can constitute small dikes and sills (plate 3.1g, h). They may attain 5 km in diameter in the lower crustal section and do not exceed a few hundreds of meters in the upper section. Picritic dikes radiating from the uppermost intrusions have been traced up to the upper extrusives (Juteau et al., 1988b) (fig. 3.10). An important feature is, at least in the deeper wehrlitic and gabbroic intrusions, the absence of chilled margins against the layered gabbros. On the contrary, one observes magmatic reactions and deformation of the gabbros, expressed by breccias, shear bands and, more commonly, by an open folding clearly induced by the intrusion (plate 3.1h).

Juteau et al. (1988b) and Benn et al. (1988) insist on the fact that the ultramafic-mafic compositional layering which belongs intrinsically to the layered gabbro unit, and is crystallized from the same melt as graded bedded sequences, is mainly present at the lowest levels of this unit. The ultramafic layers met upsection are sills belonging to the wehrlitic magma ; they show clear-cut contacts with the surrounding gabbros which can be traced locally into discordant intrusive contacts. They have been either injected as sills (plate 3.1g) or as dikes (plate 3.1h) or stocks, subsequently transposed into parallelism with layering by the very large magmatic shear flow (§ 2.4.3). Still higher in the sequence the wehrlites take preferentially the shape of intrusive plugs.The parental affinity of the layered gabbros and wehrlites is confirmed by Nd isotopes (Michard- Vitrac, unpublished), presenting similar values of εNd, respectively 7.5 and 8.2, and suggesting a slightly more depleted source for wehrlites.

The source of these intrusions would be located in the transition zone below the Moho because one looses track of wehrlite dikes and plugs in this zone, and because wehrlites have never been observed down into the harzburgites. The above-cited authors also believe that the ultramafic magma was injected within the magma chamber or close to it, in still hot gabbros. They disagree somewhat about the cause : for Juteau and his co-workers, the intrusions derive from the magma trapped in the transition zone which would be expelled during the first stage of compression at the ridge related to the detachment (§ 3.4.3), whereas for Benn and his co-workers, the wehrlitic intrusions represent a normal product of a fast ridge activity, the melt being expelled from the transition zone when the mantle flow diverging from diapirs below the ridge is squeezed laterally (§ 10.3). The ultramafic nature of the magma is ascribed to the mixing of the residual melt with fragments of the disaggregated dunites of the transition zone (§ 10.3).

*Trace elements signatures* - Geochemically, lavas $V_1$ and $V_2$ differ by the more 'primitive' characters of $V_2$ basalts relative to $V_1$ ('Geotimes'). In $V_2$, Lasail and Alley volcanics follow the same trend of fractionation from basalts to felsic lavas ; the Cpx-phyric unit (of picritic composition) represents the least fractionated term in the extrusive section (Lippard et al., 1986). Figure 3.14 shows the trace elements signatures of the extrusives (sheeted dikes and volcanics) and of the plutonic gabbros and wehrlites, obtained respectively on whole rocks and on separated clinopyroxenes. The patterns confirm the geologically established parentage of the $V_1$ 'Geotimes' lavas with the diabase sheeted dikes and underlying gabbros. These formations share a common MORB signature. The relation between the intrusive wehrlites and those associated with dunites in the transition zone is also confirmed. Interestingly, their common trace elements pattern is closer to the $V_2$ 'Alley' volcanics pattern than to the main sequence one. This is well in agreement with the timing of wehrlite intrusions and $V_2$ extrusions which both occur very close to the ridge, in a still hot crust for the wehrlites (see above). The high dispersion of REE patterns in plagiogranites (Lippard et al., 1986) may account for multiple possible origins of these differenciated intrusions, as suggested by field evidence.

*Hydrothermal alteration* - Alteration by water circulating at various temperatures is observed throughout the volcanics, sheeted dikes and the upper gabbros. Below, the hydrous activity is concentrated in the vicinity of faults and within local zones, possibly more tectonized (Nehlig and Juteau, 1988). In the uppermost peridotites, this activity is recorded by a talc-tremolite alteration of orthopyroxene and more rarely by an antigorite-serpentinization of olivine. It occurs locally within the upper 2 km below the crustal section or in the vicinity of large mylonitic shear zones (§ 3.4.3) in which the motion is accompanied by an hydrous magmatism (Ceuleneer, 1986).

A complex metamorphic history of hydrothermal alteration has been documented by Alabaster (1982), Alabaster and Pearce (1985), and Stakes et al. (1984), who extended the former oxygen-isotope studies of Gregory and Taylor (1982). These authors relate this metamorphism to hydrothermal cycles taking place in the oceanic floor at decreasing temperatures. Lippard et al. (1986) give a detailed account of the mineral parageneses met with increasing depth and estimate the thermal gradient to 150°/km for the 4 km thick crust affected by hydrothermal alteration (fig. 3.10). On the basis of a fluid inclusions study, Nehlig and Juteau (1988) envisage a 240°C/km gradient for the accretion stage and 30°C/km for the off-axis stage.

In the volcanics an increasing metamorphism is recorded from top to bottom, grading from the brownstone facies (clays), to the zeolite facies (zeolites, calcite and celadonite) and the greenschist facies (albite, epidotes, quartz and sulphides). These secondary assemblages fill the voids between pillows and the cooling fractures inside the pillows and lava flows ; they also circulated along fractures cutting through these formations. Progressive enrichment in sulphides can lead to massive sulphide deposits like in the Zuha Gossan, near Wadi Salahi, and in the Lasail, Bayda and Aarja mined districts in Wadi Jizzi. It is in the Bayda mine that Haymon et al. (1984) have discovered the fossil worms evoked in a preceding section. In such areas, due to more intense rock-water interaction, the albite-epidote-prehnite assemblage is replaced by a quartz-chlorite-sulphide assemblage which is a reaction already described in oceanic hydrothermalism.

In the sheeted dike unit, contiguous dikes are variously affected by the greenschist facies metamorphism related to water circulating in fractures. The fractures are rich in epidote with subordinate quartz and sulphides. There is also, together with this introduced water, an alteration caused by residual fluid trapped within crystallizing dikes. Nehlig and Juteau (1988) show that the fractures are preferentially oriented parallel to the dike system and are located along their margin. They conclude that the hydrothermal circuits below the oceanic ridge are characterized by a dominant along-strike vertical attitude. This attitude can be traced down into the layered gabbros where the fractures are predominantly amphibole or zoisite-bearing.

Interestingly, the plumbing corresponding to hydrothermal circulation at 200-400°C seems to be different from the most primitive one, which is responsible for the hydrous recrystallization of the upper gabbros into isotropic amphibole gabbros and diorites and for local anatexis, and which occurs around 700-800°C. Our measurements show that the patches of isotropic gabbros within the foliated gabbros are crudely shaped and internally banded horizontally when the crustal unit is restored into its ridge orientation. A subordinate orientation in these isotropic gabbros is that of the sheeted dikes. This suggests that the first hydrous circuits were closing downward in this horizon where the foliated gabbros recrystallize into isotropic gabbros and where the sheeted dike unit is rooting.

*Thickness of the units of the mafic section* - Many data have been published on the thickness of the main units of the mafic section (Pallister, 1981 ; Pallister and Hopson, 1981 ; Browning, 1982 ; Rothery, 1983 ; Dahl, 1984; Lippard et al., 1986 ; Juteau et al., 1988a ; Reuber, 1988). For the $V_1$ 'Geotimes' volcanics below the umbers level and for the sheeted dikes units the estimations are straightforward. Typical results are between 400 m and 1600 m for the Geotimes units and 1000-1700 m for the sheeted dikes unit. This gives an average thickness for these volcanic and hypovolcanic units of around 2500 m.

Estimating the thickness of the plutonic section is more difficult for several reasons : as seen above, there is no consensus on the definition of the constituent units ; the thickness would greatly vary from one section to the next, mainly in the layered gabbros which would vary in the NW-SE direction for example, from 150 m to 1800 m in the Fizh massif (Reuber, 1988) and from 300 m to 2300 m in the Haylayn massif (Juteau et al., 1988a), reflecting for these authors the variable thickness of magma chambers along ridge strike ; estimations depend on whether or not the wehrlite intrusions are considered ; eventually, the estimations are based on insufficient structural data and disputable assumptions about the layering and foliation attitude.

New estimates are now proposed which are based on our detailed structural mapping throughout the belt, and which integrate the concept of rotations in the layering-foliation attitudes due to identified causes (upwarping of layering upsection, effect of listric faults, of wehrlite intrusions,...). The layered gabbro unit varies from 1200 m to 2600 m, clustering around 2000-2500 m and the foliated and isotropic gabbro unit, from a few tens of meters to 1600 m, clustering around 500-1000 m. The plutonic section unit, including the wehrlite intrusions, would vary in thickness between 1700 m and 4100 m, with an average value around 2750 m. This variation is real, though the lowest estimate of 1700 m does not reflect the thinner possible sections (see above).

The total thickness of the crustal section in the Oman ophiolite, excluding the V2 volcanics, is around 5000 m, with expected variations of over 1000 m along ridge strike.

### *3.3.3. Ultramafic section*

*Transition zone* - Below the lowest layered mafic, the transition to the harzburgitic mantle

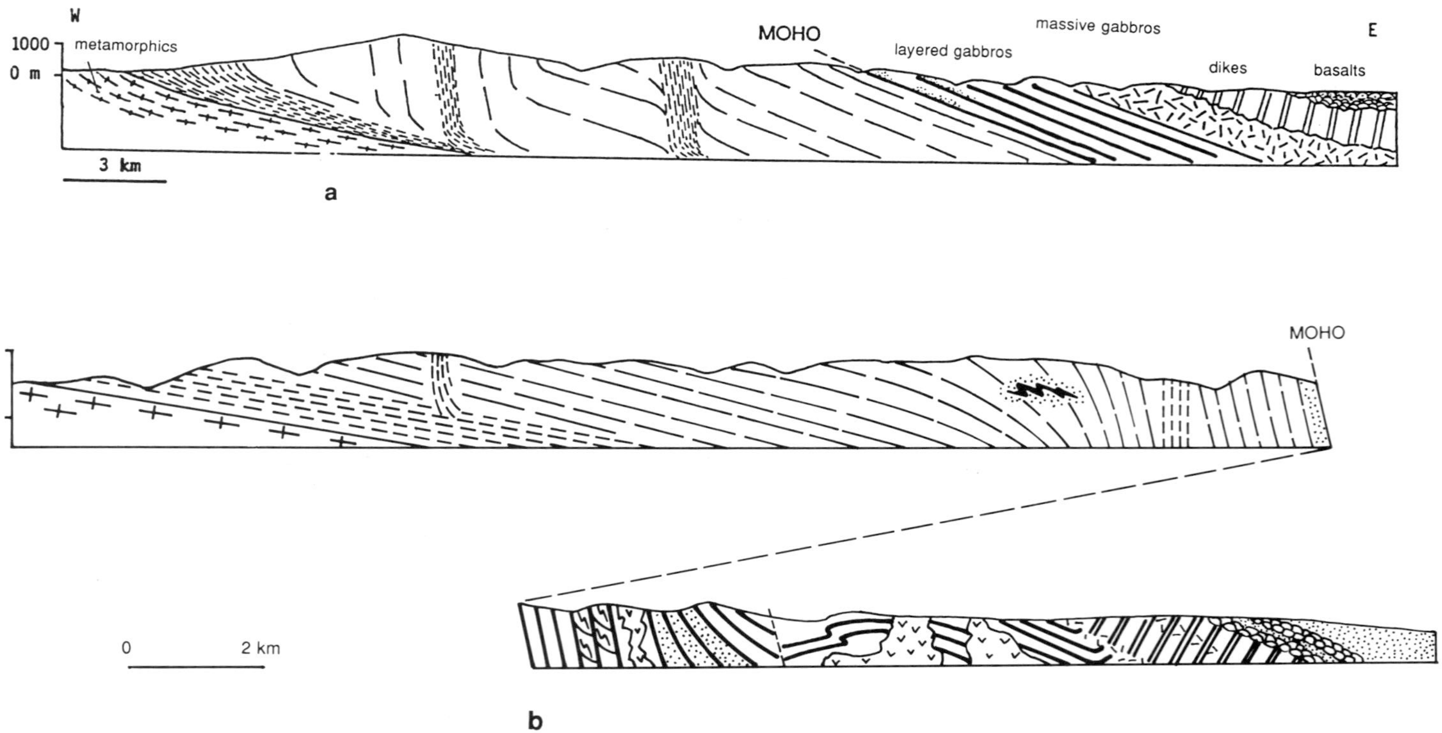


Fig. 3.15. Cross sections (location in figure 3.8b), showing the relation between high-T (wide spaced dashes) and low-T (narrow spaced dashes) structures in the ultramafic section and how the low-T shear zones rotate into the basal thrust zone. a) southern Fizh massif, b) northern Fizh massif illustrating also the large rotation in the structures occurring locally in the vicinity of the Moho (Boudier et al., 1988).

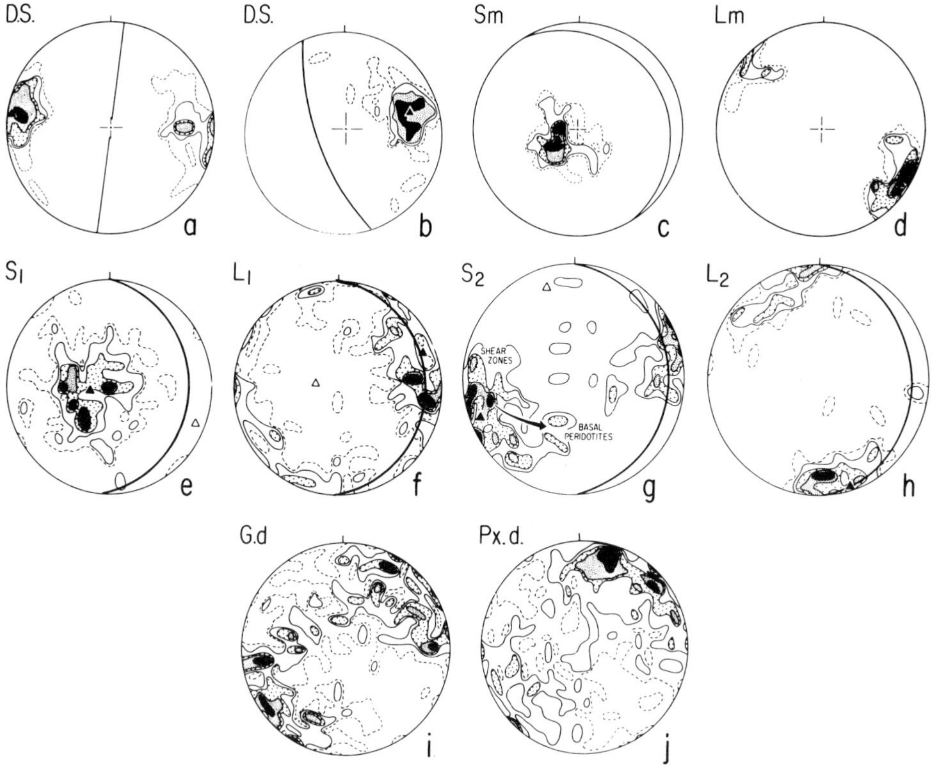


Fig. 3.16. Stereonets of the principal structural elements in the South-Fizh and Hilti massifs. a) D.S., diabase sheeted dikes in Hilti (231 measurements ; contours : 1, 3, 4, 10, 18 %). b) D.S., diabase sheeted dikes in Fizh (67 measurements ; contours : 2, 3, 6, 10 %). c) Sm, layering plane in Hilti gabbros (46 measurements ; contours : 2, 4, 6, 8, 10 %). d) Lm, magmatic mineral lineation in these gabbros (45 measurements ; contours : 2, 4, 6, 8, 10 %). e) S1, high-T foliation in peridotites from both massifs (184 measurements ; great circle : calculated paleo-Moho ; contours : 0.5, 2, 3, 4, 5 %). f) L1, mineral and aggregate lineations in these peridotites (173 measurements ; contours : 0.5, 2, 3, 4, 5 %). g) S2, low-T to mylonitic foliation in peridotites from both massifs showing the continuity from shear to thrust zones (94 measurements ; contours : 1, 2, 3, 4 %). h) L2, lineations in these peridotites (103 measurements ; contours : 1, 3, 5, 7, 10 %). i) Gd, gabbro dikes from ultramafic section in both massifs (227 measurements ; contours : 0.5, 1, 1.5, 2, 2.5 %). j) Px.d, pyroxenite dikes from same massifs (516 measurements ; contours : 0.5, 1, 2, 3, 4 %) (Ceuleneer, 1986).

section is always marked by the occurrence of plastically deformed dunites. In some sections, like Fayd or Bani Kharus, they are only a few meters thick and grade downward into strongly depleted harzburgites and, after a few hundreds of meters, into normal harzburgites. The thickness of the dunites is more commonly around 50 m with numerous exceptions : in the Batin area of the Wadi Tayin massif and in the Maqsad area of the Sumail massif, they attain a few hundreds of meters ; on the contrary, in many places, like in Wadi Bani Kharus (Rustaq massif) or Wadi Hilti (Hilti massif), they do not exceed a few meters.

In terms of mineral composition, the mafic-ultramafic transition is smoothed by the presence of ultramafic layers within the basal crustal sequence (plates 2.1b, c and 3.1e) and by that of gabbro dikes, sills and impregnations in the dunites and the uppermost harzburgites (plate 3.2 b, c, d). The magmatic impregnations, composed of plagioclase and poikilitic diopside grade from diffuse patches (plate 3.2g) to heavily impregnated zones (plate 3.3a, b) and finally to gabbro dikes which can incorporate blocks and fragments of dunite (plate 3.3c, d). These impregnations have penetrated the peridotites before or during the plastic deformation because they are usually deformed along with the olivine matrix, but locally they postdate this deformation (plate 3.2h) (§ 2.5.3). The melt responsible for these impregnations was saturated in sulfur, as shown by the local enrichment of the dunites in sulfur (up to 540 ppm S) and the occurrence of Cu-Ni-Fe sulphides, associated with the introduced plagioclase and clinopyroxene (Lorand, 1988). Incidentally, these sulphides have a higher Cu/Ni ratio that those accompanying the orthopyroxene in the depleted harzburgites.

An important discovery has been the presence of gabbro sills which are strictly parallel to the overlying layered cumulates. The distinction in the field between these intrusions and a mafic-ultramafic cumulate sequence would be impossible if the observation of dikes feeding these sills had not been made repeatedly (plate 3.2b, c). Misinterpretating this intrusive series for a magmatic sequence would result in assimilating the dunites to the cumulate sequences and in extending the base of the cumulates down to the harzburgites as commonly accepted (§ 10.4). These sills can attain 2 m in thickness ; they are internally differentiated and can grade into an impregnation zone (plate 3.2d). The numerous gabbro dikes intruding the transition zone some of which feeding the impregnation zones, are described below. More exceptionally, olivine-spinel-clinopyroxenite sills are observed interstratified with the dunite together with gabbro sills ; they range in thickness from several tens of centimeters to several tens of meters. Here again, pyroxenite dikes have been observed branching on these sills.

In contrast with the petrological continuity, structurally, the transition zone is a zone of sharp contrast. In Oman, the magmatic structures are preserved down to the limit with the dunites. Unusually, in Wadi Fayd, the plastic deformation overprints the cumulate structure for a few meters above the dunites. The plastic deformation of the dunites is usually neither apparent in the field, nor under the microscope where the texture mimicks that of an adcumulate. It is mainly revealed by the olivine lattice fabrics which are surprisingly strong (fig. 2.7a) and are characteristic of large strain at very high temperature where grain boundary migration is very active. Hyper-solidus condition in the presence of melt has been proposed (§ 2.5.5.) to explain such textures and fabrics. The foliation is, however, clearly visible in the associated harzburgites and in the impregnated dunites (plate 3.2d, h). Thus, the top of the transition zone is marked by an abrupt disappearance of plastic deformation. The absence of plastic deformation in many gabbro dikes proves that this deformation ceased earlier than the magmatic activity.

Chromite pods are restricted to the transition zone and the domain of the upper harzburgites. They are specifically considered in § 10.5.

*Peridotite section* - This section is homogeneous in terms of petrologic features along a profile penetrating several kilometers downsection within a given massif, as well as from one massif to the next (Ceuleneer, 1986).

As mentioned above, the uppermost harzburgites are strongly depleted, mainly in the vicinity of dunite veins and bodies. This progressive depletion starts a few meters from a vein where the orthopyroxene crystals or aggregates decrease in volume and the associated spinel recrystallizes, as shown in figure 10.6. The overall depleted character is thus due to the abundance of dunites and to the width of each depletion zone between them and the harzburgites. Interstitial clinopyroxene and plagioclase are observed with an irregular but pervasive distribution in both the dunites and the harzburgites ; their fraction is usually in the one percent range or below. These minerals are secondary and related to the magmatic impregnation (§ 2.5.3.).

Texturally, the top harzburgites in Oman are particularly coarse-grained with an important recrystallization of the olivine and even of orthopyroxene, giving aggregates of mosaic-shaped grains flattened parallel to the foliation (fig. 2. 7a, b). The olivine fabrics are also very strong (fig. 2.7a,b and 3.20c and d). The bulk of the peridotites displaying these coarse facies are cut at any level by vertical shear zones which reflect relatively high-T conditions (§ 3.4.3).

Downsection, the monotonous mass of harzburgites tends to be well layered, parallel to the foliation (plate 3.3e). The compositional layering is a few centimeters thick, defined by dunitic or pyroxenitic (essentially orthopyroxene) diffuse segregations. Otherwise, these harzburgites contain less interstitial diopside than the upper ones. They are now porphyroclastic, with a well recovered substructure ; the olivine fabrics are also weaker (fig.2.7c). This reflects a plastic strain probably less intense than in the upper peridotites, still produced at high-T, but in the absence of a molten phase.

*Dikes sequence* - From the layered gabbros level to the lowest peridotites, there is a remarkable organization in the nature of the dikes and their sequence of intrusion with respect to the thermal evolution of the intruded peridotites and their plastic flow history. This relation conforms with the general analysis presented in § 2.5.2. and suggests an origin of the various dikes during accretion at a spreading center. Two or possibly three distinct generations of dike and sills can be traced using structural criteria (Nicolas and Jackson, 1982). The *younger one* comprises, from the Moho downwards, a series of diabase, gabbro, websterite dikes and dunite veins possibly not continuous ; the *older one* includes a series of gabbro sills, chromite-dunite veins and websterite-dunite layers. Other dikes completely independent from these sequences are very rare ; they are typically meter-sized, diabase and plagiogranite dikes which cross-cut without any interaction the peridotites as well as the gabbros. In Wadi Tayin, diabase dikes have been observed cutting the basal peridotites affected by the detachment-related deformation ; they evidently postdate this detachment. The plagiogranite dikes have no chilled margins and are plastically deformed, indicating that they were emplaced in peridotites which were still at temperatures of several hundred degrees.

The rare mafic dikes in the layered gabbro section belong to the *younger series*. They are diabase dikes cutting the layering at high angles. A transition from such diabase dikes to fine grained clinopyroxene-hornblende-plagioclase gabbros occurs within the lowest gabbro section or within the transition zone. The microgabbro dikes can be remarkably abundant in the transition zone, mainly where it has been affected by shear zones, extending downward usually no more than 1km. They occur in swarms of dikes, each 5-20 cm across with sharp contacts and numerous inclusions of acute wall rock fragments grading into breccias (plate 2.2h). This is ascribed to hydrofracturing

(Nicolas, 1986a) in a peridotite rendered brittle by its low temperature. These dikes are always at a high angle with respect to the paleo-Moho and tend to be parallel to the dikes of the sheeted dike unit.

A contemporaneous or slightly earlier set of gabbro dikes is met only in the peridotites. They bear clinopyroxene, olivine, plagioclase and occasionally a pargasitic amphibole. They are characterized by a coarse texture with locally comb or magmatic flow structures, sharp contacts with the peridotites with neither mineral reactions nor chilled margins (plate 2.2g). Their thickness is variable in the 1-30 cm range, usually closer to 5-10 cm. These dikes have been emplaced after the episode of plastic deformation in the upper peridotites because they are not deformed. Their average orientation, like that of the microgabbro dikes is often near to that of the sheeted dikes (compare figs. 3.16i and a, b). These observations suggest that these dikes were intruded into a medium which had already undergone a large cooling, though still being under the influence of stress related to the high-T plastic flow episode.

At around 2 km below the paleo-Moho, the set of gabbro dikes grades into websterite dikes, essentially clinopyroxene-rich with minor olivine and orthopyroxene. The plagioclase disappears completely in all such dikes at around 4 km below the paleo-Moho. This zoning is a rule without any exception in Oman peridotites (as well as in other visited ophiolites). It is suggested that this vertical evolution is not due to the disappearance of plagioclase on the phase diagram of the basaltic melt which circulated through the dikes, but to the order of crystallization at the dike margins which shifted from plagioclase to diopside. The websterite dikes are altogether more contorted and statistically more loosely oriented than the corresponding gabbro dikes, with a tendency to approach the foliation orientation (compare figs. 3.16e, i and j). These observations suggest that the dikes have been affected by the plastic deformation. Moreover, 4-5 km below the Moho, these intrusive websterites dikes grade into indigenous dikes (§ 2.5.2), developing diffuse contacts with the country peridotites, with a screen of residual dunites in-between. The first dunite veins with the orientation typical of this generation of dikes also begin to appear (plate 3.2f).

These observations about reactions with wall peridotites and relation with peridotite flow lead to the conclusion that *this generation of diabase-gabbro-websterite dikes was emplaced in a cooling mantle* : at depths greater than 4 km below the Moho, they intruded peridotites at or around their solidus (1250°C) where plastic flow was still operative, whereas at the Moho depth, they intruded peridotites sensibly colder and already rigid and eventually penetrated the crustal section at even lower temperatures (transition microgabbro/diabase around 450°C, § 2.5.2).

The *older series* of intrusives consist of sills and dikes emplaced even at Moho depth during plastic flow of the peridotites. They are intrusive within the transition zone as sills parallel to the Moho (plate 3.2b, c, d), developing both sharp and diffuse contacts with the enclosing peridotites. Below, they constitute internally foliated dikes and dunites veins weakly inclined with respect to the foliation and co-zonal with the peridotites lineation. This generation of dikes is also responsible for the plagioclase-clinopyroxene impregnations locally observed in the uppermost peridotites (plates 3.2g, h and 3.3a, b, c, d). In areas of more intense deformation or in the lower peridotites, the dikes are tectonically rotated parallel or close to the foliation plane. This could explain why in the deeper section of the peridotites the compositional layering, which could at least partly derive from this type of dike by tectonic transposition (Nicolas and Jackson, 1982) is more developed than in the upper section and, on the contrary, why the cross cutting dikes are less common. These sills and dikes are, like the preceding ones, clinopyroxene-olivine-plagioclase-gabbros in the upper peridotites and olivine-websterites in the lower ones. Clearly, chromite pods in Oman belong to this dike generation (§

10.5) as shown by their similar orientation, and by the observed continuity between chromite schlieren and such gabbro dikes.

Because they invade peridotites during their plastic flow below the Moho at solidus or hypersolidus temperatures, that is in asthenospheric conditions, *these dikes have been emplaced just beneath the spreading center, and not at some distance in a cooling lithosphere like the preceding ones. They are thus believed to represent the main feeding system for the accreting overlying crust.*

*Basal peridotites* - The lower contact of the Oman ophiolite nappe is marked by a metamorphic aureole, the nature of which has been studied in the northern and central massifs by Lippard (1983) and by Searle and Malpas (1982) and in Wadi Tayin by Ghent and Stout (1981). The structural analysis conducted in most metamorphic aureoles of Oman and in the overlying basal peridotites has led to a model for the obduction of the Oman ophiolite (Boudier et al., 1985, 1988 ; Cannat, 1983) discussed in § 3.5.3.

A new tight substructure appears in the olivine of the harzburgites some 2 km above the lower contact. It is the first sign of the superimposition on the high-T, low-stress asthenospheric deformation prevailing upsection, of a low-T, high-stress lithospheric deformation (§ 2.5.5) related to the ophiolite oceanic detachment. Rapidly the strain increases and a new foliation and lineation become visible in the field. The foliation is parallel to the basal contact and to the foliation in the underlying metamorphic rocks (figs. 3.9 and 3.15). The strain increases down section and as a consequence the dikes and dunite veins are rotated into parallelism with this new foliation, thus creating the banded aspect of these lower peridotites (the 'banded unit' of the Open University maps) (plate 3.3f). Mylonitic peridotites with olivine recrystallized in 5-30 μm neoblasts appear in the last hundred meters above the basal contact (fig. 2.7e). Using the structural piezometers (§ 2.5.4), applied stress responsible for these structures seems to be in the range of 0.1-0.2 GPa. Measurements in the basal peridotites are often difficult due to the heavy lizardite-serpentinization.

### 3.3.4. Metamorphic aureoles

The metamorphic rocks in contact with the mylonitic peridotites are garnet amphibolites recording in Wadi Tayin the highest temperatures of 755°-865°C at 200 to 500 MPa estimated pressures (Searle, 1980 ; Ghent and Stout, 1981). After a few meters downsection the garnet disappears, and after a few tens of meters, fine grained epidotite amphibolites and greenschists are observed (plate 3.3g). Quartzites derived from cherts are interlayered with the amphibolites. A few hundreds of meters below, a distance difficult to estimate properly because of the low-T shear zones and faults affecting this series, one attains the schistose and weakly metamorphic cherts and argilites of the Hawasina formation. The downward decreasing temperature conditions of the metamorphic soles is not due to retrograde metamorphism, but represents an inverted thermal gradient (Ghent and Stout, 1981). The greenschist facies rocks were successively incorporated to the advancing ophiolite sheet, in cooling conditions. In the southern Oman mountains, these metamorphic aureole formations rest upon the limestones of the autochtonous Saih Hatat dome, which at the contact in Wadi Tayin, are strongly foliated with plastic deformation in calcite, and further away are deformed by stylolites and vein-filling, implying a solution-deposition mechanism of deformation.

The foliation and cleavage in all these formations remain more or less parallel to the peridotite contact but the mineral lineations show a systematic drift down section as described in § 3.5.3.

An important discovery in the highest and hottest part of the metamorphic aureole has been that of a hydrous anatexis (Searle and Malpas, 1980, 1982; Boudier et al., 1988). It is not common in the Oman part of the ophiolites, yielding locally plagiogranitic patches and dikelets, but is spectacular in the Emirates, where K-bearing phyllites are transformed into migmatites with full structural evidence of a syntectonic origin of the melt (Boudier et al., 1988) (plate 3.3h). The felsic melt issued from this anatexis is now observed in the overlying peridotites making numerous granite stocks and dikes. The origin of K-granite in the Emirates, related to metamorphic aureole melting on structural evidence by Boudier and co-workers, is on the contrary ascribed on the basis of a K-Ar biotite age of 85 ± 3 Ma, to the melting of the continental edge during obduction by Lippard et al. (1986). This conclusion is difficult to accept because the temperature in the thrust plane, 10 Ma after detachement, is only able to develop a green schist facies metamorphism, in conformity with thermal calculations. Interestingly, in the central and southern massifs where the melted aureoles are K-poor, the dikes injected into the above peridotites are dominantly K-devoid plagiogranites.

### *3.3.5. High pressure metamorphism*

High pressure parageneses have been described in the para-autochtonous formations on the eastern flank of the Saih-Hatat limestone dome where they define a gradient of increasing metamorphism in a N.E. direction (fig. 3.26) (Lippard, 1983 ; Michard et al., 1984 ; Le Métour et al., 1986 ; Goffé et al., 1988). The characteristic facies are Fe-Mg carpholite, lawsonite, glaucophane, and garnet schists, grading to eclogites in the easternmost site of As Sifah. In a complex P/T path traced by Goffé et al. (1988), the eclogites record the most severe conditions for this metamorphism :1-1.2 GPa and 400-450°C. Such a pressure requires an overload of some 30 km, possibly composed by 25 km thick ophiolitic nappes added to 5 km of folded and sliced continental cover.

Finally, many radiometric ages have been obtained in the metamorphic aureoles and in the underlying high-P metamorphic formations. They are reported in table 3.I. Hornblendic amphiboles from the high temperature metamorphic soles have consistent K/Ar isotopic ages of 93 to 101 Ma. A few Ar/Ar ages on white micas from greenschist facies rocks are bracketed between 71 and 81 Ma. Finally K/Ar determinations on high-pressure minerals from the Saih Hatat provide ages between 67 and 89 Ma for metabasites, and scattered ages up to 240 Ma for white micas in metapelites, and eclogites (Montigny et al., 1988).

## *3.4. STRUCTURE OF THE OMAN OPHIOLITE*

### *3.4.1. Introduction - main structural events.*

The present day structure of the Oman ophiolite nappe results from the superimposition of three major events : 1) accretion at an oceanic spreading center, 2) detachment and oceanic thrusting towards the Arabic margin and 3) obduction on to this margin and related post obduction deformations. For instance, in the Wadi Fayd section, layering planes in the layered gabbros and foliations in the upper peridotites are vertical respectively over 2 km and 4 km (fig.3.15). The magmatic sedimentary structures observed in this section demonstrate that these formations have been vertically tilted after their deposition on a flat-lying surface. The tilting can be alternatively produced at the ridge by the effect of listric faults, during the early detachment event or during obduction on to the continental margin, as a ramp effect.

Each of these events has a distinct structural and mineralogical signature, usually

| LITHOLOGICAL UNIT | METHOD | AGES | REFERENCES |
|---|---|---|---|
| *HAGAR Supergroup*. Subsident carbonate shelf Autochtonous basement | biostratigraphy | Mid-Permian (~ 260Ma) to Cenomanian (~ 95 Ma) | |
| *MUTI formation*. Marls and shales, conglomerates with reworked fragments of Hagar | biostratigraphy | Coniacian (~ 88 Ma) to Campanian (~ 80 Ma) | Glennie et al. (1973) |
| *HAWASINA formations*. Continental slope and abyssal sediments, reefal limestones. Imbricated thrust sheets above Hagar | biostratigraphy | Mid-Permian (~ 260 Ma) to Cenomanian (~ 95 Ma) | |
| *HAYBI volcanics*. Alkali basalts associated with Hawasina, in Fizh massif | K/Ar on biotite | 233 ± 9 to 200 ± 8 Ma clustering around 218 Ma | Lippard and Rex |
| Alkali sills intrusive in Haybi and Hawasina, in northern massifs | K/Ar on biotite | 162 ± 6 to 92 ± 4 Ma no clustering | (1982) |
| Alkali tuffs in Hawasina mélange | K/Ar on biotite | 96 ± 4 Ma | |
| *Ophiolitic layered gabbros.* | | | |
| Wadi Tayin massif | Sm/Nd on coexisting minerals | 128 ± 20 Ma<br>150 ± 40 Ma | McCulloch et al. (1981) |
| Fizh massif | | 100 ± 20 Ma | |
| *Plagiogranites* intrusive in high crustal levels, complete ophiolite belt | U/Pb on zircon | 93.5 to 97.9 Ma with clustering at 95 Ma | Tilton et al. (1981) |
| K-granite intrusive in the ophiolite Haylayn massif | K/Ar on biotite | 85 Ma | Browning and Smewing (1981) |
| *Amphibole gabbro dikes* intrusive in peridotites | K/Ar on amphibole | | |
| Wadi Tayin massif | | 114 ± 9 and 103 ± 8 Ma | Montigny et al. |
| Sumail massif | | 85 ± 5 and 99 ± 7 Ma | (1988) |
| Sarami massif | | 81 ± 6 Ma | |
| *V1 (Geotimes) volcanic unit* Fizh massif | biostratigraphy of interbedded sediments | Early Cenomanian (~ 97 Ma) | Tippit et al. (1981) |
| *V2 (Alley) volcanic unit* Fizh massif | idem | Early Cenomanian (~ 97 Ma) to Early Turonian (~ 91 Ma) | Tippit et al. (1981) |
| *Supra-ophiolite radiolarian* | biostratigraphy | Campanian (~ 81 Ma) | Schaaf and Thomas (1986) |
| *High-grade metamorphic sole* | K/Ar on amphibole | | |
| Wadi Tayin massif | | 89.2 ± 2.0 and 96.5 ± 5.7 Ma<br>97 ± 3 | Lanphere (1981), |
| Sumail massif | | 98 ± 3 and 93 ± 3 Ma | |
| Haylayn massif | | 98 ± 3, 97.5 ± 4 and 95 ± 4 Ma, 99 ± 3 Ma | Montigny et al. |
| Wadi Asjudi | | 96 ± 3 Ma | (1988) |
| Wadi Ahin area | | 99 ± 5 Ma | |
| Wadi Jizi area | | 95 ± 4 Ma | |
| Jebel Sumeini area | | 101 ± 4, 100 ± 5, 99.2 ± 3 and 97.5 ± 3.5 Ma | |
| Mafasi area (United Arab Emirates) | | 105 ± 2 to 97.6 ± 2 Ma | Moody (1974) |
| *Greenschist metamorphic sole* | | | |
| Northern Oman Mountains (UAE) | K/Ar on muscovite<br>K/Ar on biotite | 85 ± 5 Ma<br>83 ± 5 Ma | Alleman and Peters (1972) |
| Wadi Tayin massif | K/Ar on phyllite | 76.0 ± 2.0, 80.9 ± 1.3 Ma and 70.8 ± 8.6 Ma | Lanphere (1981) |
| *Blueschists* from the autochtonous basement, Saih-Hatat | K/Ar on phengite | 80 ± 2, 100 ± 4 Ma and 101 ± 4 Ma, 88 ± 3 Ma | Lippard (1983) |
| | on glaucophane | 68.5 ± 12 Ma | Montigny et al. |
| | on crossite | 76 ± 6, 79 ± 4 Ma | (1988) |
| *Eclogite* from the autochtonous basement | on glaucophane | 103 ± 4 Ma | |
| As Sifar area-Saih Hatat | on phengite | 131 ± 4 Ma | |
| *Neo-autochtonous sediments* | biostratigraphy | Maestrichian (~ 75 Ma) to Early Miocene (~ 20 Ma) | Glennie et al. (1973) |

Table 3-I. Biostratigraphic and radiometric ages of the Oman formations (compilation by G. Ceuleneer and F. Boudier)

recognizable in ophiolite specimens (chapter 2). In peridotites for instance, the asthenospheric flow below the spreading center develops a high-T foliation with melt evidence, the thrusting within the mantle during the first stage of detachment develops a mylonitic foliation in the peridotites located above the metamorphic aureole, and eventually the obduction and subsequent deformations in a colder peridotite produce a cleavage and tectonic brecciation first in antigorite-serpentinite bands and eventually in lizardite-serpentinite bands.

Using such signatures, it becomes possible to unravel the intricacies of the ophiolite history. We describe first the structures developed at the spreading center of origin in the mantle and in the crustal sections, respectively by solid state asthenospheric flow and by magmatic flow. The various asthenospheric flow patterns in the mantle section of the Oman ophiolite have been analyzed by Ceuleneer et al. (1988) and Nicolas et al. (1988a), using the foliations, lineations and shear senses reported in the structural maps accompanying these papers. These measurements are also presented under the form of simplified trajectory maps (figs. 3.8a, b). From these documents, four typical mantle flow patterns at a spreading ridge have been distinguished (Ceuleneer et al., 1988). They are briefly presented here, except the transform flow pattern which is presented in chapter 5. Together with a synthesis at the scale of the belt, these flow patterns are incorporated into ridge models in § 9.2. Magmatic flow structures in the gabbro units are also represented in the maps referred to above. How this magmatic flow relates to plastic flow in the underlying mantle and to the sheeted dike attitude has been examined thanks to a dozen detailed cross sections, eight of which are published in Nicolas et al. (1988b). The remarkably convergent results of these cross sections allow us to restrict the description to a single one, complemented by a brief comment of the maps.

Ascribing to a spreading center activity the high temperature structures evoked above, and to the oceanic detachment-obduction activity, the other lower temperature structures seem reasonable. However, some low temperature faults and shear zones appear to have been produced in the crust at the ridge itself and should be described in the corresponding section. Finally the oceanic detachment and obduction structures will be described together as being parts of a continuum.

### 3.4.2. Structures related to accretion at the spreading center

*The typical flow patterns in the ultramafic section*

*Homogeneous mantle flow away from the ridge axis in the Fizh and Salahi massifs* - The first flow pattern is by far the most common. It has been observed along about 70% of the Oman paleo-ridge segment (figs. 3.8a,b). It is exemplified by the Fizh and Hilti massifs, where very homogeneous structures along the ridge strike are still preserved on the 100 km scale. The flow plane attitudes have been constructed from those of the foliations and from petrofabric analysis (§ 2.5.4.). They dip very gently with respect to the paleo-Moho (fig. 3.15). The flow lines are at right angle to the ridge axis and follow the steepest dip line of the flow plane (figs. 3.15 and 3.16f). The shear strain increases rapidly towards the top of the mantle section : from an average shear strain g ~3 in the main of the mantle section, estimated approximately from the angle between foliation and shear flow plane (§ 2.5.4), it reaches a value of g ~10 at about 500 m beneath the paleo-Moho. The most intense strain is measured just below the layered gabbros of the mafic section which, contrasting with the peridotites, show virtually no sign of plastic strain. In this zone of very strong shear strain, the flow plane can be considered as being parallel to the paleo-Moho. Its dip increases gradually downsection. Applying the principle proposed in § 2.6, this dip should be parallel to the lithospheric front surface. Incidentally, this implies that in these massifs the ridge of origin should be sought Westwards.

The detailed kinematic study has revealed a reversal in shear sense at the top of the mantle section. The reversal zone closely corresponds to the 500 m thick zone of very strong shear strain mentioned above. Beneath this top level, the shear sense is very consistent throughout the ultramafic section : the upper peridotites flowed Eastward, presumably away from the ridge axis, at a higher rate than the deeper ones.

*Mantle flow in asthenospheric diapirs, the Maqsad, Batin and Shamah areas* - This second configuration is more exceptional. It features steep flow lines, down the dip of the flow planes, and closing flow plane trajectories. Such a pattern has only been found so far without doubt only in three areas the Oman paleo-ridge segment, namely in the Maqsad (fig. 3.17), Batin (fig. 3.20) and Shamah (fig. 3.21) areas (a fourth area in Wuqbah is still in the mapping process).

In the Maqsad area, the width of the steep flow domain is 8 km at a right angle to the local orientation of the diabase dike swarm, and about 10 km along this direction ; it is hidden in places below the crustal formations, but is no longer than 20 km. It may, therefore, be described in terms of a vertical pipe, slightly elliptical in cross-section, elongated along the ridge axis. The asthenospheric flow does not remain vertical up to the Moho ; it breaks up 200-300 m below the layered gabbro unit, within dunites of the transition zone and associated harzburgites (figs. 3.12 and 3.17c). Beyond a radius of at least 30 km around the center of the pipe, the flow lines are radial with respect to the pipe axis, although the directions parallel and perpendicular to the ridge axis are clearly preferred.

The analysis of the plastic deformation in the central domain has revealed several singularities : the transition between undeformed layered gabbros and the deformed dunites and harzburgites of the transition zone is very sharp, occurring within a few tens of meters below the Moho. This can be appreciated by reference to the chromite ore deformation in the pods. The olivine nodules in a chromite matrix constitute a strain marker, unfortunately underestimating the strain because massive chromite is much stronger than peridotite (Sécher, 1981). However, these nodules record shear strain of 2 at 200 m below the Moho in the transition zone, and of 3.5 in the concordant pods of the underlying harzburgites. The average obliquity of the foliation to the flow plane corresponds to a shear strain of 5. The olivine fabrics confirm that the strain was very large and indicate a shear flow regime with, in domains of horizontal flow, the top layer moving away from the diapir (figs. 3.17c, 3.18b,c, 3.19). This shear sense is maintained everywhere, throughout the sections in the outer domains as well as in the core of the diapir, where the foliation steepens downward. The sharp rotation of the foliation corresponds in fact to a rupture in the flow pattern (fig. 3.19). The olivine fabric in a specimen from the rotation zone is very weak indicating a weak plastic deformation (fig. 3.18a). These data introduce severe constraints on the physical models of asthenospheric flow beneath a spreading center (§ 9.4).

In the Batin area (fig. 3.20) the diapir has also an elliptical shape with a 12 km long axis, oriented NW-SE, parallel to the trend of the sheeted dike complex. The vertical flow breaks up a few hundred meters below the Moho, at a level where large tabular dunite bodies invade the harzburgitic section. The largest dunite of the Oman ophiolite (13 km x 2 km) has been mapped in this diapiric area. Interestingly, the foliation in these dunites is mainly vertical, whereas the lower dunite-harzburgite contacts are mainly horizontal (plate 3.2e). In contrast with the situation in Maqsad, the rotation occurs here entirely within the dunites. Beyond a radial distance of 5 kilometers, flow lines diverge rapidly and tend to become perpendicular to the ridge orientation. Shear sense determinations in the center of the diapir and in the radially diverging zone give the same results as in Maqsad, i.e. flow

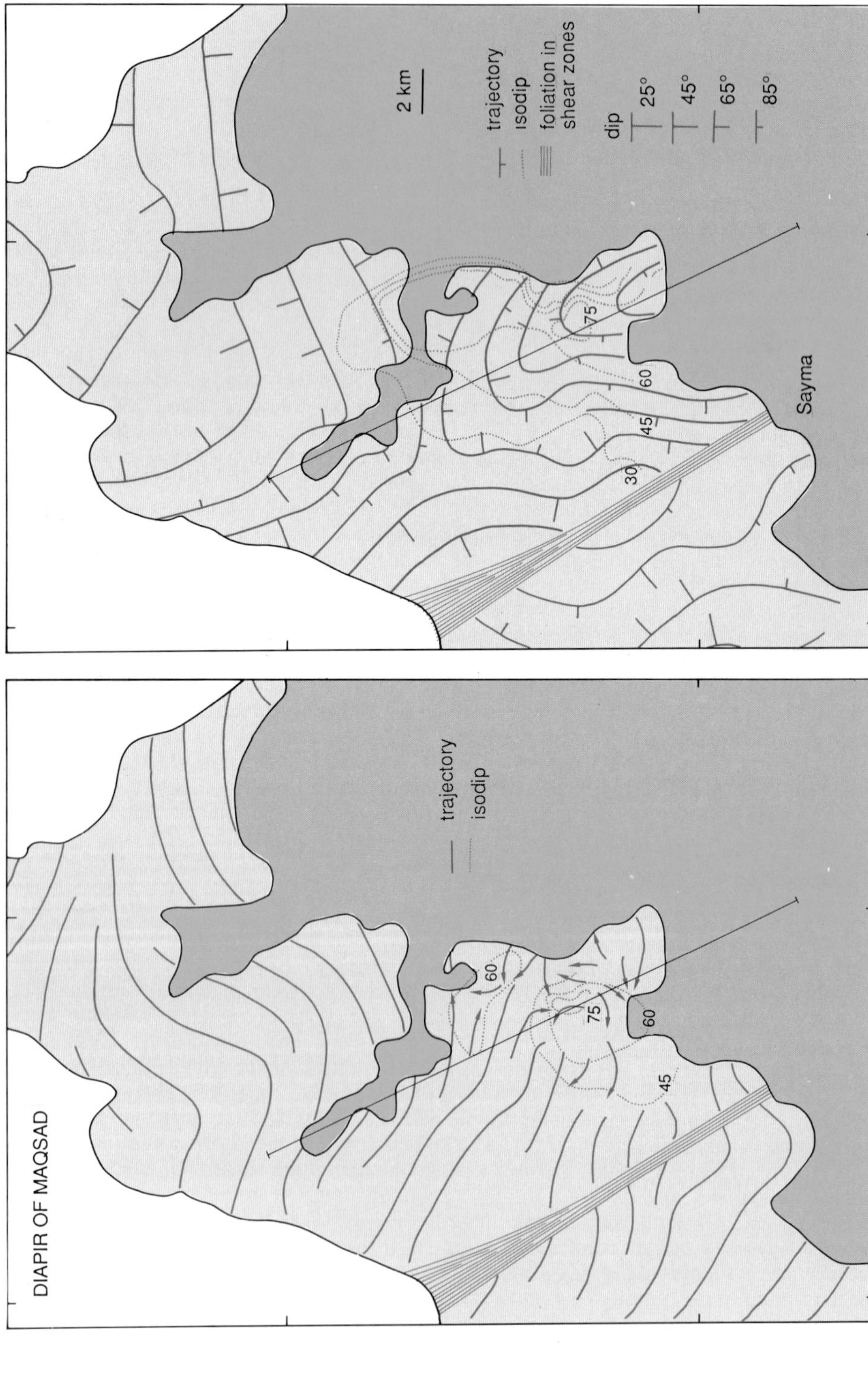


**a** LINEATIONS

**b** FOLIATIONS

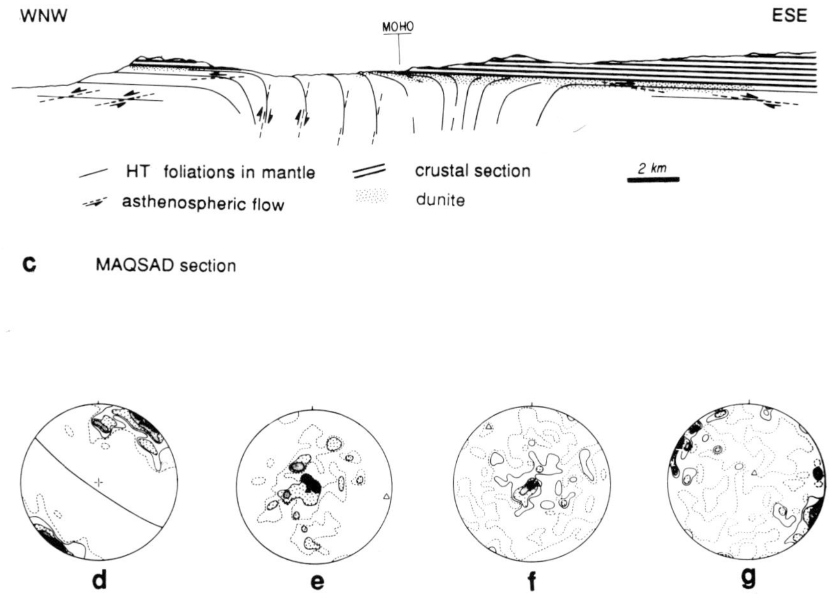


Fig. 3.17. Structure of the Maqsad diapir in the Sumail massif (see location in figure 3.8a). a and b) Simplified maps showing, beneath the undifferentiated crustal section (darker color), the diapiric pattern of lineations(a) and foliations(b) in the mantle section (lighter color). c) Cross section (location above), illustrating the sharp rotation of foliations and flow planes at the top of the diapiric intrusion. d) Diabase dike swarm (117 measurement ; contours : 1, 2, 5, 9, 15 %). e) Layering plane in gabbros (80 measurements ; contours : 1, 3, 4, 5 %). f) Foliations in peridotites (217 measurements ; contours : 1, 2, 3, 4, 5 %). g) Lineations in peridotites (213 measurements; contours : 1, 2, 2.5, 3 %) (Ceuleneer et al., 1988).

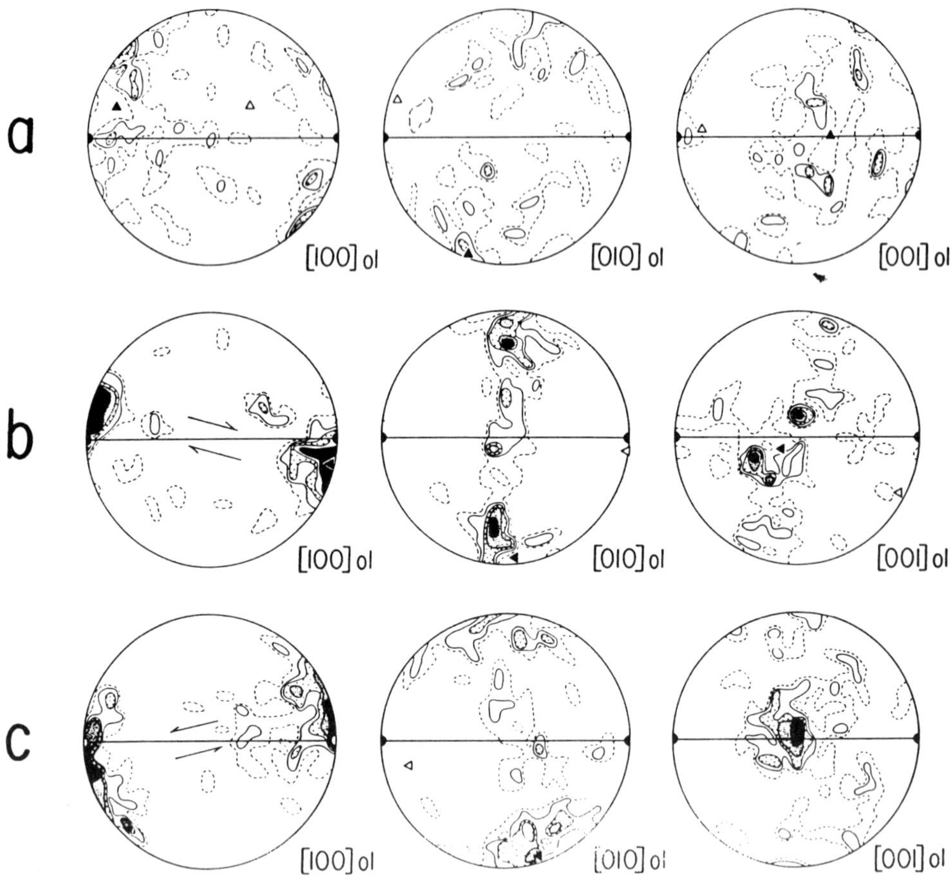


Fig. 3.18. Olivine fabrics in the top most section of the Maqsad diapir (100 measurements ; contours : 1, 2, 3, 4, 10 %). a) Fabrics in the zone of flow rupture. b and c) Respectively fabrics in the southern and northern peripheral zones. Line : foliation trace ; dot : mineral lineation. Note the opposite sense of shear indicated by the opposite obliquity of lattice fabrics, on the two opposite margins of the diapir (Ceuleneer, 1986).

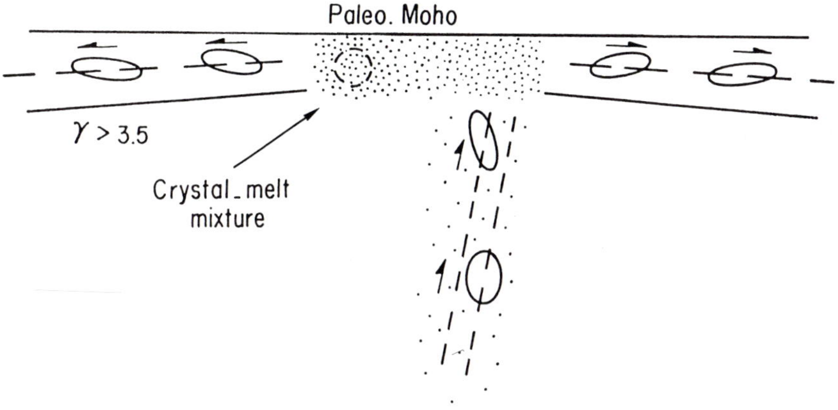


Fig. 3.19. Schematic flow structure at the top of the Maqsad mantle diapir deduced from fabric analysis. Rotation of the foliation trajectory (represented by ellipses) and rupture in flow trajectory (dashed line).

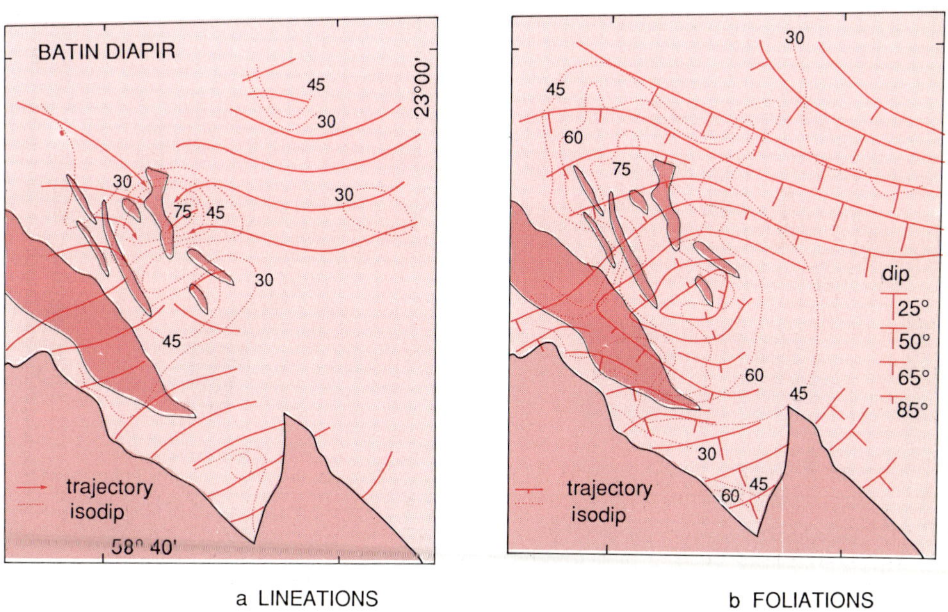


Fig. 3.20. Lineations (a) and foliations (b) trajectories in the high temperature peridotites of the Batin area, Wadi Tayin massif (location in fig. 3.8a). Colors gradation as in fig. 3.17 ; deepest tone : dunites (Ceuleneer et al., 1988).

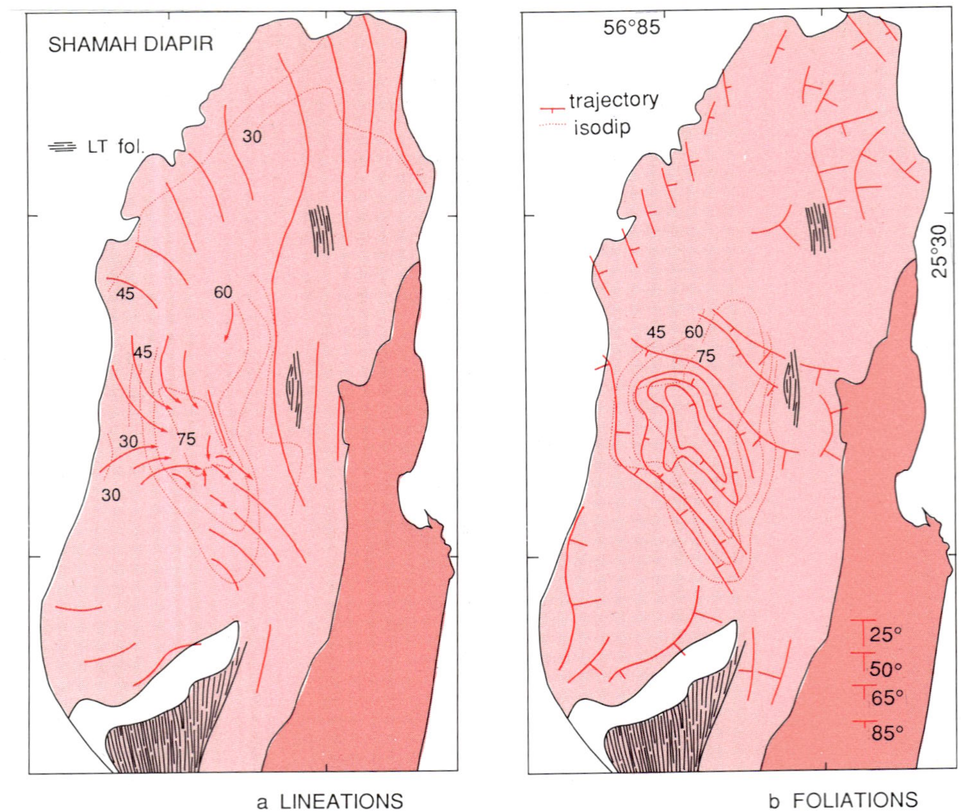


Fig. 3.21. Lineations (a) and foliations (b) trajectories in the high temperature peridotites of the Shamah area, Khawr Fakkan massif (location in fig. 3.8a). Color gradation as in fig. 3.17 (Ceuleneer et al., 1988).

away from the diapir was faster in the upper levels than in the lower ones.

In the Shamah area (fig.3.21), the vertical flow zone has an elliptical shape with its long axis (12 km) NNW-SSE, presumably parallel to the ridge trend. This diapir is truncated by the topographic surface, approximately two kilometers below the Moho, providing the opportunity of observing a diapiric structure at a lower level than in the two previous cases. Inside the pipe, the harzburgite is relatively fertile, rich in websterite layering often folded and oblique to the foliation. Dunites are rare, so are pyroxenite and gabbro dikes. Beyond a radial distance of 6 kilometers, the flow lines diverge into a flat attitude, trending dominantly parallel to the assumed ridge orientation.

The Maqsad and Batin areas are marked by the abundance of gabbro dikes and sills, grading into diffuse impregnations. They are related to chromite pods which are particularly abundant in Maqsad. As mentioned above, tabular dunites are also exceptionally thick and abundant. These characteristics evoke an exceptional magmatic activity in the transition zone above mantle diapirs. Deeper in the diapir like in the Shamah area,this activity may be expressed by the websterite layering. Nicolas et al. (1988a) have tempted to identify diapirs by using these petrologic characteristics in areas of the Oman ophiolite where, due to the steady-state drifting from the spreading center, no structural print of a diapir could be preserved into the mantle section (§ 2.6.). This will be discussed further in § 9.2.6.

*Channelling of the mantle flow along the ridge axis* - The third configuration is examplified by the areas extending from diapirs in the paleo-ridge direction, such as around the Shamah diapir, the presumed Wuqbah diapir and the Maqsad diapir (fig. 3.8.). This flow geometry has been recognized along about 15% of the Oman paleo-ridge segment. Plastic deformation is very intense and linear. The flow line is parallel to the ridge axis ; the flow plane is in a zone around the flow line. Although on average sub-horizontal and parallel to the Moho, especially in the uppermost level of the mantle, the flow plane dip is irregular, in agreement with the dominantly linear character of the deformation. Lattice fabrics are very strong ; the obliquity between the shape and lattice fabrics has an average value of 5° which corresponds to shear strain of the order of 10. Enstatite is usually entirely recrystallized also indicating intense deformation at high temperatures. The shear sense is quite constant throughout the thickness of the mantle section, and in particular, no shear sense reversal was observed on nearing the paleo-Moho, despite close sampling in such areas. Mantle flow parallel to the ridge axis has been shown to be genetically linked to the diapir flow pattern, as discussed in the previous section. In the Ragmi-Fayd area, this flow pattern was recognized without a diapir being structurally identified farther upstream (fig. 3.8b). Its existence is however suspected (fig. 9.3), on the basis of the petrological signatures envisaged in the preceding paragraph.

*Structure of the plutonic section*

*Magmatic flow structures* - Structural mapping in the uppermost ultramafic section and in the overlying layered gabbros (figs. 3.8b) and detailed cross sections through the plutonic section (Nicolas et al., 1988b ; Nicolas et al., in press) have revealed three outstanding features which can be illustrated by the Samra cross section (fig. 3.12).

i) There is a striking structural continuity between the uppermost peridotites with a tectonic fabric and the lower layered gabbros with an igneous fabric. As seen in figures 3.12 and 3.17, the plastic foliation of the peridotites is parallel to the layering and magmatic foliation plane of the gabbros. The lineations also remain parallel on both sides of the paleo-Moho.

ii) In most places, the layered gabbros have been affected by a large magmatic flow. As discussed in § 2.4.3., the orientation of gabbro layering now reflects the magmatic flow field and no longer an attitude related to an 'in situ' crystallization process, either because the layering has been directly caused by the magmatic flow (magmatic segregation) or because it has been transposed by the large rotations which can be imposed by the flow.

iii) In the upper gabbros, magmatic foliations steepen and become parallel to the dikes of the diabase sheeted dike complex (see § 3.3.2.).

In view of modelling magma chambers (§ 11.4.3), it is critical to be able to determine from which flank of the ridge of origin, the considered ophiolitic section was derived. The evidence here is based on the sense of shear in the peridotites deeper than 1 km below the Moho. As discussed in this paragraph, it is believed that the shear sense of the asthenosphere flow component away from the ridge is determined by the relative motion of the lithosphere. This criterion is not entirely satisfactory because it is model-dependent. The evidence available so far favors the curvature of the gabbro layering facing away from the magma chamber axis like in the model of figure 11.8. This is suggested from the eastward concavity of the layering-foliation planes upsection in the Samra section where the mantle flow suggests that the ridge was located to the West (fig. 3.22).

*Lower temperature structures-listric faults* - The magmatic flow structures, mainly layering and foliation of the gabbro unit are locally affected by amphibolite facies shear zones which are in continuity with shear zones of the ultramafic section ; they are examined together below, as emplacement-related structures. Otherwise, the upper part of the gabbro unit has been fractured by the hydrothermal circulation with apparently amphibolite facies circuits closing down in the zone of transition between sheeted dike and upper gabbro units and greenschist facies circuits penetrating deeper (§ 3.3.2). Finally, distinct listric faults have been observed mainly in the Sumail and Wadi Tayin massifs. They flatten down at the Moho level where they are marked by one or more meters of brecciated and sheared gabbros and antigoritic peridotites. They are responsible for the thrust contacts between the layered gabbro and peridotite units mapped in these areas by the American group and the BRGM. Directly above these contacts, the layering of the gabbros is commonly tilted to 45°. This tilting has an important consequence on the drawing of cross sections and on the estimation of the layered gabbro thickness. Maps of the Oman ophiolite show a striking abundance of faults trending parallel to the sheeted dike attitude. A number of these faults may represent listric faults. This will be difficult to prove because many presumed listric faults trending more or less normal to the obduction direction, are now inverse faults reflecting the obduction kinematics. For instance, this is the case for the NW-SE faults mapped by the OU group in the Ragmi-Zabin area.

Finally, it is tempting to ascribe to listric faulting some tectonic structures from the layered gabbros and underlying transition zone units which have been generated close to the paleo-ridge as demonstrated by the accompanying magmatic activity. Such structures are normal shear zones or brecciated zones responsible for large block rotations, injected by hydrous gabbros, and possibly some folded and brecciated zones, injected by wehrlites.

### *3.4.3. Structures related to oceanic thrusting and obduction*

*Introduction* - The new foliation observed in the basal peridotites (§ 3.3.3.) has been

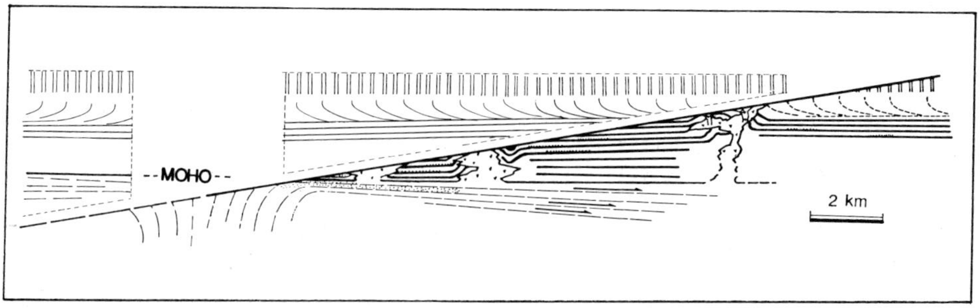


Fig. 3.22. Ridge sketch deduced from the Samra cross section, East of the Maqsad diapir, after rotation of Moho to horizontal (fig. 3.12),. Note the Eastward concavity of the layering-foliation upsection in the gabbro unit.

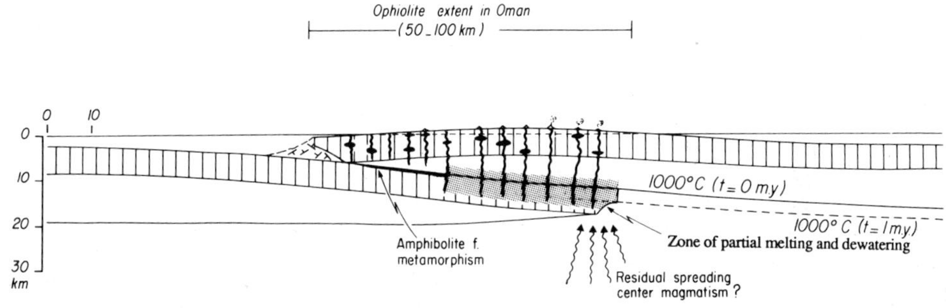


Fig. 3.23. Physical model for the oceanic thrusting at an active ridge and related magmatism. The lithosphere structure has been modelled for a 5 cm/yr spreading rate, the 1000°C surface separating lithosphere from asthenosphere at the initiation of the thrusting is assumed to remain that of the active ridge (the dashed line shows its position after 1 Ma of static cooling). This assumption may be valid if a residual spreading center activity or the effects of shear heating are taken into account or if the convergence rate rapidly attains 4 cm/yr (Boudier et al., 1988).

induced by detachment within the lithospheric mantle at temperatures around 1000°C (Boudier et al., 1988). Once the lithospheric slab thus detached begins to be thrust over the facing oceanic crust, it develops within this crust the highest granulite facies metamorphism, locally accompanied by partial melting (fig. 3.23.). The slab rapidly cools during its progression and the deformation moves downward to surfaces of decreasing temperatures first in the amphibolite and next in the greenschist facies. Locally, the piling of nappes over the Arabic margin develops a new high pressure metamorphism within the margin formations.

By ascribing P, T conditions of equilibrium to the minerals symptomatic of these events, by measuring their associated foliations and lineations, and by dating them by radiogenic means, it is theoretically possible to reconstruct the history of the motion of the ophiolite from its cradle to its present position. The obtained sketch can be referred to geographic coodinates and becomes a paleogeographic reconstruction provided that paleomagnetic data are obtained on the various rocks from the metamorphic aureole. This project is only partly achieved in the Oman ophiolite as seen below.

*Detachment and oceanic thrusting* - Shear zones and basal thrusts : large shear zones have been discovered in the ultramafic sections of the ophiolite mainly in the central part of the belt (fig. 3.24) (Ceuleneer, 1986 ; Boudier et al., 1988). Some of these shear zones can be traced into the basal thrusts (fig. 3.15) and thus bring an important information on the detachment. The shear zones are best developed in the Fizh and Hilti massifs. They are sub-vertical, strike NS or NW-SE and maintain a relatively constant orientation over tens of kilometers. They are characterized by a mylonitic foliation, which transposes high-T deformation structures, and by a sub-horizontal stretching lineation indicating a strike-slip movement.

In the peridotite section, the shear zones range in width from a few meters to 1-2 km. Locally, the shear zones can be followed up into the gabbros of the crustal sequence where they keep the same orientation and kinematics (Reuber, 1988). Flaser structures are developed in these shear zones, now split into numerous splays a few meters wide, like those described in Wadi Ragmi (Smewing, 1980). In contrast, the shear zones in the peridotite section become progressively wider and more diffuse. They may be responsible for a local tilting of the crustal section, such as in Wadi Sayjani (northern Sumail massif) where the contact between peridotites and gabbros is rotated to the vertical along a shear zone 100 m wide which extends for 7 km along strike. In Wadi Ragmi, where the shear zones are particularly common in the gabbros and further North in Wadi Fayd, a progressive rotation to the vertical affects the ultramafic and mafic sections over a total width of 6 km (fig. 3.15).

The general attitude of the major shear zones is close to that of diabase dike swarm (fig. 3.24). Most are oriented N-S and show dextral shear ; others, oriented NW-SE, show both dextral and sinistral shear. Dextral and sinistral shear zones interpenetrate each other and seem to have been active at the same time. The shear sense, deduced from the obliquity between lattice and shape fabrics in olivine and orthopyroxene (§ 2.5.4.), is confirmed by field observation of the rotation of the high-T foliation in the vicinity of the shear zone. These rotations operate on a scale of a few meters or less in the gabbros, of around 200 m at 1-2 km below Moho and of around 1 km in the deepest peridotites, indicating that at the time of shearing there was a large vertical temperature gradient from the gabbro to the peridotite. The complete geometrical continuity between vertical major shear zones and subhorizontal basal thrust zone (fig. 3.15) corresponds to a kinematic continuity. The dextral shear sense in the vertical shear zones thus grades into a southerly-directed thrusting as foliation becomes horizontal.

In the mafic crustal section, a correlation between shear zones and late intrusions of wehrlites, gabbros and plagiogranites has been reported by Smewing (1980), Browning (1982), Ceuleneer (1986) and Beurrier (1987). In the peridotite section, the shear zones are invaded by amphibole-gabbro dikes. Field relationships show that the magmatic activity responsible for these dikes was contemporaneous with the shearing episode (Boudier et al., 1988).

*Detachment at the oceanic ridge* - Boudier et al. (1988) have developed three arguments in favor of the ophiolite detachment occurring close to the paleo-ridge :

i) The nappe thickens Northeastward ; for example, thickness increases along the Bahla-Rustaq cross section from 2-4 km in Bahla to 10-12 km in Rustaq, and along a cross section through Wadi Tayin (fig. 5.20), from 1 km at the SW end to 15 km at the NE end (Hopson et al., 1981). Systematic mapping in the peridotites shows that there is no major low inclination tectonic discontinuity within these sections, i.e. they are not thickened by internal thrusting. Thus, the observed thickness probably reflects that of the original lithosphere at the paleoridge. From these field estimates, the initial slope of the thrust plane was in the range of 2-3°.

ii) The lithosphere was still hot at the time of its detachment. This is deduced from the characters of the shear zones which, as discussed above, are connected with the detachment and early thrusting. The progressive transition in plastic deformation between peridotites outside and inside the shear zones, particularly at depth in the peridotite section, indicates that the peridotites in order to deform plastically,were still above 800°C (§ 2.5.5.). Crystallization of hydrous gabbros at Moho depth during shear deformation indicates a temperature around 1000°C (Wyllie, 1980). Incorporated into thermal models for the oceanic lithosphere (Kusznir, 1980 ; Morton and Sleep, 1985), these constraints require the site of origin of the Oman ophiolite to be ocean crust no older than 1-2 Ma.

iii) The radiometric ages of the plagiogranites in the crustal section and of the amphibolites of the metamorphic aureoles completely overlap (table 3.1). Whether the plagiogranites are considered as the products of magma chamber differentiation or as wet anatexis of the gabbros during ridge hydrothermal activity, they would date, in either case the ridge processes, with the exception that some of them may have been generated through partial melting of the metamorphic aureoles as proposed by Boudier et al. (1988). The age overlap between plagiogranites and amphibolites proves that the thrusting occurred while the ridge was still active, or immediately afterward.

*Kinematics of the oceanic thrusting* - Boudier et al. (1985) have deduced the kinematics of the detachment phase from the study of structures and fabrics in the deformed peridotites and that of the subsequent oceanic motion, from the study of quartzites from amphibolites and greenschist facies soles. The kinematic analysis conducted over the entire belt, in connection with radiogenic dating, shows that thrust directions recorded in basal peridotites and in high grade metamorphic rocks can be divided into a NNW-SSE group, parallel to the sheeted dikes, and an E-W group (fig. 3.24). Shear motions are respectively to the SSE or to the West. Although both directions have been observed in the same area, the regional distribution shows that the NNW-SSE thrusting direction is practically absent in the southern massifs of Sumail and Wadi Tayin. K/Ar ages of amphiboles in high grade metamorphics cluster between 101 and 90 Ma, with some tendency for younger ages to characterize the E-W trend and the southern domain (fig. 3.24). It is also observed that ages of 100 Ma characterize metamorphic soles in a frontal position to the ophiolite nappe, whereas ages between 90 and 95 Ma correspond to metamorphic soles in a rear position ;

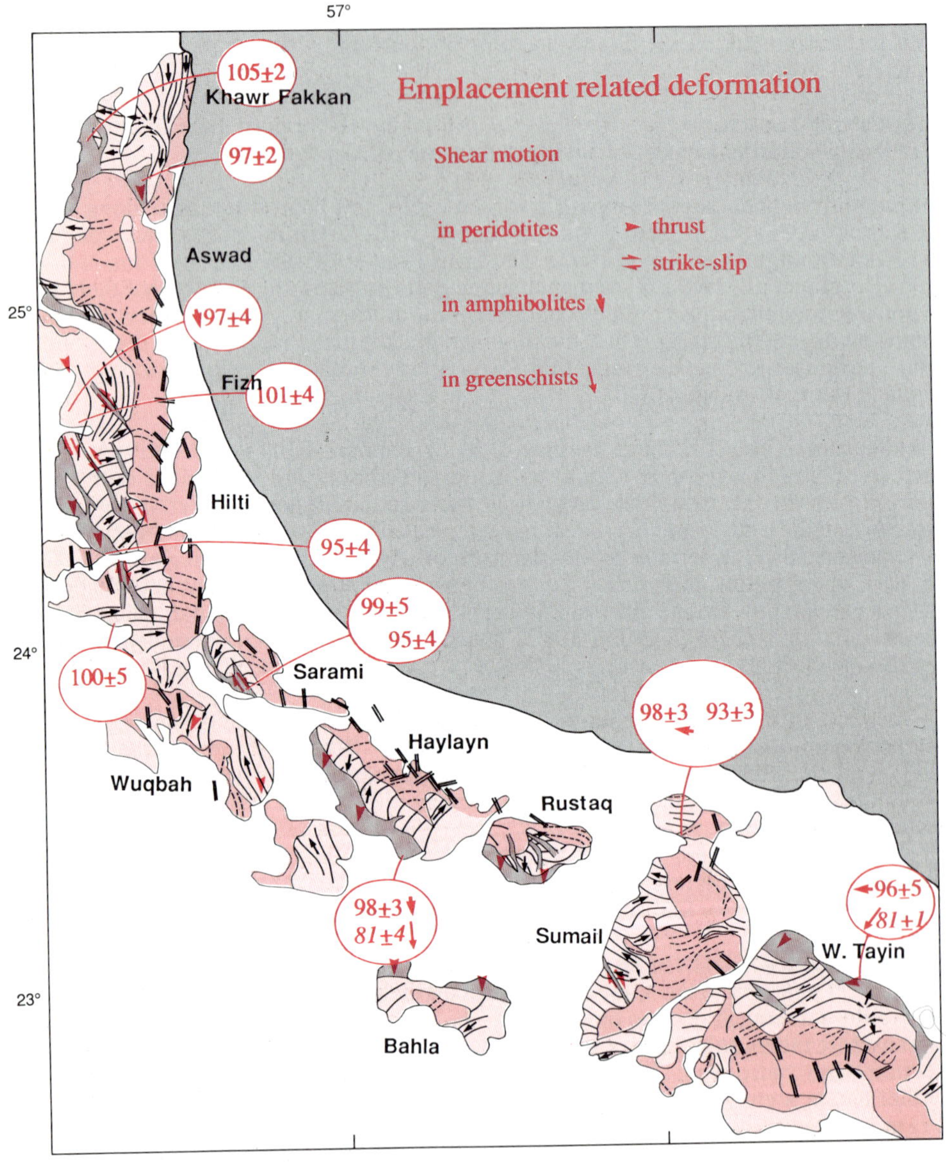


Fig. 3.24. Map of the emplacement-related deformation in the Oman ophiolite, with corresponding ages represented in color, upon the background of fig. 3.8b (modified from Boudier et al., 1985, 1988).

the time lapse would correspond to that necessary for thrusting the 50 km wide nappe over the point of emergence of the thrust which is the hottest point, where the metamorphism should be produced.

Kinematic analysis in greenschist facies rocks (fig. 3.24) shows a dominantly NE-SW trend of the transport direction in the southern massifs (Sumail and Wadi Tayin). There the greenschist formations, situated in a rear (NE) position relative to the ophiolite nappe, are thrust directly upon the Arabic margin limestone. In the massifs North of Sumail, the transport direction in greenschist formations is generally parallel to that recorded in the overlying high grade metamorphics. Downsection, motion directions become disoriented as greenschist rocks grade into the unmetamorphosed and disordered Hawasina formations.

*Obduction and high-pressure metamorphism* - The high-pressure metamorphism developed locally in the autochtonous formations of the Saih Hatat has been assigned to the obduction process, more precisely to a subduction of the Arabic passive margin beneath the ophiolitic nappes (fig. 3.28) (Lippard, 1983 ; Michard et al., 1984 ; Goffé et al., 1988). This is justified by the age of the metamorphism : 80-75 Ma, based on avalaible stratigraphic data (Goffé et al., 1988), and on isotopic dating in amphiboles from the metabasites (table 3.I). The large scale ductile deformation is comparatively homogeneous. It induces mineral lineations striking N 30 E (fig. 3.25a). A few shear senses recorded by quartz fabric in quartzite bands point to a SW directed motion of the nappes (Michard et al., 1984). Le Métour (1987) has produced some evidence of an inverse sense, possibly associated with retrograde conditions.

The unmetamorphosed autochtonous formations of the Jebel Akhdar dome are homogeneously structured with development of a penetrative cleavage slightly dipping to the West and South and of a stretching lineation trending NE-SW (Fig. 3.25b). This deformation has been also assigned to the passive margin subduction episode (Rabu, 1987).

The high pressure metamorphism and associated deformation are followed by SW directed large scale decollements (Glennie et al., 1974 ; Bayley, 1981 ; Bernoulli, 1982 ; Michard et al., 1984 ; Searle, 1985 ; Le Métour, 1987 ; Rabu, 1987) affecting the autochtonous and the base of the Hawasina formations. These thrusts are interpreted by Bernoulli and Veissert (1987) as the marks of ramp tectonics, responsible for the initiation of the Jebel Akhdar up-doming (Fig. 3.26). Subsequent gravity sliding of the ophiolite fragmented nappe represents the last trace of the obduction process.

*Post-obduction history* - The main movements subsequent to the obduction of the ophiolite nappe over the Arabian margin are the pursuing of Jebel Akhdar and Saih Hatat up-doming. This uplift is traced during Oligo-Miocene, by the tilting of the post orogenic Maestrichian to Lower Tertiary formations (Glennie et al., 1974). The domes have also tilted the peridotite nappe at their contact. They are still rising, whereas the coastal plain is subsiding. This explains the eastern dip of the structures in the ophiolites of the central massifs. The doming and uplift of the calcareous autochtonous massifs have been alternatively interpreted as a diapiric effect due to the ophiolite load upon lighter continental formations (Andrews-Speed and Johns, 1985). It seems however that excessive temperatures are required at the base of the nappe (1300°C) in order to heat up and activate diapirism in the underlying formations.

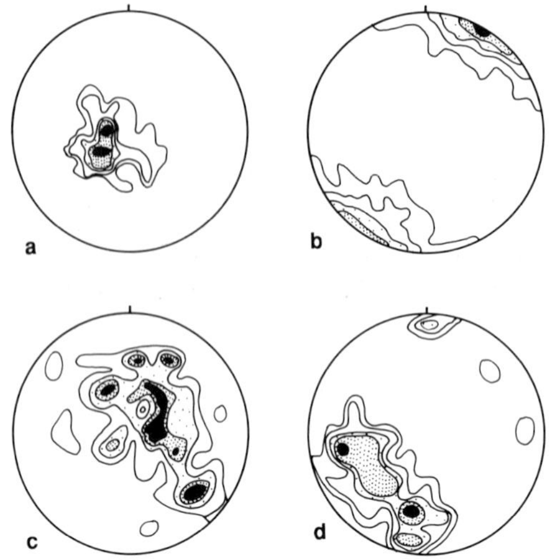


Fig. 3.25. Stereonets of foliations and mineral lineations in Arabic margin formations induced by the ophiolite nappe obduction. a and b) Respectively foliations and lineations in high P schists of the Saih Hatat (a : 252 measurements ; contours : 1,2,3,4,8 % ; b : 252 measurements ; contours : 1,2,3,4 %). c and d) Respectively foliations and lineations in low grade phyllites of the Djebel Akhdar (c : 74 measurements ; contours : 1,2,3,4 %); d : 82 measurements ; contours : 1,2,3,4,8 %) (Boudier et al., 1985 and unpublished data).

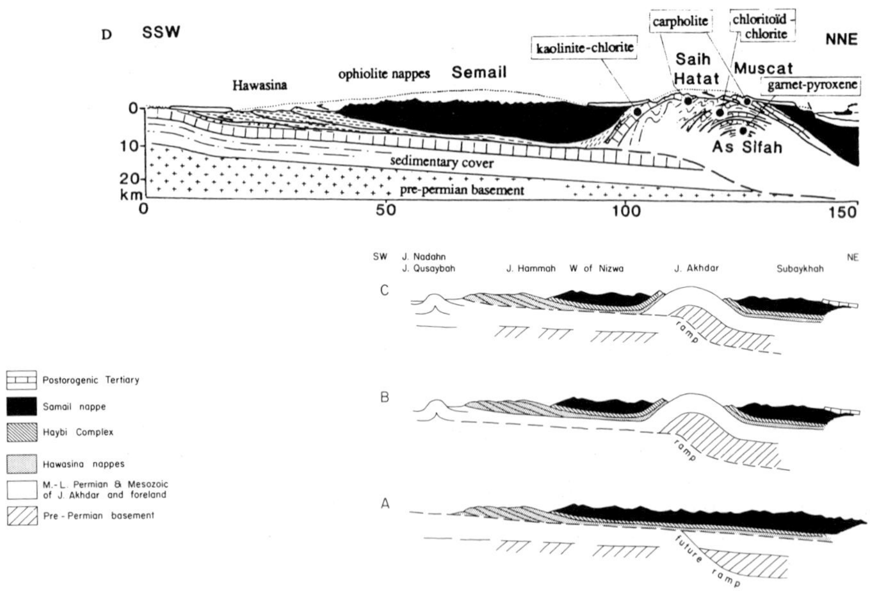


Fig. 3.26. a, b, c) Kinematic model of ophiolite obduction and ramp tectonics in the Arabic margin. d) Enlarged c section, illustrating the high pressure gradient in the Saih Hatat (a,b,c, from Bernoulli and Weissert, 1987 ; d, from Goffé et al., 1988).

## *3.5. GENERAL INTERPRETATION OF THE OMAN OPHIOLITE*

### *3.5.1. Introduction*

The single most influencial parameter in the study of oceanic lithosphere and ophiolites seems to be the spreading rate (§ 8.2.3 and chapter 9). Hence, we discuss this parameter in the case of the Oman ophiolite first. In the study of an ophiolite, one may have a thematic or a regional interest. The most controlled features of the Oman ophiolite which are relevant to a thematic study of the functioning of oceanic ridges will be used in the appropriate place in this book : large scale mantle flow pattern, in § 9.2 ; melt extraction in relation with mantle diapirism, in § 7.5 and 10.3 ; and eventually magma chamber structure and dynamics, in § 11.4. Consequently, we will restrict ourselves here to a brief discussion of the regional aspects : paleo-environment of origin of the ophiolite and obduction history in the frame of a closing Neo-Tethys.

### *3.5.2. Spreading rate estimation*

Estimating this important parameter directly, even in a large piece of ophiolite like that of Oman is very difficult because of the insufficient accuracy of radiogenic dating (Tilton et al., 1981). However, several factors favor a medium to fast spreading rate in the case of the Oman paleo-ridge (Nicolas et al., 1988a).

i) The continuous and thick character of the layered gabbro sequence suggests a permanent magma chamber (Pallister and Hopson, 1981) which seems tenable only in the case of fast spreading rates (§ 11.4).

ii) No transform fault has been found with the exception of the central domain of Wadi Tayin. The absence of such structures for over 300 km along the ridge axis is compatible only with an East Pacific Rise type of environment (fig. 3.1). Moreover, the transform fault of Wadi Tayin would correspond to a fast spreading environment (§ 5.4.3).

iii) The foliations and associated flow planes in the uppermost peridotites are parallel to the Moho surface and apparently remain flat-lying within a few kilometers downsection, except in diapiric areas. In accordance with the discussion of § 2.6., this implies that the isotherms in the uppermost mantle were only gently dipping which is characteristic of a fast spreading ridge.

### *3.5.3. Paleo-environment of origin and obduction history*

Reinhardt (1969) and Coleman (1981) considered the Oman ophiolite as a piece of oceanic lithosphere formed at a mid-oceanic ridge and derived from the Cretaceous spreading of the Neo-Tethys. Coleman also proposed that the ophiolite detachement occurred at the ridge, along a NE dipping thrust in response to the closing of the Tethys. Boudier and Coleman (1981) analysed the kinematics of the early thrusting as corresponding to a Westward transport direction. For these authors, the ophiolite was derived from the NE flank of the ridge.

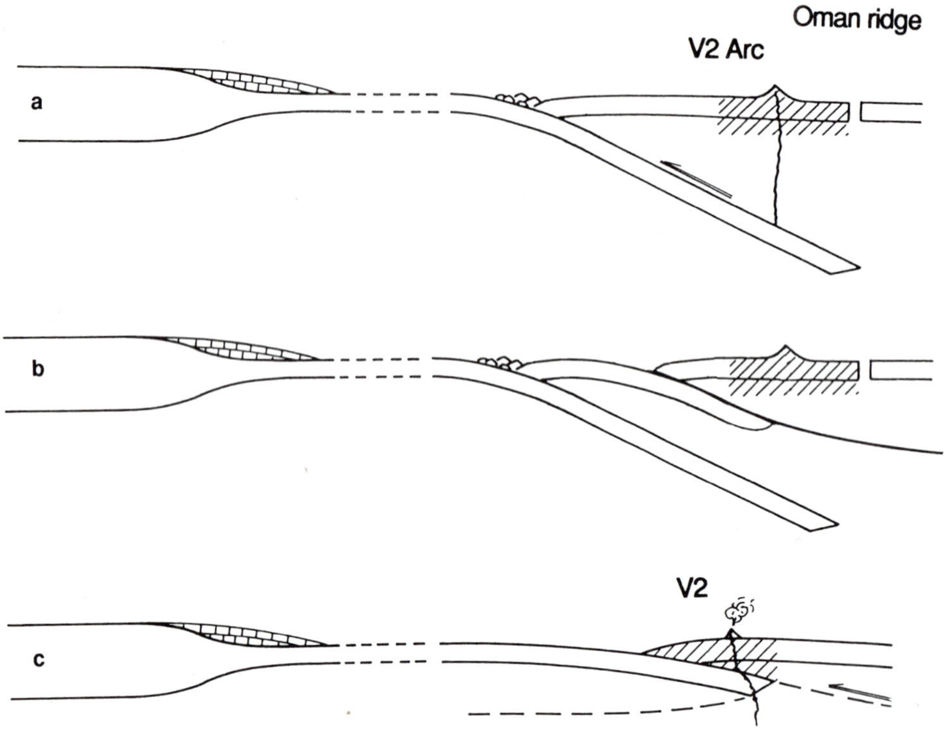


Fig. 3.27. Proposed schemes for the origin of the Oman ophiolite (shaded area) and of the secondary volcanism. a and b) The ophiolite is generated in a back-arc and contaminated by the Lasai immature island-arc volcanism ; a) The basal sole is the subduction surface ; b) The basal sole is an intra-oceanic thrust ; c) The ophiolite is generated at a ridge ; the thrusting occurs at the ridge, inducing by crustal contamination the secondary volcanism; the basal sole is the young lithosphere asthenosphere boundary. (a) Alabaster et al., 1982 ; b) Lippard et al., 1986 ; c) Boudier et al., 1988).

This interpretation has been challenged by the discovery, mainly in Central Oman, of the secondary magmatism (§ 3.2.). Its geochemical affinities led to the conclusion that the Oman ophiolite had been formed in an island arc environment or in a back-arc basin with superimposition of an island arc volcanism (Pearce et al., 1981 ; Alabaster et al., 1982 ; Searle and Stevens, 1984 ; Lippard et al., 1986 ; Beurrier, 1987 ; Le Métour, 1987). In this new interpretation, there is a subduction zone between the spreading center and the Arabic passive margin (fig. 3.27a,b) which is responsible for the generation of the Oman oceanic lithosphere at a supra subduction spreading center, either an island arc or a back-arc. In an earlier model, both the oceanic thrusting and the obduction were taking place along the subduction plane ; the mylonitic lower peridotites and the metamorphic aureole formations derived from the dynamico-thermal processes taking place along the subduction plane. The high pressure metamorphism and, in particular, the recently discovered eclogitic one, are easily explained by the subduction of the Arabic margin formations below the lithosphere wedge from which the ophiolites are issued. In the more recent model of Lippard et al. (1986), the subduction related origin is maintained, but the thrusting of the ophiolite nappe occurs along a new plane at the limit between Tethyan and Neo-Tethyan lithosphere (fig. 3.27b).

This model of an island-arc and back-arc accretion followed by a subduction-related obduction has been applied to several ophiolites (§ 8.3), in particular in the case of the Tethyan ophiolites (Moores et al., 1984). Although it accounts well for the hydrous andesitic magmatism and for the high pressure metamorphism, this model predicts that trench and island arc formations should be found associated with the ophiolite because both are located between the passive margin and the ophiolite locus of origin and should therefore be thrust with it ; such formations are virtually unknown. This difficulty, encountered in Oman and in other ophiolites, has been dealt with by inferring that the ophiolites are generated in an immature island-arc environment. It remains difficult to explain the typical oceanic nature of the sediments in contact with the Oman ophiolite crust (fig. 3.10) and the absence of any trench series in the Hawasina Nappes below the ophiolite.

For this reason, with Boudier et al. (1985, 1988) and Ernewein et al. (1988), we favor the interpretation of accretion at a mid-ocean ridge (fig. 3.27c). This model is supported by the relation presented above (§ 3.4.3) between basal thrusting in the ultramafic section and the activity of large shear zones in a still hot and consequently very young lithosphere and by the age coincidence of lithospheric accretion and oceanic detachment (table 3.I). The flat thrusting, modelled after field data (fig. 3.23) is indeed a form of subduction which, coinsidering the particularly young age of the subducting plate, occurs along a shallow and flat plane as predicted by subduction models (Uyeda and Kanamori, 1979). Ernewein et al. (1988) dispute the arc-related origin of the various volcanisms and propose to ascribe the hydrous $V_2$ magmatism to a dying ridge activity. Boudier et al. (1988) suggest that the hydrous melting of the metamorphic aureoles during early detachment could contaminate the secondary magmatism, explaining its island arc affinity. The wehrlitic intrusions and associated picritic extrusives, whether related to the expulsion of residual trapped magma during the first compression (Ernewein et al., 1988), or injected in the cooling crust near the ridge (Benn et al., 1988) seem to be related by their trace elements patterns (fig. 3.14) to the hydrous $V_2$ magmatism. Finally, in this interpretation the high pressure metamorphism is explained by the ephemeral subducting of the Arabic platform and shelf formations (Goffé et al., 1988) (fig. 3.28). This would occur at around 80 Ma (table 3.I), after the oceanic lithosphere has been entirely doubled by the oceanic thrusting. The subduction of continental shelf rapidly creates a blockage situation which, in turn, triggers a more stable oceanic subduction in the Makran. The initiation of subduction in the Makran, dated by the age of the oldest subduction-related magmatic intrusions in the Bazman area, would be between 80 and 70 Ma (Berberian and Berberian, 1981). The age coincidence between the end of obduction of the ophiolite nappe and the beginning of the Makran subduction, should be noted (table 3.I). This succession of events is outlined in figure 3.28. In this regard, the fact that the accretion prism in the Makran subduction zone is partly above sea level could be due to the subducting of a doubled oceanic crust.

At this stage, it should be noted that there is some convergence between the interpretations in figure 3.27. In their last model, Lippard et al. (1986) retain the idea of the flaking in a young lithosphere with a shallow thrust surface, distinct from the subduction surface, whereas the other authors prefer an Arabic margin subduction to explain the high pressure metamorphism. The difference is that for the former authors (fig. 3.27b), the subduction is first marine, active over 20 Ma, and precedes the Oman ophiolite generation. For the latter authors (fig. 3.28), the subduction is solely continental, operating only during a few Ma when the ophiolite is being obducted over the Arabic margin.

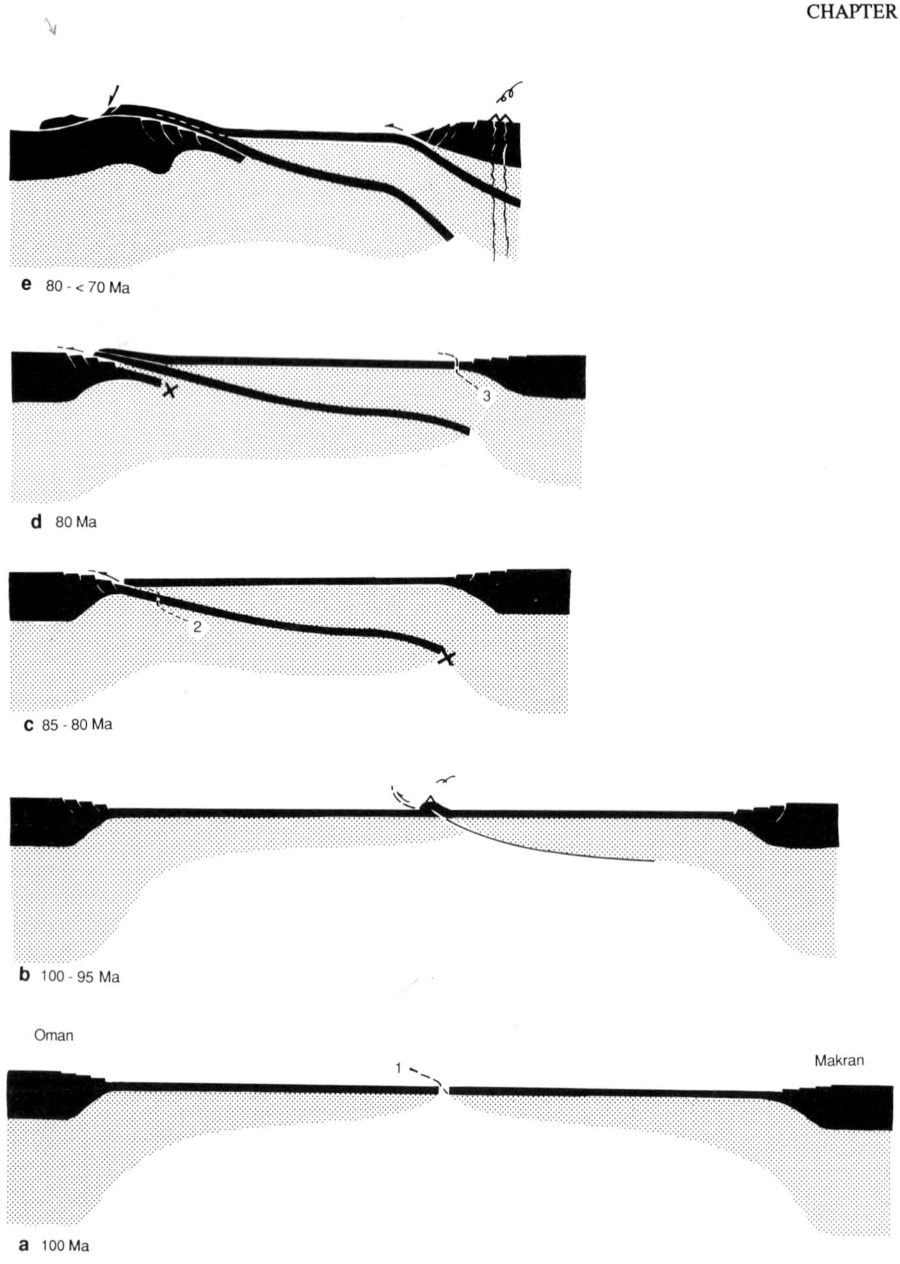


Fig. 3.28. Scheme of a possible history of the Oman ophiolite emplacement; 1), 2) and 3) : successive detachment surfaces. a) Oceanic accretion ; b) Detachment in the ridge vicinity and secondary volcanism ; c) Oceanic thrusting leading to a blockage (X) due to the meeting with older and thicker lithosphere ; d) Arabic margin subduction, progressively choking (X) ; e) Initiation of a new subduction in the Makran, isostatic rebound of the Arabic margin and gravity sliding of the ophiolite nappe.

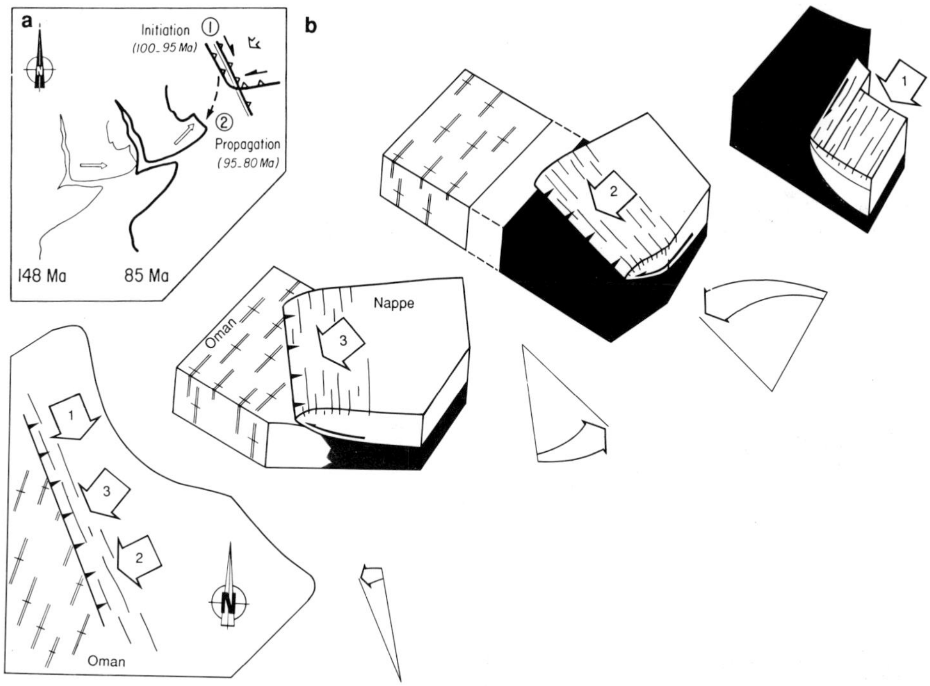


Fig. 3.29. Two models of oceanic thrusting and obduction of the Oman ophiolite in relation with the motion of Africa with respect to stable Eurasia (Patriat et al., 1982). The thrusting is initiated along the ridge. In model a), there is no block rotation and the first motion is either normal or parallel to the presumably NNW-SSE ridge. The motion normal to the ridge is now recorded in the Westward shear lineations, and the parallel motion, in the Southward lineations. Eventually, in response to the change in motion of Africa at 85 Ma from Eastward to Northeastward, the thrusting and obduction path becomes Southwestward as recorded by the low-T lineations. In model b) there are large rotations (curved arrows) between the ophiolite lithosphere and the Arabic margin and the motion is assumed to maintain a fairly constant direction ; arrow 1 : high T kinematics in peridotites and amphibolites, dated 101-95 Ma ; arrow 2 : kinematics in the same formations, dated 100-90 Ma ; arrow 3: low T kinematics dated 80-70 Ma (a) Boudier et al., 1985 ; b) Thomas et al., 1988).

In the second model, thermal constraints require a rapid shift from oceanic spreading to oceanic convergence occurring within one to a few Ma (Boudier et al., 1988). Such a rapid shift can be ascribed to various causes, such as a change in plate velocity or in spreading direction. The first case can be illustrated by the India-Eurasia convergence, which corresponds to a paleo-geographic situation comparable to the Africa-Eurasia convergence under consideration. When India and Eurasia first collide at 50 Ma, the convergence rate falls by some 10-20 cm/yr (Patriat and Achache, 1984) within 2 Ma. A collision of Eurasia with any block located between it and Africa could similarly produce a rapid drop in subduction rate, inducing in the Oman paleo-ridge located South of the Eurasia subduction zone, a rapid shift from expansion to compression (chapter 13). The second case has been documented by Casey and Dewey (1984) in relation with ophiolite emplacement. In the Oman case, these authors propose an along-strike shift from spreading to convergence due to the fact that the pole of rotation of the interacting plates was located close-by.

The oceanic detachment and first thrusting motion would have occurred along two directions : E-W or NNW-SSE (in the present day referential), except in the Sumail and Wadi Tayin massifs, where only the E-W motion has been recorded (fig. 3.24). The shear senses deduced from the kinematic analysis are systematically Westward or Southward. The NNW-SSE thrusts are in structural continuity with the NNW-SSE shear zones (fig. 3.15) which are parallel to the ridge (§ 3.4.3). These thrust directions are thus interpreted as being controlled by the ridge topography and thermal structure. In the Fayd area, the rotation to the vertical affecting several kilometers of lithosphere (fig. 3.15) possibly took place during this longitudinal Southward motion. Where the ridge control did not operate, for instance because the detachment emerged at some distance from the ridge, the first motion would be directly from East to West.

The situation seems simpler dealing with the subsequent oceanic thrusting recorded in the oceanic formations of greenschist facies (age span : 85-71 Ma) and with the obduction motion recorded in the margin formations of high pressure metamorphic facies (average age : 87 Ma). The dominant motion is from NNE to SSW, with the exception of the Dibba metamorphic aureole and a few other small thrust zones which could have been rotated during the final emplacement.

Boudier et al. (1985) have proposed that the ophiolite thrusting is related to the drift of Africa with respect to Eurasia, and that the EW to NNE-SSW change in kinematic path of the ophiolite nappe between 100 Ma and 90-80 Ma is related to the change in convergence direction between Africa and Eurasia occurring during this time span (fig. 3.29a). Thomas et al. (1988) have obtained new paleomagnetic data on umbers and radiolarians associated with the accreting crust of origin of the Oman ophiolite. Integrating these data to those previously published on diabase dikes and gabbros (Luyendyk and Day, 1982 ; Luyendyk et al., 1982 ; Shelton, 1984), they propose a model implying large rotations of the ophiolite nappe (fig. 3.29b). In this model the ridge of origin was oriented NNE-SSW. Interestingly, their model predicts that during an assumed constant NE drift of Africa, the motion of the ophiolite nappe keeps a steady SW direction ; the apparent rotations in thrusting directions presented above would in fact reflect rotations of the nappe itself. In spite of its coherence, this model must be considered as speculative, because it is still insufficiently constrained by paleomagnetic data.

# Chapter 4

# XIGAZE AND TRINITY OPHIOLITES - PLAGIOCLASE LHERZOLITE MASSIFS : THE LHERZOLITE OPHIOLITE TYPE

## *4.1. INTRODUCTION*

A few ophiolite massifs depart from the Oman type and seem to define a specific trend. In contrast with the Oman ophiolite, which is characterised by a depleted harzburgitic ultramafic section and defines the Harzburgite Ophiolite Type (HOT), these massifs which, among several other pecularities, have a lherzolitic or sublherzolitic ultramafic section, define the Lherzolitic Ophiolite Type (LOT) (Boudier and Nicolas, 1985). The differences between HOT and LOT are summarized and discussed in chapter 8.

The Xigaze ophiolite of Tibet, which for many reasons can be regarded as intermediate between the harzburgite and the lherzolite types is described first, followed by the Trinity ophiolite complex of Northern California, which is representative of the lherzolite type, although the complexity of its general setting hinders understanding some of its characteristic issues. Moving toward the extremity of the HOT-LOT spectrum, one next encounters plagioclase lherzolite massifs, which may or may not be covered by an ophiolitic crust and which are illustrated by the western Alps ophiolites ; finally they are plagioclase and spinel lherzolite massifs which, although they are associated with continental crust, still show many affinities with LOT.

## *4.2. XIGAZE OPHIOLITE*

### *4.2.1. Introduction*

Our presentation here of the Xigaze ophiolite of Tibet is justified by its remarkable geodynamic location along the Himalayan suture and by some peculiar internal features. The ophiolite belt which underlines this suture between the Indian and Eurasian plates is strongly deformed, being commonly reduced to meter-sized slivers of serpentinite. Large massifs have been exceptionally preserved in the Xigaze area, and among these the largest and least dismembered is located just South of this locality (fig.4.1) ; it is referred to as the Xigaze massif. Being representative of this part of the ophiolite belt, this massif has been the most studied (Nicolas et al., 1981 ; Xiao Xuchang, 1984 ; Wang Xibin et al., 1984 ; Deng Wanming et al. 1984 ; Girardeau et al., 1984 ; Gopel et al., 1984 ; Girardeau and Mercier, 1985, 1988 ; Girardeau et al., 1985a and b) and will be dealt with in detail here. This massif displays all the characteristic units of an ophiolite and, in spite of locally sheared contacts, it seems possible to restore these units in their natural order and to estimate their initial thickness.

### *4.2.2. Geological setting*

The Xigaze ophiolite belongs to the Indus-Yarlung Zangbo suture, marked by

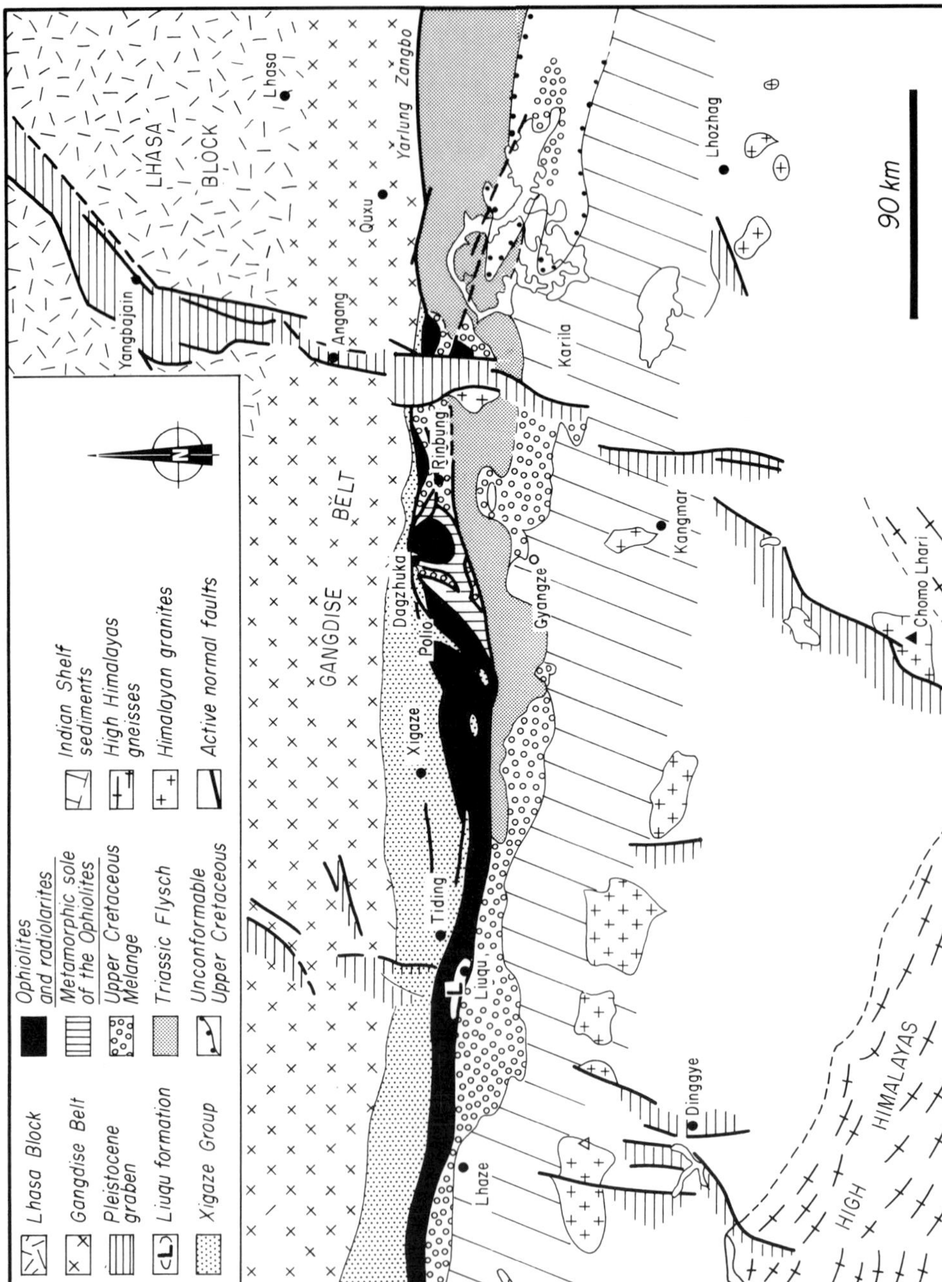


Fig. 4.1. Simplified geological map of the Indus-Yarlung Zangbo suture in Tibet (after Tapponnier et al., 1981).

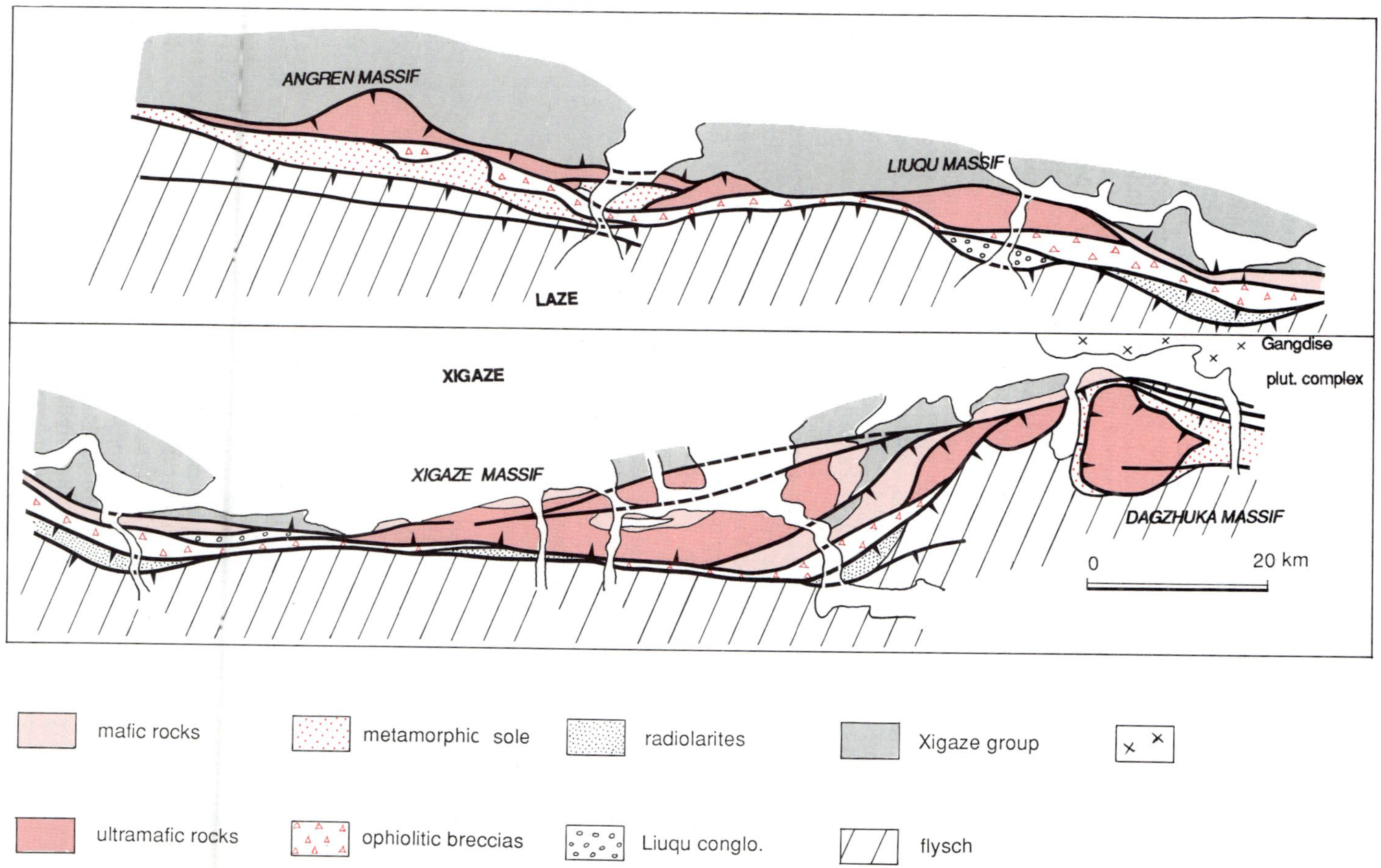


Fig.4.2. Structural map of the Xigaze ophiolite between Dagzhuka and Angren (simplified from Nicolas et al., 1981 and Girardeau et al., 1984).

discontinuous ophiolite bodies disseminated from Burma to Ladakh (fig. 3.2). In the Xigaze area, this suture extends E-W and separates the Indian plate from the Lhasa block to the North (Chang et al., 1977 ; Gansser, 1977 ; Xiao Xuchang et al., 1980).

The Indus-Yarlung Zangbo ophiolites were formed in the paleo-Indian Ocean separating India from the Lhasa block during upper Albian-lower Cenomanian (110-100 Ma) as shown by the age of radiolarian cherts directly covering the Xigaze ophiolites (Marcoux et al., 1982). A $^{238}U/^{206}Pb$ age has given 120 ± 10 Ma (Göpel et al., 1984) and a Nd/Sm one, 109 ± 21 Ma (Prinzhofer, 1987). The collision of India with Eurasia, to which the Lhasa block had been accreted during late Jurassic-early Cretaceous (Girardeau et al., 1984 ; Allegre et al., 1984), began in Cenozoic at 50 Ma as shown by paleomagnetic data (Patriat and Achache, 1984). Since then, approximately 2000 km of crustal material has been absorbed in a N-S direction along a North dipping subduction zone, whose trace is the Indus-Yarlung Zangbo suture. The ocean was completely resorbed and the ophiolites obducted before the end of the Eocene (before 40 Ma) (Tapponnier et al., 1981). The subductions of oceanic crust induced island-arc magmatism in the Gangdise mountains of the Lhasa block. Accordingly, the youngest granodiorites dated at 41 Ma (Shärer et al., 1984) record, with a delay of a few million years, the closure of the ocean and thus the end of oceanic subduction. Thereafter, the shortening occurred by continental collision and subduction (Mattauer, 1983 ; Burg and Chen, 1984).

The well-developed ophiolite in the Xigaze area has been mapped in detail (fig.4.2) (Nicolas et al., 1981 ; Girardeau et al., 1984 ; Girardeau and Mercier 1985). A N-S cross section (fig.4.3) through the Xigaze massif illustrates the comparatively simple structure in this area, where the ophiolite is essentially tilted between the Gangdise plutonic complex and its Xigaze Group sedimentary cover to the North, and the Tethyan flysches of the Himalaya to the South. The steep attitude of the ophiolite massif relates to the fact that, due to the collision along the suture, the overall structure of the suture fans from a Northward increasing dip in the South, to a steep southward dip in the North. East of the Xigaze massif, the large Dagzhuka ophiolite massif has been squeezed out of the suture and is now partly covering it.

### *4.2.3. Description*

From the numerous cross sections established in the Xigaze ophiolite (Girardeau et al., 1984), it is easy to draw two typical logs , one through the Xigaze massif (fig. 4.4a) and the other through the Dagzhuka massif (fig. 4.4b), the later differing mainly by some development of gabbros in the mafic section and by the presence of basal formations (mylonitic peridotites and amphibolites) typical of a metamorphic aureole (chapter 12).

The volcanics on top of the ophiolite are covered, in stratigraphic continuity, by cherts and marine pelagic sediments belonging to the Xigaze group and are thus attached to the Lhasa block (Nicolas et al., 1981 ; Girardeau et al., 1984). Below the cherts, either pillow-lavas or lava-flows are observed. Their deposition plane is parallel to the chert bedding plane except locally in the pillow lavas and in strongly brecciated areas. Polarity in the pillow-lavas points to the top Northward.They are either largely variolitic, with 1 cm varioles, or massive with only a finer-grained rim. Their matrix is not abundant and exclusively constituted by volcanic debris and glass. Below a few hundred meters of volcanics, the first diabase dikes and sills appear. They typically present chilled margins and have been observed cutting through the pillows. They progressively

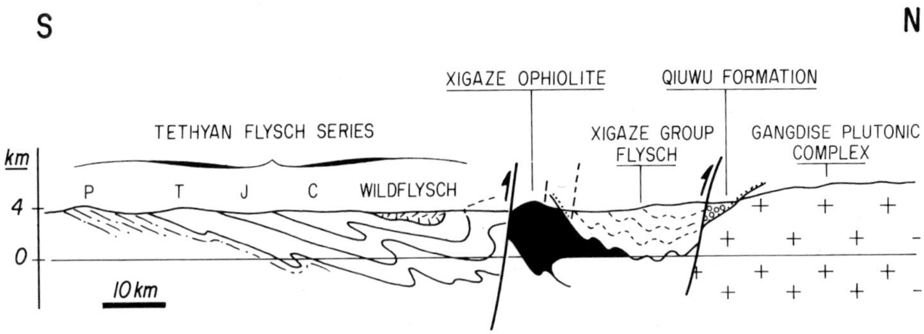


Fig. 4.3. Cross section through the Yarlung Zangbo suture in the Xigaze area (after Burg, 1983).

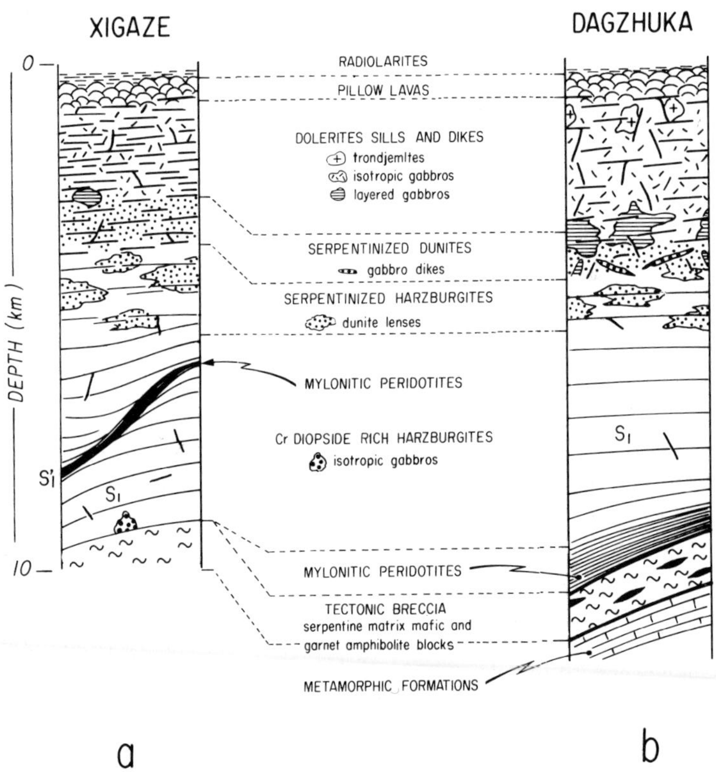


Fig. 4.4. Schematic logs through a) the Xigaze and b) the Dagzhuka massifs (after Nicolas et al., 1981 and Girardeau et al., 1985a).

become more abundant until they constitute a sill complex with mutually intrusive sills. The sill rather than dike nature is deduced from their general parallelism with the overlying volcanic lava-flows and sediments. A few dikes, normal to the sills, are also present. The thickness of individual sills is in the meter range. When they reach a few metres, microgabbro textures can be obtained in their center. New sills intrusive in such rocks can create the illusion that they are intrusive in a gabbroic formation. True plutonic rocks are uncommon between diabase sills with the exception of a few isotropic gabbros and trondjhemite screens.

In the Dagzhuka (fig.4.4b), Bainang, Tiding and Angren massifs, small bodies of layered gabbros and ultramafic rocks are locally present at the base of this sill complex (Wang Xibin et al., 1984). In these massifs, thickness of the layered olivine gabbros does not exceed 350 m, except in Angren, where a thickness of 2500 m has been locally measured (Prinzhofer et al., 1984). In the Dagzhuka massif, the walls of the chambers are made of troctolites and dunites which can be interpreted as the first cumulates or, alternatively, as residual dunite walls impregnated by feldspathic liquid, feldspar coming from the chamber (§. 10.4.3). The shape of the magma chambers is very irregular and the total thickness mentioned above could be achieved by the addition of smaller chambers. Total thickness of the mafic unit does not exceed 3 km.

This mafic section lies over a serpentinized harzburgite and dunite formation penetrated by numerous diabase sills. Sills constitute 50 % of the rock volume in the upper part of this formation, the fraction continuously decreasing downwards. Locally, they are particularly thick, up to 7 m. They have been observed to branch into perpendicular dikes which probably represent their feeders. In the various massifs, the diabase sills and dikes have a similar composition whatever their level of intrusion. A few rodingitized gabbro dikes also intrude the ultramafic rocks ; they are cut by the diabase dikes. In contrast with the gabbros, the diabase dikes are not rodingitized. As rodingitization accompanies serpentinization of the peridotite in greenschist facies conditions (§. 2.5.6), it can be concluded that the gabbro dikes were emplaced before the metamorphic serpentinization, and the diabase dikes, after or later during the process. The peridotite structure is not altered by the serpentinization except locally, where low-temperature deformations have disrupted and boundinaged the dikes and developed a slickenside cleavage in the serpentinites. As a consequence, the coarse porphyroclastic texture of the peridotite, typical of high-temperature flow, is still visible and its orientation can be mapped. Thickness of the harzburgite-dunite formation containing the sills is 500 m on average with large local variations. Downwards, it grades into a 500 m-thick unit formed of serpentinized foliated harzburgites and dunites with a decreasing amount of gabbro and diabase dikes. Lizardite has been identified as the dominant phase in the serpentinites, but antigorite is also present. After 1-2 km, the degree of serpentinization rapidly decreases in these peridotites which become fresh diopside-bearing harzburgites with a crude layering resulting from the alternation of Cr-rich and Cr-poor diopside facies. The rocks are well foliated with discordant shear zones of mylonitic peridotites a few meters thick. Outside these zones, porphyroclastic structures related to high- temperature plastic flow are observed. Girardeau and Mercier (1988) give a detailed petrological and textural description of this ultramafic section.

The southern E-W contact of the Xigaze ophiolite with the sedimentary formations is tectonic. More than 500-800 m away from this contact, ultramafic rocks are transformed into an ophiolitic breccia or mélange with blocks of rodingitized gabbro and diabase floating in a schistose serpentinite matrix. On the other side of this major tectonic contact, either Jurassic radiolarites or a coarse red conglomerate separate the ophiolites from the Triassic Wild Flysch series. They are also strongly deformed and schistosed at the contact but, like the serpentinites, are devoid of any suture-related metamorphism.

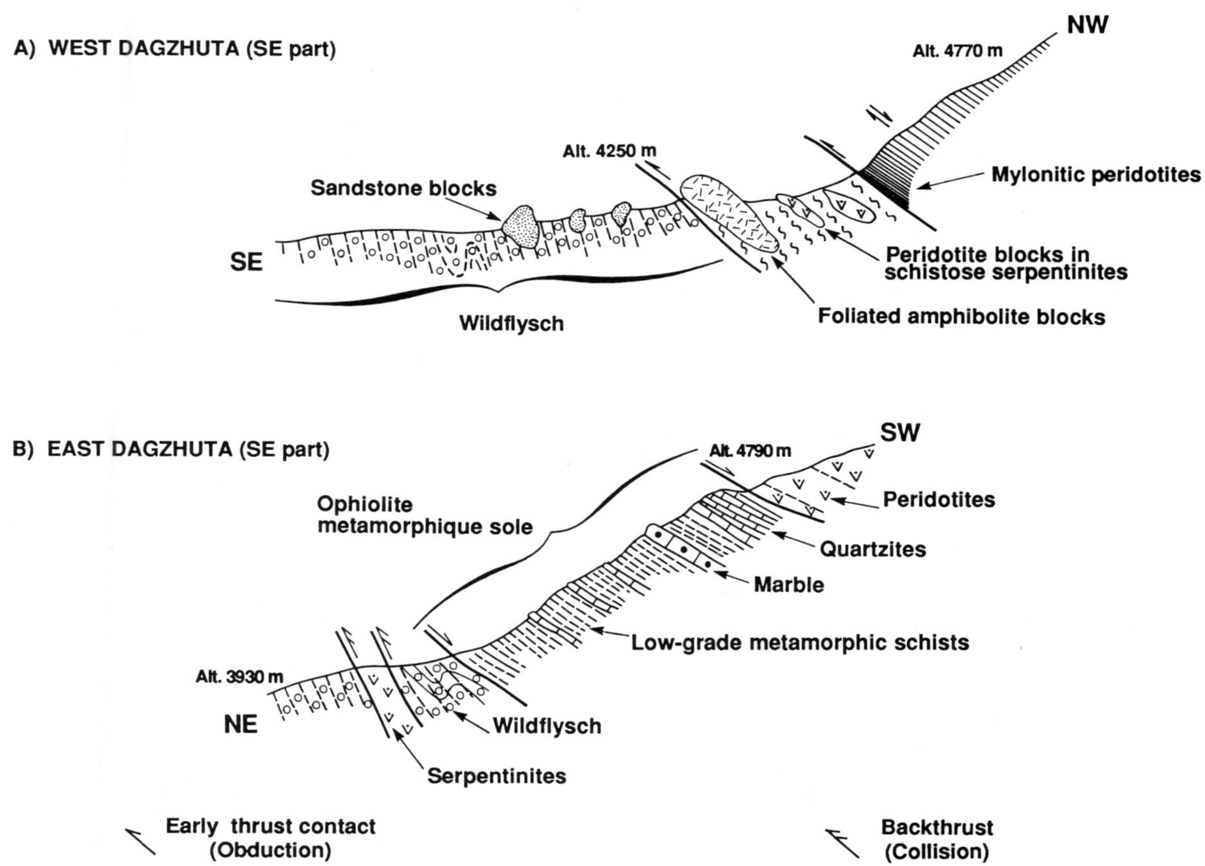


Fig. 4.5. Basal contact in the southern part of the Dagzhuka massif between fresh mylonite peridotites and the southerly Wild Flysch (after Girardeau et al., 1984).

This type of metamorphism marked by amphibolites, garnet amphibolites and quartzites has been observed as inclusions in the serpentinite mélange of several localities (Liuqu, Xigaze, Bainang and Dagzhuka). In the Dagzhuka massif, mylonitic peridotites are developed over more than 300 m above the basal contact which contains blocks of amphibolites in a serpentinite matrix (fig.4.5). Below, a metamorphic formation of quartzites, phyllites and ophicalcites, crops out locally over 300 m. This has been interpreted as the basal part of a metamorphic aureole (Girardeau et al., 1984).

### *4.2.4. Structural analysis*

A detailed structural mapping has been conducted in the Xigaze, Dagzhuka and Liuqu massifs (Girardeau and Mercier,1985 ; Girardeau et al. 1985b). In the Xigaze massif, the maps of planar and linear elements respectively shown in figures 4.6a and 4.6b illustrate the orientations, density and variability of measurements in the field.

These data are reported on the stereonets of figure 4.7, where their average attitudes have been computed. The vertical E-W contacts are of tectonic origin due to secondary shearing. The original contacts between the units in this massif are oriented NE 60°with a 60° NW dip. This attitude is also that of the deposition plane of radiolarites and lava flows, and that of the diabase sills. The diabase dikes are oriented on average at a right angle. A few lamination surfaces and associated mineral lineations have been measured in the gabbro screens between the diabase intrusives. The surfaces are roughly parallel to the general attitude and lineations plunge weakly to the NE.

The contact between the mafic and ultramafic sections is commonly sheared. It seems to be irregular but retains the common NE 60°, 60°NW orientation, with the thick diabase sills also parallel to that orientation. Below, the high temperature S1 foliation (fig.4.7b) in the fresh harzburgites and the associated L1 spinel lineation (fig.4.7c) have nearly constant orientations. They record a large plastic flow produced in non-coaxial regime with a dominant shear sense indicating an upward and northward motion of the northern block relative to the southern one. From the analysis of the substructure, high to intermediate temperature conditions (as defined in §2.5.5) are predominantly recorded in the ultramafic section. There are however a number of several meters-thick shear zones displaying low-T porphyroclastic to mylonitic textures; a sinistral shear sense is associated with these new S'1 foliations (fig.4.7d) and L'1 lineations (fig.4.7e). In the Dagzhuka massif, similar textures appear at the base of the massif in connection with the development of the metamorphic aureole (fig.4.5).

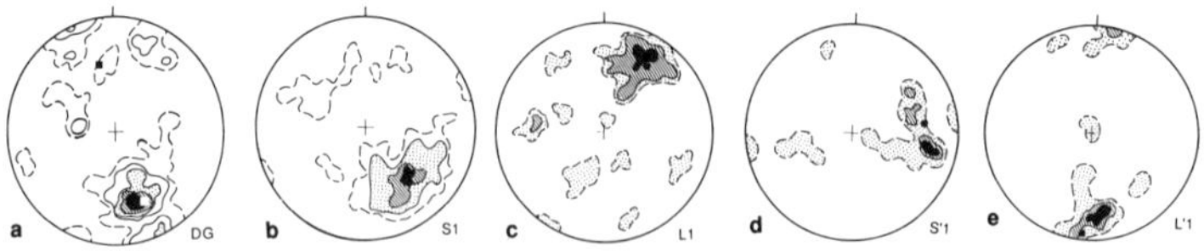


Fig. 4.7. Stereonets of the structural elements in the Xigaze massif (equal-area projection ; lower hemisphere ; solid squares represent the computed average poles of planar structures and directions of linear structures). a) DG dolerite intrusives in the dolerites and gabbros unit (67 measurements ; contours : 1, 3, 7, 12, 17 %). b) S1 foliation plane in peridotites (148 measurements ; contours : 1, 3, 5, 9 %). c) L1 lineations in peridotites (58 measurements; contours : 2, 3, 5 %). d) S'1 mylonitic foliation plane in peridotites (17 measurements ; contours : 18, 12, 6 %). e) L'1 mylonitic stretching lineations (14 measurements ; contours : 8, 14, 21 %). (After Girardeau and Mercier 1985).

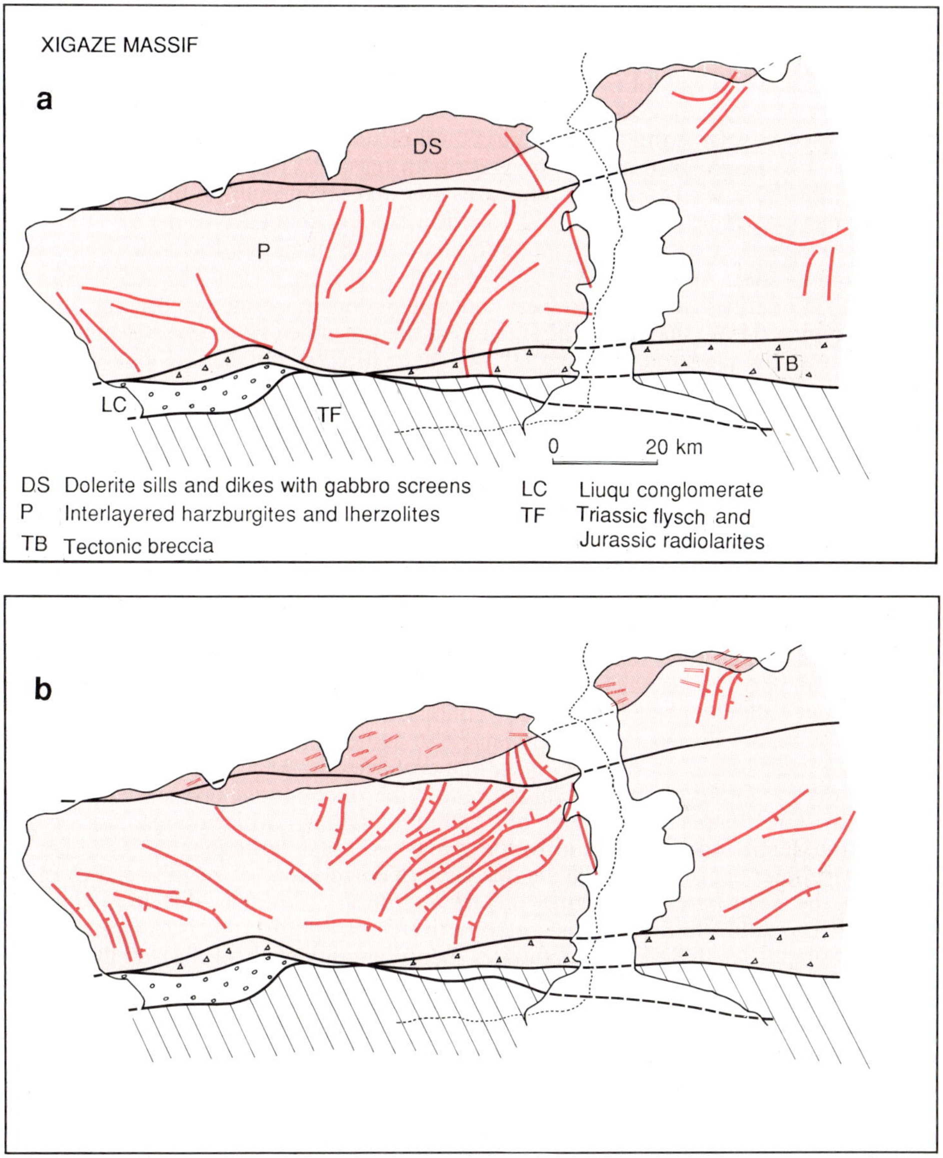


Fig. 4.6. Structural maps in the Xigaze massif.Trajectories of a) lineations and b) foliations in peridotites. Short parallel lines in b) : diabase sills (modified from Girardeau and Mercier, 1985).

These data have been used by Girardeau and Mercier (1985) and Girardeau et al. (1985b) to restore this ophiolite in its oceanic situation following the method proposed in § 2.2.1. For this purpose, the cherts and lava flow deposition planes which are parallel to the Moho separating the mafic and ultramafic sections have been rotated horizontally. It has been assumed that the strike of the diabase dikes after this rotation may indicate the ridge direction. Throughout the Xigaze ophiolite these dikes strike N 60°-80°. The ridge direction defined in this way does not take into account the rotations of the whole ophiolite deduced from paleomagnetic studies which would occur during the closure of the Indian paleo-ocean (Pozzi et al., 1984 ; Girardeau et al., 1985b). The computed average attitudes of the structural elements described above appear in figure 4.8 for the Xigaze and West Dagzhuka massifs.These models suggest that, below a thin crust with flat-lying diabase sills and steep diabase dikes, the mantle structure has a nearly horizontal high-T foliation dipping about 20° to the SE (fig.4.8b). The asthenosphere flow plane orientation would be dipping some 20-30° to the SE, considering its shear sense (see § 2.5.4) ; the flow lineation would be at 50° from the ridge trend as defined above. The shear zones in mylonitic peridotites become oriented N120°, dipping 64°W, that is at a high angle to the presumed ridge direction and fairly steep, with a flow lineation plunging 60°W. Interestingly, the diabase sills which are nearly horizontal (dip of 10° to the SE) in the crustal section become progressively more inclined in the mantle section with a 18° dip at Moho depth and 28° in the deeper section.

### *4.2.5. Geochemistry*

Petrology and major-element chemistry of the Xigaze ophiolite mafic section has been studied by Den Wanming et al., (1984), Girardeau et al. (1985a) and the trace-elements by Wang Xibin et al., (1984), Den Wanming et al.(1984), and Prinzhofer et al. (1984). These authors conclude that the lavas and the dolerites have the same MORB signature. The dolerite sills and dikes retain the same composition from the top of the mafic section down into the peridotites where they become scarce ; this is illustrated by major and trace element bulk compositions and by clinopyroxene compositions (Girardeau et al., 1985a). Some isotropic gabbros have the same composition as the dolerites, confirming their origin as coarser crystalline products in the center of large dolerite sills and dikes. Other isotropic gabbros and the uncommon layered gabbros are derived by fractional crystallization of a tholeiitic magma, which on the basis of lead isotope measurements is shown to be the same as the dolerite magma (Göpel et al., 1984). Prinzhofer et al. (1984) explain the lack of correlation between Ni and Cr of the volcanics by their parental magma undergoing only a limited differentiation by fractional crystallization and some orthopyroxene resorption ; this is supported by the paucity of cumulates in the whole ophiolite. Den Wanming et al. (1984), on the basis of the relative abundance of transition metals in the ultramafic tectonites, conclude that they are the residue of a comparatively small degree of melt extraction.

Hydrothermal alteration, thought to be related to spreading center activity (§. 4.2.3), has been studied in detail for its petrological aspect by Girardeau et al. (1985a) and for the oxygen and hydrogen isotopes by Agrinier et al. (in press). The relation with a spreading center activity is confirmed and it has been shown that there was first a high-T, low-P, hydrous and static metamorphism creating conditions of the amphibolite facies (430°-700°C) down to the base of the crustal section, followed by a greenschist-zeolite facies metamorphism (300°), the latter being responsible for the high $^{18}O$ ratio in albitized plagioclase. This hydrothermal alteration is ascribed by Agrinier et al. (in press) to sea-water circulation during the slow cooling of the oceanic crust.

A puzzling problem has arisen with lead isotope measurements which showed much higher $^{207}Pb/^{204}Pb$ values in the peridotite section than in the mafic section (Göpel et al., 1984). The lead discrepancy between mafic and ultramafic sections is not induced by a secondary alteration as shown by careful analyses on clinopyroxene separates and the good agreement between radiometric and paleontologic age determinations. The data shown in figure 4.9 point to island arc affinities for the ultramafic tectonites and

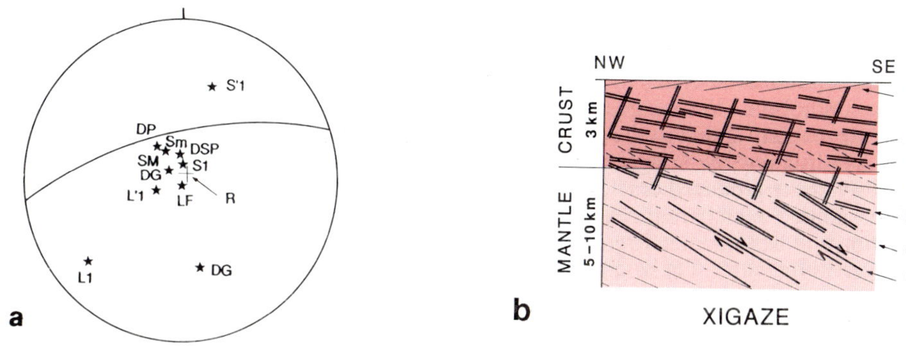


Fig. 4.8. a) Ridge referential obtained from computed best poles of planar structures and best directions of linear structures for the Xigaze massif,which have been derived from the data in figure 4.7, after rotation of radiolarite beds to the horizontal. The ridge direction is assumed to be parallel to the cyclographic representation of DG diabase dikes. b) Reconstructed model. LF, lava flows ; DG, DSP and DP : dolerite intrusives respectively in dolerites and gabbros, in serpentinized peridotites, and in fresh peridotites ; S1, foliation in fresh peridotites ; L1 : lineation in fresh peridotites ; S'1, mylonitic foliation in sheared peridotites ; L'1 : stretching lineation in sheared peridotites ; Sm, lamination plane in gabbro ; C, shear plane in peridotites (Girardeau and Mercier,1985).

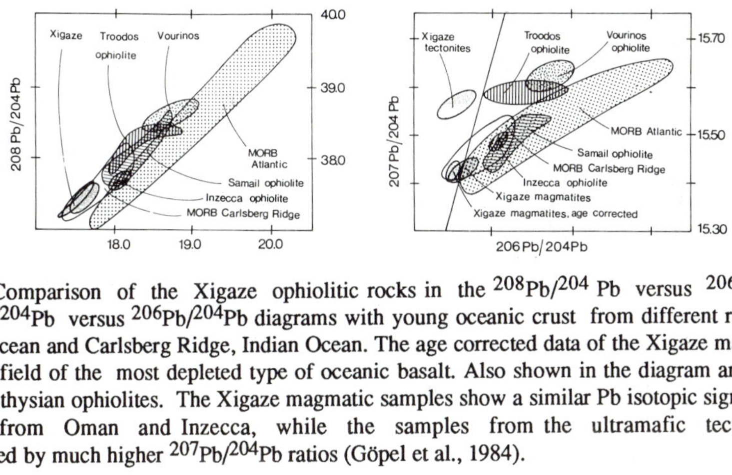


Fig. 4.9. Comparison of the Xigaze ophiolitic rocks in the $^{208}Pb/^{204}Pb$ versus $^{206}Pb/^{204}Pb$ and $^{207}Pb/^{204}Pb$ versus $^{206}Pb/^{204}Pb$ diagrams with young oceanic crust from different regions, the Atlantic Ocean and Carlsberg Ridge, Indian Ocean. The age corrected data of the Xigaze magmatic part plot in the field of the most depleted type of oceanic basalt. Also shown in the diagram are data from various Tethysian ophiolites. The Xigaze magmatic samples show a similar Pb isotopic signature as the ophiolites from Oman and Inzecca, while the samples from the ultramafic tectonites are characterized by much higher $^{207}Pb/^{204}Pb$ ratios (Göpel et al., 1984).

depleted MORB affinities for the mafic formations. A fortuitous association of the mafic and ultramafic sections due to the regional tectonics can be excluded because the dolerite dikes cutting the ultramafic tectonites cannot be distinguished structurally, chemically or isotopically from those constituting the mafic section. To explain the lead isotope discrepancy, Göpel and co-workers propose the interpretation of a new oceanic opening responsible for the mafic formations in an older interarc basin represented by the ultramafic formations. This interpretation is difficult to accept because structurally and petrologically the mafic and ultramafic formations are homogeneous, and no sign of a mixed origin for neither of them can be found. They are also uniformily covered by Albian-Cenomanian sediments with an extension of several hundred kilometers (Girardeau et al., 1985a) ; this does not militate in favor of a dual origin for the ophiolites. We suggest that the geochemical discrepancy between crustal and mantle formations is due to the heterogeneous nature of the melting mantle on a small scale. As observed by Allègre and Turcotte (1986), this can result in a great isotopic variability of the produced crust, mainly in the case of a limited melt extraction from the mantle source which reduces the opportunity for homogenization. Limited melt extraction and isotopic heterogeneity are also correlated by Allegre and Turcotte with a slow spreading rate. These conclusions correlate well with the present situation (§. 8.2.3).

### 4.2.6. Discussion

*Specific characters of the Xigaze ophiolite*

Compared to more common ophiolites which will be discussed more fully in chapter 8, the Xigaze ophiolite has some particular features, which can be thus summarized :

i) The mafic part of the sequence is nearly devoid of plutonic rocks except for the small layered gabbro bodies found in a few massifs. The mafic section is composed of basaltic volcanics overlying diabase and dolerite swarms with a few isotropic gabbro and scarce trondjhemite screens.

ii) The diabase unit is a sill complex rather than a dike complex, with sills intrusive one into the other, and only a few branching dikes at a right angle to the sills.

iii) This mafic part of the sequence seems remarkably thin, compared with that measured in non-dismembered ophiolites. It is in the range of 3 km with local variations in thickness. These pecularities had already been noted by Bally et al. (1984) in their preliminary report. It should however be remembered that the E-W shear zones may have modified the original thickness.

iv) The upper harzburgites and dunites are invaded over a thickness of around 1 km by thick diabase sills which become progressively less abundant down-section. In ophiolite massifs, diabase intrusive in the harzburgites are very scarce.

v) Cr-diopside-rich harzburgites are the dominant facies of the ultramafic unit and appear as shallow as 2 km beneath the mafic unit, and thus 5 km beneath the sedimentary cover of the ophiolite sequence. They grade downward into more lherzolitic peridotites.

vi) Local thin shear zones with low-temperature plastic flow structures cut through the mass of the peridotites. They are distinct from those observed in many ophiolites, including Dagzhuka massif, at the base of the ultramafics where they are thicker and related to thrusting over a metamorphic sole.

vii) The uppermost peridotites are heavily serpentinized in lizardite, more exceptionally in antigorite, which is uncommon in ophiolites. From the presence of a rodingitic alteration of the gabbro and its absence in the dolerite dikes cross-cutting these ultramafics, it can be demonstrated that an episode of serpentinization occurred close to the spreading axis. It has been ascribed to a high temperature sea-water circulation penetrating at depth of 3-4 km below the ridge (§. 8.3.1). Girardeau et al. (1985a) have documented the effects of this hydrothermalism in the mafic section, where it develops a greenschist to amphibolite facies metamorphism. Its existence is also supported by stable isotope studies (Agrinier et al., in press).

viii) Eventually, the lead isotopes reveal a large discrepancy between the signature of the ultramafic tectonites and the mafic volcanics and intrusives ; this discrepancy is incompatible with the latter being formed by direct partial melting of the former.

*Geodynamic environment of origin*

The Xigaze ophiolite is conformably covered by the Xigaze Group flysch which was deposited in an oceanic basin at the southern edge of Eurasia, and thus represents the oceanic basement of this basin. Mylonitic deformation in basal peridotites and, just below, remnants of metamorphic aureoles tectonically intermingled with formations of the Xigaze Group demonstrate that the ophiolite has been overthrust within this basin. The simultaneous formation of the andesitic Gangdise arc just North of this oceanic basin shows that it was a fore-arc basin, being located between an arc and a North dipping subduction zone. This leads to a scenario evoking the Sierra Nevada-Great Valley ophiolite situation of California as proposed by Bally et al. (1980), Shackleton (1981) and Nicolas et al., (1981) (fig. 4.10). This view is supported by paleomagnetic data on the pillow lavas and radiolarites of the Xigaze ophiolites (Pozzi et al., 1984) showing that the ophiolite formed at a latitude of about 10°-20°N, thus close to the Southern margin of Eurasia during lower Cretaceous time. Integrating the other available data, the following history can be traced.

i) Formation of the ophiolite at 110-100 Ma during Albian-Cenomanian in an oceanic basin with MORB affinities. This basin is at least as old as upper Aptian-Albian as shown by the age of the lowest formations of the Xigaze Group (Cherchi and Schroeder, 1980 ; Wang Naiwen et al., 1983). It could have been opening as early as the Lias in relation with the Eastward drift of Africa with respect to Eurasia (Girardeau et al., 1985b). Nicolas et al. (1981), Wang Xibin et al. (1984), Girardeau et al. (1985 a and b) and Girardeau and Mercier (1988) envisage for this ridge a slow-spreading environment on the basis of the singularities of the Xigaze ophiolite (chapter 8).

ii) Initiation of subduction around 110 Ma, which is the oldest age of the Gangdise arc magmatism (Coulon et al., 1986 ; 96 Ma for Schärer et al., 1984). This arc seems to have been built upon an oceanic crust as indicated by presence of metamorphic ophiolite screens between the granodioritic plutons (Proust et al., 1984). Such a crust would belong to the same basin as the Xigaze ophiolite (fig.4.10). The subduction could result from the change in convergence vector between Africa and Eurasia, which at 110 Ma rotates from an ESE to a ENE direction (Patriat et al., 1982).

iii) The beginning of convergence which has stopped the oceanic spreading in the Xigaze basin induces a thrusting of this young lithosphere over the basin formations, responsible for the mylonitic deformation at the base of the Dagzhuka massif and the metamorphic aureole in garnet-amphibolite facies conditions in the formations below the thrust. This overthrusting of the ophiolite has promoted its later survival during collision.

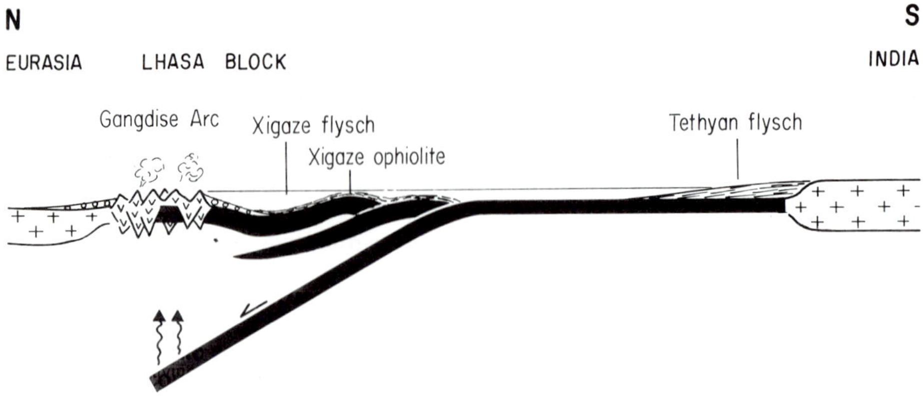


Fig. 4.10. Model of formation of the Xigaze ophiolite with respect to the Gangdise subduction. After the formation of this subduction zone from oceanic thrusts like that of Dagzhuka, the ophiolite becomes incorporated to the oceanic lithosphere of a fore-arc basin. From there, it is obducted onto the Indian continent during the final collision.

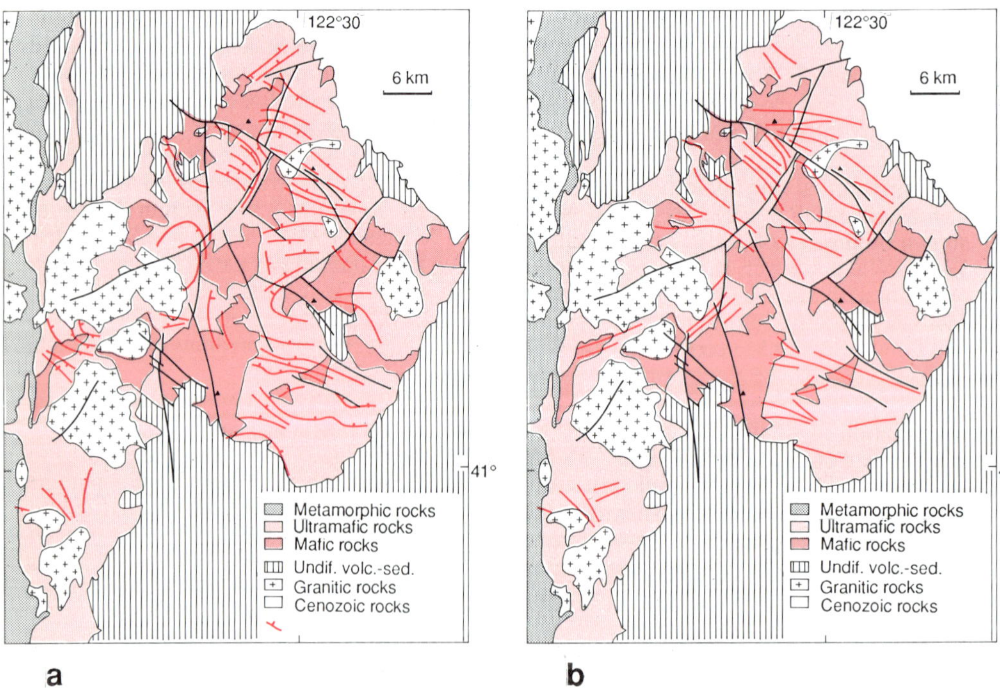


Fig. 4.11. Structural map of the Trinity massif in the eastern Klamath Mountains (northern California). Trajectories of a) lineations and b) foliations in the ultramafics. (Compilation of works cited in the text and new structural data by Boudier et al., in press).

iv) Since 85 Ma, the rapid NE motion of India towards Eurasia (Patriat and Achache, 1984) induced an active subduction and consequently an island arc magmatism along the southern margin of Eurasia. The magmatism dies off at 50 Ma with the beginning of continental collision (Patriat and Achache, 1984). For the ophiolites of the suture zone, squeezing, block rotation and shearing are the main consequences of this collision.

## *4.3. TRINITY OPHIOLITE*

### *4.3.1. Introduction*

The Trinity ophiolite is exposed in the Klamath Mountains of California (location in fig. 4.11b). As stated in the introduction to this chapter it is, to our knowledge, the best example of the lherzolite subtype of ophiolites, showing more radical characters than the Xigaze ophiolite. Beyond this type, it seems that the ophiolite concept starts to disaggregate and that a new type of mantle-crust association appears (§ 4.4.).

Four recent field studies have been devoted to the Trinity massif, each covering restricted areas and each with a specific objective. Lindsley-Griffin (1977) mapped the northwestern quadrangle, including mainly mafic parts of the complex. She drew conclusions, after Hopson and Mattinson (1973), as to its ophiolitic nature, tying together the ultramafic bodies with the gabbros, sheeted diabase complex and basaltic vocanics. This view, gaining support from the petrological and geochemical studies of Lapierre et al. (1987) and Brouxel and Lapierre (1988), conflicted with that of earlier workers (Lipman, 1964 ; Irwin, 1966) who interpreted the mafic formations as intrusives into the ultramafics. Goullaud (1977) provided the first structural survey of the ultramafic formations with a detailed study of the Coffee Creek mafic body. Quick (1981 a,b) mapped in detail the ultramafics in the China Mountain - Mt Eddy quadrangle within the frame of a petrological and geochemical study. Finally, Boudier and her co-workers (Le Sueur et al., 1984 ; Cannat and Boudier, 1985 ; Boudier et al., 1989) covered structurally the entire massif. They support Lindsley-Griffin's conclusion about the ophiolitic association of the mafic and ultramafic sections. More specific studies dealing mainly with age data (Lanphere et al., 1968 ; Mattinson and Hopson, 1972 ; Jacobsen et al., 1984) show some discrepancy between ages in the mafic and ultramafic sections. For this reason Jacobsen and his co-workers tend to consider the mafic and ultramafic formations as independent intrusions. This central problem is addressed below.

### *4.3.2. Geological setting*

The Trinity complex crops out over 3 500 km$^2$ in the eastern Klamath Belt. This belt appears to have been accreted to continental North America during the Nevadan collision (Roure, 1982, 1984 ; Ketner, 1984) ; a related magmatism was responsible for the granodioritic intrusions which occur in the area. The belt underwent a ninety degree clockwise rotation between Triassic and Upper Jurassic (Mankinen et al., 1982). During Early to Middle Devonian a subduction-arc system was active in this belt (Dickinson, 1981 ; Lapierre et al., 1985). It induced a collision with the Central Metamorphic Belt, West of the eastern Klamath Belt, with thrusting of the Trinity complex over the Central Metamorphic Belt during Devonian at about 380 Ma (Lanphere et al., 1968 ; Cashman, 1980).

Gravity, magnetism and seismic data (La Fehr, 1966 ; Griscom, 1977 ; Fuis et al., 1987), supported by the regional mapping cited above, suggest that the Trinity complex is a relatively thin (2-4 km) easterly dipping sheet, overlying a less dense basement. To the West it rests upon formations of the Central Metamorphic Belt whose metamorphism is at least partly caused by overthrusting of the peridotites (Lipman, 1964; Cashman, 1980 ; Cannat and Boudier, 1985). The latter authors describe a typical metamorphic aureole (chapter 12) with a high grade dynamo-metamorphism in both the basal peridotites and the underlying metamorphites which are derived from oceanic basalts and sediments. They also studied the kinematics of the oceanic thrusting responsible for development of aureole, and related it to an easterly dipping subduction zone in the present day geographical frame. To the East and the South, the Trinity complex is unconformably overlain by volcanic and sedimentary rocks ranging in age from Devonian (Boucot et al.,1974) to Jurassic (Roure, 1984) and belonging to the Reding formations. To the North, the Ordovician-Silurian Duzel and Gazelle formations are thrust over the complex and over a mélange unit lying stratigraphically upon the complex, which has been dated by its Devonian matrix (Lindsley-Griffin, 1977).

Radiometric and paleontological ages point to an Ordovician origin for the complex, however with some serious discrepancies. A first group of ages suggests formation of the complex at 470-480 Ma. They comprise 480 Ma and 455 Ma U/Pb ages on zircons from a gabbro and a trondhjemite of the mafic section (Mattinson and Hopson, 1972) and a 472 ± 32 Ma Sm/Nd age in the lherzolites (Jacobsen et al., 1984) which would correspond to the crystallization of a trapped feldspathic liquid, thus dating the intrusion of a mantle diapir at a shallow depth. A second group corresponds to younger ages : 418 ± 17 Ma and 439 ± 18 Ma K/Ar age on amphiboles from a gabbro dike in the ultramafic section (Lanphere et al., 1968) and 435 ± 21 Ma Sm/Nd age in a microgabbro, unfortunately from a float (Jacobsen et al., 1984).

It is suggested that the complex was formed around 470 Ma and that younger ages represent either reset ages for the K/Ar data with relation to the granitic Nevadan intrusions which are abundant where these ages have been obtained, or correspond to secondary basaltic intrusions not directly related to the formation of the complex.

### *4.3.3. Description*

In the Trinity complex, the mafic formations constitute independent bodies surrounded by peridotites (fig. 4.11). They had been considered as intrusions into the peridotites, like the granodioritic Nevadan plutons, because of locally observed steep contacts with the peridotites and of gabbro dikes cutting the peridotites. The latter evidence rather militates in favor of a common origin because, in the case of a gabbro dike emplacing in cold peridotites, chilled margins and doleritic textures should be expected. The former evidence is not necessarily conclusive. In SE Oman and New Caledonia where the horizontal trend of the major contacts is obvious at the landscape scale, contacts are often steep at the outcrop scale mainly where serpentinization is important. This is the case in many places in the Trinity massif where the gabbro-peridotite contact is deformed with the coeval development of antigorite and greenschist facies assemblages (Goullaud, 1977; Boudier et al., 1989). When it is primary, this contact is flat-lying like at Deadfall Meadow or South China Moutain (fig. 4.12) and the gabbro section rests on a transition zone (see below), as observed in typical ophiolites. The various mafic bodies have lithologies and structures somewhat diverse compared to the uniformity of the underlying ultramafic section.

The pillow lavas are only locally preserved and their relation with the mafic formations has been questioned (Irwin, 1981) ; however the geochemical similarity of these volcanics with the sheeted dike diabases attached to the complex favors their consanguinity (Lapierre et al., 1985 ; Brouxel and Lapierre, 1988). The continuous sequence of the mafic bodies starts with a sheeted complex of mutually intrusive dikes upsection and with isotropic gabbro screens downsection. This unit has commonly a large development, for example

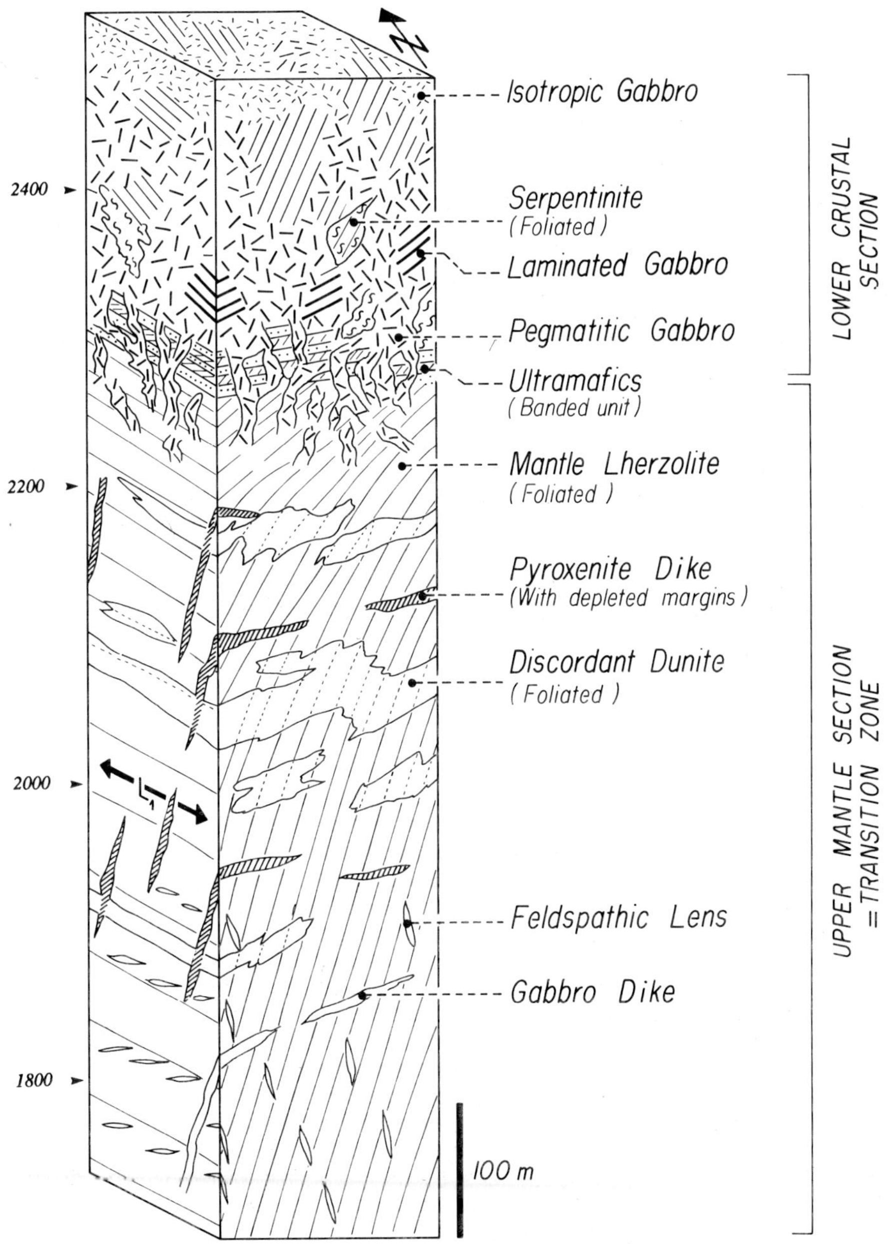


Fig. 4.12. Synthetic log in the mafic and ultramafic sections of the Trinity massif showing in particular the complexity of the transition zone of South China Mountain (Boudier et al., in press).

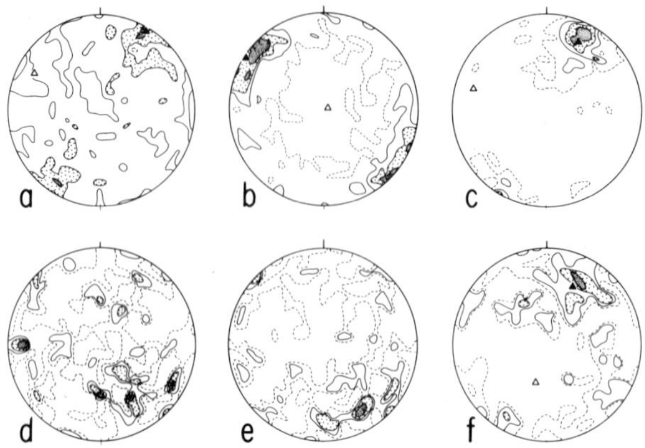


Fig. 4.13. Stereograms of the structural elements in the Trinity Complex. a) Foliations in peridotites (547 measurements). b) Mineral and aggregate lineations in peridotites (374 measurements). c) Diabase dikes in the sheeted dikes unit (92 measurements). d) Pyroxenite dikes in peridotites (142 measurements). e) Pegmatitic gabbro dikes (116 measurements). f) Diabase dikes in peridotites (153 measurements). Equal area projection, lower hemisphere. Contours approximately 0.2 %, 1 %, 2 %, 3 %, 4 %, 5 % per 1 % net area, solid triangle : best computed axis ; open triangle : pole of best computed plane (Boudier et al., in press).

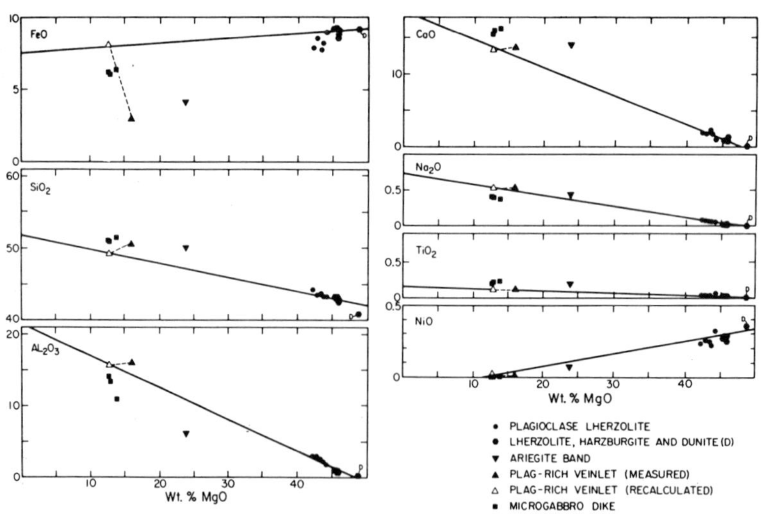


Fig. 4.14. Oxides versus MgO diagrams for calculated bulk compositions of the Trinity peridotites, gabbro veinlets and microgabbro dikes. Solid line to emphasize trend defined by the recalculated composition of the gabbro veinlet and that of the plagioclase lherzolite and harzburgite (Quick, 1981b).

in the Bonanza King area, where its thickness exceeds 600 m. The nature of the dikes varies widely, including diabases with magmatic amphiboles and keratophyres with a quartz matrix and spilites ; their average orientation however is well defined (fig. 4.13c), suggesting a common origin. Downsection, the dikes become less numerous and more dispersed in orientation (fig. 4.13f). They remain fairly abundant in the ultramafic section where they can attain a few meters in width, displaying a microgabbro texture in their center.

The gabbro section is exposed over a thickness of approximately 300-500 meters. Its internal organization and relation with the ultramafics are illustrated by figure 4.12. The normal components of the gabbro section in ophiolites, namely layered and foliated isotropic gabbros, are poorly organized and extensively invaded by gabbro pegmatites grading to coarse grained-gabbros (plate 4.1a). The layered gabbros have a one centimeter-thick mineral banding composed of various proportions of plagioclase and green clinopyroxene plus subordinate primary hornblende. Associated foliated gabbros generally have a larger amount of green hornblende. The layered and foliated gabbros occur as relicts, a few meters to a few hundreds of meters in size within the network of isotropic gabbros. The layering or magmatic foliation plane has a poor preferred orientation striking NW-SE with a moderate SW dip. Mineral lineations are also poorly marked.

The pegmatitic gabbros composed of clinopyroxene, hornblende and plagioclase appear abruptly upsection in the ultramafics of the transition zone, where they form vertical dikes, several meters thick . Above, these pegmatitic gabbros invade the layered gabbros in a more diffuse way, with magmatic reactions occurring at contacts. In the vicinity of pegmatitic dikes, coarse-grained and amphibole- bearing isotropic gabbros are generated by porphyroblast growth within the magmatic foliation plane of the gabbros (Plate 4.1 a,b). The pegmatitic dikes and coarse-grained gabbros represent approximately fifty per cent of the mafic section.

Scarcity of feeder dikes in the mantle section and the abrupt occurrence of pegmatitic gabbro dikes in the transition zone (fig. 4.12) suggest that the initial liquid did not originate from depth. The pegmatitic and coarse-grained gabbros and the related trondhjemites seem to originate from the layered gabbro section. A possible origin could be related to sea water convecting down to the Moho through still hot gabbros and producing a hydrous anatexis of these gabbros at temperature conditions of amphibolite facies. Silica oversaturated melts (trondhjemitic veinlets) could represent the differentiated product of this melting (plate 4.1.b).

Time of emplacement of the pegmatitic gabbros is constrained by two facts. On the one hand, they are not affected by the high-T plastic deformation when included in the uppermost lherzolites ; on the other hand, their average orientation is perpendicular to the average peridotite foliation (fig 4.13e), suggesting that their intrusion, occurring soon after the end of the plastic deformation, was controlled by the stress field responsible for the plastic flow.

As seen above, in many places the contact between the ultramafic and mafic rocks has been tectonically activated as steep shear zones related to greenschist metamorphism with development of antigorite, chlorite, talc, tremolite and piedmontite. The ultramafic section begins just below the mafic section with a banded unit which is generally flat-lying, 50-100 m thick and composed of alternating dunites, wehrlites and pyroxenites (fig. 4.12). Fabric studies showing preferred orientations typical of high-T plastic flow demonstrate that this banded unit records a large plastic flow parallel to the banding. The banded unit could originally represent either ultramafic cumulates deposited at the base of a magma chamber and subsequently deformed by plastic flow, or the upper part of the mantle section. In the second interpretation, the banding would be thus a consequence

***Plate 4.1. : structures in the Trinity massif***

a - Poorly layered gabbros, partly recrystallized into isotropic and pegmatitic amphibole-gabbros (Boudier et al., 1989).

b - Pyroxenites from the Bonanza King transition zone, first recrystallized into pegmatitic amphibole gabbros and next brecciated by trondhjemite (marker is 10 cm long) (Boudier et al., 1989).

c,d - Indigenous pyroxenite dikelet with symetric borders of dunite grading into harzburgite, depleted lherzolite and, at around 50 cm distance, into undepleted plagioclase lherzolite (foliation parallel to hammer) (Boudier et al., 1989).

a b

c d

of large horizontal plastic flow, tectonically transposing the numerous pyroxenite sills and dikes which intrude the residual and impregnated dunites located just below this unit.

The banded unit grades downsection into the plagioclase lherzolites interlayered with large dunite bands of the transition zone. The pegmatitic gabbros rapidly disappear and the number of diabase dikes with microgabbro differentiations become less abundant. In contrast, the number of indigenous and intrusive dikes with an olivine-clinopyroxene-plagioclase composition increases. The high temperature foliation also steepens (figs. 4.12 and 4.13a). Plagioclase and spinel lherzolites, with a discrete websterite banding represent in the massif approximately 30 % and 10 % respectively of the mapped surface ; depleted peridotite facies is 40 %, with 15 % harzburgites and 25% dunite-wehrlite. Dunites form rather flat bodies up to 100 m thick, extending over distances up to several kilometers ; they are especially abundant within 500 meters below the mafic section. Werhlites form irregular patches associated with the dunite bodies. Finally, approximately 15 % of the ultramafic section is totally serpentinized to antigorite plus lizardite.

#### *4.3.4. Structural analysis*

The peridotites microstructures are porphyroclastic with evidence of high-T conditions during plastic flow (Goullaud, 1977). Plagioclase lenses, oblique with respect to the foliation, provide evidence of melt being extracted during this flow (§ 2.5.3) which thus corresponds to hypersolidus conditions. Foliations and mineral lineations have a regular pattern throughout the massif except for local rotations related to shear zones active at somewhat lower temperature (fig. 4.11). Outside such zones the foliation is on average steep, striking NW-SE and the lineation, horizontal (figs. 4.13a and b). The indigenous and intrusive gabbro- pyroxenite dikes have no preferred orientation (fig. 4.13 d) in contrast with the pegmatitic gabbro dikes (fig.4.13e) and diabase dikes (fig. 4.13f) of the top of the ultramafic section which are respectively perpendicular to the mineral lineation of the peridotites, and parallel to both the mineral lineation and the orientation of the diabase dikes in the mafic section (fig. 4.13c).

Finally, as mentioned above (§ 4.3.2), in the Trinity Alps along the western contact of the massif, there is at the base of the ultramafic section a zone of increasingly deformed lherzolites which grade over 200 m downward into mylonites. These mylonites are in contact with a series of metamorphic rocks presenting an inverse gradient with garnet-amphibolites at the contact and greenschists below, following the general scheme of a metamorphic aureole (chapter 12). The peridotites near the contact are heavily serpentinized with development of greenschist facies mineral (antigorite, chlorite, tremolite, talc, magnetite) accompanying the recrystallization of olivine into centimeter-sized grains. These mineral reactions may be related to the abundant Nevadan intrusions along this eastern contact of the Trinity complex. Deformation also imparted by these intrusions obscures the origin of the metamorphic aureole.

#### *4.3.5. Melt extraction and melt reaction*

The Trinity plagioclase lherzolite massif shares with other plagioclase lherzolite massifs, and with the Lanzo massif in the Western Alps in particular (Boudier and Nicolas, 1972, 1977 ; Boudier, 1978) some remarkable features indicative of a melt activity, principally the occurrence of plagioclase lenses, of dunite bodies and, genetically related to the latter, of indigenous gabbro-pyroxenite dikes (§.2.5.3). The plagioclase lenses are a few tens of centimeters in length and are slightly oblique to the high-T foliation. The orientation of these lenses track that of the shear flow plane (§

2.5.3) ; they have been ascribed to fluid- assisted fracturing.

Relationships of the Trinity dunite bodies with the enclosing plagioclase lherzolite have been described by Quick (1981b). The rather flat contacts of the dunite bodies are discordant with the vertical foliation measured inside the dunite bodies and in the enclosing lherzolites. Occasionally, the primary websterite banding of the lherzolite which is parallel to the foliation can be traced as spinel relicts in the dunite (Boudier et al., 1989) ; this demonstrates the dominantly residual character of the dunite, a conclusion confirmed by the irregular shape of the dunite contacts with the lherzolite and their relation with the depletion halo bordering indigenous dikes (plate 4.1c,d). Inside the dunite bands, gabbro dikes, clots and diffuse aggregates of feldspar and clinopyroxene are commonly observed, producing locally wehrlites of very inhomogeneous composition and random distribution. Such features have been ascribed to the peridotite being impregnated by and reacting with a melt (§. 2.5.3).

As proposed by Quick (1981b), the process responsible for the formation of these dunites and the related indigenous dikes is the incongruent melting of the orthopyroxene (§ 10.4.4). Quick, however, restricts this process to the marginal few meters of the large dunite bodies, and ascribes to the central dunites a cumulative origin by fractionation from a picritic melt. The above observation of banding being traced through the dunite bodies shows that at no time did the solid dunites lose their coherence. It is concluded that if there is olivine fractioning in the dunites, it is a small fraction, and that most olivine is either residual or in a subordonate fashion derived from the incongruent melting of the orthopyroxene. This is more amply discussed in § 10.4.4.

### *4.3.6. Petrology and geochemistry*

The most complete petrological study of the Trinity massif is that of Quick (1981 a,b) which focuses mainly on the ultramafics with a special attention paid to the partial melting of the plagioclase lherzolites and to the origin of the dunites. This author notes the similarity in mineral and bulk compositions of the Trinity ultramafics with other 'alpine-type' peridotites, notably the Lanzo (Boudier, 1978) and Lizard (Green, 1964) plagioclase lherzolites and the Tinaquillo spinel lherzolites (Green, 1963). The Trinity ultramafics are equilibrated at temperatures estimated by pyroxene-pyroxene geothermometers (Wells, 1977) between 1160°C and 1275°C and pressures around 500 MPa as estimated by the orthopyroxene-plagioclase geobarometer (Obata, 1976). Reequilibration in porphyroclasts rims and neoblasts record cooling down to a temperature of 960°C. Geochemical study of the partial melt products points to their low pressure origin (100-1000 MPa), their tholeiitic affinities and the importance of chemical exchanges with the wall rocks, a conclusion which agrees with structural evidence (depleted margins and dunite reaction zones at the indigenous dikes margins). From the calculated composition of the melt it can be proposed that the spinel lherzolites and the harzburgites are refractory residues produced by partial melting of the plagioclase lherzolite (fig. 4.14). As in the case of Xigaze (§ 4.2.5), Nd and Sr isotopic studies of the Trinity peridotites (Jacobsen et al., 1984) show a discrepancy between the Nd ratios in the peridotite and in the cross cutting gabbro or pyroxenite dikes, demonstrating that they are not genetically related. On the basis of major and trace elements and Sm-Nd isotopic data, a multistage melting is described with the stage at 472 ± 32 Ma, corresponding to 10-15 % melt extraction, which is comtemporaneous with the diapiric intrusion. The gabbro and pyroxenite dikes would be intruded at 435 ± 21 Ma with a lower Nd ratio, more typical of oceanic mantle at this time. This conclusion raises a difficult problem : occurring 40 Ma after the peridotite emplacement at shallow depth, the gabbro and pyroxenite intrusions should develop chilled margins because the wall-rock peridotite should have cooled ; such thermal reactions are unknown.

In terms of mineral chemistry, the ultramafics of the banded unit just below the mafic section and the layered gabbros of this mafic section are similar to the dunites-pyroxenites of the transition zone and to the gabbro-pyroxenite dikes intrusive into the ultramafic section respectively. This confirms the co-genetic character of mafic and ultramafic sections. In the mafic layered sequence, orthopyroxene is present as a cumulus phase.

The chemistry of the microphyric dikes and volcanic flows presumably capping the mafic section is more diverse, with a series of low-K, low-Ti, and LREE depleted tholeiitic suites (Lapierre et al., 1985 ; Brouxel and Lapierre, 1988).

### *4.3.7. Discussion*

*Specific characters of the Trinity ophiolite*

Detailed structural mapping has shown that most components of an ophiolite are present in the Trinity Massif, and that they are related according to the ophiolite model, confirming previous conclusions (Hopson and Mattinson, 1973 ; Lindsley-Griffin, 1977). In particular, in locally preserved sections, the peridotite section grades into the mafic section through a transition zone of flat-lying layered dunites, pyroxenites and wehrlites ; pegmatitic gabbro dikes have intruded both sections while they were still hot. This is difficult to explain if interpretating the gabbros as independent intrusions through older peridotites, as proposed by authors who do not accept the ophiolitic nature of the Trinity complex.

The reconstruction of the Trinity complex in its spreading center geometry (fig. 4.15) is based on reference to the preferred orientations of all major structures, to the peridotite/gabbro limit (paleo-Moho) regarded as a paleo-horizontal and to the trend of the sheeted dike complex, regarded as giving the azimuth of the spreading axis (§ 2.2.1). In this reference system, the high-T mantle plastic flow in peridotites just below the crustal formations follows a vertical plane parallel to the ridge trend with a horizontal flow direction.

Comparing the model of figure 4.15 with that of a typical ophiolite (fig. 9.1), one is struc by many singularities of the Trinity ophiolite which are discussed in chapter 8.

i) Layered gabbros in the mafic section of Trinity are either totally absent as componants of the crustal section or reduced in thickness and extension. They are partly replaced by recrystallized gabbros and magmatic brecciae, emphasizing the importance of hydrous recrystallizations, possibly accompanied by anatectic reactions.

ii) The mafic section has a comparatively reduced thickness with respect to other ophiolites. The difference in thickness is difficult to estimate precisely because the section is incomplete, but the reduction occurs at least partly in the gabbro unit, where it is ascribed to the quasi-absence of layered gabbros.

iii) The ultramafic section is dominantly composed of plagioclase lherzolites with the development of thick dunite bodies in its upper part.

iv) No chromite pods have been reported.

v) The foliation in the ultramafic section is essentially vertical and the lineation parallel to the ridge azimuth as defined by the sheeted dike trend.

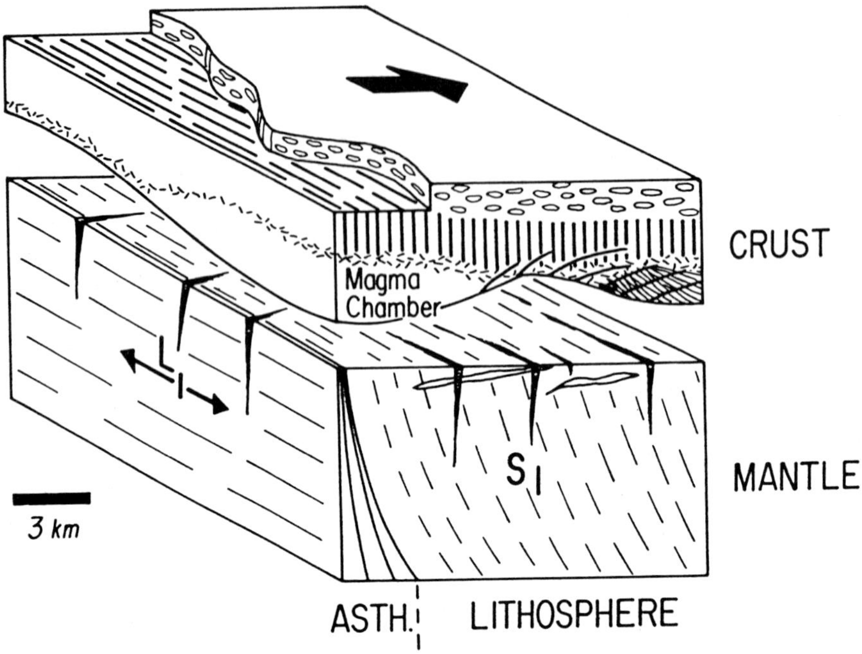


Fig. 4.15. Reconstruction of the Trinity complex in its presumed oceanic spreading situation based of the structural data of figure 4.13 (Le Sueur and Boudier, 1986).

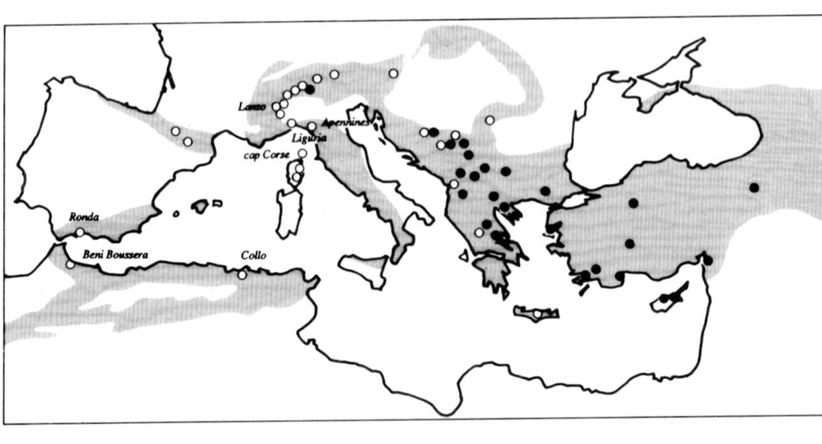


Fig. 4.16. Simplified map of the Mediterranean belts with location of the lherzolite and LOT massifs (open symbols) and of the HOT massifs (closed symbols) (modified from Nicolas and Jackson, 1972).

vi) Serpentinization to antigorite and related rodingitization occur in the upper ultramafic section.

*Geodynamic environment of origin*

As discussed extensively in Chapter 8, the specific characters of the Trinity ophiolite have led Boudier and Nicolas (1985) to propose that it formed along an oceanic spreading center characterized by a slow spreading rate, perhaps an environment similar to a modern island arc or a back-arc basin as proposed by Quick (1981a) and Lapierre et al. (1985), or perhaps an oceanic rift of the Red Sea type (Dickinson, 1981) or of its Miocene precursor (Boudier and Nicolas, 1985). After its formation during Ordovician time, the Trinity piece of oceanic lithosphere has been incorporated into island arc formations of Devonian and younger ages. It was obducted some 100 Ma after its accretion, from its marginal basin or trapped rift of origin, over the marine formations of the Central Metamorphic belt.

## *4.4. THE WESTERN ALPS OPHIOLITES*

The ophiolites of western Alps are highly tectonized and metamorphosed, making them inappropriate for structural studies ; in a few external areas like Mont Genèvre, and Queyras in French Alps, northern Apennines and Liguria in Italia (fig. 4.16), the metamorphism is less severe and some cross sections, limited in extension but well preserved, have permitted us to classify them as LOT and to make important discoveries. Typical characteristics are, 1) the plagioclase lherzolite composition of the ultramafic unit found everywhere inasmuch as identification is allowed by the serpentinization (see next section for their specific study), 2) the absence or limited extent of layered gabbros, and 3) the abundance of diabase dikes at any level, including the lherzolites.

The major discovery made in these ophiolite is that they derive from a reduced ophiolitic crust in the Piedmont-Liguria oceanic basin of origin, with local exposure of feldspathic lherzolites on the sea-floor (Decandia and Elter, 1972 ; Grandjaquet and Haccard, 1977 ; Abbate et al., 1980 ; Lemoine, 1980 ; Cortesogno et al., 1981).This is illustrated by figure 4.17 a and b, showing the exposure of serpentinites in relation with respectively listric faulting at the European passive margin and with normal or strike-slip faulting within the oceanic basin.

The reconstitutions of figure 4.17 rely on careful sedimentological and tectonic studies, showing in particular that some serpentinite breccias, the ophicalcites, were either due to sediments filling fractures opened in the serpentinized ocean floor or to sedimentary reworking of this floor. It was also demonstrated that locally gabbros were strongly sheared in amphibolite facies conditions at the ridge of origin. This follows from the evidence that diabase swarms cross cut them (Mével et al., 1978 ; Steen et al., 1980).

All authors agree on a small rifted ocean, transected by numerous transform faults, as the oceanic site of origin (fig. 4.18). Views differ on how to explain the thin crust and mantle exposed as sea-floor : intrusions in transform faults (Gianelli and Principi, 1974 ; Messiga and Piccardo, 1974 ; Abbate et al., 1980 ; Ishiwatari, 1985), rifting-mediated denudation (Bortolotti et al., 1976 ; Piccardo, 1977 ; Lombardo and Pognante, 1982 ; Nicolas, 1984 ; Lemoine et al., 1987) or some combination of the two (Lagabrielle et al., 1984 ; Tricart and Lemoine, 1986). This question is addressed in very general terms in § 8.2.3.

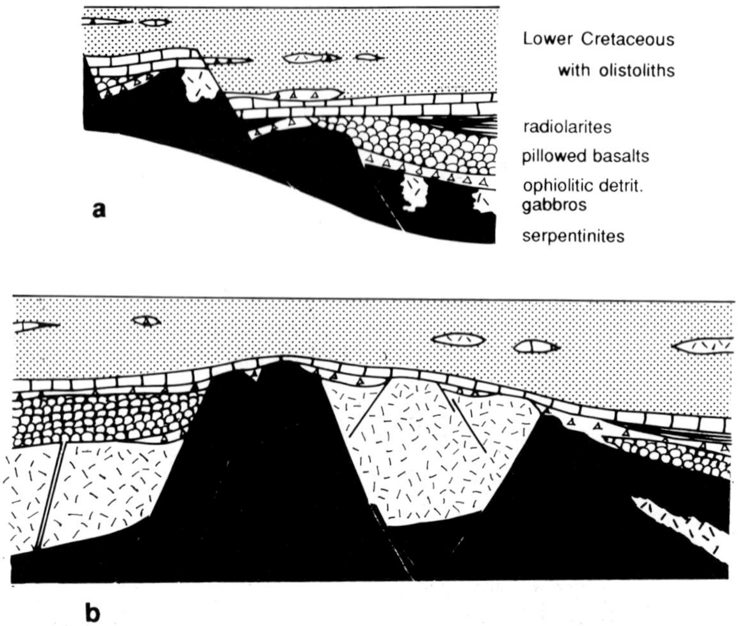


Fig. 4.17. Reconstructed cross sections illustrating the nature of the oceanic crust in the Piedmont-Liguria basin and in particular the tectonic denudation of the ultramafic floor (black). a) Listric faulting along the European margin in Queyras. b) Ultramafic protrusions within the Liguria basin. (a) Lemoine, 1980, b) Tricart and Lemoine, 1986).

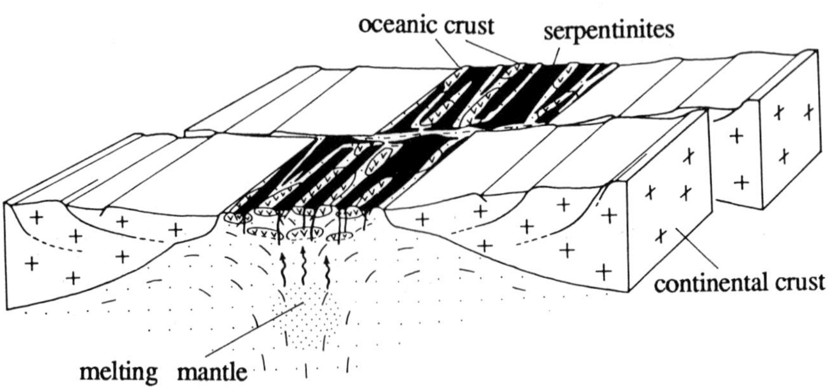


Fig. 4.18. Largely accepted model of the Piedmont-Liguria basin during late Jurassic in western Alps, as a rift transected by numerous transform faults between European and Southalpine margins (Lombardo and Pognante, 1982).

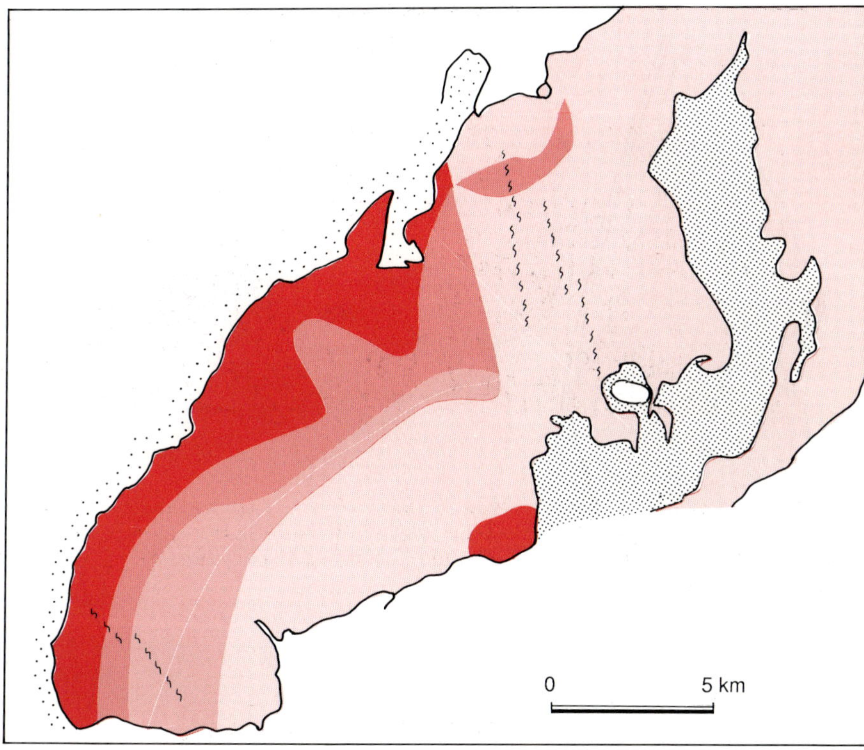


Fig. 4.19. Map of the peridotite facies in the Sierra Bermeja massif. From darker to lighter colors : garnet lherzolite ; garnet pyroxenite and spinel pyroxenite facies in spinel lherzolites ; plagioclase lherzolites. Tight and spaced dots, high and low grade metamorphic rocks respectively. 'S' serpentinized corridors (Obata, 1980).

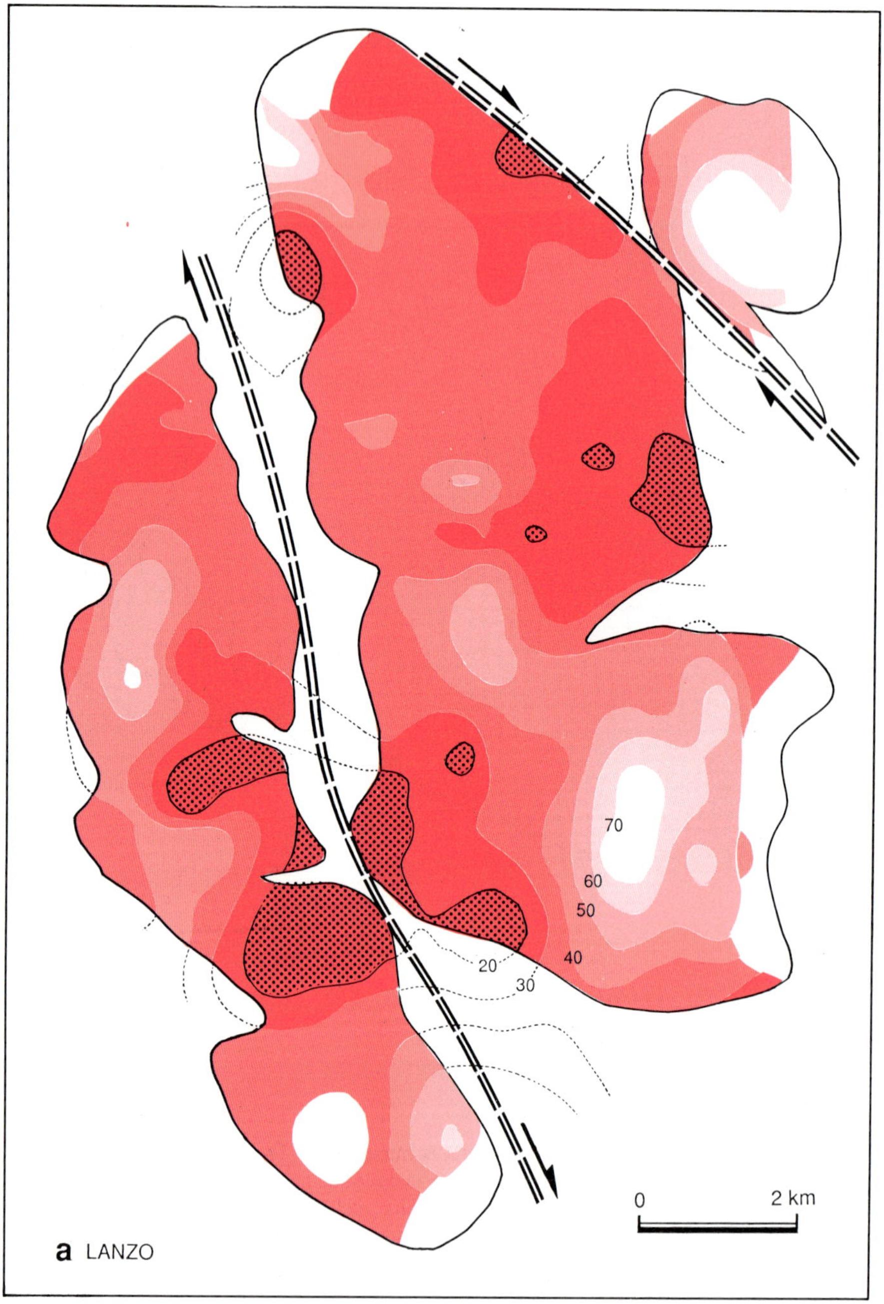


Fig. 4.20. Plastic strain maps in a) Lanzo (the three parts of the massif have been restored into their presumed original position), b) Sierra Bermeja and c) Beni Bousera massifs. Maps a) and b) are based on grain size which reflects the amount of dynamic recrystallization which is itself related to strain (contours : grain size in 1/100 mm) ; map c) is based on strain estimates based on the shape of spinel and orthopyroxene, which tends to underestimate the total strain. (a) Boudier and Nicolas, 1980 ; b) Darot, 1973, 1974 ; c) Reuber et al., 1982).

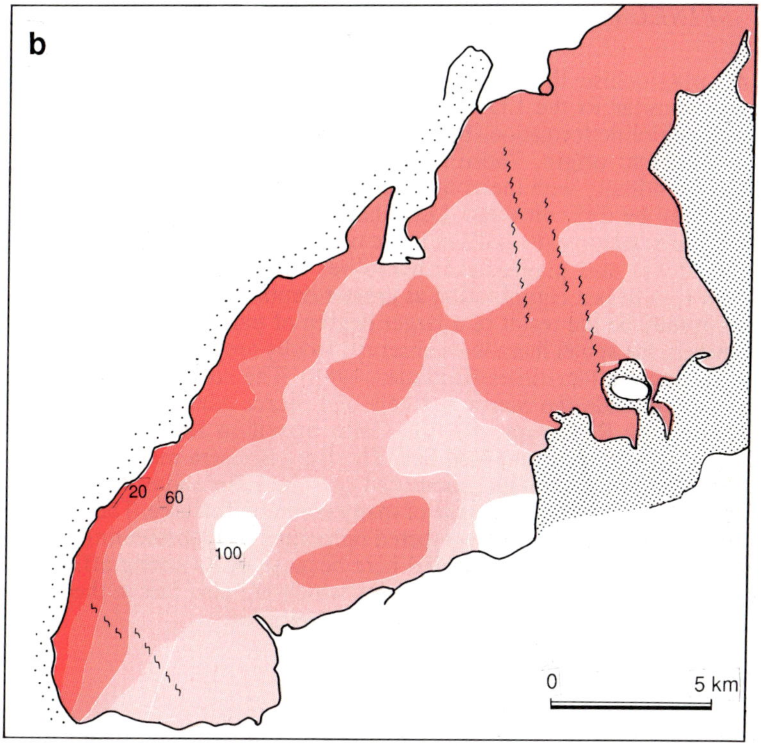
b
20
60
100
0
5 km

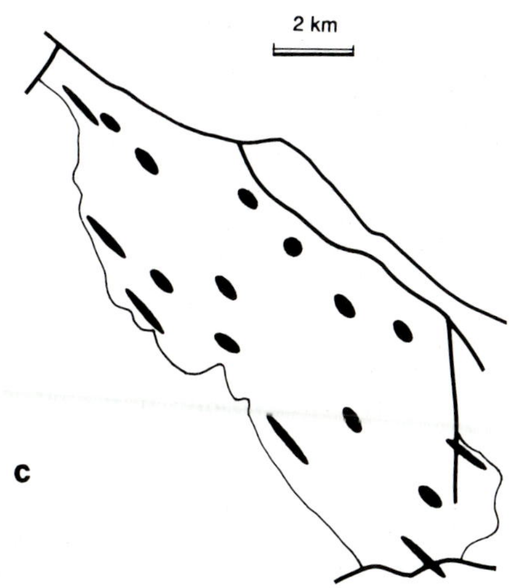
2 km
c

## 4.5. THE SPINEL-PLAGIOCLASE LHERZOLITE MASSIFS

A few spinel-plagioclase lherzolite massifs of western Mediterranean (fig. 4.16), otherwise very similar to the LOT (Lherzolite Ophiolite Type) peridotites, are not associated with ophiolitic formations but are on the contrary, associated with granulites of apparently continental origin. These massifs are nevertheless briefly described here because of their continuity with lherzolite massifs typical of a LOT situation. Distinction between them may be difficult when LOT lherzolites are no more associated with an ophiolitc crust due to tectonic denudation in the ocean of origin, as seen above. The typical association of spinel and plagioclase lherzolites in the massifs considered now, is interpreted as a sign of last equilibration at greater depth than in the LOT case (§ 7.2.2). Alternatively, it may be the result of a larger degree of partial melting when, like in the LOT Lanzo massif, the spinel lherzolites observed locally are also more refractory than the surrounding plagioclase lherzolites (Bodinier, 1988). The considered spinel-plagioclase lherzolite massifs include the Sierra Bermeja and Sierra Alpujata massifs in southern Spain, also called the Ronda massifs, the Collo massif in Kabylie (Algeria) and the Zabargad Island occurrence in the Red Sea. The Beni Bousera massif on the Moroccan side of the Gibraltar arc, although composed of spinel lherzolites, will be considered together with the Ronda massifs because of many common features. The Lanzo massif in the western Alps, surrounded by dismembered ophiolites and representing the floor of the Piedmont basin (§ 4.4), is also considered here because of its strong affinities with the above- cited massifs. It best displays structures which are also met in the Liguria (pers. obs.) and Cap Corse (Jackson, 1979) plagioclase lherzolites.

### 4.5.1. Petrological zonation

Typically, these massifs are less refractory than LOT lherzolites ; the spinel lherzolites are generally more fertile than the plagioclase lherzolites, indicating a lesser degree of melt extraction. In Sierra Alpujata (Tubia and Cuevas, 1987) there is a zonation with plagioclase lherzolites constituting the bulk of the massif and spinel lherzolites, the margins. In Sierra Bermeja, there is a rough zonation with more residual plagioclase lherzolites located Eastward in the presumed center of the intrusion, and more fertile spinel lherzolites located Westward where they grade locally at their contact with garnet granulites into a film of garnet lherzolites (fig. 4.19) (Hernandez Pacheco, 1967 ; Dickey, 1970; Darot, 1973 ; Obata, 1980 ; Frey et al., 1985). This zonation has been interpreted by Obata and Frey and his co-workers as the result of cooling along the margins of a mantle diapiric intrusion which would result in the preservation of a marginal lherzolite facies equilibrated at the highest pressures. For Obata, the intrusion occurred at subsolidus temperatures, a conclusion based on pyroxene thermometry which indicates temperatures of equilibration not higher than 1200°C, whereas for Frey et al. (1985), the intrusion was accompanied by decompression melting (§ 7.2.2). Rapid cooling along the margins limits the melting and thus explains the more fertile character of the garnet lherzolites compared to the more slowly cooling plagioclase lherzolites in the interior of the massif ; this conclusion is based on trace elements geochemistry and is supported by personal observations made in the center of the massif where we have discovered a few gabbro dikes emplaced during plastic flow.

Fig. 4.21. Structural maps of Sierra Bermeja massif. Trajectories of a) lineations and b) foliations (Darot, 1973).

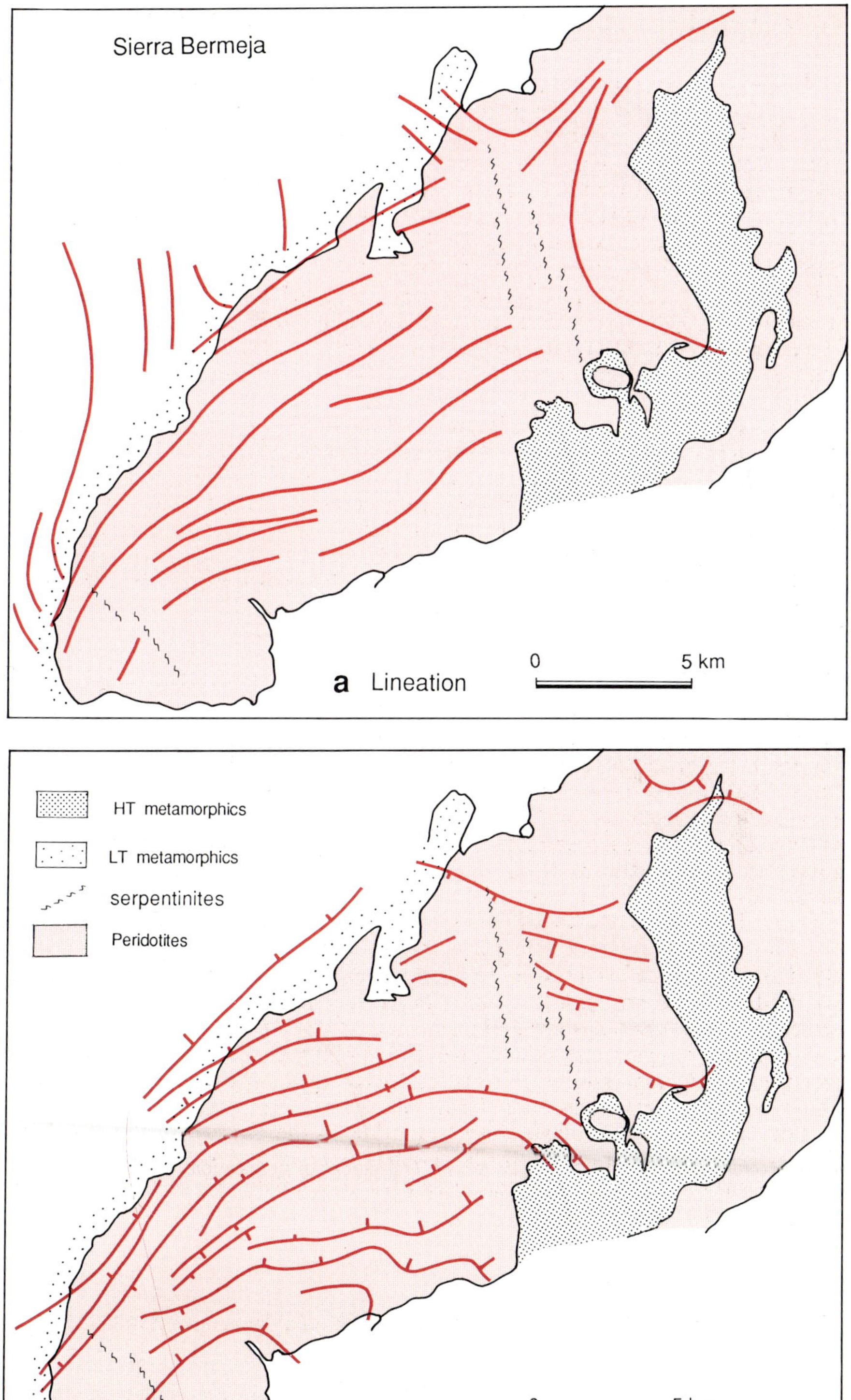
Sierra Bermeja
a Lineation
0
5 km
HT metamorphics
LT metamorphics
serpentinites
Peridotites
b Foliation
0
5 km

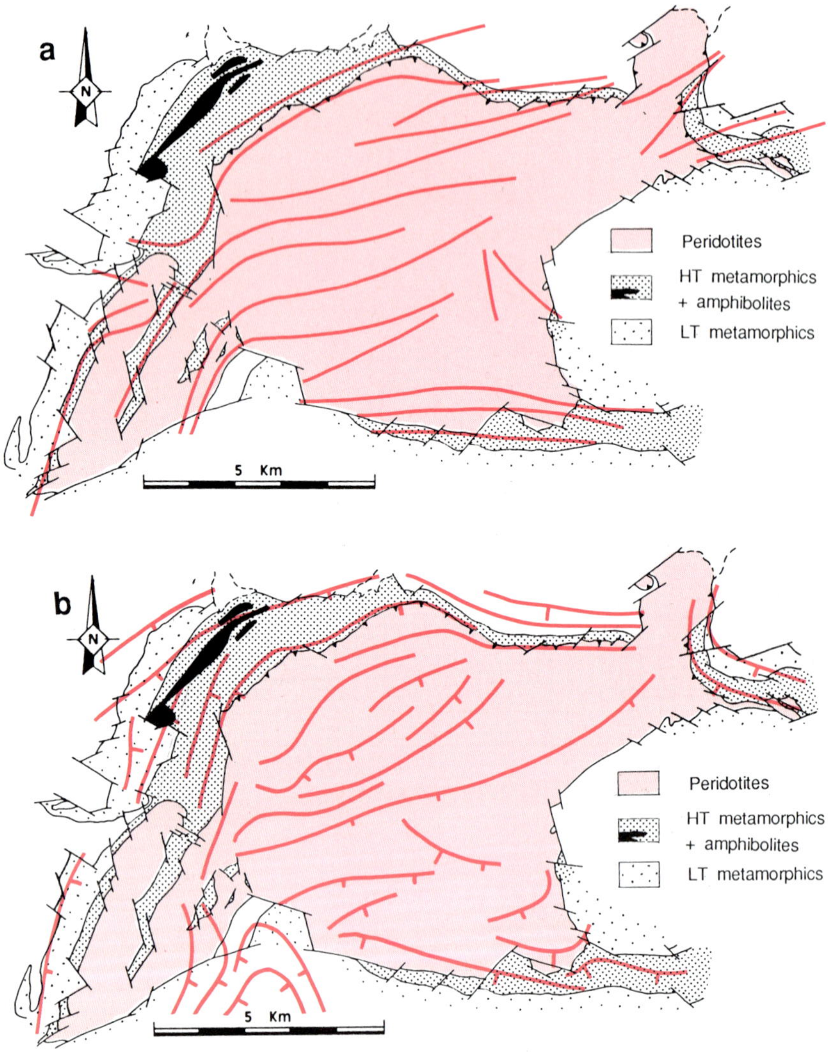


Fig. 4.22. Structural maps of Sierra Alpujata massif. Trajectories of a) lineations and b) foliations (Tubia and Cuevas, 1987).

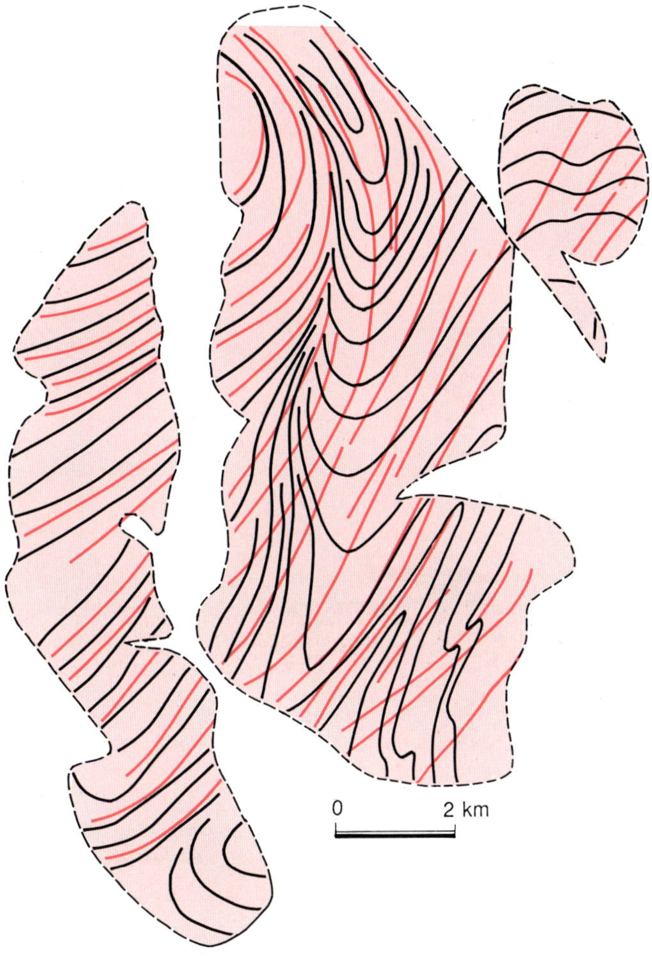


Fig. 4.23. Structural map of Lanzo massif. Trajectories of $S_0$ compositional layering (black lines) and $S_1$ foliation (colored lines) (Boudier and Nicolas, 1980).

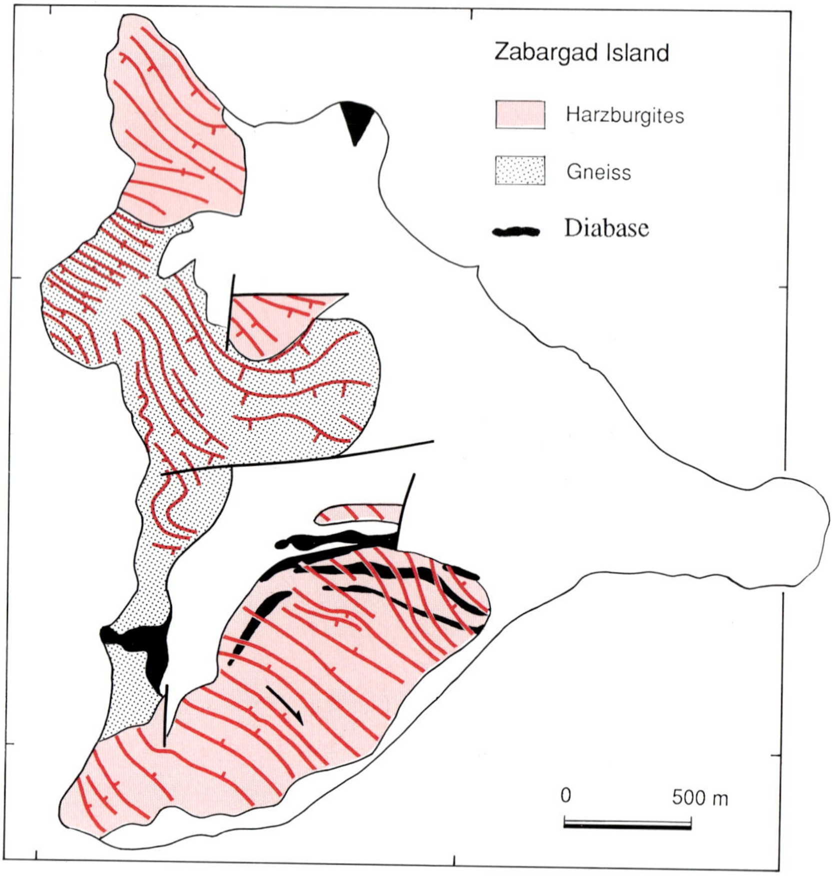


Fig. 4.24. Structural map of Zabargad Island. Trajectories of foliations (Nicolas et al., 1987).

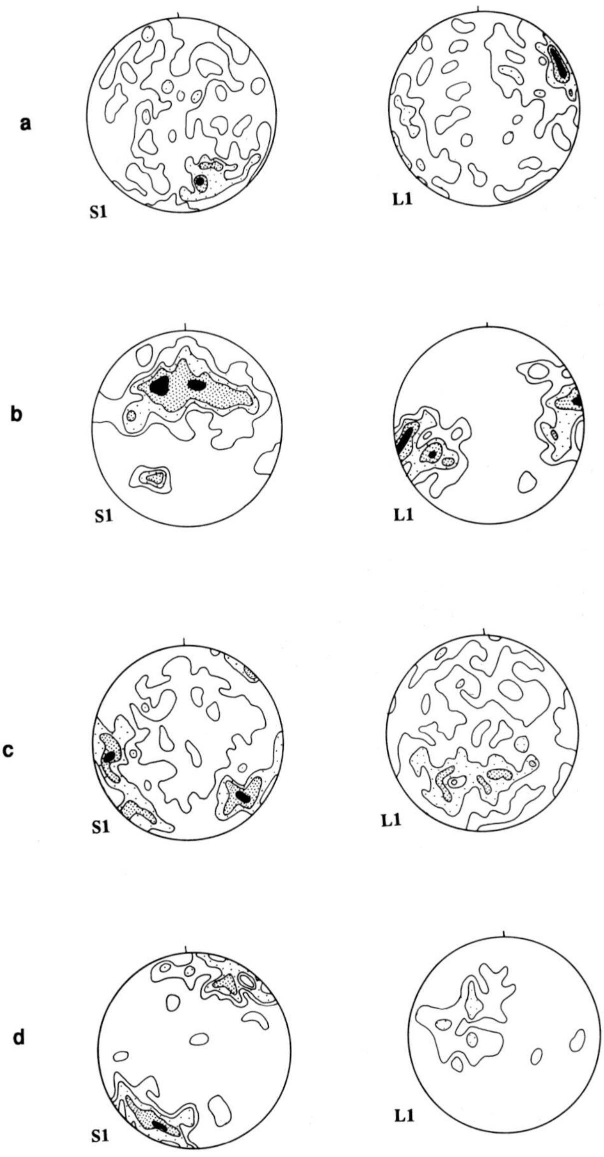


Fig. 4.25. High temperature foliations (S1) and mineral lineations (L1) in various massifs (lower hemisphere stereographic projections, contours for 0.45% net area). a) Sierra Bermeja (S1 : 225 measurements, contours : 1, 4, 7, 10% ; L1 : 181 measurements, contours : 1, 4, 6, 8 % ) ; b) Sierra Alpujata (S1 : 72 measurements, contours : 1, 2, 3, 5 % ; L1 : 45 measurements, contours : 1, 2, 3, 5 %); c) Lanzo (S1 : 525 measurements, contours : 0.5, 1, 2, 4 % : L1 : 389 measurements, contours : 0.5, 1, 2 %) ; d) Zabargad (S1 : 106 measurements, contours : 1, 2, 4, 8 % ; L1 : 60 measurements, contours : 1, 2 %). (a) Darot, 1973 ; b) Tubia and Cuevas, 1987 ; c) Boudier, 1978 ; d) Nicolas et al., 1987).

### *4.5.2. Structural zonation*

The microstructures in these massifs are similar to those of the LOT lherzolites. In Lanzo (fig. 4.20a), Sierra Bermeja (fig. 4.20b), Beni Bousera (fig.4.20c) and Sierra Alpujata (Tubia and Cuevas, 1987) there is a structural zonation superimposed on the petrological zonation such that the microstructures indicate increasing strain and stress and decreasing temperature towards the margins of the intrusion, with the development of mylonites in the contact areas locally. Similar observations are made in the northern outcrops of Zabargad Island (Nicolas et al., 1987), where in the vicinity of granulite country-rocks a syntectonic pargasite invades the spinel lherzolite and a new low temperature-high stress foliation develops (fig. 4.25.d). In Zabargad, Beni Bousera and Ronda massifs, the granulites-kinzigites which are in contact with the mylonitic peridotites show the same deformation as the peridotites.

### *4.5.3. Structure and geodynamic environment*

The structural studies conducted by the above cited authors in these various massifs lead to the same general conclusions as in LOT, namely that foliations are dominantly steep and lineations moderately inclined. It is puzzling to observe that this tendency is recognized in spite of the fact that some of these massifs are rootless, and that probably all of them have experienced some crustal deformation during their emplacement. Lanzo and Zabargad could be the least displaced because both sit on large positive gravity and magnetic anomalies with, in the case of Lanzo, the additional evidence of an important mantle uprise well shown by seismic profiling and gravity modelling ( ECORS-CROP Gravity Group, in press). Both show vertical high temperature foliations and stretching lineations plunging at 40°-50° (figs. 4.23, 4.24 and 4.25). These structures are parallel to the regional geodynamic trend. In the case of Lanzo, which is derived from the lithosphere of the Piedmont-Liguria ocean (fig. 4.18), being later squeezed between the colliding European and Southalpine plates, the high temperature foliation attitude is parallel to the plates boundary. In the case of Zabargad, the foliation attitude is parallel to the Red Sea trend. The low temperature deformations in these massifs vary in orientation ; in Lanzo, they bend the high temperature foliations in a new NW orientation of strike-slip motion and, in Zabargad, they mold the variously oriented contacts with the surrounding granulites. The Gibraltar arc massifs (Beni Bousera and Ronda) and Collo massif (Misseri, 1987), although they are displaced with respect to the gravity anomaly high, show the same general structure with generally steep foliations and flat lineations also parallel to the regional geodynamic trend.

### *4.5.4. Contact metamorphism and nature of metamorphosed formations*

With the exception of Lanzo which is in an ophiolite environment, all of these massifs are in contact with granulites and kinzigites which have been deformed and thermally affected by the hot intruding peridotite. These effects are quite obvious in the Ronda massifs (Loomis, 1972, 1975 ; Lundeen, 1978 ; Tubia and Cuevas, 1987) and in Zabargad (Boudier et al., 1988). In Zabargad, there has been an important contamination of the country rock by mafic material added at various stages during the course of deformation and metamorphism induced by the peridotite intrusion (Nicolas et al., 1987 ; Bonatti et al., 1987 ; Boudier et al., 1988). Mafic granulites deriving from gabbros, amphibolites deriving from diabase dikes and tonalitic intrusions due to intrusion-related anatexis can be so abundant that it becomes difficult to identify the country rock. This is now a felsic granulite equilibrated at fairly high pressures, which suggests a deep crustal origin (Bonatti et al., 1987 ; Boudier et al., 1988) (see however § 8.2.3). In the Ronda massifs the mafic intrusions are much more restricted and the kinzigites at the contact grade away into formations representing a continental crust basement and its sedimentary cover affected by a regional metamorphism.

# Chapter 5

# BOGOTA PENINSULA AND NE DISTRICTS OF NEW CALEDONIA - WADI TAYIN IN OMAN COASTAL COMPLEX OF NEWFOUNDLAND : POSSIBLE ORIGIN IN TRANSFORM FAULTS

## *5.1. INTRODUCTION*

Ophiolite massifs or districts in ophiolitic massifs which are good examples of oceanic fracture zones are not common. Ophiolites with crustal features evoking this environment are found in Cyprus, the Arakapas Valley (Moores and Vine, 1971 ; Simonian and Gass, 1978) and the adjacent Limassol Forest (Murton, 1986 ; Murton and Gass, 1986), Gibbs Island in the Shetlands (De Wit et al., 1977), the northern Apennines (Abbate et al., 1980) and the Coastal Complex of Newfoundland (Karson and Dewey, 1978 ; Karson, 1984). A situation somewhat comparable to this last example has been reported from the SW Sierra Nevada Foot Hills of California, although structural evidence and in particular the steep plunge of high temperature lineations (Saleeby, 1978) suggests a more complex origin (Saleeby, 1982). In these massifs the evidence for an oceanic fracture zone environment is mainly found in the mafic section. In particular in the Arakapas valley of Cyprus, it is still possible to observe the seafloor and shallow formations of a paleo-transform fault. The fault is indicated by the valley morphology, the presence of polymict breccias representing relief screes over the basaltic floor and a complex E-W to NE-SW diabase dike pattern compared to that of adjacent domains, regularly N-S oriented.

In other cases, such as the Antalya ophiolite of Turkey (Reuber, 1984), the Ingalls Complex of Washington (Miller, 1986, 1987), the Bogota Peninsula (Prinzhofer and Nicolas, 1980) and the Tiebaghi-Poum- Belep districts of New Caledonia (Sécher, 1981), the evidence for a transform environment is found in the ultramafic section. The central part of the Wadi Tayin massif in Oman, already discussed in § 3.4.2, corresponds to the only case reported so far where this evidence is clearly displayed both in the ultramafic and in the mafic section of a single ophiolite (Nicolas et al., 1988a). We will describe it briefly after the New Caledonia districts and the Coastal Complex of Newfoundland. These three occurrences have been selected because of the extensive structural evidence available.

## *5.2. BOGOTA PENINSULA AND NE OPHIOLITIC DISTRICTS OF NEW-CALEDIONA*

### *5.2.1. Introduction*

The Bogota Peninsula and the NE districts of New Caledonia have been selected because they illustrate well the geometry of a large-scale shear zones in the mantle section contemporaneous with magmatic activity, and, in the NE districts, the association of lherzolite facies with such environments.

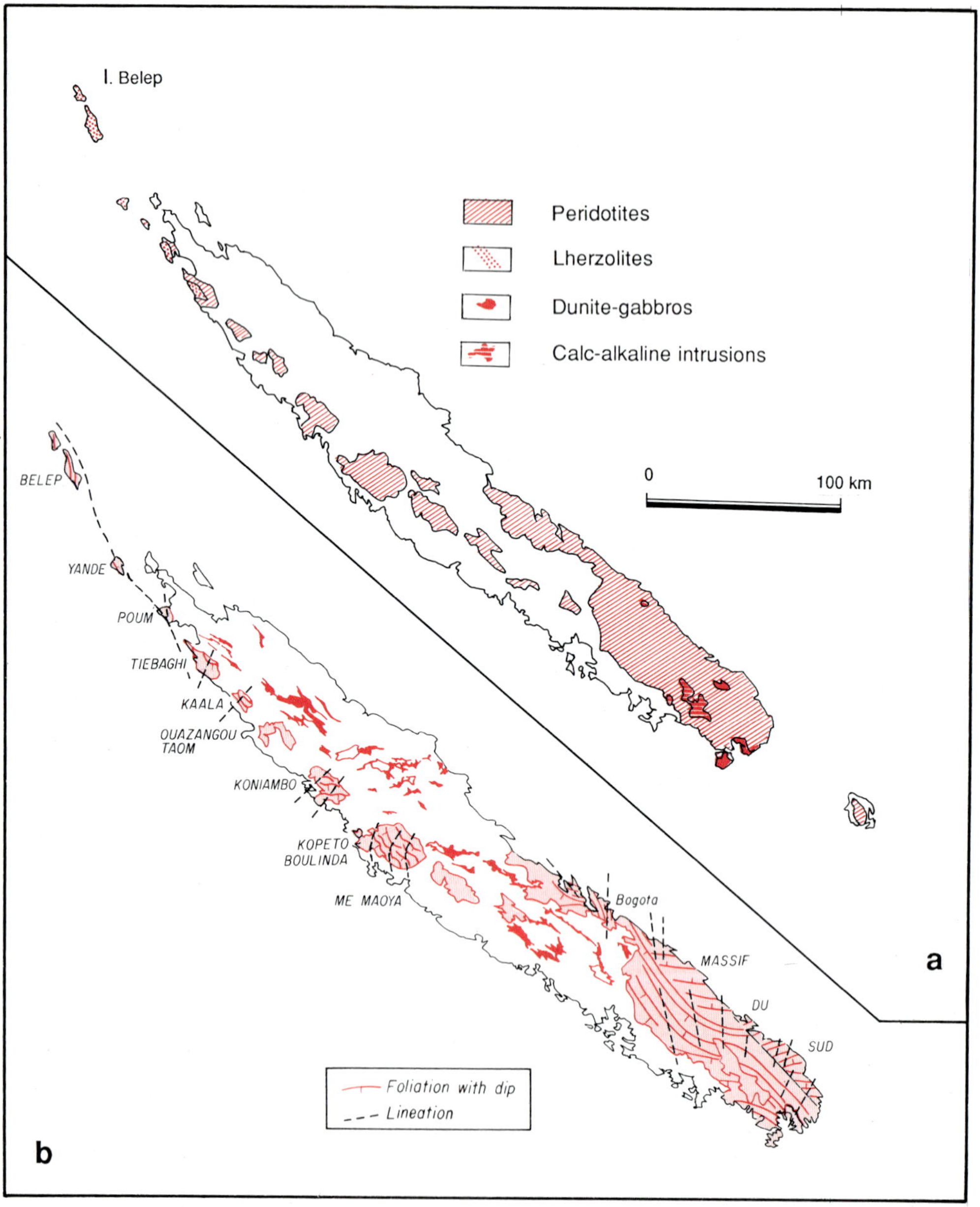


Fig. 5.1. The ophiolite nappe of New Caledonia. a) Petrological facies. b) Trajectories of high-T plastic deformation in peridotites ; solid lines : foliations trajectories ; dashed lines : lineation trajectories; black : serpentinite stringers below the nappe. (Prinzhofer et al., 1980).

### *5.2.2. Geological setting*

Ophiolitic, essentially ultramafic formations, cover 7000 km$^2$ in New Caledonia, that is 40% of the surface of the island (fig.5.1). They crop out predominantly in the Massif du Sud and in a series of massifs along the SW coast, extending to the Belep islands, some 50 km from the NW tip of the main island.

This suggests a much larger extension of the ophiolite before erosion. These massifs clearly form an horizontal nappe, only a few kilometers thick, at the scale of the island (Glasser, 1903 ; Routhier, 1953 ; Prinzhofer et al., 1980). Serpentinite stringers (fig.5.1) pinched into the underlying formations are interpreted as remnants of the base of the nappe. The ophiolite nappe is incomplete, being composed of ultramafics, essentially harzburgites, except in the Massif du Sud where it is locally capped, above a transition zone of dunites and werhlites, by small outcrops of layered gabbros ; the contacts between these units are horizontal (Prinzhofer et al., 1980). The trace element geochemistry of the harzburgites and dunites has been studied by Prinzhofer and Allègre (1985), who point to their residual character with some melting occurring in the feldspar field. Prinzhofer (1987) has also dated the layered gabbros by Nd/Sm isotopes at 131 ± 16 Ma.

The nappe rests on various sedimentary and volcanic formations, including basalts and dolerites probably derived from oceanic crust (Challis and Guillon, 1971) which, along the NE coast, could belong to the same piece of lithosphere as the main nappe. The basalts of the SW coast, however, have a Nd/Sm age of 58 + 29 Ma (Prinzhofer, 1987) which rules out their parenthood with the ophiolite. The youngest sedimentary formations below the nappe have been used to date the nappe emplacement as Upper Eocene (Paris et al., 1979). The gravimetric data (Collot et al., 1987) indicate that in the Massif du Sud the nappe thickens seaward with respect to the NE coast. The structural mapping shows that the foliation planes in the peridotites which are on average horizontal (fig.5.4a) tend to dip to the NE along this coast. Finally, a blueschist metamorphism affects the autochtonous formations along the coast at the NE of the Island (Espirat, 1963 ; Brothers, 1974). Its age at the Eocene-Oligocene limit (Coleman, 1967; Blake et al., 1977) confirms the age for the nappe emplacement proposed above. All of these data suggest that the nappe is rooted in a NE direction. Taking into account also the geophysical data available for this part of the SW Pacific, Aubouin et al. (1977) have proposed a model for the origin of the New Caledonian ophiolite in which it originated above an Eocene subduction zone and was emplaced when New Caledonia entered the subduction zone (fig.5.2). The subduction being choked by this piece of continental lithosphere would have migrated to the NE with respect to its present position. This model is compatible with the available results except that the dip of the first subduction plane needs to be very flat in the first hundred kilometers from the trench to explain the thin and flat character of the ultramafic nappe (see § 12.4.2).

Mapping the high temperature foliations (fig. 5.3b and 5.4) and lineations (fig. 5.3a and 5.4) in the Massif du Sud reveals a surprisingly regular pattern, considering the area covered and the thinness of the nappe. The plastic flow pattern deduced from these structural elements (see § 2.5.4) has the following characteristics : a flat flow plane, a NS flow direction and a remarkable shear sense inversion at around 1 km below the Moho.

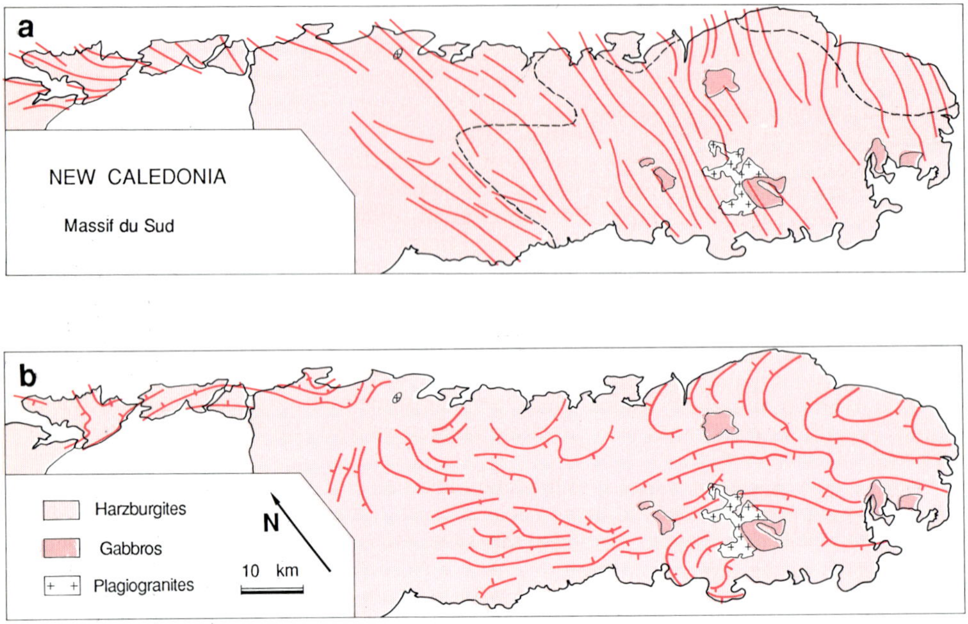


Fig. 5.2. Model of obduction of the New Caledonian ophiolite, inspired by Davies' (1971) model for New Guinea. a) Before Eo-Oligocene. b) After Eo-Oligocene. (Aubouin et al., 1977).

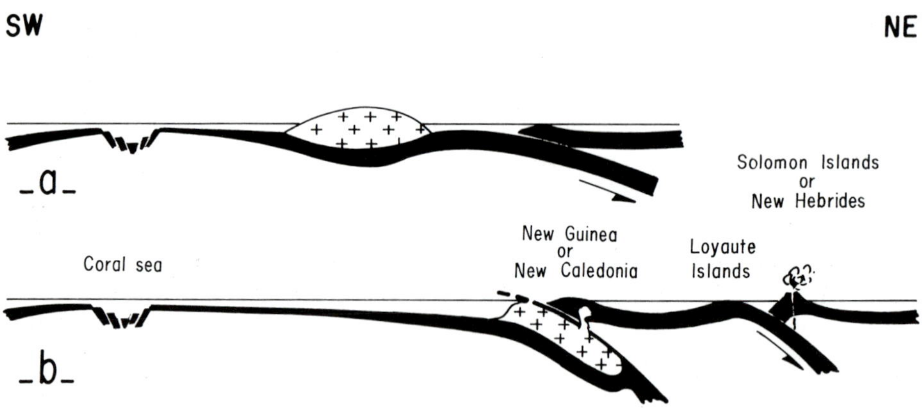


Fig. 5.3. High-T plastic flow structures in the Massif du Sud peridotites. Trajectories of a) mineral lineations and b) foliations. The dashed line in fig. 5.2a separates an upper mantle sheet within which the shear sense is such that it moved Northward, from the underlying mantle formations moving Southward (Prinzhofer et al., 1980 ; Podvin, 1983).

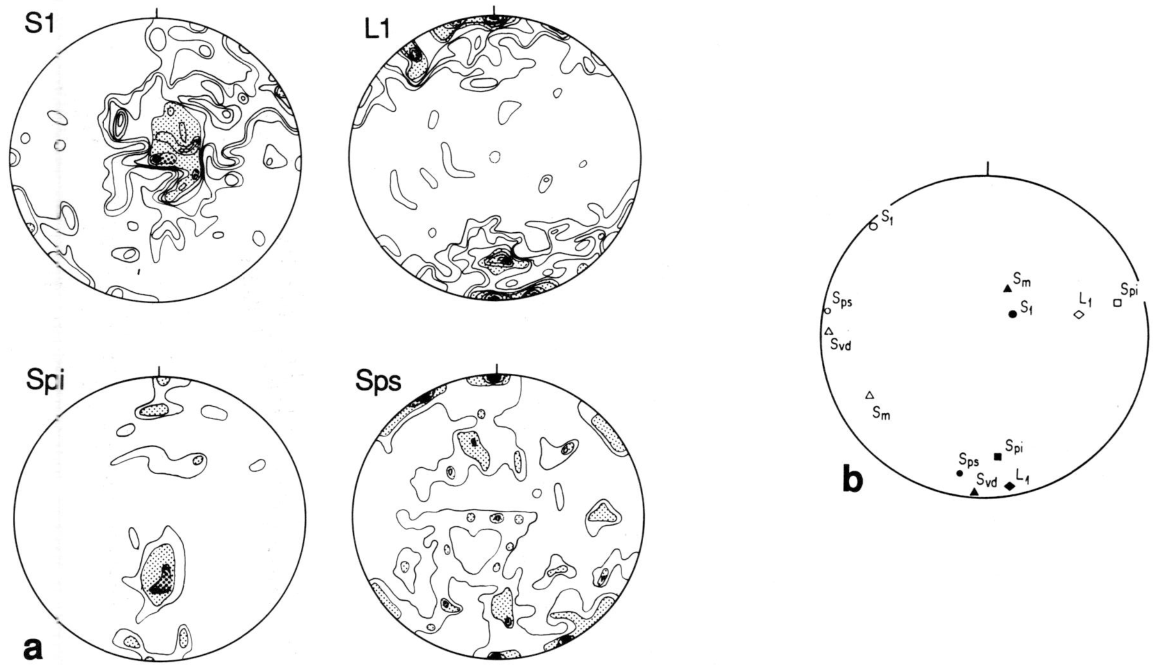


Fig. 5.4. a) Stereograms of the main structures in the Massif du Sud peridotites. S1 : high-T foliation (375 measurements ; contours : 0.5, 1, 1.5, 2, 2.5, 3, 3.5, 4, 4.5 %) ; L1 : associated mineral lineation (325 measurements ; contours : as for S1) ; Spi : indigenous pyroxenite dikes (30 measurements ; contours : 3, 6, 9 %) ; Sps : intrusive pyroxenite dikes (124 measurements ; contours : 0.8, 1.6, 2.4, 3.2%). b) Computed best poles and axes with as additional symbols : Sm, gabbro layering plane; Svd, dunite veins. Black symbols : field data ; open symbols : constructed axes of girdles or of open folds (Prinzhofer et al., 1980).

### 5.2.3. Description of the Bogota Peninsula shear zone

*Structure*

The Bogota Peninsula belongs to the Massif du Sud (fig.5.1b). It has been studied specifically by Prinzhofer and Nicolas (1980). Structurally, the Bogota Peninsula can be divided into two distinct domains (fig.5.5). The domain along the tip of the Peninsula (northern domain) has steeply dipping foliations ; the peridotites are strongly deformed, little serpentinized, and are cut by numerous mafic dikes. In contrast, the peridotites on the southern part of the Peninsula have flat-lying foliations ; they are moderately deformed and highly serpentinized and only a few dikes have been observed. Both domains are composed of harzburgites intercalated with dunite lenses.

The peridotites of the southern domain seem to rest on the strongly deformed peridotites of the tip of the Peninsula. The contact, following a near-horizontal limit locally underlined by a sharp morphological difference, is interpreted as a surface of a late thrust. Concerning the thrusted formations of the southern domain, it is sufficient to observe that the foliations are moderately inclined toward the North but their attitudes are irregular in detail. The mineral lineation has a dominant northwestern trend. The rest of this description will focus on the northern domain.

The northern domain contains a 3 km wide mylonite zone outcropping along part of the coast (fig. 5.5). The mylonite is texturally recognizable by its strongly developed banding originated by heterogeneous straining ; some bands are composed of trails of enstatite and olivine neoblasts ascribed to superplastic flow, whereas others, deformed by plastic flow s.s., contain enstatite porphyroclasts with an elongation ratio up to 33/1. Within the mylonite zone the vertically dipping foliation trends North (fig. 5.6a) and the mineral lineation is horizontal (fig. 5.6b).

East and West of the mylonitic band, the strain progressively decreases with some local inclusions of moderately deformed peridotites. The foliation rotates toward a NNW azimuth with decreasing dips. The lineation remains essentially horizontal. The peridotites are fine-grained with elongated pyroxenes. Locally, the pyroxene layering displays tight isoclinal folds with axes parallel to the mineral lineation. Altogether the domain of intense straining is 15 km wide. Finally the divergence of foliation away from the axis which is a typical feature of shear zones is more clearly observed to the East than to the West. Although the mapping West of the zone is still insufficient, it seems that the overall structure is more irregular with possibly another shear zone.

From the study of microstructures and fabrics in the peridotites, a systematic dextral shear sense has been determined. Considering the obliquity between lattice and shape fabrics (§ 2.5.4), shear strain increases from 1.7 in the least deformed facies, to 4 in more deformed ones and to 8 in the mylonite.

*Hydrothermal alteration and magmatic intrusions*

An interesting feature of the peridotites in the shear zone is the presence of numerous dikes and the evidence of a high temperature hydrothermal contamination in the bulk of the rocks. Such a hydration is exceptional outside this zone, the harzburgites of the Massif du Sud being remarkably dry, except for the late, low temperature and static serpentinization in lizardite.

Fig. 5.5. Structural map of the Bogota Peninsula. The high T and low T units are separated by the dashed line. Trajectories of a) lineations and b) foliations presented only in the shear zone unit (modified from Prinzhofer and Nicolas, 1980).

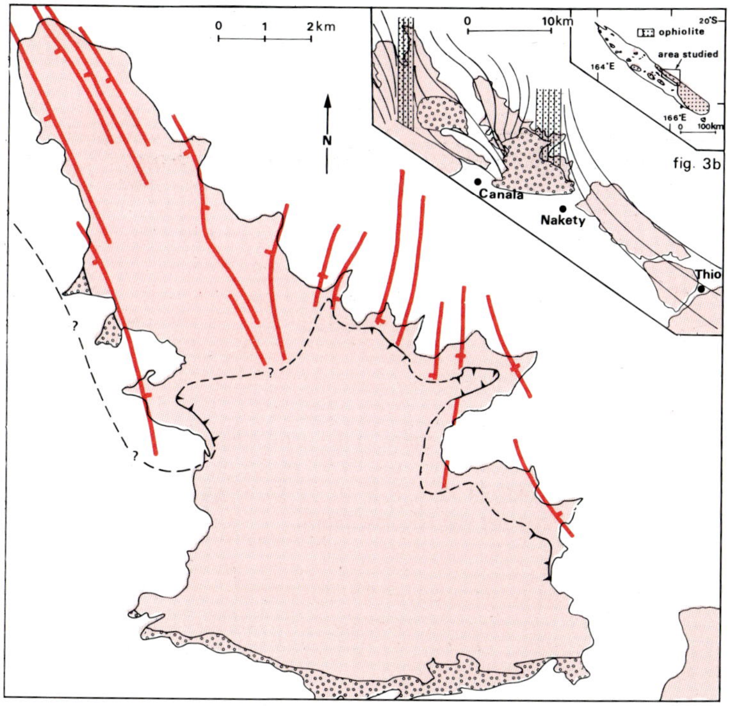
0 1 2km
N
0 10km
20°S
ophiolite
area studied
164°E
166°E
0 100km
fig. 3b
Canala
Nakety
Thio

a

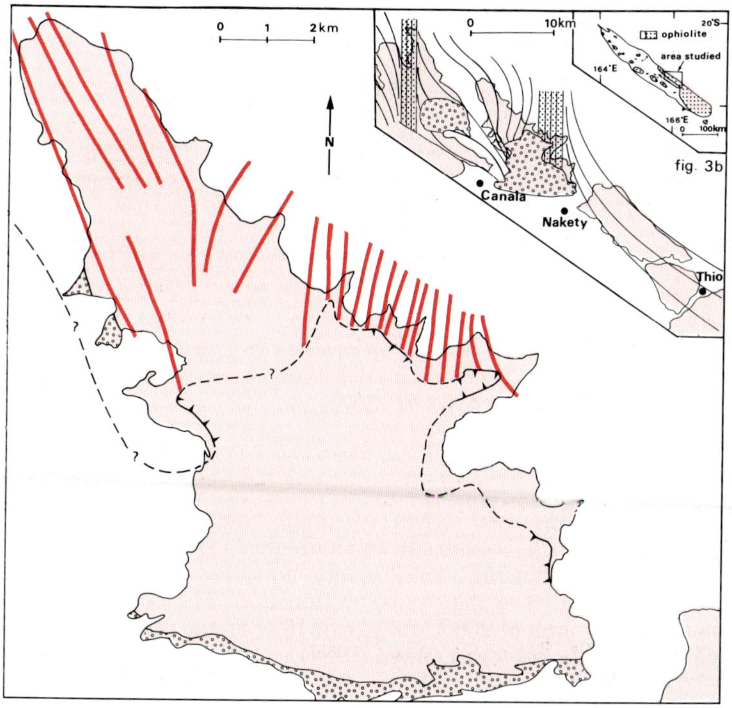
0 1 2km
N
0 10km
20°S
ophiolite
area studied
164°E
166°E
0 100km
fig. 3b
Canala
Nakety
Thio

b

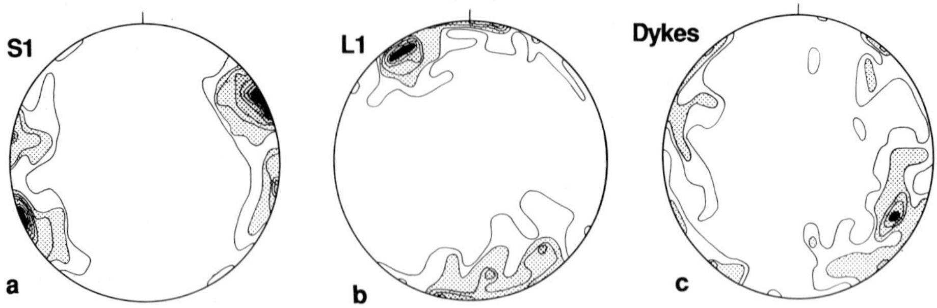


Fig. 5.6. Stereograms of the main structures in the peridotites of the Bogota shear zone. a) S1 foliation (70 measurements ; contours : 1.5, 4.5, 6, 7.5, 9, 10.5, 12, 13.5 %). b) L1 mineral lineations (58 measurements ; contours : 2, 4, 6, 8, 10, 12 %). c) All types of dikes (71 measurements ; contours : 1.5, 3, 6, 9 %). (Prinzhofer and Nicolas, 1980).

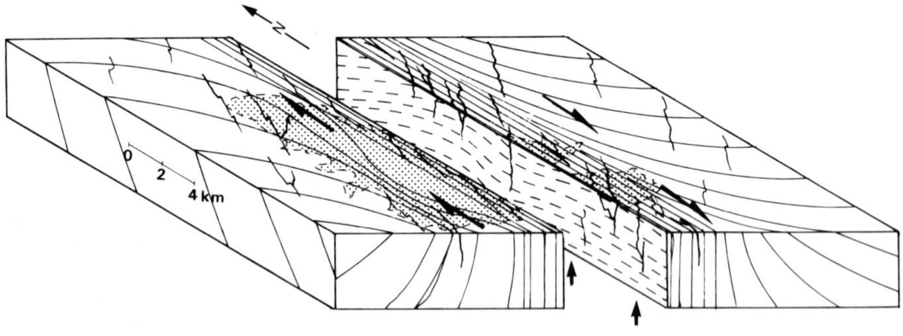


Fig. 5.7. Diagram of the northern domain, interpreted as a dextral shear zone (the Bogota Peninsula corresponds to the decorated area) ; foliations are continuous lines ; lineations, dashed lines ; dikes, wavy lines. (Prinzhofer and Nicolas, 1980).

The harzburgite contains two types of amphiboles : rare poikiloblasts of green edenitic and chromium-rich hornblende and individual crystals and clusters of tremolite replacing pyroxenes and showing a substructure indicative of plastic strain. Amphiboles tend to be oriented parallel to the pyroxene lineation. The occurrence of syntectonic tremolite puts an upper limit of 900°C for plastic flow in the peridotites, the temperature above which tremolite is not stable (Boyd, 1954).

The chronological sequence of dikes, as seen from cross cutting relations, is pyroxenite, followed by feldspathic pyroxenite and hornblende gabbro, and finally diabase. Locally a network of deformed plagiogranite has been observed. It predates the emplacement of the diabase. The amphibole from gabbros is a black magmatic hornblende, pointing to the hydrous character of the magma. The pyroxenite and the gabbro dikes are sheared, boudinaged, and rodingitized in the greenschist facies.

Several small copper occurrences in lenticular bodies 10 to 60 cm thick and up to 100 m long, with chalcopyrite in magnetite and pyrrhotite have been described (Guillon and Saos, 1971). These lenses are oriented parallel to the average orientation of the dikes.

The dikes have a dominant 30°E vertical orientation (fig. 5.6c) independent of their nature ; thus, no discrimination is introduced in figure 5.6. Since diabase dikes are undeformed, they were injected after the plastic deformation. In contrast, feldspathic pyroxenite and hornblende gabbro dikes are often deformed. Prinzhofer and Nicolas (1980) report field evidence showing that they were emplaced during the plastic flow, with, in the central mylonitic band, rotation due to tectonic transposition toward an attitude nearly parallel to the foliation.

*The Bogota Peninsula as an oceanic fracture zone*

Figure 5.7 summarizes the data presented above and clearly shows that the Bogota Peninsula is a major transcurrent shear zone with a 15 km width comparable to that of oceanic transform faults. It was deformed with a dextral sense of movement which is deduced from the following observations : (1) rotation of foliation on each side of the highly deformed zone, (2) consistent dextral shear sense measured in specimens from the shear zone, and, (3) preferred direction of injection of the various dikes, including undeformed diabase dikes. This preferred direction of injection is thought to correspond to the tension orientation ($\sigma_1, \sigma_2$ plane) related to the flow. This orientation of tension plane with respect to the foliation is typical of a dextral shear sense. This topic is further discussed in § 5.5.2.

Common orientations of the dikes and structural evidence indicate that magmatic activity occurred during movement in the shear zone and, considering the absence of deformation in the diabase dikes, continued after the movement ceased. Black magmatic amphiboles in the gabbro dikes and metamorphic-syntectonic green amphibole in the peridotites show that this was a hydrous magmatism and indicate temperatures of 1000°C-900°C for this magmatism and plastic flow in the shear zone. Hydrous circulation continued after plastic flow in greenschist facies conditions. It is probably responsible for the copper sulfide deposition.

From the average N-S trend of plastic flow lineations over such a large area as the Massif du Sud it can be assumed that the azimuth of the oceanic spreading center of origin was E-W (ignoring here possible rotations introduced by paleomagnetic information). The NNW-SSE horizontal lineations and the vertical foliation in the Bogota Peninsula shear zone, which laterally rotate progressively into the general orientation of the Massif du Sud, are then entirely compatible with what is expected for an oceanic fracture zone.

### *5.2.4. Description of the Tiebaghi-Poum-Belep shear zone*

This shear zone extends over 120 km in a NNW direction at the NW tip of New Caledonia, covering part of the Tiebaghi massif, the Poum massif and a series of small islands, Yandé, Art and Pott, the last two belonging to the Belep

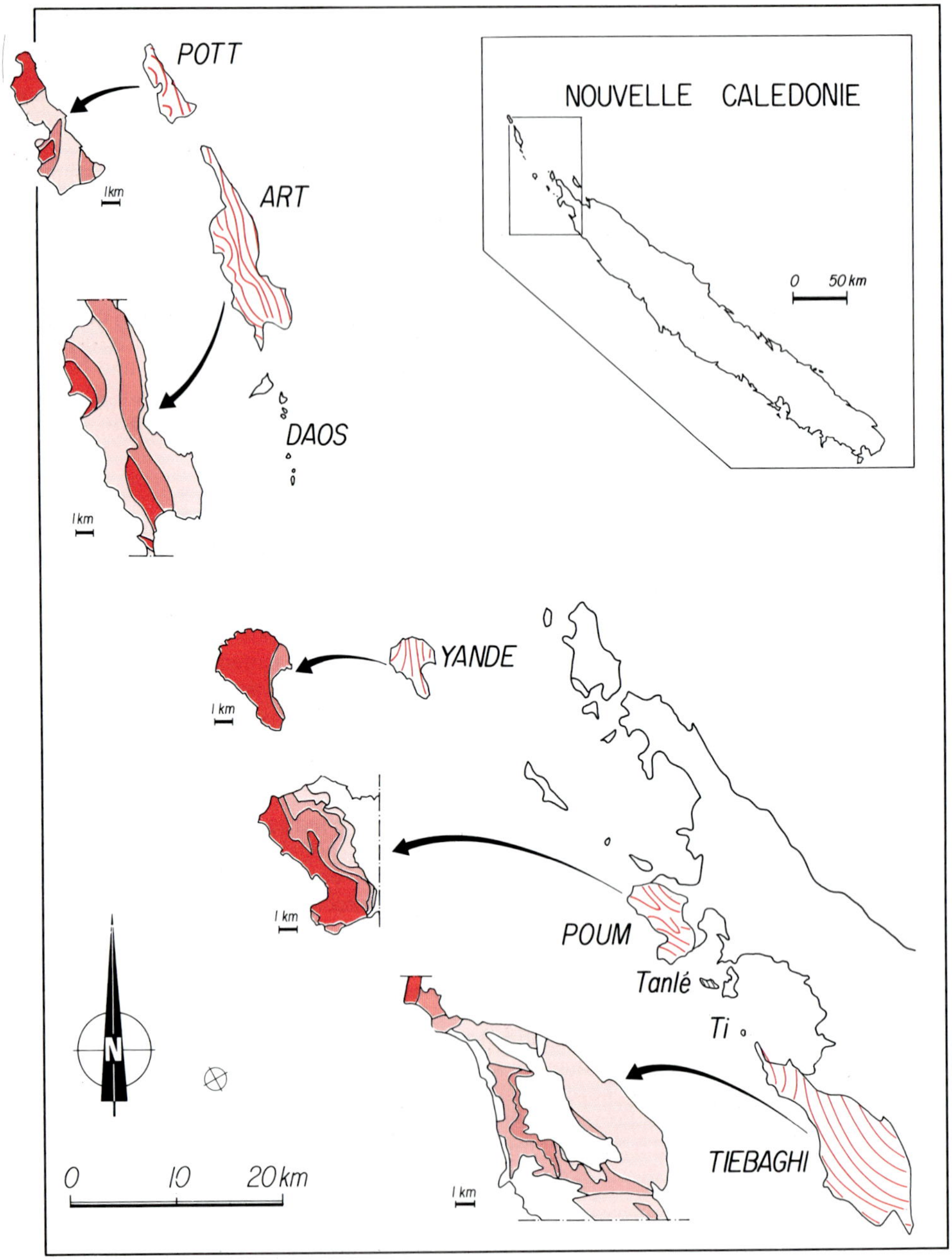


Fig. 5.8. Structural and petrological map of the Tiebaghi-Poum-Belep shear zone. The continuous lines are the trajectories of the peridotite foliation ; Decreasing color tones from plagioclase lherzolites, spinel lherzolite, harzburgites and finally dunites (After Moutte, 1979 and Sécher, 1981).

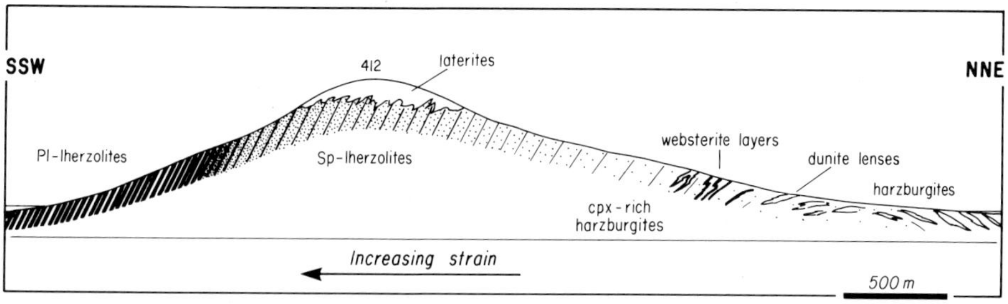


Fig. 5.9. Cross section through Poum district showing the evolution of the peridotite facies to increasingly fertile facies in relation with the proximity of the shear zone. The largest strain is measured in the steeply dipping plagioclase lherzolites. (After Sécher, 1981).

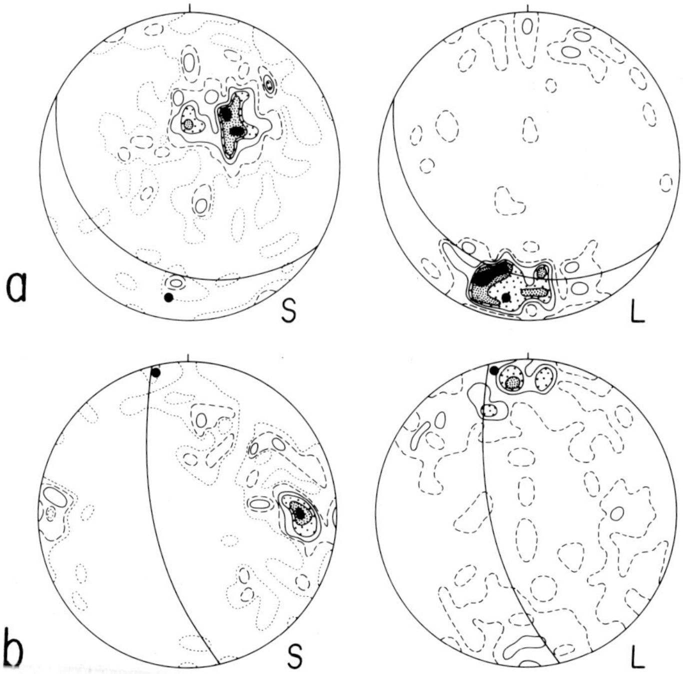


Fig. 5.10. Stereograms of high strain foliation (S) and mineral lineation (L) in the peridotites of Poum (a) and Art Island (b). a) S (616 measurements ; contours : 0.5, 1, 1.5, 2, 2.5, 3 %), L (187 measurements ; contours : 1, 2, 3 ,4, 5 %). b) S (302 measurements ; contours : 0.5, 1, 1.5, 2, 3 %), L (107 measurements ; contours : 1, 2, 4, 6 %). Line : best computed plane ; dots : best computed line. (Sécher 1981).

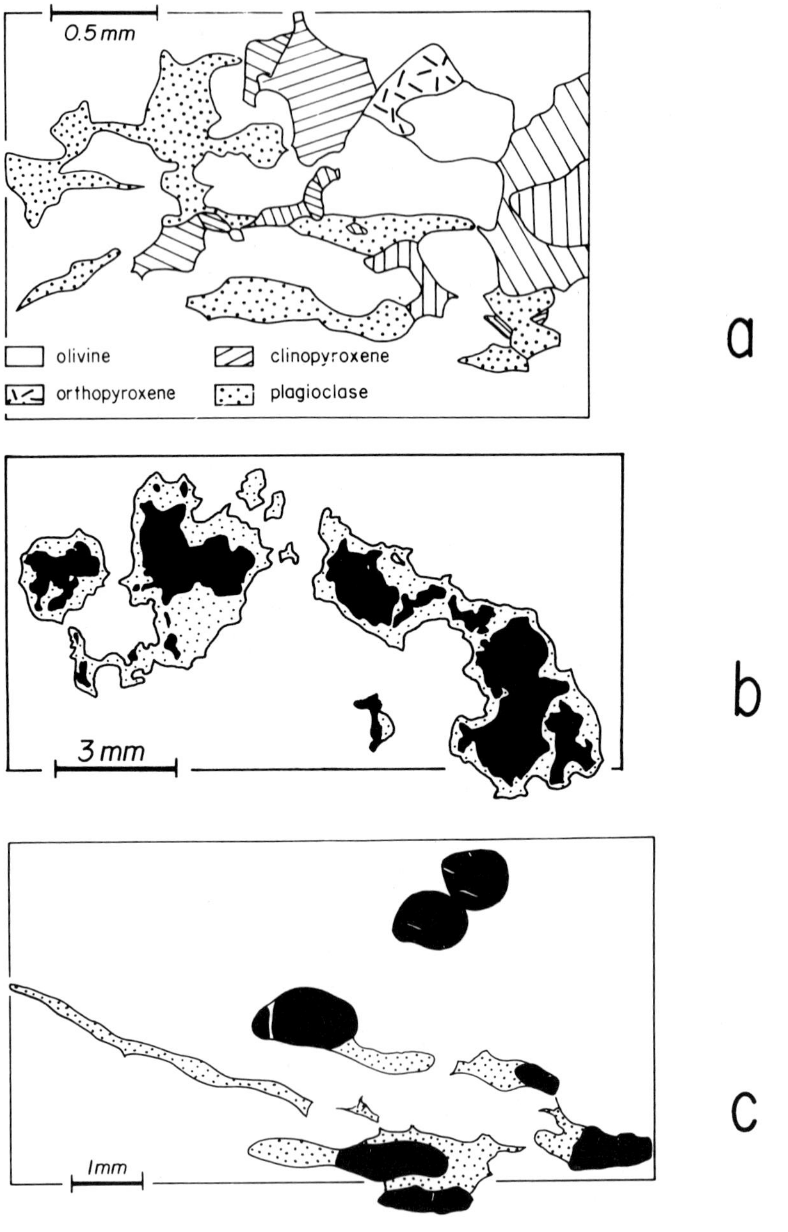


Fig. 5.11. Plagioclase and clinopyroxene microstructures in the Tiebaghi-Poum- Belep shear zone. a) Interstitial habit of the two minerals in a plagioclase lherzolite interpreted as the result of trapped melts. b) Plagioclase corona around spinel in a plagioclase lherzolite. c) Plagioclase both interstitial , between olivine crystals, and in corona around chromite in a dunite. (Sécher, 1981).

archipelago (fig. 5.8). This district has been specifically studied by Sécher (1981), with more detailed studies on the chromite-rich massif of Tiebaghi by Moutte (1979, 1982).

In these northwestern massifs, the plastic flow structures in the peridotites are less coherently organised than in the Massif du Sud, possibly due to late deformation and rotation (Leblanc, 1980) (fig.5.1). However, the foliation remains flat-lying and the spinel lineation trends N-S on average.

*Structure*

The plastic flow structures in the axis of the shear zone and the orientation of the shear zone itself are turned slightly more toward the North than the alignment of the ultramafic massifs. The shear zone is centered on Yandé, where the structures are predominantly mylonitic, whereas only the western parts of Poum and Tiebaghi and the eastern parts of Art and Pott are affected by this deformation (fig.5.8 and 5.9). The divergence of foliations typical of a shear zone is best observed in Tiebaghi. In the other massifs, the structures outside the mylonitic band are irregular ; locally the foliation warps in 1-3 km arcs with steep lineations (fig.5.8).

The fact that the structure pattern on maps is less regular than in the Bogota area is not surprising considering the difference in scale and the possibility of some rotation from one massif to the next. Structures are also less regular at the scale of a single massif, as shown by maps of foliations and lineations (Sécher, 1981 ; Moutte, 1979) and by stereonets of these structures (fig.5.10). However, the microstructures are quite similar to those in Bogota Peninsula. The shear sense deduced from their study is dominantly, but not systematically, dextral as in Bogota. Interestingly, the sinistral shear sense becomes as important as the dextral one in the Yandé and Belep massifs which are located on the western side of the shear zone.

*Petrology*

The most remarkable feature of this shear zone is its association with lherzolites. On the East side of the shear zone where the transition to the normal high-temperature, flat-lying foliations can be followed, there is a complete gradation from harzburgites with abundant dunites outside the shear zone to clinopyroxene-bearing harzburgites, spinel lherzolites and plagioclase lherzolites. Spinel and mainly plagioclase lherzolites are found in the most deformed domains of the axis of the shear zone. This is visible on the map (fig. 5.8) and in cross section (fig. 5.9).

Moutte (1979, 1982) and Sécher (1981) have analysed these lherzolites. Their average modal composition is olivine 61.5, enstatite 24.5, diopside 7.4, plagioclase 5 and spinel 1.6 for the plagioclase lherzolite and olivine 67, enstatite 25, diopside 6.2 and spinel 1.3 for the spinel lherzolite. The plagioclase lherzolite has a major element composition compatible with pyrolite (Ringwood, 1966) or with lherzolites assumed to be directly derived from the mantle (Maaloe and Steel, 1980), but the spinel lherzolite has a depleted signature.

The origin of lherzolites is not unique. They can represent mantle rocks, either pristine or little depleted by partial melting, or alternatively harzburgites secondarily impregnated by a basaltic melt introducing the clinopyroxene and eventually the plagioclase. This question is debated more fully in §.2.5.3 where criteria to distinguish these two possibilities are presented.

It so happens that in this case, these criteria are somewhat ambiguous, resulting in diverging opinions. Moutte (1979,1982) favors a direct mantle origin but does not exclude an entirely magmatic origin, explaining the large scale zonation mentioned above by differentiation during crystal settling. Sécher (1981) observes that structurally, the clinopyroxene seems secondary because it is interstitial between the other minerals (fig. 5.11a) and is not specifically associated with enstatite and spinel as in mantle lherzolites (fig. 2.5) ; the plagioclase is interstitial (fig.5.11a), indicative of an impregnation origin, and forms coronas around spinel (fig. 5.11b), as typically observed in mantle plagioclase lherzolites, although some intermediate habits are also reported (fig. 5.11c). Mineral composition and chemical zonation are not considered by this author to be conclusive either one way or the other. Nicolas and Dupuy (1984), considering the REE patterns in the various peridotite facies of this area, note their discrepancy with the major and transition element trends. Thus, the major and transition elements in the plagioclase lherzolite indicate a relatively fertile composition whereas the REE indicate a refractory character. They conclude that the various facies are refractory residues locally enriched, during a stage subsequent to partial melting by a liquid which, in conformity with the chemistry of the basic rocks of the ophiolite (Dupuy et al., 1981) would be itself produced by a depleted LREE source. Alternatively, the plagioclase lherzolites may reflect a lesser degree of melt extraction as discussed below.

### *5.2.5. Discussion*

*An oceanic model for the New Caledonia ophiolite*

From the overall N-S trend of lineations in the various massifs, we assumed above that the oceanic ridge where the New Caledonia ophiolites originated had an E-W azimuth with respect to present coodinates. Speculating further, one can infer on which flank along this ridge the Massif du Sud lithosphere originated, by considering the pattern of the shear sense inversion observed in the peridotites of this massif and referring this pattern to the model of figure 2.2. This piece of lithosphere would thus derive from the southern flank of the ridge. Figure 5.12 is a sketch of the ridge-transform structure based on the data in the Massif du Sud and Bogota Peninsula. The data on the Tiebaghi-Poum- Belep transform fault are compatible with this model and it would represent another structurally similar fault located 150-200 km to the West.

*Kinematics in the Bogota and Tiebaghi-Poum-Belep fracture zones*

It has been mentioned that the attenuation zone of the Tiebaghi-Poum-Belep transform fault presented some heterogeneous structures and inverse shear sense along its western side. This suggests that fracture zones may differ somewhat from typical shear zones in keeping with the expected complexity of the kinematics of mantle flow in the vicinity of a fracture zone. In figure 5.13 it is shown that a *transform fault is a true shear zone only between the two ridge segments.* Outside this domain, the trace of the transform motion, which is now frozen, is preserved only along the older lithosphere, probably as fragments of a shear zone. Moreover, the asthenospheric flow, supposed to diverge from a mantle diapir and to be channelled by the transform wall (chapter 9), induces a new shear motion whose sense is that of the transform fault only between the ridge segments. Outside, this new flow imposes an opposite shear sense along the older sheared lithosphere. It is expected however that the new shear strain be much smaller than the one produced within the active transform and that its interference be modest. This could be what is observed on the western side of the Tiebaghi-Poum-Belep transform fault.

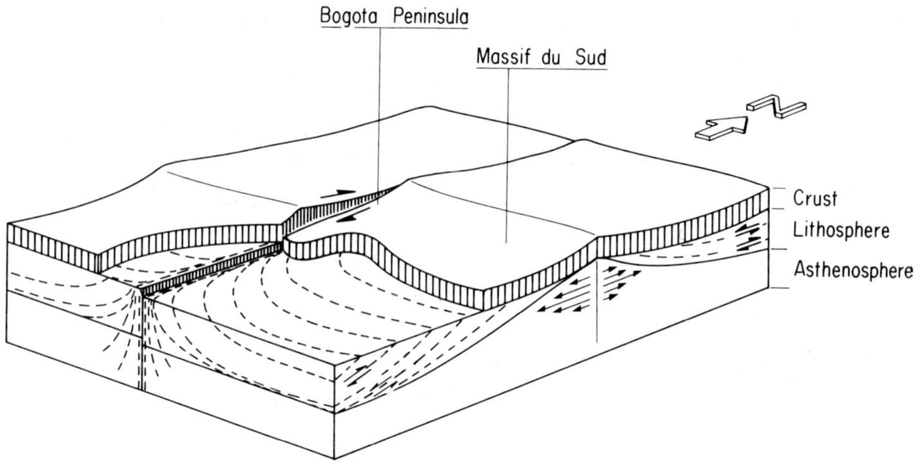


Fig. 5.12. Ridge-transform system modelled after the structural data of the Massif du Sud (Fig. 5.4) and of the Bogota Peninsula (fig. 5.7).

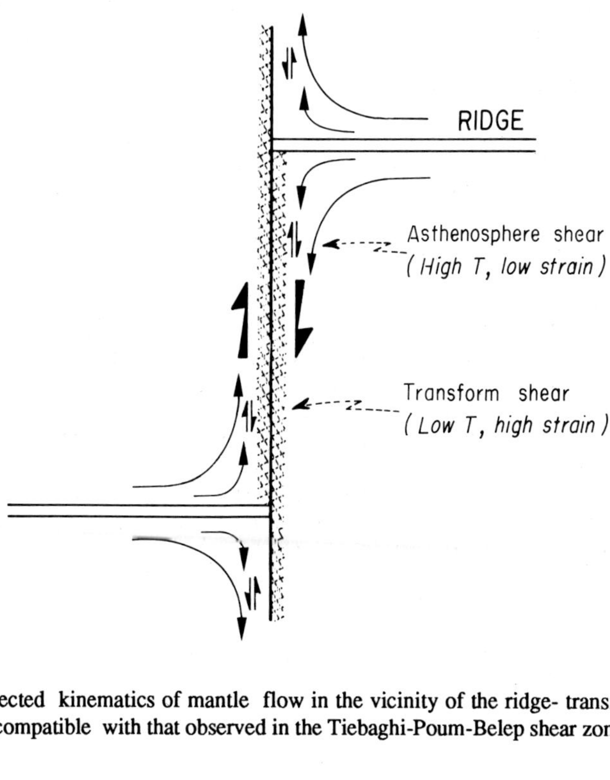


Fig. 5.13. Expected kinematics of mantle flow in the vicinity of the ridge- transform intersection. The complexity is compatible with that observed in the Tiebaghi-Poum-Belep shear zone.

*Petrological signatures of the Bogota and Tiebaghi-Poum-Belep fracture zones*

In spite of their structural similarities and their relative proximity, the Bogota Peninsula and the Tiebaghi-Poum-Belep fracture zones have very distinct petrological signatures. In the former, the peridotites are harzburgites similar to the country rocks except for their high temperature contamination by water and the abundance of hydrous magmatic intrusions. In the latter, the peridotites are lherzolites increasingly fertile toward the axis of the zone ; numerous pyroxenite gabbro and diabase dikes are also present but they have dry parageneses ; also this zone is rich in chromite pods with, in its vicinity, the Tiebaghi district which is the largest chromite deposit of New Caledonia and one of the largest ophiolitic deposits in the world. These lherzolites have been ascribed, at least partly, to the impregnation by a basaltic melt of a depleted peridotite, harzburgite or lherzolite (fig. 5.11). Similar impregnated lherzolites are common in oceanic transform faults (Nicolas and Dupuy, 1984 ; Dick et al., 1984 ; see also § 8.2.3). They have, however, some features evoking an origin as comparatively pristine mantle (fig. 5.11b). Considering the expected pecularities of an oceanic transform fault environment, Sécher (1981) suggests that the origin of the lherzolites could be a compromise between these two possibilities. The new lithosphere accreting in the vicinity of a transform fault is cooled by the older lithosphere. As a consequence, a mantle wedge uprising in this environment will be less depleted by partial melting than the mantle uprising away from it, where the thermal structure is not altered. This mantle wedge would thus retain a lherzolite composition (Boudier and Nicolas, 1985) ; furthermore, when it attains a shallow depth below the Moho it is cooled well below 1000°C and thus tends to trap the melts issued from below and migrating upward, producing an impregnated lherzolite (Fox and Gallo, 1984). Figure 5.14 illustrates these two possibilities which together could explain the ambiguous petrological characters of this shear zone. This dual origin of the plagioclase lherzolites is also retained for a similar ophiolitic situation described by Miller and Mogk (1987) in the Ingalls Complex (North Cascades, Washington). The mylonitic textures are also ascribed to an oceanic fracture zone environment. Interestingly, these authors describe, in the band of highest strain, both lherzolites and hornblende peridotites, a situation which juxtaposes here the two peridotite facies met separately in New Caledonia fracture zones.

It is not understood why the two New Caledonia fracture domains have such distinct petrological signatures. Sécher (1981) has looked for an explanation in the possibility that in a transform fault there is a minor component of motion normal to the fault plane, making it a compressive or extensive structure depending on the sense of this motion (Bonatti, 1978). In the present case, considering the importance of its diffuse and intrusive magmatism, the Tiebaghi-Poum-Belep domain should be distensive compared to Bogota. Other explanations can be proposed. For instance, it may be important to know which site within the transform system has been sampled. If it derives from inside the domain between the ridge segments, a simpler shear zone structure and a weaker thermal contrast can be expected (Bogota ?) than if it derives from outside these segments (Tiebaghi-Poum-Belep ?) (fig. 5.13).

## 5.3. COASTAL COMPLEX OF NEWFOUNDLAND

### 5.3.1. Introduction

The Coastal Complex of western Newfoundland may be, among ophiolites, the best example of a crustal section representative of an oceanic fracture zone and in this way it is complementary to the mantle sections illustrated by the New Caledonia cases . The general structure and petrology have been studied by Karson (1977, 1984) and Karson and

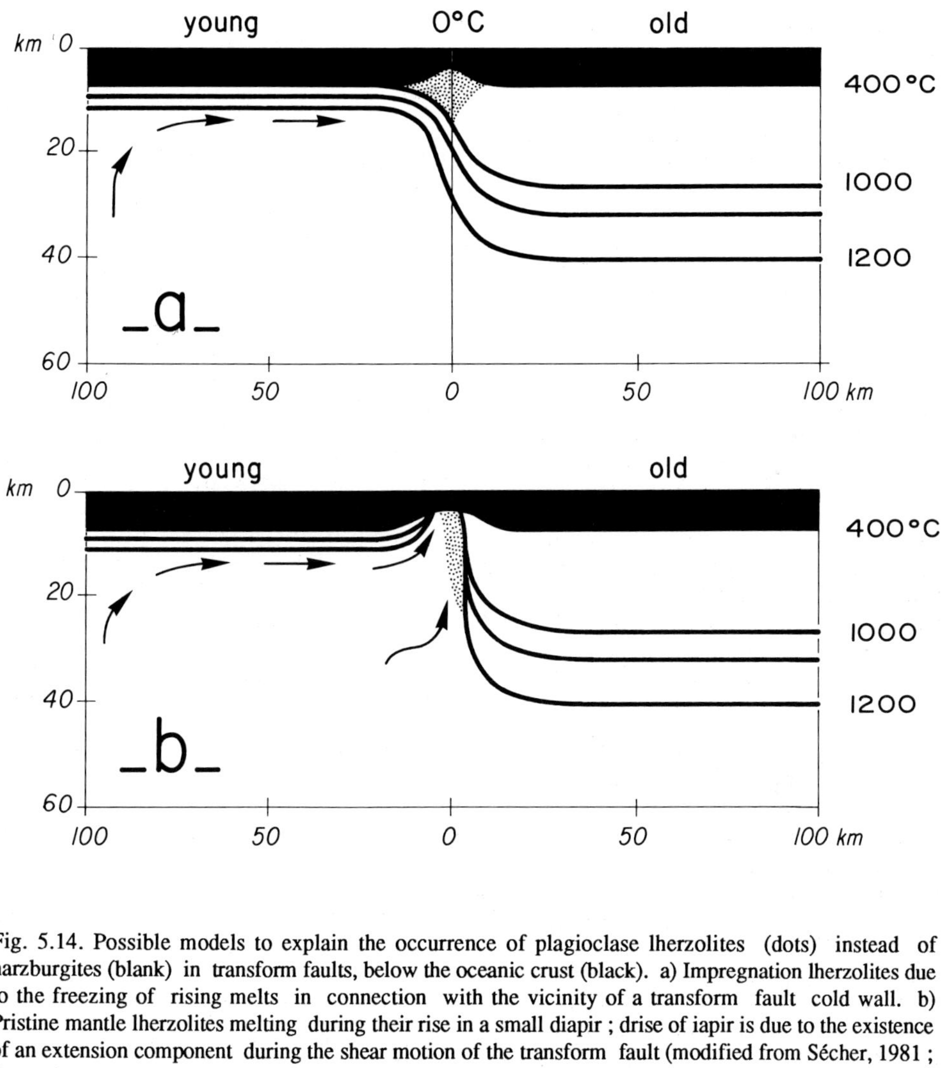


Fig. 5.14. Possible models to explain the occurrence of plagioclase lherzolites (dots) instead of harzburgites (blank) in transform faults, below the oceanic crust (black). a) Impregnation lherzolites due to the freezing of rising melts in connection with the vicinity of a transform fault cold wall. b) Pristine mantle lherzolites melting during their rise in a small diapir ; drise of iapir is due to the existence of an extension component during the shear motion of the transform fault (modified from Sécher, 1981 ; thermal structure in figure 5.14a from Forsyth and Wilson, 1984).

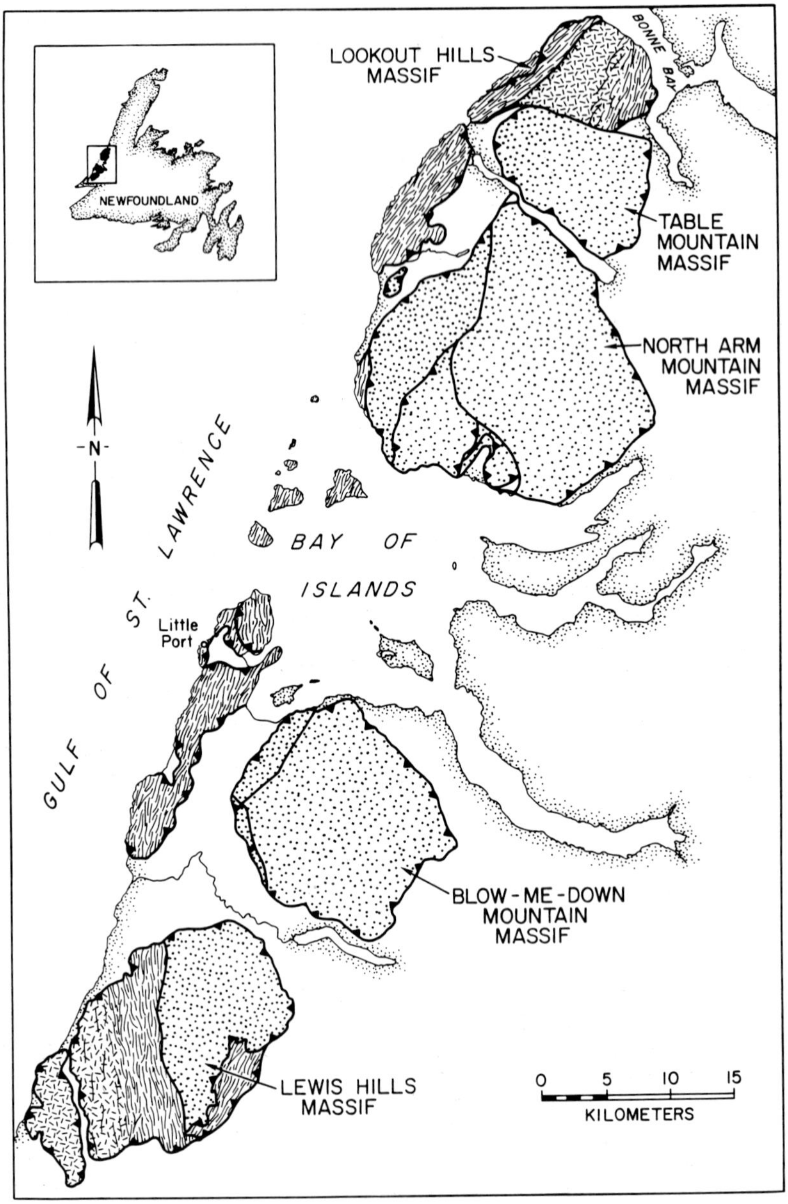


Fig. 5.15. Bay of Islands ophiolites. The ophiolite thrusts lie upon allochthonous Cambro-Ordovician sedimentary rocks (blank). Bold lines are major faults and thrust faults are indicated by barbs on the upper slices. Partial ophiolite sequences (random dashed pattern) grade into highly deformed assemblages (discontinuous lines) in the Coastal Complex. Solid lines with marks indicate the igneous contact at the western edge of the Bay of Islands complex (heavy stippled). (Karson, 1984).

Dewey (1978), and more specific problems regarding ultramafic formations or the seismic structure, by Karson et al. (1983) and Karson (1982) respectively.

### *5.3.2. Geological setting*

The Coastal Complex constitutes a linear band 2 to 7 km wide and about 100 km long along the western edge of the Bay of Islands ophiolite complex of western Newfoundland (fig. 5.15). This ophiolite, isotopically dated between 501 and 508 Ma (Mattinson, 1976 ; Jacobsen and Wasserburg, 1979), was obducted over a Cambro-Ordovician continental margin during the middle Ordovician, around 460 Ma ago (Strong et al., 1974 ; Williams, 1975). Preceding the obduction, a phase of oceanic thrusting, producing a mylonitic deformation in basal peridotites (Girardeau and Nicolas, 1981) and below, a high grade metamorphic aureole (Williams and Smyth, 1973 ; Malpas, 1979 ; Jamieson, 1981), has been dated at 469 Ma (Dallmeyer and Williams, 1975). The oceanic thrust-obduction phase has been related to the activity of a subduction zone dipping East. The ophiolites are considered by many authors as deriving from the fore-arc basin between the trench and the Notre Dame island arc massif (Strong, 1973 ; Strong et al., 1974 ; Casey and Dewey, 1984), but other interpretations have been proposed (Searle and Stevens, 1984).

The Bay of Islands ophiolite is a classical one, evocative of the Oman ophiolite. It has been extensively studied (Williams, 1973 ; Malpas, 1977 ; Church and Riccio, 1977 ; Casey et al., 1981 ; Girardeau and Nicolas, 1981). The sequence of units is complete and little dismembered, with a relatively flat-lying general attitude. The well developed crustal section has in particular a thick layered gabbro unit. The ultramafic section is composed of harzburgites except in the transition zone where locally the dunites are exceptionally thick, and along the base of the complex which is marked by lherzolitic facies . Their mylonitic texture contrasts with the high-temperature texture of the overlying harzburgites ; a similar deformation extends into the high-grade metamorphic aureole just below the mylonitic lherzolites (Malpas, 1979, Girardeau and Nicolas, 1981). From the attitude of the diabase sheeted dikes, it has been concluded that the ophiolite originated from a ridge striking WNW-ESE, in present coordinates and probably from its northern flank (Casey et al., 1983). A mid-oceanic situation has been proposed from the MORB affinity of the lavas (Casey et al., 1985).

### *5.3.3. Description*

The contrast between the simpler Bay of Islands structure and that of the Coastal Complex can be seen in the Lewis Hills massif (fig. 5.16). There the flat-lying formations from the transition zone of the Bay of Islands ophiolite abut against the steaply dipping deformed metamorphics of the Coastal Complex.

Along this contact, plutonic rocks of the Bay of Islands Complex display a distinctive fragmental facies of polymict igneous conglomerates produced by disruption and redistribution of previously crystallized layered gabbros. The map-scale character of this section is also peculiar in its unusual thickness of dunites (3 km) interpreted by Elthon et al. (1982, 1984) as cumulates (see § 10.4) and in the presence of layered mafic/ultramafic megalenses within the dunites.

It is also in the Lewis Hills massif of the Coastal Complex that the deepest structural levels with the widest extension (10 km) are exposed. The following description is borrowed from Karson (1984). The westernmost part of the Lewis Hills massif is comprised of a 5-6 km wide assemblage of mainly gabbroic rocks that include isotropic and layered gabbros, greenschist facies metagabbros, trondhjemite bodies up to a few

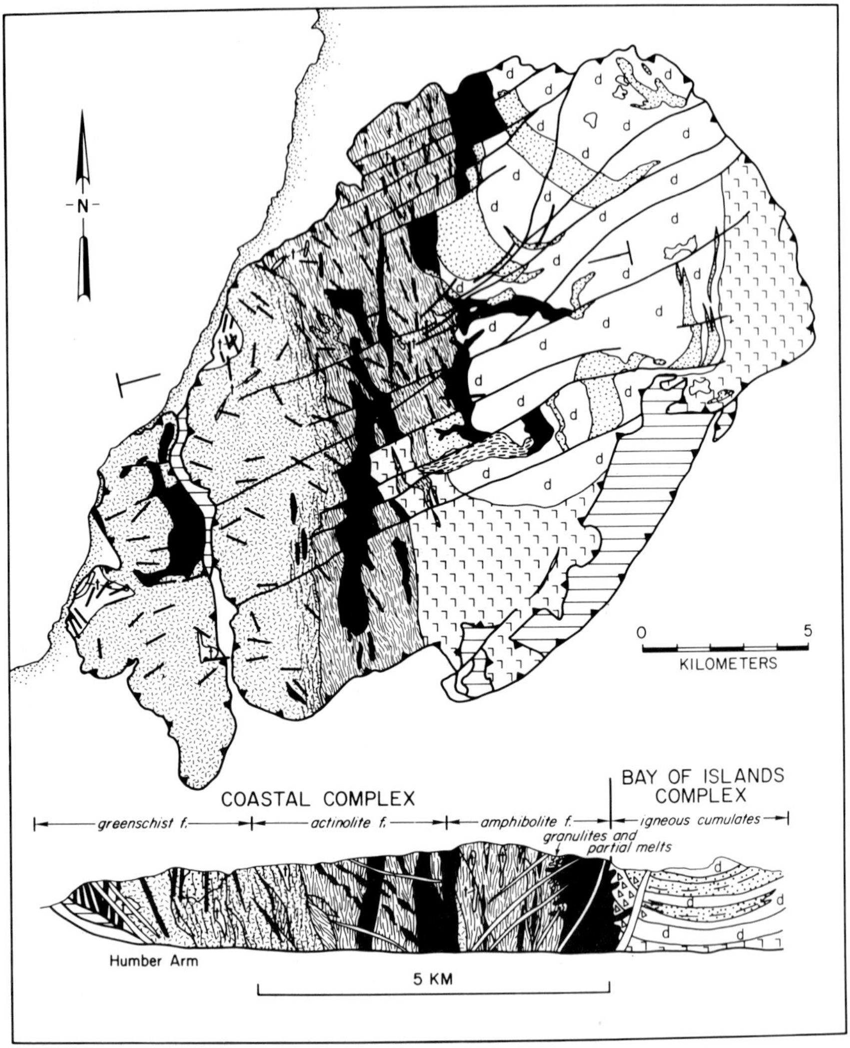


Fig. 5.16. Geological map and cross-section of the Lewis Hills Massif. Map symbols as follows : inverted L : harzburgite ; d : dunite ; stippled : interlayered gabbros and wehrlites ; bold dashed lines : mylonitic wehrlites and dunites ; random dashed lines : gabbros and metagabbros ; black : peridotitic crystal mush intrusions ; short bold lines : representative dike orientations ; blank with bold lines : sheeted dike complexes ; t : masses of trondhjemite to quartz-diorite with mafic inclusions (variably deformed and metamorphosed with amphibolites in the northern part of the Lewis Hills) ; discontinuous line pattern : highly deformed amphibolites and mafic gneisses ; v : basaltic to silicic volcanic rocks and minor sediments ; inverted v : alkalic volcanic rocks and volcano-clastic sediments ; open dots : basaltic pillow lavas ; s : slaty argillite and melange ; horizontal lines : small ophiolitic slices underlying the main allochthons ; blank : Humber Arm Supergroup sedimentary rocks ; bold lines : faults (barbs on upper slice for thrust faults) ; wavy line : brecciated non-conformity. Symbols are the same in the cross-section except : open triangles show position of fragmental cumulates in the Bay of Islands Complex and open lines show latest generation of mafic dikes. (Karson, 1984).

tens of meters across, and vertical diabase dikes. The dikes occur individually or in sheeted swarms and strike dominantly ENE. This lithological assemblage is typical of mid-crustal levels of the nearby Bay of Islands Complex. In contrast to the latter, numerous plugs and larger megadikes (up to a few hundreds of meters across) of peridotite intrude the gabbroic rocks and are themselves cut by diabase dikes.

To the East, these lithologies become progressively more deformed and metamorphosed over a lateral distance of 3 to 4 km. The first signs of strain are the appearence of isolated actinolite facies shear zones, up to about 1 m wide. Further to the East, these lithologies grade into strongly lineated and foliated amphibolites. The strong linear fabric is defined by elongate hornblende crystals and stretching of initially equant igneous features. These lineations plunge gently to the NW. Deformed diabase dikes in this region strike NW.

The easternmost 1 km consists of an extremely complex belt of clinopyroxene-bearing amphibolites and subordinate granulite facies gneisses. Throughout most of the area, highly deformed igneous layering and dikes are still recognizable. The layering dips steeply to the East and the dikes strike NNW. These are cut by vertical, NNW-striking, high-strain zones with parallel metamorphic layering. The well developed stretching lineation plunges very gently to the NNW. Despite the lack of obvious deformation, some mafic dikes in this area have amphibolite to granulite facies mineralogies. This, and the observation that many strongly lineated rocks have almost optically strain-free mineralogies, suggest a relatively late, high-temperature, static thermal event. In some places, extremely complex migmatitic rocks occur. These generally have numerous undeformed ptygmatic folded leucocratic veins in a strongly lineated, mafic granulite host. Throughout this domain, outcrop-scale relations indicate a complex history of syntectonic magmatism and polymetamorphism. For example, many outcrops display multiple generations of cross-cutting igneous features and shear zones. The northernmost part of the area is exceptional in that it includes a 2-3 km wide area of highly deformed amphibolite facies metagabbros to metadiabases invaded by a variably deformed trondhjemitic net-vein assemblage which includes some trondhjemite bodies up to a few tens of meters wide.

The contact between the Coastal and the Bay of Islands Complexes is obscured by a family of ultramafic intrusions up to a few hundreds meters wide (Karson, et al., 1983). Undeformed diabase dikes that strike ENE cut across the contact between the two complexes and the high-grade metamorphics of the Coastal Complex.

North of the Lewis Hills a series of correlated massifs are aligned along a NE direction. They correspond to progressively shallower sections in the same type of sheared crust (fig. 5.17). The nature of the deformed rocks varies from predominantly gabbros in the Lewis Hills to volcanics, minor volcano-clastic sediments and serpentinites in the northern massifs.

Compared to the Lewis Hills, the metamorphic grade decreases from granulite to amphibolite-greenschist facies in the northernmost Look-out Hills massif. Simultaneously, the deformation becomes less penetrative and less homogeneous, tending to concentrate in retrograde greenschist facies shear zones enclosing less strained amphibolite facies lensoid masses.The motion between these lenses involves substantial vertical shuffling superimposed on the general dextral shear affecting the entire belt. Possibly due to the same cause, the strain in Look-out Hills is partitioned into two nearly parallel belts (2 to 3 km wide) separated by a much less deformed layered section (8 km wide). Finally,differences exist as to the nature of syntectonic intrusive rocks. Apart from the ubiquitous diabase dikes injected in several episodes, the main syntectonic intrusives in Lewis Hills are ultramafic bodies which in the northern massifs are transformed into sheared serpentinites. In these latter massifs, the main intrusions are abundant felsic bodies up to several kilometers wide. They have been emplaced during the general shear motion as shown by their locally gneissic texture together with the preservation of intrusive contacts against the tectonites. They are extremely heterogeneous in composition, including diorite, quartz-diorite and tonalite and could derive from the anatectic melting of the lower crustal formations. This interpretation is deduced from the occurrence of variably deformed net-vein assemblages of felsic migmatites in Lewis Hills. Alternatively, they could represent highly fractionated magmas from a spreading- center source. This question is discussed further below.

### *5.3.4. Petrology and geochemistry*

There is a large amount of literature on the petrology and geochemistry of the Bay of Islands ophiolites (Irvine and Findlay, 1972 ; Williams and Malpas, 1972 ; Malpas, 1977, 1978 ; Church and Riccio, 1977 ; Suen et al., 1979 ; Jacobsen and Wasserburg, 1979 ; Coish and Church, 1979 ; Siroky et al., 1985 ; Elthon et al., 1982, 1984) with a few papers dealing more specifically with the Coastal Complex (Elthon et al., 1986 ; Smith and Elthon, 1988). These studies generally point to an origin of this ophiolite at a mid-oceanic ridge, although the island arc environment has also been envisaged (Upadhyay and Neale, 1979 ; Dick and Bullen, 1984). This subject has been since reviewed by Casey et al. (1985), who argue convincingly for the mid-oceanic interpretation.

Considering the continuity of the Coastal Complex with the Bay of Islands complex, a similar environment should be retained for it. This is also proposed by Casey et al. (1985), who discuss Malpas' (1979) and Searle and Stevens' (1984) earlier conjectures of an island arc origin. Considering the large amount of plagiogranites within the northern part of the Coastal Complex, these latter authors had proposed that the Complex represented a deformed island arc assemblage.

This indeed correlates with the principal petrological singularity of the Coastal Complex ophiolite, namely the abundance of differentiated magmatic intrusions of trondhjemite-plagiogranite, diorite and tonalite. Their relation with the high-T migmatitic metamorphism evoked above, has lead us to interpret them as the anatectic products of this metamorphism. More recently, Elthon et al. (1986) have described two groups of dikes in the diabase dike unit of the Coastal Complex. One group, whose intrusion is prekinematic with respect to the Complex deformation, is similar to mid-oceanic ridge basalts. The other group, whose intrusion ranges from pre- to postkinematic, has a peculiar geochemical signature, being altogether silica-undersaturated and depleted in incompatible trace elements. This second group is regarded as the product of remelting of the layered gabbros and troctolitic rocks of the lower crustal unit. Although rocks of the plagiogranite stem are in higher proportions in present-day fracture zones than in adjacent oceanic areas as pointed out by Casey et al. (1985), this is not the case for the silica-undersaturated diabases which are not known in this environment. Elthon et al. (1986) relate this to their emplacement in the lower crustal section of the Lewis Hills, a horizon which is not sampled in present-day fracture zones.

The problem of the heat source for these anatectic processes is discussed below.

### *5.3.5. Interpretation*

The internal structure of the Coastal Complex has been interpreted (Karson and Dewey, 1978) as the result of the juxtaposition of two segments of oceanic

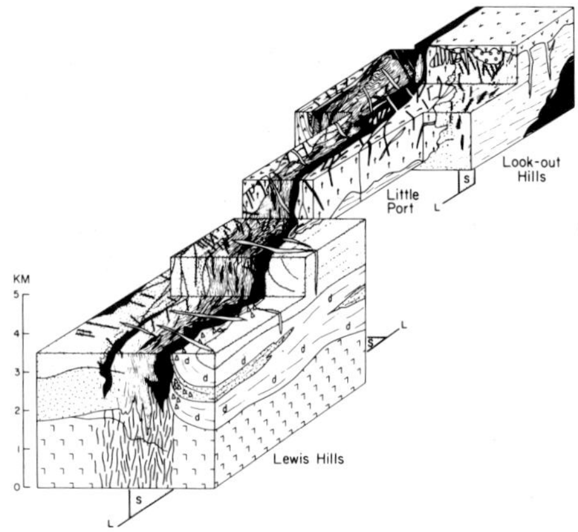


Fig. 5.17. Schematic block diagram depicting geological relationship between the three exposures of the Coastal Complex. Each exposure represents a sample of the crust derived from different crustal depths and lateral positions along the obducted fracture zone assemblage. The vertical sections overlap significantly, and therefore different exposures of the same crustal level are exposed at different points. From this reconstruction both vertical and lateral variations may be determined. Symbols are the same as in figure 5.16. (Karson, 1984).

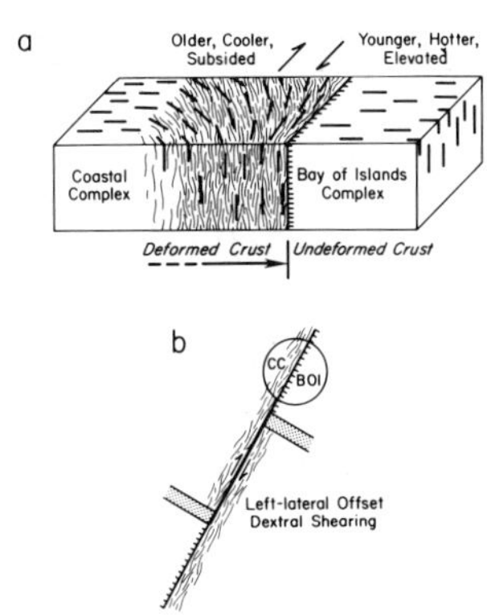


Fig. 5.18. Schematic diagrams showing the general relationship between the Coastal Complex and the Bay of Islands Complex. a) Block diagram illustrates how undeformed crust grades laterally into a steeply dipping high-strain zone (discontinuous lined pattern). The Bay of Islands Complex is undeformed and has an igneous contact (ticked line) against the Coastal Complex. Short bold lines show dike orientations (assumed to be ridge-parallel where undeformed) in both complexes relative to the contact between them. b) Left-lateral offset transform fault (bold solid lines) between spreading axes (stippled). Non-transform extensions are contacts between older, deformed lithosphere (discontinuous lined pattern) and younger undeformed lithosphere (solid line with ticks on younger side). Asymmetric deformation, dextral strike-slip history, and younger side to the East relationships are consistent only with the region circled in the left-hand diagram. (Karson, 1984).

lithosphere of somewhat different age at a ridge-transform intersection (fig. 5.18).

The western part of the Coastal Complex is a segment of an older oceanic lithosphere which is increasingly deformed and metamorphosed Eastward. The undeformed WNW-ESE dikes of the western part are assumed to have an orientation approximately parallel to that of the spreading center as it is defined by the attitude of the sheeted dike unit in the Bay of Islands Complex. Eastward, the dikes rotate nearly 90° in a clockwise sense into parallel trend to the shear belt. The trend of foliations across the deformed region has an asymmetrical fanning relationship with respect to the contact between the Coastal and the Bay of Islands Complexes, similar to that of one-half of a ductile shear belt (fig. 5.18). This and other observations on the sense of offsets across mesoscopic-scale shear zones are all consistent with a dextral shear sense across a steeply dipping shear belt. The nearly horizontal trend of stretching lineations indicates a dominantly horizontal direction. All this suggests a dextral transform fault origin for this shear belt and would locate the Coastal Complex with respect to Bay of Islands as shown in figure 5.18b. However, in contrast with what is observed in Bogota (§ 5.2.3) and predicted in oceanic fracture zones (Fox and Gallo, 1984), the rotation of the diabase dikes in the vicinity of the transform fault would be induced by their straining and would not be under the control of the rotating principal stress direction.

In the eastern part of the Coastal Complex and in its deepest crustal levels observed in Lewis Hills the metamorphism may have attained the stage of anatexis ; the felsic intrusions observed at shallower depths may be rooted in this zone. The high temperature syn- to post-tectonic metamorphism and the ultramafic intrusions of the eastern part of the Coastal Complex are ascribed to the contact, at the ridge-transform intersection, of this older and deforming crust with the magma chamber below the ridge where the younger Bay of Islands ophiolite was being formed (fig. 5.18).

Finally, several features evoke those described in oceanic fracture zones such as the serpentinite protrusions observed in the northern massifs of the Coastal Complex, where the crustal section could also be abnormally thin, the occurrence of late dikes emplaced after these protrusions and their composition which grade from mafic to silicic. The silicic dikes are probably related to the felsic intrusions located below. As in the Bogota district of New Caledonia (§ 5.2.3) and in oceanic fracture zones, high temperature hydration is pervasive ,in the shear belt, in contrast with the surrounding undeformed domains which are dry.

Fig. 5.19. Structural map of the Wadi Tayin central area (scale : 1/400,000). Trajectories and direct representations of a) lineations and b) foliations ; colored lines : HT deformation ; blank lines : LT deformation. Decreasing color tones from extrusives and sheeted dike units to plutonic gabbros and tectonic peridotites. Moiré in gabbros : wehrlites, and in peridotites : dunites (Nicolas et al., 1988).

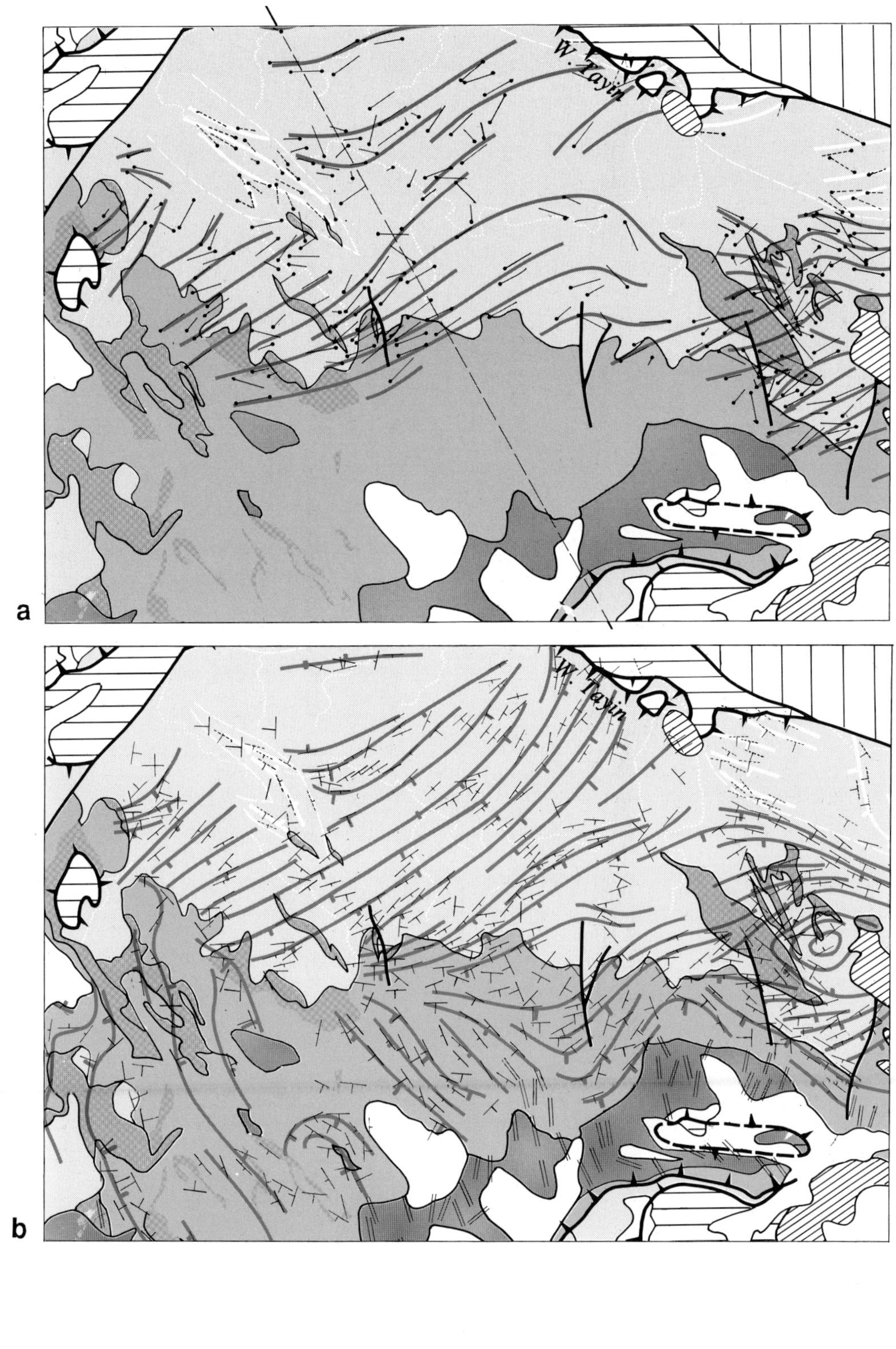
W. Tayin
a
W. Tayin
b

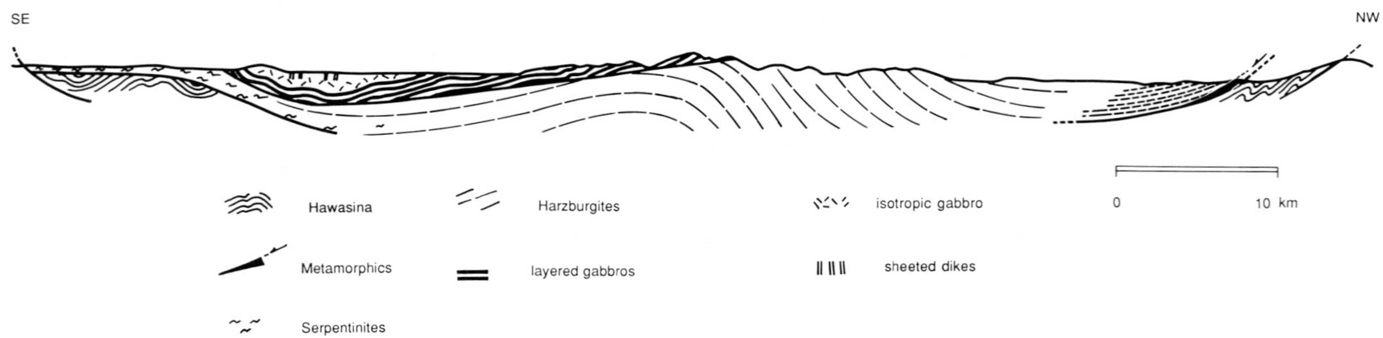


Fig. 5.20. Cross section through the Wadi Tayin massif, showing the rotation of HT foliation within the transform domain (location in fig. 5.19).

## *5.4. WADI TAYIN MASSIF IN OMAN*

### *5.4.1. Introduction*

Compared to the two examples treated above, the central part of the Wadi Tayin massif in Oman presents both the ultramafic and mafic units and differs structurally in a striking way, although it has also been interpreted as formed in a transform fault environment (Misseri, 1982 ; Ceuleneer et al., 1988 ; Nicolas et al., 1988).

The setting of the Wadi Tayin massif in the Oman ophiolite is described in chapter 3, and more detailed descriptions can be found in the 'Oman ophiolite' volume 86 of the Journal of Geophysical Research (1981). In particular, Boudier and Coleman's (1981) cross section throughout the Wadi Tayin massif closely followed the presumed transform domain. We report here only on the characteristics considered as specific of a transform environment. They are observed in the central part of the Wadi Tayin massif at some 20 km from both the Batin mantle diapir in the eastern part of the same massif and the Maqsad diapir in the Sumail massif (figs. 3.8, 5.22).

### *5.4.2. Structural description*

The structural map (fig. 5.19) and cross section (fig. 5.20) in the central part of the Wadi Tayin massif shows a 20 km thick zone in the peridotite section, where the foliations are oriented NE-SW. This is at right angles to the presumed NW-SE ridge axis as determined by the regional sheeted dike trend (fig. 5.21e). These foliation planes dip steeply to the SE (fig. 5.21a) ; lineations are sub-horizontal, indicating a strike-slip movement (fig. 5.21b). The shear direction, deduced from the obliquity between shape and lattice fabrics (§ 2.5.4), is consistent along the full length of the zone and indicates a sinistral sense, also confirmed by the rotation of the foliation on each side of the zone. The harzburgite found here has the classical coarse-grained structure typical of asthenospheric deformation. Numerous dunite bands are parallel to the foliation. Gabbro dikes and plagioclase-clinopyroxene magmatic impregnations are abundant in the upper part of the peridotite section, being relayed by pyroxenite dikes, deeper. The sheared domain is also comparatively rich in chromite pods, some of which are found abnormally deep within the peridotite section (see § 10.5.2).

The foliation becomes progressively flat-lying in the top few hundred meters of the ultramafic section, and it is parallel both to the paleo-Moho separating this section from the overlying mafic section and to the layering plane in the basal gabbros (fig. 5.21c). Interestingly, the overlying layered gabbros show no sign of plastic deformation and do not differ from the gabbros of other sections in Oman, except that they are intensely strained in the viscous state with a very strong magmatic lineation which is parallel to the plastic flow lineation in the peridotite section (fig. 5.21d). Finally, the sheeted dike unit, regionally oriented NW-SE (fig. 5.21e), tends to rotate into a N-S strike in the vicinity of the considered domain.

### *5.4.3. Discussion*

The flow pattern recorded by the high temperature peridotites of the central part of the Wadi Tayin massif is exceptional. With the exception of the zones of diapiric ascent, we have concluded (§ 2.6) that the flow plane attitude is controlled by the thermal lithosphere-asthenosphere boundary orientation. In Wadi Tayin, the flow plane deduced from the foliation (§ 2.5.3) is surprisingly steeper than elsewhere in Oman and is at a right

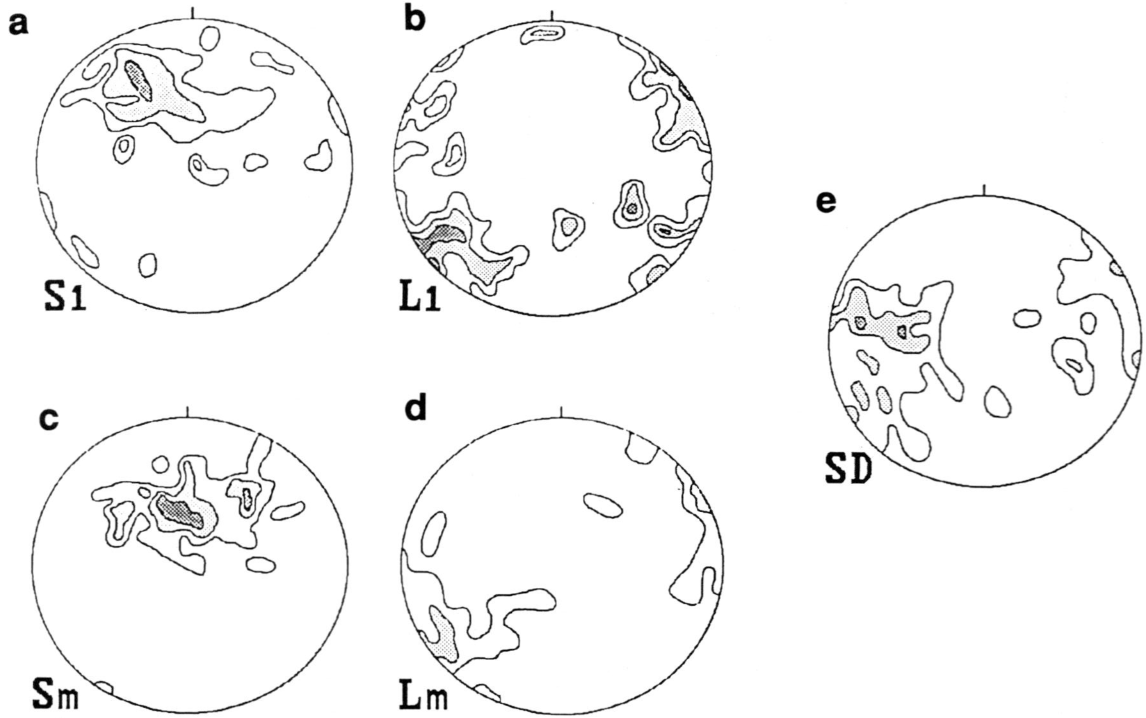


Fig. 5.21. Preferred orientations of the field structures in the Wadi Tayin transform structure, except 5.20e, measured outside this domain ; a) S1 high-temperature plastic flow foliations, 224 measurements ; contours : 1, 2, 5 %. b) L1 lineations in the ultramafic section, 90 measurements ; contours : 1, 1.5, 2.5%. c) Sm magmatic layering planes, 273 measurements (including Pallister and Hopson's data, 1981) ; contours : 1, 2, 3.3 %. d) Lm lineations in the overlying layered gabbros of the mafic section, 20 measurements ; contours : 5, 10 %. e) SD diabase dikes in the sheeted dike unit, 100 measurements ; contours : 1, 3, 6 %. Lower hemisphere, equal-area projection (After Nicolas et al., 1988).

angle to the ridge axis with a horizontal flow line, suggesting that the asthenosphere there flowed along a pre-existing lithospheric wall. Shearing at a right angle to the ridge trend suggests a transform origin (Misseri, 1982 ; Ceuleneer et al., 1988). The transform fault interpretation in the case of the Wadi Tayin shear zone presents a few difficulties. In New Caledonia (§ 5.2) and Antalya (Reuber, 1984), the two other massifs where the peridotite structure has been studied, the rotation of the mantle flow structures into a transform orientation coincides with the development of higher stress-lower temperature microstructures in the peridotites. Such a microstructural evolution does not occur in the Wadi Tayin shear zone where the deformation structures of the mantle peridotites remain of the asthenospheric type. The sheared domain is also 20 km across, that is substantially larger than in the other ophiolites ascribed to transform situations which have been described or referred to above. The foliations in Wadi Tayin are nowhere vertical as in all these transform faults ophiolites ; they keep a steady 60° SE dip (fig. 5.20).

If the Wadi Tayin asthenospheric shear zone actually represents a piece of upper mantle deformed in a transform fault, one must admit that the classical 'cold edge effect' (e.g. Sleep and Biehler, 1978) is attenuated here. The observed thermal configuration might be consistent with a fast-spreading transform zone (Fox and Gallo, 1984 ; Forsyth and Wilson, 1984). Young ages and a small difference in age on each side of the transform would explain the high temperature character of the deformation and the moderate slope of the cold wall. To explain the SE dip of foliations, the cold wall would have to be to the SE. Incidentally, this conclusion, related to the sinistral sense of motion in the shear zone, imposes a general scheme like that of figure 5.22.

Finally, a major contrast with other ophiolites considered as representative of transform faults is the evidence of a normal crustal accretion and the absence of plastic shearing in the crustal section. This indicates that the transform motion was accompanied by the extension component responsible for crust generation. The extension is also a feature which is more consistent with the broad transform zones found along fast-spreading ridges (e.g., Madsen et al., 1986) than with the narrow transform zones of slow-spreading ridges where the continuity of magma chambers is frequently impeded and crust generation considerably reduced (Fox and Gallo, 1984 ; Whitemarsh and Calvert, 1986 ; Potts et al., 1986). In this respect it should be noted that the Wadi Tayin shear zone coincides with a clockwise rotation of the sheeted dike complex. This pattern of sheeted dikes implies a dextral ridge offset which is also consistent with the scheme of figure 5.22. Finally, in the crustal section of central Wadi Tayin, the strength and orientation of magmatic lineations and the strike of the sheeted dike complex are the only evidence for the transform origin.

## *5.5. CONCLUSION*

### *5.5.1. The diversity of ophiolitic transforms*

In this chapter, we have described three ophiolites displaying evidence of origin in oceanic transform faults. The striking contrast emphasized just above between the Wadi Tayin and all the other ophiolites, presented or referred to in this chapter, seems to reflect different spreading rates in the oceans of origin. A fast spreading rate has already been proposed for the Oman ophiolite (§ 3.5.2) which seems to be confirmed by the characteristics of the Wadi Tayin transform ; these can be summarized by stating that, in both the ultramafic and mafic sections, there is no difference (same petrology and structures) between the transform domain and outside, except for the plastic flow pattern in the ultramafic section, the trend of magmatic lineation in layered gabbros and the local rotation of the sheeted dikes. In contrast, the other ophiolitic transforms would reflect slow spreading situations,

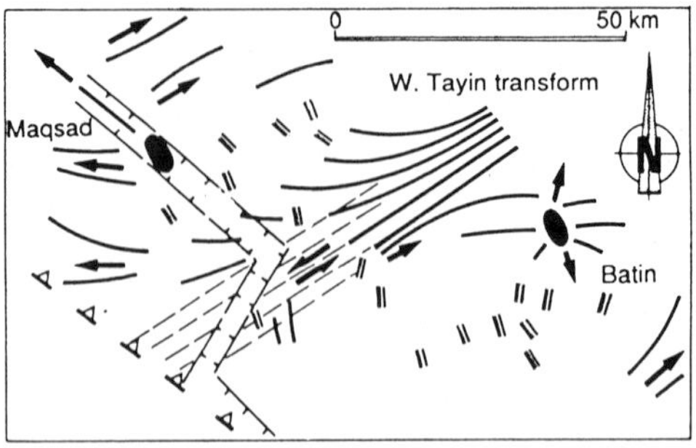


Fig. 5.22. Interpretative map of the Wadi Tayin transform zone. The full lines represent measured features, and the thin lines, inferred ones. Lines : flow trajectories with arrows pointing to relative motion of upper mantle compartment ; NW and SE black spots : Maqsad and Batin diapirs respectively ; short double lines: average orientation of sheeted dikes complex ; facing barbed lines : presumed ridge segments ; dashes with triangles : front of the nappe and presumed detachment locus (Nicolas et al., 1988).

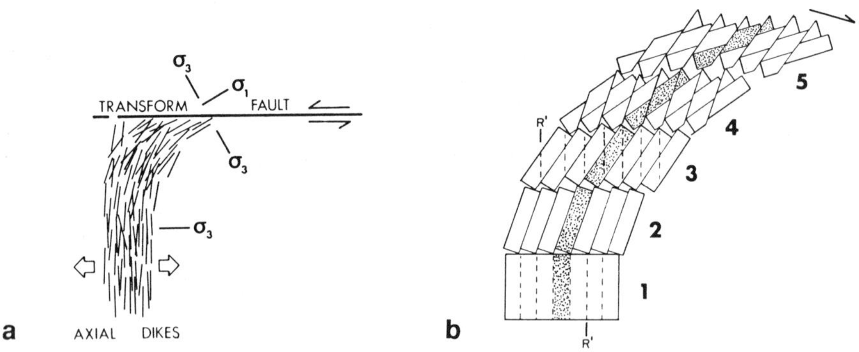


Fig. 5.23. Possible dike trend pattern near a ridge-transform intersection. a) Stress-controlled rotation in a sinistral transform : dikes progressively change trend because of interplay of ridge and transform stress patterns. b) Block rotation in a dextral transform : dashed lines equal incipient antithetic Riedel R' faults. Stippled blocks represent once continuous dike or block with shear strains as follows : Step 1, $\gamma = 0$, undeformed mass, incipient faults forming parallel to dike ; step 2, $\gamma = 0.34$ ; step 3, $\gamma = 0.70$ ; step 4, $\gamma = 1.2$ ; and step 5, $\gamma = 2.0$. Faults are assumed to 'lock' at an arbitrary angle of 66° to $\sigma_1$ (steps 3 and 5), assuming an angle of internal friction = 30°, and a shear direction at 75° to R' (Young et al., 1985).

a conclusion already proposed by Miller and Mogk (1987) for the Ingalls ophiolite. In such situations, the greater age difference between the two plates would explain the steep foliations, the low temperature plastic deformation and the strain localization. The more heterogeneous temperature field around the fault, compared to the Wadi Tayin case, may also create a more complex flow pattern (fig. 5.13). This could explain the more complex structure in the Tiebaghi-Poum-Belep district (fig. 5.8) than in Wadi Tayin (fig. 5.22). In the presumed slow spreading ophiolites considered in this chapter, the crustal section is highly sheared, suggesting the absence of an extensional component in the transform domain. Colder and expectedly thinner crust could favor water penetration down to the mantle as recorded by the hydrous character of some peridotites. Elsewhere, these colder conditions in the upper mantle could account for the presence of plagioclase lherzolites (§ 5.2.4).

### *5.5.2. Dike orientation in transform zones*

Contradictory interpretations regarding the mechanisms responsible for the sheeted dikes sygmoidal orientation across transform faults have been envisaged in this chapter. Two mechanisms have been proposed, with for a given orientation of dikes, opposite directions of displacement on the transform : local rotation of the stress field responsible for dikes opening in a new direction (fig. 5.23a), or tectonic rotation either of blocks (bookshelf tectonics) (fig. 5.23b) or, in ductile conditions, of individual dikes by shear transposition. Both mechanisms are supported by observations in active rifts, in oceanic fracture zones and in ophiolites. From field studies in rifts, Courtillot et al. (1974) in the Afar of eastern Africa infer a stress-related orientation of dikes and Young et al. (1985) in North Central Iceland, a block rotation (fig. 5.23b). From sonar and Seabeam observations in oceanic fracture zones, both interpretations are again inferred (see Searle, 1981, 1983 or Fox and Gallo, 1984 for the former and Cowan et al., 1986 for the latter). As seen in this chapter, dikes would track the $\sigma_1$ direction in Bogota and Wadi Tayin and be transposed by shear in Coastal Complex. The sheeted dike orientation in the Arakapas-Limassol Forest of Cyprus has been interpreted as stress-controlled (Varga and Moores, 1985 ; Murton, 1986; Murton and Gass, 1986) and, on the basis of paleomagnetic data, as controlled by tectonic rotation (Bonhommet et al., 1988).

Obviously both mechanisms can operate separately or in a combined fashion (Karson, 1987). The Bogota and Coastal Complex data indicate that in deep crust and upper mantle, the attitude of dikes is stress-controlled, provided that, after their intrusion, the intruded domain has not been strongly sheared, in which case dikes would be transposed toward the shear direction. Such dikes would be now strongly foliated. In shallower crust, it is suggested that dikes intrusive into young transform zones would preferentially track the stress field ; they would be cut by strike-slip faults and with aging, a decollement surface may appear at depth, possibly in serpentinized mantle, allowing block rotations within the transform domain.

# Chapter 6

# CANYON MOUNTAIN OPHIOLITE : POSSIBLE ORIGIN IN AN ISLAND ARC

## *6.1 . INTRODUCTION*

Until Miyashiro (1973), ophiolites were regarded as fragments of oceanic lithosphere generated at midoceanic spreading centers. Now island arc basalt signatures have been detected in many ophiolitic volcanics and the marginal basin-island arc origin for ophiolites has become increasingly popular (§ 8.3). Considering the importance of this issue, it seems useful to include in the description of selected ophiolite complexes, the Canyon Mountain case for which concerned authors postulate that the geodynamic environment of origin was an island arc. Striking differences appear with a typical ophiolite like Oman ; however, it would be premature to claim that they reflect the differences between ophiolites formed in an island arc and in a mid-oceanic ridge.

The Canyon Mountain complex in eastern Oregon has only recently been recognized as an ophiolite (Avé Lallemant, 1976 ; Thayer, 1978). This is not surprising since it possesses some singular features : the mafic sector is abnormally silicic and the overall structure is more complex than in typical ophiolites.

It is in this complex that pioneering studies on structures of the ultramafics were conducted by Thayer and his co-workers starting in 1956, who drew attention to flow structures in chromite pods (Thayer, 1963, 1964). One of the first structural maps in an ultramafic massif was also achieved here by Avé Lallemant (1976), since complemented by the work of Misseri (1982) and Misseri and Boudier (1985). Petrological and geochemical studies were performed by Himmelberg and Loney (1980), and Gerlach et al., (1981 a, b). Thayer (1978) and Avé Lallemant (1984) have reviewed the available literature on this massif.

## *6.2. GEOLOGICAL SETTING*

The Canyon Mountain ophiolite belongs to the Blue Mountains province of eastern Oregon comprising marine volcano-sedimentary formations from Devonian to Cretaceous with a complex subduction-island arc history. The province can be subdivided into four distinct terrains ; each of which has its own depositional and tectonic history (Valier et al., 1977 ; Dickinson, 1979). From North to South (fig. 6.1) they are : 1) Wallowa terrain, which could represent a Permian-Triassic island arc ; 2) Baker terrain, representing Paleozoic to Jurassic dismembered oceanic formations uplifted during the activity of a subduction zone ; 3) Izee terrain, derived from a Late Triassic to Late Jurassic forc-arc basin and ; 4) Olds Ferry terrain, a Triassic island arc formation (Avé Lallemand, 1984). Most ophiolite fragments including the Canyon Mountain ophiolites belong to the Baker terrain (fig. 6.1). All these formations are largely covered by Tertiary volcanic rocks.

The Canyon Mountain complex is the largest and most complete ophiolite body in the Blue Mountains, covering an area of 200 $km^2$ (fig. 6.2). Its northern and eastern contacts are steep faults. On the West, it rests tectonically on mélange formations without evidence of a metamorphic sole which would be indicative of emplacement by intra-oceanic

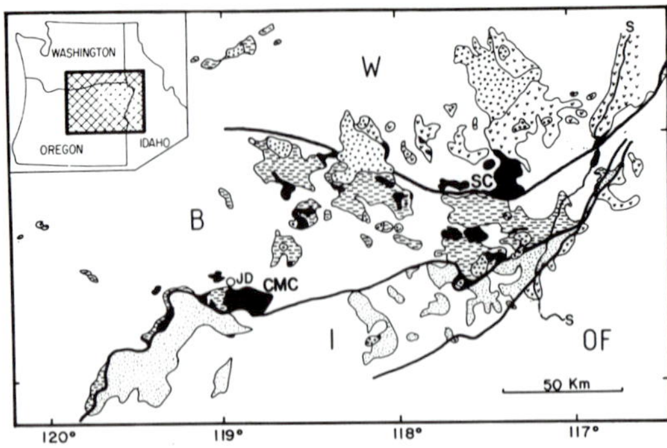


Fig. 6.1. Simplified map of NE Oregon. W : Wallowa ; B : Baker ; I : Izee ; OF : Olds Ferry. Black areas: ophiolites and fragments thereof (CMC : Canyon Mountain Complex ; SC : Sparta Complex) ; V-pattern: Permian and Triassic formations ; dashed pattern : chert-argilites ; finely-stippled pattern : Triassic and Jurassic flysch ; coarsely-stippled pattern : Jurassic-Cretaceous intrusives. J.D : John Day locality ; S : Snake River (after Dickinson, 1979 and Avé Lallemant, 1984).

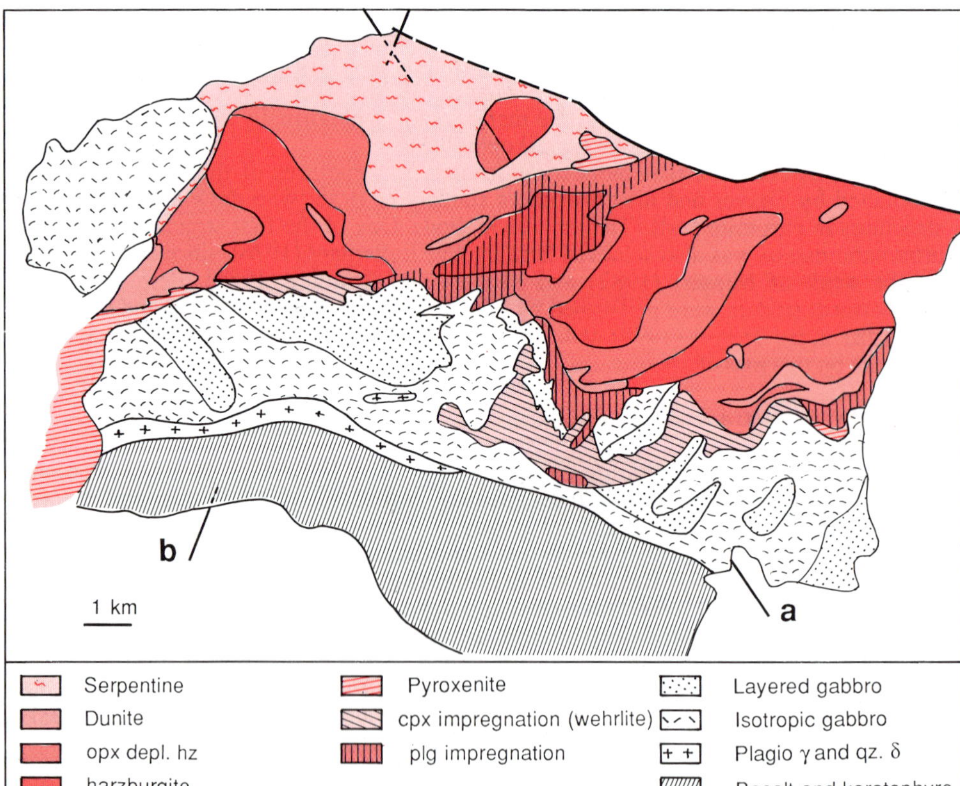


Fig. 6.2. Simplified petrologic map of the Canyon Mountain Complex . a and b) traces of the cross sections of figure 6.4. (compiled on the basis of works referred to in the text by Misseri, 1982).

thrusting. On the South, the complex is discordantly overlain by Tertiary volcanics (fig. 6.3). The formation of the Canyon Mountain complex has been dated at 278 Ma by U/Pb in zircons from gabbro (Walker and Mattinson, 1980) and plagiogranite (Walker, 1981) ; $^{40}Ar/^{39}Ar$ ages on amphiboles range from 269 to 262 Ma (in Avé Lallemant, 1984).

## 6.3. DESCRIPTION

Structurally, the highest formation exposed in the complex is what appears as a sheeted dike sequence. This unit consists of vertical E-W sheets of basalt or diabase, keratophyre and plagiogranite (fig. 6.6) which individually do not exceed more than a few meters in thickness and are mutually intrusive, although the keratophyres seem to be altogether older. The unit has a thickness between 0.8 and 4 km and a gross internal stratification (fig. 6.3.) ; the plagiogranite dikes are concentrated in a belt 500 to 1 200 m wide, next to the gabbro ; Southward, they are mixed in various proportions with keratophyres. The basalt sheets are widely distributed. Petrologically, this is a complex unit with hydrous alteration and deformation making recognition of original rocks types often impossible. The dominant facies is that of the keratophyres ; they consist of high temperature quartz and sodic plagioclase phenocrysts in a fine-grained trachytic matrix. The plagiogranite facies covers a series of $K_20$-deficient rocks, ranging from albite granite to hornblende diorite. Thayer (1978) mentions a progressive transition from keratophyres to plagiogranites.

No pillow lavas nor definitely extrusive tuffaceous material were found in this unit, leading Thayer (1978) and Misseri and Boudier (1985) to interpret it as a sheeted dike unit. However, Avé Lallemant (1976, 1984) interprets it as a sheeted sill unit, because the basalt sheets sometimes contain vesicles. This argument is not entirely convincing because vesicles can be present in shallow dikes.

North of this unit (fig. 6.4), occurs an E-W striking gabbro unit composed of layered, pegmatitic and isotropic gabbros. The layered gabbros are not systematically in contact with the ultramafics ; neither are the isotropic gabbros, which represent over 50 % of the mafic section, restricted to the vicinity of the sheeted complex. As stressed by Avé Lallemand (1976), the main difference in the gabbro unit is not to be found in a N-S cross section but between the eastern and western part of the unit (fig. 6.4). The layered gabbros are well developed in the West . They always contain clinopyroxene, basic plagioclase, and locally olivine or orthopyroxene and a brown magmatic amphibole.The only tectonites observed in the West are late-stage mylonites in shear zones equilibrated in amphibolite facies. There the transition to the ultramafic tectonites is abrupt, marked only by a band of olivine-websterite above chromite-bearing dunites. In the eastern part of the gabbro unit, the transition to the ultramafics is gradational with digitated facies of gabbro, olivine-websterite, wehrlite, dunite and harzburgite (see detailed map in Himmelberg and Loney, 1980). The gabbros are foliated and folded with a mineral lineation parallel to the fold axes. The layering is tectonically transposed. This deformation extends from high-temperature dry conditions, probably just subsequent to the gabbro crystallization, to conditions of the hydrous amphibolite facies. The local brecciation and intrusion of the gabbros by plagiogranitic melt occurred during the syntectonic hydrous contamination of the gabbros (Gerlach et al., 1981). Thayer (1978) provides evidence of such a water contamination at 600°-700°C in the upper ultramafic tectonites, by describing foliated and amphibolitized basaltic dikes in otherwise fresh harzburgites.

The northernmost unit is dominantly composed of harzburgites and serpentinites. The harzburgites have a comparatively heterogeneous composition (fig. 6.2) following bands subparallel to the foliation. These bands are variously enriched in clinopyroxene, plagioclase and locally amphibole ; on the contrary, they may be depleted in these minerals

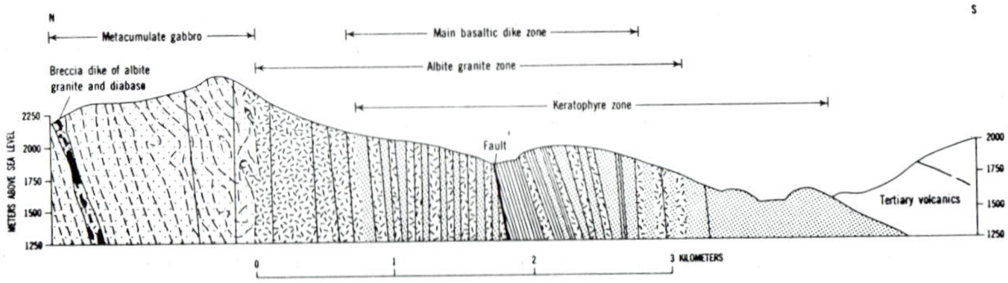


Fig. 6.3. Cross section through the contact between the gabbro and the dikes units in the Canyon Mountain Complex (Thayer, 1977).

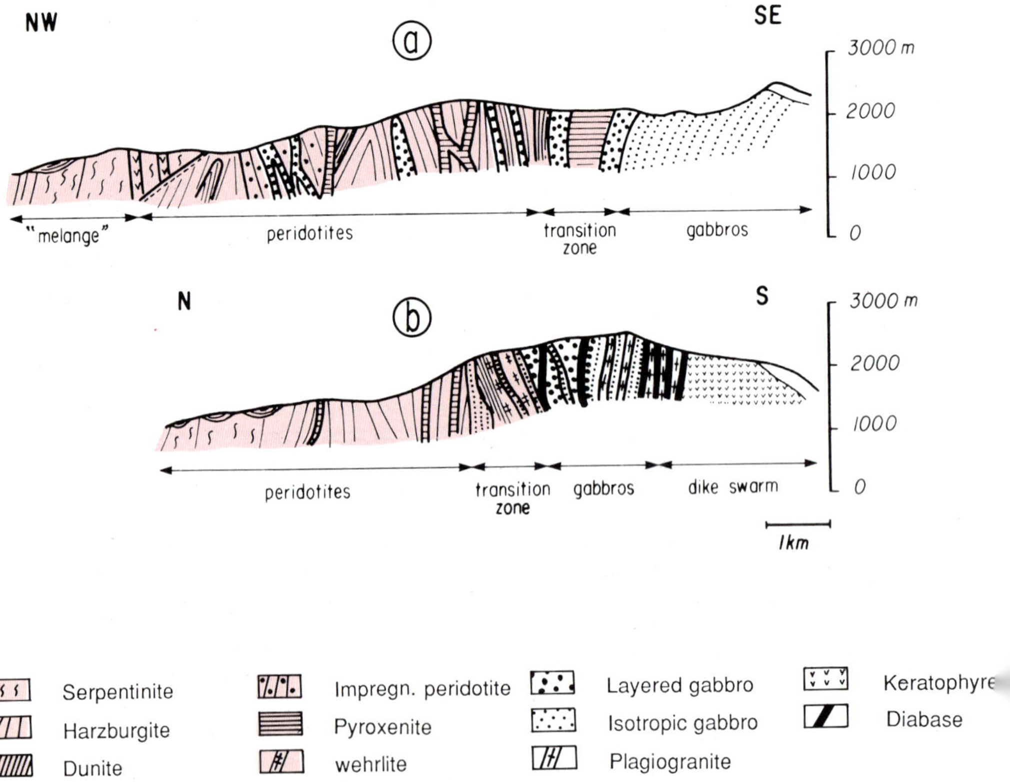


Fig. 6.4. Cross sections through the entire Canyon Mountain Complex (location in figure 6.2) showing the contrast between the eastern (a) and the western (b) domains (Misseri and Boudier, 1985).

and in orthopyroxene, thus trending towards dunites. As already mentioned, in the East the transition to the gabbro unit is gradational over a distance of several kilometers. Several chromite deposits have been mined from the dunites of this transition zone. On the basis of textural observations (§ 2.5.4), Boudier and Misseri (1985) interpret the heterogeneous composition of the harzburgites, with in particular the abnormal development of a banded transition zone and the existence of a plagioclase-rich zone in the Pine Creek area (fig. 6.2), as a local impregnation of harzburgites and dunites by a melt crystallizing clinopyroxene , plagioclase and amphibole. The melt-trapping process occurred at different times during the course of the high-T plastic flow affecting the peridotites, because locally the impregnation minerals present evidence of plastic deformation along with their matrix, while elsewhere they do not. This is indirect evidence demonstrating that the penetrative foliation and lineation in ultramafic tectonites have been produced by plastic flow at solidus or hypersolidus temperatures. Accordingly, the microstructures are well recovered and the petrofabric data (Avé Lallemant, 1976 ; Misseri, 1982) point to the activation of high-temperature slip systems in olivine ; large deformation in shear regime is also deduced from these data. At the NW margin of the massif, low-temperature mylonitic deformation is locally recorded.

Pyroxenite dikes, gabbro dikes and dunite veins are numerous in the ultramafic unit. The gabbro may contain a magmatic amphibole showing that the circulating magma was hydrated.

## *6.4. STRUCTURAL ANALYSIS*

The first structural map of the Canyon Mountain Complex including data in the ultramafic section has been published by Thayer in 1956 and is updated in his 1978 publication. A new map focusing on the solid state flow and metamorphic foliations has been published by Avé Lallemant (1976). Finally, a third map, by Misseri (1982) extends the preceding works mainly by incorporating more measurements of mineral lineation. Figure 6.5 represents the compiled results of these studies as a map of trajectories of planar structures; the lineations cannot be properly represented because they are steeply plunging, as shown by the stereonets of figure 6.7b and d.

As already seen in cross sections (fig. 6.4), the contacts between the main units are EW striking, and steep. The internal structures are discordant with these contacts except in the dike unit where the dikes are on average parallel to the contact with the gabbro unit (figs. 6.5 and 6.6). In fact, the dikes retain also this average orientation in the gabbro unit ; this is why dikes from both units have been plotted together in figure 6.6g. The gabbro dikes in the ultramafic section are steep, but present a dispersed azimuthal orientation (fig. 6.6e) and pyroxenite dikes in the same unit (figs. 6.6f and h) tend to orient subparallel to the host rock foliation, as a result of progressive rotation due to plastic strain.

The magmatic layering and foliation in the gabbros are often difficult to distinguish from the tectonically transposed layering and foliation of gabbros deformed at high temperature. Both types of structures have been grouped on the stereonets of figures 6.7c and d. The magmatic structures are mainly present in the western part of the unit (fig. 6.5), where they are generally steeply dipping and forming large scale folds with steep axes, like the tectonic structures. Interestingly, these tectonic structures are in continuity with those of the ultramafic unit, indicating that the same high-T deformations affected both units (fig. 6.5). The foliation in the peridotites also makes folds visible from the outcrop scale to that of the map. Their axes are parallel to the spinel lineation (figs. 6.7a and b). There are, however, domes where the foliation becomes horizontal. The complex structural pattern is sketched in figure 6.5. The overall structure evokes that of nested diapiric intrusions. In keeping with the analysis of folding in the Lanzo diapiric intrusion (Nicolas and Boudier,

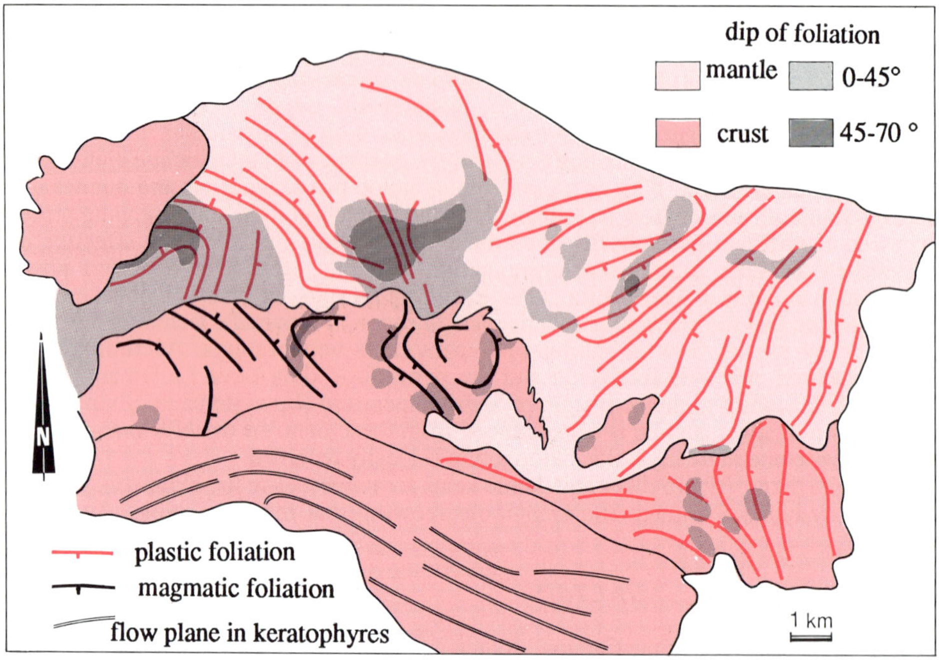


Fig. 6.5. Trajectories of plastic foliations (color) and magmatic foliations (black) in the Canyon Mountain Complex. Colors refer to the petrologic facies and superimposed grey tones, to the regional dip of foliations (compiled from Thayer, 1956 ; Avé Lallemand, 1976 ; Misseri and Boudier, 1985).

1975), the superposed folds observed in the field need not represent independent episodes of folding but, considering the persistence of the same physical conditions throughout the folding sequence, they are more easily interpreted as the result of the continuation of a heteregeneous deformation in rising diapirs.

## *6.5. PETROLOGY AND GEOCHEMISTRY*

Petrological description and mineral chemistry of the peridotite-gabbro units of Canyon Mountain are given by Himmelberg and Loney (1980). They confirm the ophiolitic signature of the complex. In conformity with the result of textural analysis, the local presence of wehrlite, lherzolite and plagioclase-bearing harzburgite is ascribed to a trapped mafic melt reacting with a residual harzburgite at low pressure.

Gerlach et al. (1981 a and b) have reported on the geochemistry, including trace elements in the various facies of the dike unit. Regarding the origin of the keratophyres which appear to be the original volcanic or hypovolcanic component of the mafic section of the complex, they conclude that the keratophyres are very similar to dacites from island arcs. These keratophyres could thus represent the subvolcanic portion of a young island arc. The plagiogranite series which is contemporaneous with the keratophyre series is ascribed to hydrous partial melting of the mafic crust, either in relation with sea water

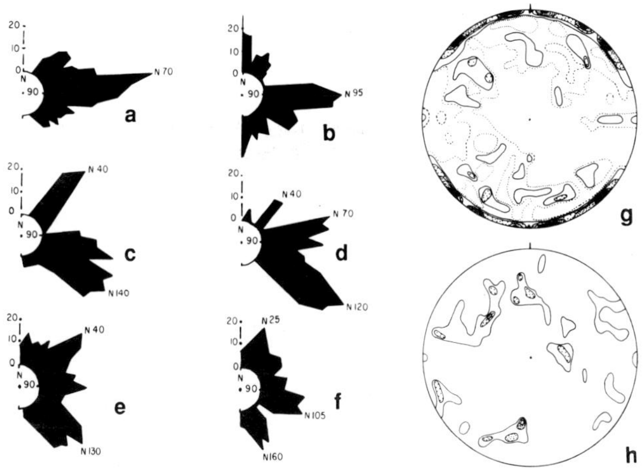

Fig. 6.6. Dike orientations in the Canyon Mountain Complex : a, b, c, d, e, f : rose diagrams ; g, h : lower hemisphere, equal-area projections. a) Aphyric diabase dikes (70 measurmements). b) Porphyric diabase dikes (23 measurements). c) Flow attitude in keratophyre dikes (25 measurements). d) Plagiogranite dikes (41 measurements). e) Gabbro dikes (49 measurements). f) Pyroxenite dikes (42 measurements). g) All dikes except pyroxenites (210 measurements). h) Pyroxenite dikes (50 measurements) (Misseri and Boudier, 1985).

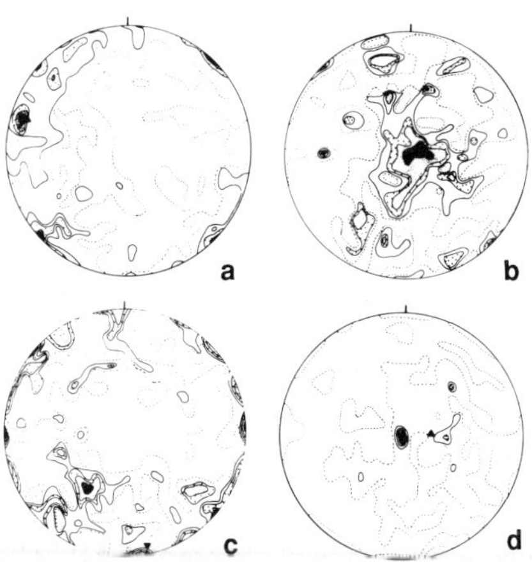

Fig. 6.7. Stereonets of the penetrative structures in peridotites and gabbros of the Canyon Mountain Complex. a) Foliation in peridotites (250 measurements). b) Spinel lineation in peridotites (210 measurements). c) Foliations and magmatic planes in gabbros (195 measurements). d) Tectonic and magmatic lineations in gabbros (134 measurements). Lower hemisphere, equal-area projection : contours at 1, 2, 3, 4, 10 % per 1 % area (Misseri and Boudier, 1985).

circulation in a cooling crust, or with water ascending through the island arc formation as a result of dehydration of the underlying subducted lithosphere. The presence of magmatic amphiboles in the harzburgites and gabbros is however a direct evidence of a water vapor or a water-rich melt percolating through the Complex formations, and tends to support the hypothesis of a subduction zone-related origin of the water.

## 6.6. DISCUSSION

### 6.6.1. Specific characteristics of the Canyon Mountain ophiolite

A few characteristics of the Canyon Mountain ophiolite are distinctive.

i) The main singularity is the keratophyric composition of the volcanic or hypovolcanic unit. Pene-contemporaneous with the keratophyres are the plagiogranite intrusives which here have an exceptional development. Both the field evidence and the geochemical characteristics point to the origin of the latter by hydrous melting of the mafic formations.

ii) Also remarkable is the importance of a magmatic melt impregnation in the harzburgites. This is responsible for their overall heterogeneous composition and possibly for the exceptional development of the transition zone in the eastern part of the massif. This melt was locally hydrated as shown by the occurrence of magmatic amphibole associated with the impregnation minerals. The hydrous metamorphism and anatexis in the mafic section may be related to the same source.

iii) Whatever the adopted structural model (see next section), the peridotite and gabbro structures are typical of diapiric intrusions penetrating through the crustal formations, tilting and deforming them, first penetratively at high temperature, next along shear zones at decreasing temperature.

### 6.6.2. Structural models

The structural data on Canyon Mountain are difficult to reconcile with the simple stratification of the ophiolite model. The simplest interpretation is that a 90° rotation around an WNW horizontal axis as proposed by Avé Lallemant (1976). The structure obtained would be that of mantle diapir intrusions with a western diapir stopping its intrusion at Moho depth and allowing the development of a magma chamber above it, and an eastern diapir rising through the crystallizing gabbros of this magma chamber. The main difficulty with this model is that the dikes of the sheeted dike unit, the flow lineations and the fold axes in the diapir are also rotated and become, on average, subhorizontal. It has also been noted (§ 6.3) that there is no evidence that the dike unit was initially a sill swarm mixed with flows.

Alternatively, the dike unit has been interpreted as a dike swarm, implying no or little rotations about an horizontal axis (Thayer, 1978 ; Misseri, 1982). The peridotites section has characteristics of nested mantle diapirs (Misseri and Boudier, 1985). The deformation cutting across the peridotite-gabbro contact would also be produced by a mantle diapir locally penetrating through the base of a crystallized magma chamber and deforming the layered gabbros. The steep westerly contacts in a domain where the gabbro unit is only deformed by secondary shear zones are more difficult to explain by this model. It is speculated that here the mantle diapir may have penetrated and tilted the floor of a magma chamber. This interpretation seems possible if this ophiolite forms within the crustal environment of an island arc.

### *6.6.3. Geodynamic environment of origin*

The regional environment and the petrological and geochemical characteristics of the hypovolcanic unit both indicate that the Canyon Mountain ophiolite originated in an island arc environment. Such an environment could explain the hydrous contamination visible throughout the complex as a consequence of dehydration taking place in the underlying subduction zone. This seems more probable than a per descensum hydration due to a spreading center hydrothermalism, since in this case, the water hardly penetrates in the mantle and when it does, it tends to generate a lower temperature metamorphism (§ 2.5.6).

The nested diapiric character of the mantle upwelling points to a small spreading rate (§ 9.3). In this case, comparatively fertile peridotites should be expected and not the depleted peridotites which characterize the Canyon Mountain Complex outside the zones of melt impregnation. The association with important chromite deposits also indicates a large degree of melt extraction (§ 10.5.2). This could be due to the hydrous character of the melting which, lowering the solidus temperature, would increase the melt production for a given temperature. It is also speculated that the low viscosity, estimated from the recovered substructure of the peridotites rising through the Moho surface, is due to the water weakening of olivine ductility (Mackwell et al., 1985). The lithosphere around the diapiric intrusions remaining thick would account, as in the LOT situation (§ 9.3), for the steep foliations measured in Canyon Mountain.

# PART III

# ACTIVITY OF OCEANIC SPREADING CENTERS AND THE ORIGIN OF OPHIOLITES

## *INTRODUCTION*

The variety of ophiolites described in part II evidently reflects a similar variety of oceanic situations where new lithosphere is created. The variety in oceanic situations arises from the effect of independent or combined factors such as spreading rate, geodynamical environment (for instance mid-ocean or back-arc ridge) and local conditions of spreading (vicinity of a hot spot, of a transform fault, tip of a propagating rift, anomalously elevated zone, etc.). One object of part III is to try to correlate the main ophiolite types described in part II with the main oceanic spreading situations. With the structural approach favored here, characterization of ophiolites will be based mainly upon geological and structural features. In this respect, the harzburgite and lherzolite ophiolite types are distinguished, and their characteristics are described in chapter 8. These two types can be related to the spreading rate, which is the most influential physical parameter of oceanic spreading. The distinction between oceanic environments, which is generally made on geochemical grounds, will nevertheless be discussed in chapter 8.

Whatever their differences, ophiolites possess several common features which reflect general processes also common to the spreading systems of origin. Structural studies of various ophiolites, including the ultramafic section, provide invaluable tools for understanding the general physical processes taking place at spreading centers. In chapter 7 we consider the melt extraction processes occurring in the mantle beneath spreading centers by adiabatic decompression. The mantle flow patterns beneath spreading centers are described in chapter 9. How they are coupled with the accreting crust and identification of the magmatic processes occurring in the critical transition zone between mantle and crust are discussed in chapter 10. Moving upsection in chapter 11, we finally consider the magmatic processes which give birth to a new crust.

# Chapter 7

# MELT GENERATION AND EXTRACTION IN MANTLE DIAPIRS

## *7.1. INTRODUCTION*

The oceanic lithosphere, its physical, petrological and geochemical nature, are largely a function of the magmatic processes which take place at oceanic spreading centers. An understanding of the processes of mantle melting and melt extraction is therefore central to models for the development of oceanic lithosphere. Such models may need to be adapted and modified because of the specific features of the accretion site : mid-oceanic ridge, island arc or back-arc environments, but the basic principles still hold. The processes of melt extraction and magma segregation need to be considered from both physical and geochemical points of view, taking into account the observations from the ocean basins, and from ophiolite complexes and the experimental results on peridotite and basalt compositions. The data on ophiolite complexes have an important role to play here since information on the deep structure and deep processes in the ocean basins is largely an extrapolation from surface observations. In ophiolite complexes there is the potential to extend these observations down to as much as 15 km into the underlying peridotitic mantle. Observations in such peridotite massifs can be complemented by those made in peridotite xenoliths from alkali basalts, of which some derive from the deep horizons where basalts are generated.

## *7.2. MELT EXTRACTION FROM THE ASTHENOSPHERE*

### *7.2.1. Conditions of adiabatic melting*

Melt extraction from the mantle can take place at various depths and not necessarily by a single physical mechanism. For instance, kimberlites appear to be generated at great depths by small degrees of melting in the presence of $CO_2$ and $H_2O$. In contrast, melting of the mantle below mid-ocean spreading ridges occurs at much shallower depths and is associated with much larger amounts of extracted melt. This latter type of melting is the main concern of this chapter. Essentially, sea-floor spreading implies upwelling of the asthenosphere ; melting, by adiabatic decompression, occurs as the peridotite solidus is crossed during uprise (fig.7.1).

During uprise, a low temperature solidus corresponding to melting in the presence of free $CO_2$ and $H_2O$ is first met at depths greater than 3OO km (Wyllie, 1980). The fraction of first melt thus produced is controlled by the fraction of available fluids in the melting peridotite. Various estimates of $H_2O$ and $CO_2$ fractions in the mantle where basalts originate have been made relying on experimental petrology (0.02 - 0.4 % $H_2O$, 0.4 % $CO_2$, Wyllie (1980)), on modelling following measurements in glassy basalts (0.1 - 0.5 % $H_2O$ ; Moore, 1970 ; Bryan and Moore, 1977 ; data reduced to 0.08-0.2 %, more recently, by Michael, 1988) and on stable isotope evidence (0.3 - 0.5 $H_2O$ ; Javoy et al., 1986). Considering that in the pressure interval of interest here, the solubility of water in melt is large and increases rapidly with pressure (20 % $H_2O$ at 1GPa and 50 % $H_2O$ at ≥ 3GPa ; McMillan and Holloway, 1987), it is clear that a water-saturated melt, obtained for a solidus below 1000°C (fig. 7.1), would represent less than 1% of the peridotite. In order

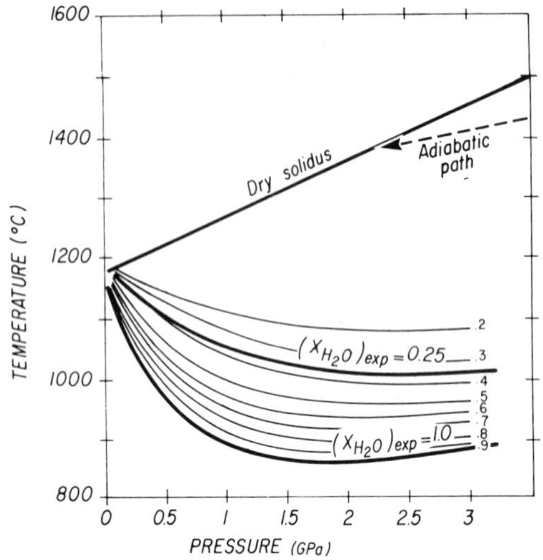


Fig. 7.1. Dry and wet peridotite solidi illustrating the large temperature drop associated with the melting in wet peridotites, and the temperature increase with decreasing mol fraction of $H_2O$ in the vapor ($XH_2O$). Thick curves are experimentally determined ($XH_2O$) and thin curves, extrapolated values (Mysen and Boettcher, 1975).

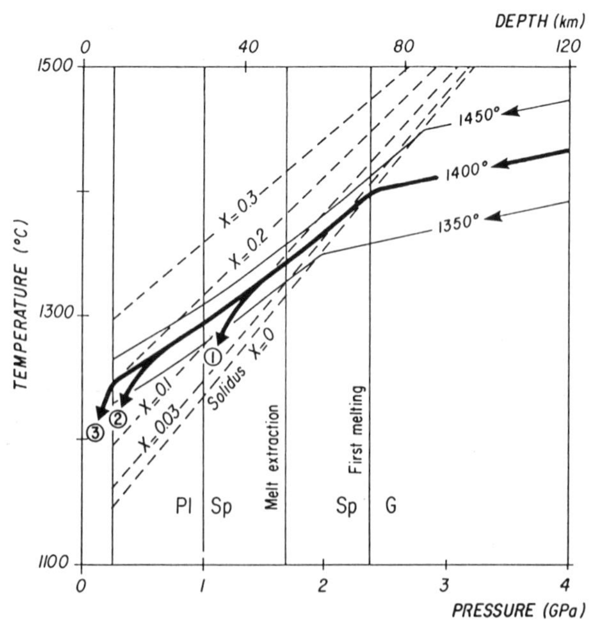


Fig. 7.2. P,T path (heavy line) of rising diapirs in McKenzie's (1984) lherzolite partial melting diagram. 1 - Spinel lherzolite path ; 2 - plagioclase lherzolite path ; 3 - harzburgite path (Nicolas, 1986a).

to produce alkali and tholeiitic basalts which result from an increasing melting fraction (8-20 % for a MORB chemistry according to Klein and Langmuir, 1987), the rising asthenosphere must intersect solidi corresponding to increasingly dry melting which are shifted to increasingly high temperatures (fig.7.1). Finally the *main tholeiitic melting can be considered as dry*. This conclusion has already been reached by Bottinga and Allègre (1978) and Presnall et al. (1979) using a similar reasoning. This simplifies the analysis of mantle melting, because the lherzolite dry solidus is well known (Takahashi and Kushiro, 1983) and because the melting mantle is a comparatively simple system, being composed of a four phases lherzolite. This conclusion may also have important bearings on melt connectivity, and thus on the amount of melt stable in a rising mantle diapir (see § 7.3.1). With the intersection of the mantle diapir adiabatic path with this solidus taking place at shallow depths (fig.7.1), there is a possible control on the melting processes by inspection of peridotite massifs and xenoliths carried by basalts.

### *7.2.2. Asthenospheric path and the meeting with lithospheric conditions.*

The asthenosphere is equated here with the convective upper mantle. Heat is carried solely by convective flow, meaning that the system is adiabatic and that therefore temperature follows an adiabatic gradient with depth :

$$\left(\frac{\delta T}{\delta z}\right)_S = \frac{g\,\alpha\,T}{Cp} \approx 0.3°K/km$$

at the considered shallow depths with T, z, g, α, Cp being the temperature (°K), the depth, the gravitational constant, the thermal expansion and the specific heat at constant pressure, respectively.

The lithosphere considered here is the thermal lithosphere which corresponds to those superficial horizons through which heat is transported by conduction. Due to the sluggishness of diffusive processes, conduction is a far less efficient way to carry heat than convection, resulting in much steeper thermal conduction gradients. In the following discussion, we adopt the simplistic view that the temperature in the asthenosphere considered is entirely defined by the vertical adiabatic gradient. Large horizontal gradients, like those expected in the vicinity of a mantle plume or of a subducting slab are ignored (for a more complete discussion of these situations, see Klein and Langmuir, 1987). In this perfectly adiabatic mantle, whatever its depth of origin, a mantle diapir crosses the lherzolite dry solidus at the same depth (fig.7.1). Its subsequent P,T path is modified due to the latent heat of fusion, with as a result a steeper slope, less steep, however, than those of the solidus and melt fraction (X) curves : $(\delta T/\delta z)_{S,X} \sim 3°/km$ which in figure 7.2 are taken from McKenzie's (1984) calculations. Thus, the diapir keeps rising with increasing degrees of melting until it meets the lithosphere front. The rising may continue above this front, although it is increasingly slowed because of the strong T-dependence of the peridotite rheology (Nicolas and Poirier, 1976), but the melting rapidly ceases because the new conductive gradient is even steeper than the melting curves (fig.7.2). Thus, depending on the depth where the lithosphere is met, the degree of partial melting, the composition of the melt (Allègre and Bottinga, 1974) and the residual nature of the melted mantle will be different. In turn, the meeting with the lithosphere depends on the ascent rate. Let us suppose that the asthenosphere diapir has a fast ascent rate. In a transient situation, it will penetrate the lithosphere, but since conductive cooling is slow, the diapir will be able to pursue its ascent to a comparatively shallow depth. In a steady-state regime, it will attain the surface of what has been described as the 'asthenosphere geoid' (Turcotte and McAdoo, 1978 ; Le Pichon et al., 1982) which is capped by a thin lithosphere only a

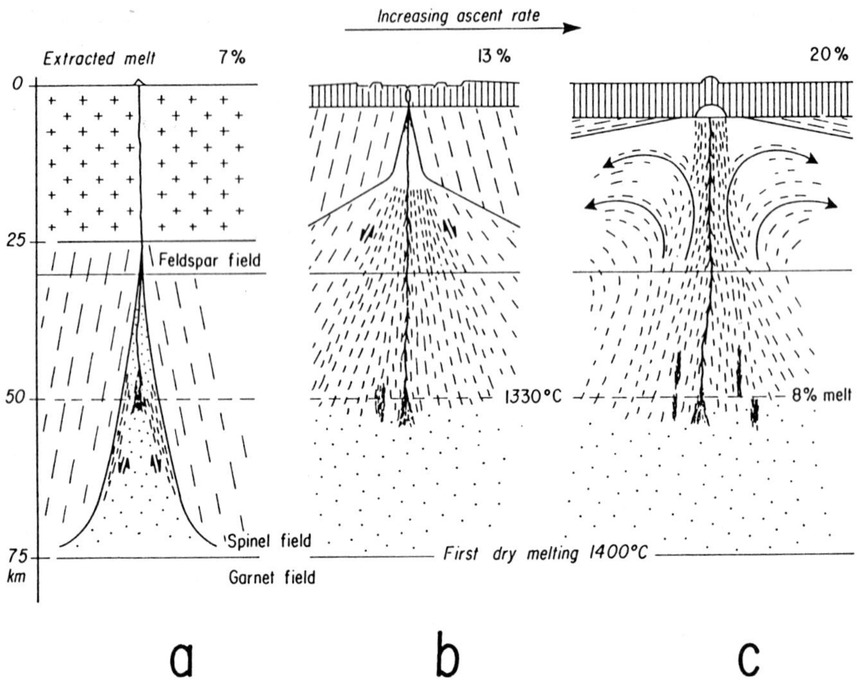


Fig. 7.3. Sketches of mantle diapirism and melt extraction in the various plate divergence situations. a) continental rift situation with associated spinel lherzolites, b) slow spreading situation with associated plagioclase lherzolites and c) fast spreading situation with associated harzburgites (a) Nicolas et al., 1987 ; b) Boudier and Nicolas, 1985 ; c) Rabinowicz et al., 1984).

few km thick (fig.7.3c). On the other hand, if the asthenosphere diapir has a slow ascent rate it will hardly penetrate into a thicker lithosphere (Nicolas, 1986a) (fig.7.3a,b).

This reasoning has been applied to melt generation at oceanic spreading ridges by Bottinga and Allègre (1978), Reid and Jackson (1981), Lewis (1983), Boudier and Nicolas (1985). These authors, relating the ascent rate to the spreading rate, show that for spreading rates larger than 1-2 cm/yr, the fast rising asthenosphere diapir can attain its highest level, thus causing maximum melting and producing the normally 6 km thick oceanic crust, while its top peridotites become particularly depleted. Below this rate, the diapir stops melting at greater depth due to thermal loss at the contact with a thicker lithosphere, thus generating less melt, a thinner crust and less depleted peridotites (Boudier and Nicolas, 1985 ; Nicolas, 1986a, b). This is shown by the trajectories in the P,T diagram of figure 7.2 where mantle diapirs meet the lithosphere at depths of around 40 km, 20 km and 6 km. In figure 7.3, they are respectively ascribed to continental rifts, to oceanic rifts or to oceanic ridges spreading at a rate below 1-2 cm/yr and finally to oceanic ridges spreading at rates in excess of 1-2 cm/yr. The mantle signature of these situations would be spinel lherzolites, plagioclase lherzolites and harzburgites respectively. Hence, some rationale appears between systems of diverging plates and the nature of melts and residual peridotites. This will be dealt with further in chapter 8. It should be stressed again that a more sophisticated model should integrate the effects of melting of heterogeneous mantle sources in terms of composition and temperature as undertaken by Klein and Langmuir (1987) in the case of oceanic ridges.

### *7.2.3. Depth of first melting*

In the diapiric situation considered here, geochemical modelling based on rare earth elements indicates that the peridotite dry solidus is crossed in the garnet field (Loubet et al., 1975 ; Prinzhofer and Allègre, 1985 ; Feigenson, 1986), that is at pressures greater than 2.5 GPa (Takahashi and Kushiro, 1983), corresponding to around 75 km depth (fig.7.1).The occurrence of dry melting has been justified in §7.2.1. Hydrous melting occurring in the diapir at a greater depth corresponds to a small melt fraction and does not significantly affect the following scenario.

The depth of first adiabatic melting in a rising diapir can be independently estimated by determining the following : 1) the adiabatic emergence temperature in tholeiitic basalts before any heat loss or gain by conduction exchange, oxidation, etc. ; this temperature, which is not known precisely, is thought to be about 1200°C, possibly not exceeding 1215°C (Wright and Okamura, 1977) and 2) the P,T path of the diapir as represented by figure 7.2. A surface temperature of 1200°C in basalts corresponds in figure 7.2 to 1300 - 1350°C solidus temperatures. For these temperatures, the solidus pressure is 2 GPa, which is well into the spinel lherzolite field, and not in the garnet field as required by geochemical data. In spite of various uncertainties in this approach (departure from adiabatic conditions in melt close to surface, insufficient knowledge of the T (X,P) function, of the dependence of spinel - garnet transition on peridotite composition, and of the effects of minor hydrous melting), this nevertheless encourages one to choose a solidus intersection as shallow as possible in the garnet field, in order to obtain a realistic emergence temperature. This is the 1400°C/2.4 GPa point in figure 7.2, which corresponds to an emergence temperature of 1230°C. In a hot spot situation like Hawaii, the depth of first melting may be deeper (60-100 km for Feigenson, 1986), due to the higher temperature in a mantle plume than in a convecting cell. McKenzie (1984) calculated that a diapir crossing the solidus at 1400°C can potentially create 10 km of crust by partial melting, which is too much for an oceanic situation. However,

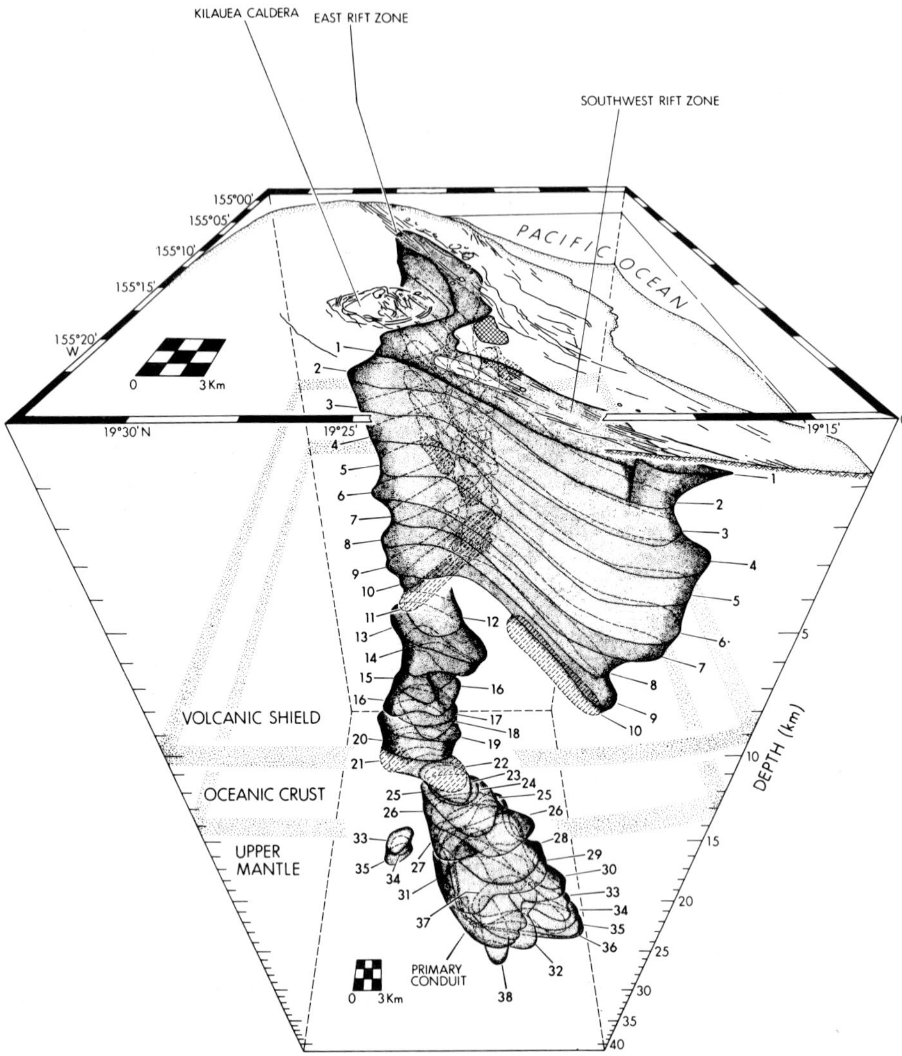


Fig. 7.4. The three-dimensional internal structure of Kilauea in an Eastward directed view. The summit magma reservoir occupies the 2- to 6-km depth interval beneath Kilauea caldera. Numbers on the leaders are cross-section depths beneath the local volcanic surface in kilometers. The conduit cross sections are dashed where uncertain. The isostatic warping of the oceanic crust produces a dip toward the island center (i.e.. toward the lower left in this view) (Ryan, 1988).

physical and numerical models (Bottinga and Allègre, 1978 ; Rabinowicz et al., 1984) show that a significant fraction of the melt in the diapir (70 % in former model and 30 % in the latter) does not contribute to crust formation because it remains trapped in the residual peridotites. Thus, the 6 km thickness of the oceanic crust is compatible with the chosen P,T path for the diapir.

In the P,T path of the figure 7.2, melting increases with uprise. It becomes sufficiently large at a depth of 60 km to produce a distinct seismic attenuation below oceanic spreading centers (Forsyth, 1977).

### *7.2.4. Maximum depth of melt extraction*

Melt extraction as envisaged below involves hydrofracturing and rapid withdrawal of melt from source rocks. These are dynamic processes which should result in earthquakes and tremors below active volcanoes (Shaw, 1980 ; Ryan et al., 1981). Their maximum depth probably reflects that of melt extraction. In Tahiti, it is about 50 km (Tallandier and Bouchon, 1979) and in Hawaii, 40 - 60 km (Aki and Koyanagi, 1981 ; Ryan, 1988). Between 40 and 60 km depth levels, the more diffuse seismicity (Koyanagi and Endo, 1977 ; Ryan, 1988) suggests that the magma conduit roots into a magma-charged volume. Above the 40 km depth level, Ryan (1988) has been able to locate a conduit zone (fig. 7.4). This is a zone of intense hydrofracturing defining a network of dikes, concentrically zoned, which do not necessarily rise to the surface and can stagnate between periods of intense activity.

In Hawaii, the crust stands 10 km above the Moho. This is the elevation of a melt column above the Moho which is necessary to counterbalance the pressure created in the mantle by a 50-60 km high melt column, with a buoyant force of 5 MPa/km created by the difference in density between basaltic melt and peridotite (§.7.4). This simple reasoning suggests that the maximum depth at which melt is evacuated by dikes is about 50 km. However, Hawaii as well as Tahiti correspond to hot spots where melting can occur deeper than below a normal ridge, as a result of higher temperatures.

## *7.3. PHYSICAL MECHANISMS OF MELT EXTRACTION*

Addressing more specifically the problem of melt extraction in a diapir melting by adiabatic decompression, there are many aspects which can be considered from theoretical, experimental or field-oriented points of view. First, if permeability is required for melt extraction, how is it achieved and for what melt fraction ? Next, how does melt concentrate, if it does, starting from an interconnected melt network in the peridotite matrix (shape and volume of melt segregation, rate-controlling process). Finally, how is melt transported to the earth's surface ? Many of these problems have been reviewed by Spera (1980).

### *7.3.1. Fraction of stable melt in a peridotite*

Dealing with the first point, there are a few theoretical and experimental contributions on how silicate liquids wet grain boundaries, a process controlling the fraction of melt necessary to create permeability in the matrix. Permeability depends on the dihedral angle shown by a liquid groove between crystals with which the liquid is in equilibrium (Bulau and Waff, 1979) (fig. 7.5). If this interfacial angle is lower than 60°, permeability is theoretically obtained for any melt fraction. Experimental results (Waff and Bulau, 1979 ; Vaughan and Kohlsted, 1982) show that in a molten mixture of olivine and basalt, this angle is less than 60° ; the liquid (1-2%) is confined in equilibrium conditions to grain corners and edges forming a three dimensionally interconnected network (fig. 7.5). However, Maaloe (1981), in more lengthy experiments on a natural spinel lherzolite, observes that the liquid wets only a fraction of grain edges. Moreover, bearing in mind that the first melt forms only at meeting points between the four phases of the lherzolite, this author concludes that interconnection throughout the medium cannot be achieved for 2 % liquid and that a fraction between 10 and 20 % melt is probably necessary to obtain a complete permeability. Fujii and Scarfe (1985) in an experimental study on a synthetic peridotite (olivine, enstatite, plagioclase) also find that for a dry melt fraction of over 8 %, connectivity may not be achieved because of the high dihedral angle of enstatite. Toramaru and Fujii (1986) emphasize the influence of the modal composition and relative grain size of the various phases on connectivity and concluded that *connectivity is more difficult to achieve in a fertile peridotite (olivine-poor) than in a depleted one (olivine-rich).* Thus it is estimated that in a fertile and dry lherzolite, a melt fraction of 7% is necessary to obtain connectivity ; it would attain 29% in a websterite. These results help to reconcile the apparent contradiction seen above between the low critical melt fraction for connectivity claimed by the authors dealing with olivine-rich systems and the high critical fraction supported by Maaloe's work on a lherzolite. It should be stressed again that these results are valid for dry peridotites, as the presence of small amounts of water would reduce the dihedral angle of enstatite (Von Bargen and Waff, 1988). Provided that these experimental results apply to melting in a mantle diapir, they would indicate that connectivity, and consequently melt extraction, can be achieved for low degrees of incipient hydrous melting of a lherzolite, and that the connectivity and melt extraction threshold would shift to the 7% ratio, proposed above, during subsequent dry melting.

In lherzolite massifs (Nicolas, 1986a) and xenoliths (Nicolas et al., 1987), structural evidence can be found which indicates that 5-10 % melt remained trapped during the process of melt extraction. In harzburgites and dunites, trapped melt is usually not detectable, except in the transition zone where the volume of melt impregnation can be very large (§ 7.5, see also § 3.3.3). In conclusion, it is suggested that a fraction of several percent melting is to be expected before connectivity is achieved in the fertile lherzolite of a rising diapir.

### *7.3.2. Melt extraction*

Once permeability is attained, physical models for melt expulsion call for percolation of the liquid through the solid frame following Darcy's law (Sleep, 1974 ; Turcotte and Ahern, 1978 ; Walker et al., 1978 ; Waff, 1980 ; Maaloe, 1981 ; McKenzie, 1984; Richter and McKenzie, 1984 ; Scott and Stevenson, 1984 ; Ribe, 1985). In particular, models differ on the rate controlling mechanism which, for the last authors, is viewed as the deformation of the matrix ; the melt once segregated migrates by itself through the compacting matrix. For other authors, the melts thus expelled constitute static reservoirs (stratae for Waff, 1980 and Maaloe, 1981) which are episodically tapped, the melt being transported from there to the surface via conduits opened by the melt pressure (hydrofracturing). Hydrofracturing is the essential process for another group of authors (Fedotov, 1978 ; Shaw, 1980 ; Nicolas and Jackson, 1982 ; Artyushkov and Sobolev,1984 ; Spence and Turcotte, 1985 ; Ryan, 1988). Melt extraction through conduits opened by hydrofracturing requires a melt pressure which exceeds the peridotite yield stress. In a permeable peridotite, this is created at the top of a static or quasi-static melt column of height h, due to the density contrast $\Delta\rho$ between the matrix (3.3 g/cm$^3$) and the melt (2.8 g/cm$^3$) at the comparatively shallow depths considered here. The fluid overpressure $\Delta P = g\,\Delta\rho h$ is 5 MPa for a melt column 1 km high, where g is the

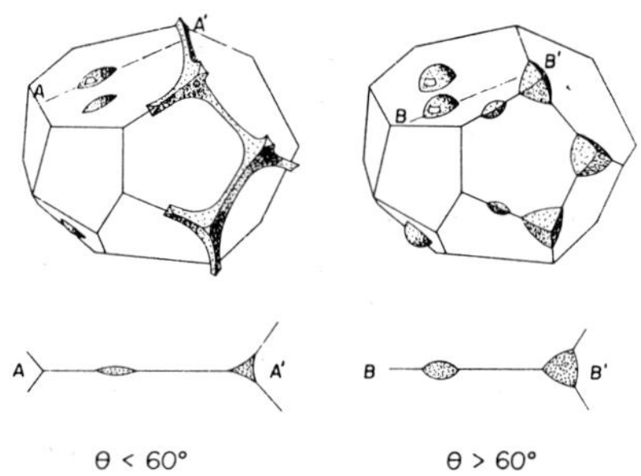


Fig. 7.5. Perspective drawings illustrating the distribution of melt (stippled) around a single grain in a rock for values of θ less than (left) and greater than 60°. Note the three-dimensional continuity of the melt in the drawing at the left, and the presence of isolated pores at the grain corners in the right-hand figure. Cross sections A-A' and B-B' are along the dashed lines drawn on the grains, perpendicular to the grain surface (Watson and Brenan, 1987).

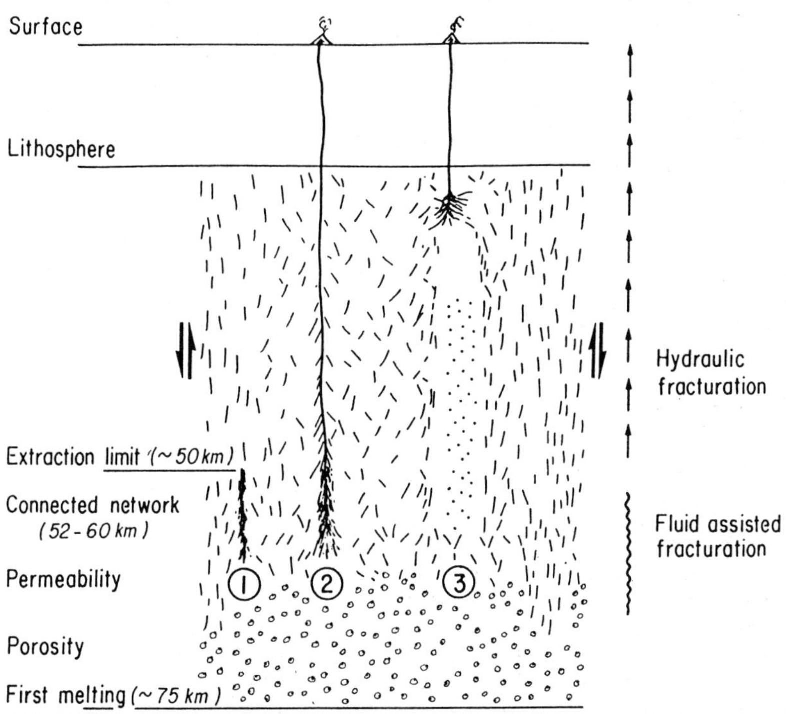


Fig. 7.6. Model of melt extraction in a mantle diapir. Circles : isolated melt drops ; dashes : melt veins ; dots : residual peridotites. The melt conduits (solid lines) follow the $(\sigma_1, \sigma_2)$ surface trajectories which are not necessarily vertical as represented here for the sake of simplicity. 1-2-3 : successive stages of melt extraction : 1 - creation of a connected melt network on the critical vertical extension ; 2 - newly formed melt conduit, draining melts mainly from deep horizons ; 3 - dying melt conduit draining shallow melts and leaving a wake of depleted peridotites. The 1-2-3 sequence is periodically achieved in the various parts of the diapir (Nicolas, 1986a).

gravitational constant. The fluid overpressure falls to 0 at Moho depth as overlying crust has a density close to that of a melt.

The discussion on melt extraction mechanisms centers around hydrofracturing and fluid migration, or percolation, through a compacting matrix. We opt below for a model calling on hydrofracturing because of several difficulties met with the percolation mechanism.

i) In the permeable matrix required for melt extraction by both mechanisms, the hydrostatic pressure gradient of the melt (5 MPa/km) in the vertical direction, responsible for hydrofracturing, exceeds by two orders of magnitude the dynamic pressure gradient (gradient of the non deviatoric part of the stress tensor, estimated at 0.05 MPa/km), which is driving the melt through the matrix. Phipps Morgan (1987) compares the efficiency of the two mechanisms through a dimensionless ratio and concludes that unless the asthenosphere viscosity, taken at $2 \times 10^{19}$ Pa.s, is raised to $10^{21}$ Pa.s, which seems unrealistically high for a melting peridotite, porous flow is not competitive with transport through dikes. As seen above, beyond a vertical distance such that the hydrostatic pressure of the melt exceeds the matrix yield stress, hydrofracturing seems unavoidable. Of course, in a partially molten horizon whose vertical extension remains below the critical height for hydrofracturing or in the crust, where there is no or little density difference between melt and solid matrix, percolation becomes possible (§ 7.5.).

ii) The presence of xenoliths in basalts heavier than the melt implies a fast moving melt; to be compatible with volcano outpour budgets, this requires that the transport occurs in narrow dikes (§ 7.4.2). Dealing with spinel lherzolite xenoliths in alkali basalts, this reasoning also implies that the same transport mechanism operates from their locus of origin (70-30 km) to the surface.

iii) Studies in peridotites, mainly plagioclase lherzolite massifs in which partial melting structures have been frozen (Boudier and Nicolas, 1972, 1977), emphasize the role of dikes in melt extraction and suggest that these dikes were opened by hydrofracturing (Nicolas and Jackson, 1982). Common observation of dikes in spinel lherzolite massifs and xenoliths equilibrated at depths to 70 km indicate that fracturing is not restricted to the shallowest mantle. On the other hand, at Moho depth, field evidence from ophiolites suggests that compaction becomes effective as a melt extraction mechanism. This will be considered after presenting the hydrofracturing mechanism.

## *7.4. MODEL OF MELT EXTRACTION BY HYDROFRACTURING*

### *7.4.1. The model*

The model illustrated by figure 7.6 comprises the following successive stages during adiabatic uprise of a mantle diapir :

i) Melt formation with only local segregation, starting around a 75 km depth. At this early stage, no interconnected melt network is created.

ii) Creation, around 60 km depth, of an interconnected network of melt veins due to increased adiabatic melting during uprise. Melt veins are preferentially oriented as tension fractures. Formation of melt gashes and veins is greatly helped and controlled in orientation by plastic flow (Boudier and Nicolas, 1972, 1977). Flow through a porous medium with solid compaction intervenes actively at this stage.

iii) For a critical vertical extension of this network, hydraulic fracturing of the overlying peridotites and propagation of a fissure through which the melt is expelled from the system.

The critical height for hydrofracturing depends on the yield strength of a partially molten peridotite which is unknown. From short-term hardness measurements in olivine at comparable high temperature conditions, Evans and Goetze, (1979) propose a crude estimate of 50 MPa. It probably represents an upper value because in long-term loading of elastic lithospheric plates (T = 600°C), the same 50 MPa threshold is estimated for the elastic-plastic transition (Menard and McNutt, 1982). With this 50 MPa value, the critical height of the column of interconnected melt is 10 km and melt extraction by hydrofracturing would be initiated around 50 km depth.

iv) If the fissure reaches the surface, drainage of the melts filling an interconnected network along the fissure path, mainly around its root.

Theoretically, the buoyant force of the melt column 50 km high is 250 MPa close to the surface, which is considerable. A similar 250 MPa depression is exerted at the roots of the column, draining very efficiently the connected melt present in the surrounding rocks. This analysis is essentially valid for the static situation, when the melt is quasi stagnant in the dike. During a steady-state flow through the dike, it is modified by the effects of the pressure drop due to melt flow in the conduit and the discharge rate is controlled by the plastic/elastic relaxation in the matrix of the melt network. Nevertheless, it can be predicted that during the transient regimes *the depression will be very efficient in draining and expelling the melt..*

The rock pressure squeezing the conduit, the main drainage moves upward until all melt in wall rocks connected to the conduit is removed and the conduit closes, except for lenses of trapped melt, constituting the dikes now observed. Such dikes, called in § 2.5.2. 'indigenous dikes', are typically rimmed at shallow depths by very depleted peridotites, in particular dunites. The origin of these dunites is more fully discussed in § 10.4.4. At greater depths in the spinel lherzolite field, such dikes are less conspicuous because, in these fertile lherzolites, the melt retention in the wall rocks becomes important (§ 7.3.1).

v) Initiation of a new hydraulic instability may occur anywhere within the rising diapir provided that a melt network attains the critical vertical extension for hydraulic fracturation of the overlying mantle.Thus, a continuous process, diapiric uprise and melting, can generate the discontinuous and episodic processes of melt extraction and volcanism.

### *7.4.2. Melt velocity within dikes, episodicity and duration of episodes of melt extraction*

Melt velocity within the dikes extracting melt from the mantle diapir can be estimated in different ways. From seismic tremors triggered by melt ascent, the velocity at 20-30 km below the Tolbachik volcano of Kamchatka was at 3-4 cm/s (Fedotov, 1978). The same author deduces possible velocities from a physical model of melt ascent in conduits based on hydrostatic forces which also integrates the thermal aspects of cooling along the walls and of heating by shear dissipation. His results introduce, for the conditions considered here, rather lose constraints on velocities (5-50 cm/s) and dike diameters (40-200 cm). Lago et al. (1982), applying Stokes' law to chromite nodules sustained by the melt within the basaltic conduits, estimate velocities of melt in the 2-20 cm/s range for dikes 5 to 25 cm in diameter. Considering peridotite xenoliths carried by alkali basalts, Spera (1980, 1984) obtains ascent rates of about 10-100 cm/s from kinetic constraints, and probably greater than 5-50 cm/s from settling rates of xenoliths, 20 cm in diameter. Rates are probably smaller in tholeiites, as suggested by the absence of peridotite xenoliths. However, tholeiites often contain olivine xenocrysts in large quantities (§ 8.3.1) and, in mantle dikes, cm-sized residual dunite xenoliths (Ceuleneer and Nicolas, 1985) proving that the rates are not considerably lower than in alkali basalts. From this discussion, it seems reasonable to propose 5 cm/s as a conservative value for the velocity in dikes of about 20 cm width, an average deduced from field observations. The linear extraction rate is thus 100 $cm^2/s$.

When applying this 100 $cm^2/s$ rate to the creation of the 6 km thick oceanic crust, a major discrepancy with sea-floor spreading rate is observed. A continuous melt discharge to build up this crust would result in a spreading rate of $2 \times 10^4$ cm/yr, instead of the expected 1-10 cm/yr, assuming that a dike emerging from the mantle feeds the same length of crust along the ridge strike. Models of ridge formation (§ 9.4) suggest that oceanic lithosphere is created, from mantle diapirs about 15 km in diameter and spaced one from the other by 50-100 km. It is thus probably more realistic to estimate that along strike, a dike feeds a longer crustal section than its own length in the mantle (fig. 11.13). The sketch of figure 7.4, based on seismological evidence indicates a ratio of 1/3 between the conduit largest dimension at depth and the extension of dikes close to the surface. The corresponding rate can be thus reduced to around $5 \times 10^3$ cm/yr, which is still incompatible with observed spreading rates. In conclusion, *melt extraction from the mantle is necessarily a discontinuous process* as already deduced from the analysis of volcanic eruptions (Wright and Tilling, 1980) and from geochemical arguments (O'Hara, 1977).

This episodicity is also deduced from the existence of a sheeted dike complex in the oceanic crust. Each dike corresponds to one rapid melt discharge, creating in a short time one meter of new crust. Periodicity of melt discharge from the mantle can be deduced from this. Depending on spreading rates bracketed between 2 and 20 cm/yr (double rate), *the periodicity is between one melt discharge every 5 years for fast spreading ridges and every 50 years for slow ones*. Based on an analysis of trace elements ratios in the Réunion basalts, Albarède and Tamagnan (in press) have estimated that for the last 50 years melt was tapped from the mantle source with a 17 years periodicity, an estimation which is within our bracketing. With the melt discharge rates estimated above, it takes between 1 and 8 weeks to create 1 m of new crust with its full 6 km thickness. Hence, *1-8 weeks measures the duration of a melt discharge event*, a duration which is compatible with the independently estimated time spans necessary to cool a 1 m thick dike of the sheeted dike complex and to form a chromite pod (§ 10.5.5).

### *7.4.3. Geochemical implications*

Two main geochemical implications are derived from this model.

*Nature of primary melts extracted from mantle sources* . The model implies that extracted melt is a mixture of melts produced by various degrees of partial melting at various depths between 75 km and 6 km in an oceanic environment (the shallowest depth depends on the environment, (§.7.2.2). This conclusion conforms to Prinzhofer and Allègre's (1985) model of desequilibrium melting based on trace elements analysis, and to Klein and Langmuir's (1987) model, which also considers major elements. These models predict that the integrated 'primary' melts are tholeiitic and not picritic, because the fraction of melt originating at deep levels remains small and is mixed with shallower tholeiitic melt.

It is still debated (see § 7.4.1) whether primary mantle melts are picritic or tholeiitic. The picritic melt would segregate at higher pressures (1.5-3.0 GPa) and fractionate a large amount of olivine on its way to the surface ; the tholeiitic melt would segregate at lower pressures (0.7-1.16 MPa) and suffer limited fractionation, resembling oceanic lavas in composition. The tholeiitic model seems in conflict with evidence presented here that first melting and melt extraction take place at pressures and depths superior to those required by tholeiite generation. The picritic model is supported by the idea that the abundant dunite bodies in the transition zone of ophiolites were formed by olivine accumulation, a view which is contested in § 10.4. It should be remembered that both picritic and tholeiitic models rely on batch (equilibrium) melting experiments which, as proposed here and also shown by trace element evidence (Langmuir et al., 1977), is an effective process in the mantle diapirs only at the early stage of fusion. In a modelling based on major elements variations during the large melting (10-20 %) imposed by the residual nature of the peridotites, Klein and Langmuir (1987) show that the discrepancy between the results of batch melting and fractional or desequilibrium melting can explain why primary melts are directly tholeiitic, notwithstanding a deep source.

*Nature of MORBs* : In this model of a discontinuous and dynamic delivery of melt in the crustal magma chamber through a narrow dike, it is conceivable that this new melt is hotter and lighter than the resident melt, and mixes poorly with it. The composition of MORB outpoured during the main course of an extraction event could then reflect that of the melt issued from the mantle. One could explain in this way the fairly constant chemical composition of MORBs overall (Cann, 1971) whatever the oceanic environments (spreading rates, presence or not of magma chambers,...), with the noteworthy exception of the tips of certain ridge segments (§ 8.3.1).

## *7.5. MELT EXTRACTION BY SOLID COMPACTION AND MELT PERCOLATION IN TRANSITION ZONES OF OPHIOLITES*

In the uppermost peridotites, a few hundreds of meters below the Moho, melt may segregate by compaction of a solid impregnated with melt and migrate upward by percolation. This is suggested by field studies in the transition zone of ophiolites, mainly in the Maqsad area of Oman (§.3.4.2). This area has been described in some detail in chapter 3 because of its peculiar mantle flow pattern (fig. 3.17), evoking a diapiric asthenosphere intrusion, and of the abundance of melt impregnations in the thick dunites and depleted harzburgites of the transition zone. In this horizon, where the vertical flow pattern attributed to upward motion in the rising diapir is broken and transposed into a horizontal flow diverging away from the diapir, the plastically induced mineral fabrics tend to disappear (figs. 3.18 and 3.19). This is ascribed to a local dismembering of the peridotite frame due to the importance of the melt fraction exceeding around 35 % (Van der Molen and Paterson, 1979 ; Wickham, 1987) (plate 3.3b,c,d).

A numerical model (Rabinowicz et al., 1987) shows that a viscosity drop of three orders of magnitude, say from $10^{18}$ to $10^{15}$ Pa.s is necessary to explain the sharp rotation of flow lines at the top of the diapir. This viscosity drop achieved by the dismembering of the peridotite generates a discontinuity of 5 bars in dynamic pressure along the high/low viscosity interface ; the higher pressure is set within the top low viscosity layer. This in turn induces melt compaction at this interface. By a feedback effect, the melt segregated by the compaction process helps to dismember the solid peridotite and to entertain the viscosity drop and pressure discontinuity.

Thus at the top of an asthenosphere diapir in the area where the mantle flow diverges, melt can accumulate by solid compaction producing segregation into gabbro sills (plate

3.2b, c ; see also § 3.3.3). This process operates here rather than hydrofracturing because locally there is a discontinuity in the dynamic pressure and because mechanically the medium is too soft to make fracturing possible. A fracture progressing upward from below is probably unable to propagate through this zone and should discharge its melt content within it. This is illustrated in the field by the observation of gabbro dikes grading into diffuse impregnation zones (plates 3.2g and 3.3a). This effect would still increase the melt-solid ratio temporarily.

Such particular conditions are necessary to promote compaction rather than hydrofracturing as the dominant melt extraction mechanism in an ascending asthenosphere diapir. They would operate only at the top of an asthenosphere diapir, possibly temporarily.

Recent studies (§ 8.3.1) emphasize the importance of olivine xenocrysts in picrites and basalts which had often been mistaken for cumulate phenocrysts. The source for the olivine xenocrysts is sought in mantle peridotites and dunites. It is suggested here that the horizon of the transition zone where the solid frame is dismembered by the percolating melt (plate 3.3c,d) is as excellent a candidate for the source of the xenocrysts as it could be for the wehrlitic intrusions so common in the crustal section of the Oman ophiolite (§ 10.3).

### *7.6. FOCUSING OF MELT EXTRACTION BELOW OCEANIC RIDGES*

It is now necessary to investigate how the melt extraction model developed above, namely hydrofracturing in a melting asthenosphere diapir combined, at least locally, with melt percolation in the transition zone just below the Moho applies to a ridge system. One of the strongest constraints on any model is the existence in fast spreading ridges of a narrow, 2 km wide, strip along the ridge axis where volcanic activity concentrates (§ 11.5.2). The 2 km width reflects that of the weak roof of crustal magma chambers and is not quite representative of the width of melt delivery through the mantle-crust boundary below the ridge. This latter width should be that of the base of the magma chamber(10-20 km, fig. 11.8) and, since a crustal magma chamber represents a rheologically weak zone, it is no surprise that tectonic activity also tends to be present over 10 - 20 km on each side of the ridge (fig. 11.12).

Melt delivery to the crust within some 5-10 km on each side of a ridge has been explained in the frame of two mantle flow models below the ridge. The first one invokes a largely diverging uprise in a constant viscosity asthenosphere, with melting over a ~ 100 km distance (Phipps Morgan, 1987) (fig. 7.7). This model has been discarded on the basis of observations of mantle flow patterns in various ophiolites (chapter 9) in favor of a model of small, low viscosity diapirs, buoyantly rising from the large asthenosphere uprise diverging at greater depth (figs. 7.3 and 9.5). Moreover, the convergence of the melt flow toward the ridge in the large scale uprise relies on disputable evidence. Spiegelman and McKenzie (1987) base their model on the melt percolation mechanism which is found to be unsatisfactory, at least in this situation (§ 7.3.2.). Phipps Morgan (1987) envisages the channelling of melt parallel to the foliation plane of peridotites which is in contradiction with observations in peridotites (see § 2.5.2. and below).

The small diapirs modelled by Rabinowicz et al. (1984, 1987) and Scott and Stevenson (in press), following those discovered in ophiolitic peridotites (fig. 9.1), have the right size below the ridge (fig. 9.5). They concentrate 70-80 % of the uprising mantle flow and consequently a similar fraction of the melt produced. Except in the shallowest transition zone, where melt is extracted by compaction in the dunites (§ 10.4), it is extracted by hydrofracturing at all depths. The trajectories of these fractures must be discussed further because they control melt delivery to the ridge.

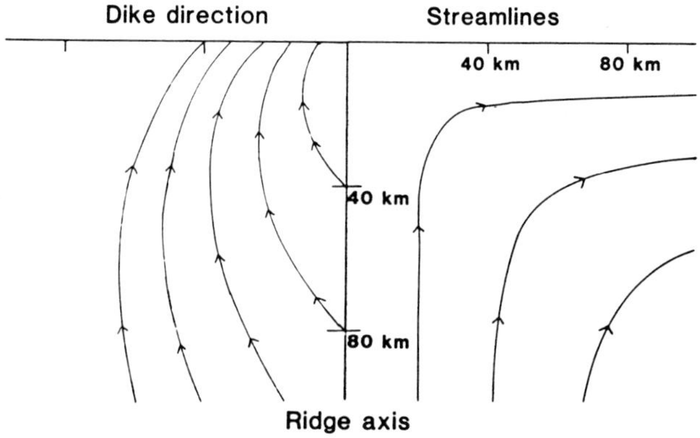


Fig. 7.7. Divergence (right) of a constant viscosity asthenosphere cell and $\sigma 1$ trajectories parallel to the presumed diking directions (left) below an oceanic ridge (Sleep, 1984).

Fractures in an isotropic medium follow the ($\sigma_1$, $\sigma_2$) plane, whereas in an anisotropic medium they can take advantage of surfaces of weakness. In peridotites, a conspicuous anisotropy is determined by the parallel orientation of the tectonic layering and foliation (§ 2.5.2.). However, repeated field observations in peridotite massifs and xenoliths show that tension fractures are not parallel to the foliation (§ 2.5.2.) but are controlled by the ($\sigma_1$, $\sigma_2$) surface (Nicolas and Jackson, 1982). If the trajectory of this ($\sigma_1$, $\sigma_2$) surface is horizontal, the melt forms sills in which the melt discharge remains stable. On the contrary, if this surface is inclined, the increasing vertical extension of the dike creates an increasing melt pressure and generates a runaway expelling melt to the surface.

It is thus critical to be able to predict the stress trajectories in the melting mantle. In the model of the large divergent cell, the $\sigma_1$ trajectories indicate that only superficial melt would reach the crust below the ridge (fig. 9.5). More primitive melt extracted from greater depths is either captured in the mantle flowing away from the ridge or expelled at distance as off-axis alkali basalts (Sleep, 1984 ; Rabinowicz et al., 1987). Stress trajectories are more difficult to predict in and around the buoyant diapirs (fig. 9.5). The root zone of the diapir at the top of the divergent cell corresponds to the situation just analyzed. Above, possibly in the 30-10 km range, a zone of horizontal $\sigma_1$ can appear where the melt would be retained, filling sills ; finally at shallower depths, $\sigma_1$ should become vertical, releasing the melt as vertical dikes. This analysis is, however, two dimensional and as observed by Sleep (in press), in the direction along the ridge strike, $\sigma_2$ and $\sigma_3$ could exchange regionally or locally in areas where the stress field is perturbed by diking. If $\sigma_2$ becomes vertical in areas of horizontal $\sigma_1$, dikes transverse to the ridge are formed which can release the melt so far trapped. In conclusion, this expected complexity of the stress field in a buoyant diapir which is reflected by that of the diking system in peridotite massifs (Nicolas and Jackson, 1982 ; § 3.3.3), seems compatible with melt delivery by tension fracturing in the vicinity of a ridge.

# Chapter 8

# THE VARIOUS OPHIOLITES AND THEIR OCEANIC ENVIRONMENTS OF ORIGIN

## *8.1 INTRODUCTION*

The structure and expected functioning of accretion centers differ markedly in various marine environments. This chapter attempts to relate these variations to those of ophiolites which also display a great variability. This variability is illustrated by the descriptions of chosen ophiolitic districts (part II). The case of ophiolites formed in, or affected by, transform zones will not be considered separately here. Transform faults are met in most oceanic environments and the differences between for instance Bogota (§ 5.2) and Wadi Tayin (§ 5.4), as discussed below, probably reflect differences in spreading rates. Indeed, spreading rate seems to be the single most influencial parameter explaining the diversity of oceanic lithospheres and ophiolites (Boudier and Nicolas, 1985). The difficulty with ophiolites is evidently that indications on spreading rate are only indirect and are therefore open to discussion. They are derived from evidence for varying degrees of partial melting in the residual peridotites associated with different ophiolites (see § 7.2.2). For this reason, we will refer to the factual classification proposed by the above mentioned authors, distinguishing the harzburgitic and the lherzolitic types of ophiolites (HOT and LOT) and will contrast the characteristics of the Oman ophiolite (chapter 3), to those of Trinity (chapter 4). The Oman ophiolite belongs to the harzburgite type to which is ascribed a high degree of partial melting and it is thought to represent a fast spreading ridge. The Trinity ophiolite, which belongs to the lherzolite type and is ascribed to a lower degree of partial melting, is thought to represent a slow ridge or rift. On the basis of petrological criteria, Ishiwatari (1985) has proposed an ophiolite classification of three types, referring to Liguria, Yakuno and Papua ophiolites. The Liguria type coincides with our lherzolite type, the Papua with our harzburgite and the Yakuno with some intermediate ophiolites. Ishiwatari uses also this relation with increased degree of partial melting to explain the differences between these ophiolites.

The geodynamic environment is the other obvious factor of diversity in oceanic spreading and presumably in ophiolites. Implicitly, in this book we refer to 'normal' oceanic spreading conditions, that is spreading due to uprise of an asthenosphere diapir from a spherically homogeneous mantle in terms of temperature and composition (§ 7.2.2.). The proposed relation of HOT/LOT with oceanic spreading rates would not hold in the case of an abnormally hot or cold mantle. For example, in a hot spot situation both the crust and the harzburgite layer in the underlying mantle are expected to be abnormally thick, whatever the spreading rate. Another problem is trying to determine whether the considered ophiolite derives from a mid-oceanic ridge, from a back-arc basin or from an island arc ; so far this has been one of the main points of discussion in ophiolite studies because the geochemistry of lavas seemed to provide criteria to solve it. Although our object here is essentially structural, we will address this point briefly. The sharp difference between Canyon Mountain (Chapter 6), thought to be a good example of an island arc ophiolite, and other ophiolites suggests that island arc ophiolites are exceptional. In contrast, back arc basins are obvious candidates for ophiolite generation, considering the dynamics of these basins and their vicinity to emerged areas. If, as proposed here, the HOT-LOT distinction mainly reflects physical conditions of spreading, other criteria, including geochemistry, are required to distinguish mid-oceanic from back arc ophiolites.

## *8.2. HARZBURGITE AND LHERZOLITE TYPES OF OPHIOLITES - ROLE OF SPREADING RATE*

The Harzburgite Type of Ophiolite (HOT) examplified by the Oman case is the most common type and includes other well known ophiolite massifs such as Bay of Islands in Newfoundland, Zambales in the Philippines, New Guinea, New Caledonia, Vourinos, Troodos, several massifs in Turkey and in the polar Urals, etc... The Lherzolite Type of Ophiolite (LOT), illustrated by the Trinity case is more restricted than the preceding type, and includes mainly the western Alps ophiolites such as Lanzo, Liguria, Apennines and Corsica, a few massifs in Yougoslavia and Othris in Greece, which is a vast district where harzburgites massifs are also present. In the Mediterranean, the earlier distinction between a western lherzolite province and an eastern harzburgitic province (Nicolas and Jackson, 1972 ; fig. 4.16) has been since revised. Pamic (1983) locates in the Southern Dinarides of Yougoslavia the transition between the two provinces, whereas Koepke et al. (1985) extend it to Crete, in the Aegean Sea. These authors contrast a western lherzolitic or sublherzolitic belt of Jurassic age with an eastern belt, definitely harzburgitic in composition and of cretaceous age. The origin of these belts is discussed in chapter 13.

LOT grades into HOT through massifs of intermediate character such as Xigaze (§ 4.3) or Yakuno (Ishiwatari, 1985) which have a clinopyroxene-harzburgite to lherzolitic peridotite section. On the other hand, the main LOT feature which is the plagioclase lherzolite nature of the tectonic ultramafic section is also seen in a few massifs like Zabargad in the Red Sea or Sierra Bermeja and Sierra Alpujata in Southern Spain which have been briefly described in § 4.5. In these massifs the plagioclase lherzolites are associated with spinel lherzolites. Such massifs, where no ophiolitic crustal section has been reported, could represent a transitional stage between continental and oceanic rifting.

### *8.2.1. Distinctive characteristics*

The distinctive characteristics between the two types of massifs are discussed below and are summarized in the following table and in figure 8.1.

Table 8-I - Distinctive characters of the Harzburgite Ophiolite Type (HOT) and the Lherzolite Ophiolite Type (LOT). The massifs considered in this table are only those for which sufficient information is available, in particular structural data in the ultramafic section. They are for HOT : (1) Oman, (chapter 3), (2) Bay of Islands (Newfoundland) (Girardeau, 1979 ; Girardeau and Nicolas, 1981 ; Church and Stevans, 1971 ; Suen et al, 1979), (3) Zambales (Philippines) (Hawkins and Evans, 1982 ; Violette, 1980), (4) Troodos (Cyprus) (George, 1978 ; Violette, 1980 ; Benn and Laurent, 1987), (5) Antalya (Turkey) (Juteau, 1979), (6) Pozanti-Karsanti (Turkey) (Cakir, 1978), (7) Massif du Sud (New Caledonia) (Prinzhofer et al., 1980 ; Cassard, 1980) and for LOT : (8) Trinity (California) (Chapter 4), (9) Piedmont-Liguria (western Italy) chapter 4, and Monte Maggiore (Corsica) (Rocci et al., 1979 ; Jackson, 1979), (10) Lanzo (western Italy) (Boudier, 1978), (11) Othris (Greece) (Menzies, 1976 ; Menzies and Allen, 1974 ; Ferriere, 1982 ; pers. obs.).

| | Harzburgite ophiolite type (HOT) | Lherzolite ophiolite type type (LOT) |
|---|---|---|
| *Environmental formations* | | |
| cover | marine sediments and volcanics (1,2,3,4,5,6,7) | marine sediments and volcanics (8,9,10,11,) breccias (8,9) |
| sole | sole of metamorphic oceanic crust (1,2,5,6) | sole of metamorphic continental (10,11) or oceanic (8) crust |
| *Mafic section* | | |
| thickness | 2-3 km (3,4), ~7 km (1,2,5,6) | 0-1 km (9), 2-3 km (8,11) |
| layered gabbros | thick and usually continuous (1,2,3,4,5,6) | thin, absent or in restricted areas (8,9,11) |
| nature of intrusives | rare diabase dikes (1,2,3,4,5)<br>wehrlite bodies (1,2,4) | numerous diabase dikes and sills (8,9,11) |
| nature of basalt directly associated with the ophiolite | tholeiitic (1,2,4,5,6) | tholeiitic (8,9,11) alkaline (9,11) |
| Fe-Ti-gabbros/ Mg-gabbros | small (1,2,3,4,5,6) | large (9,10) |
| LT plastic deformation | absent, except local shear zones (1, 2, 3, 4, 5, 6, 7) | very common (flaser gabbros) (8,9, 10, 11) |
| *Ultramafic section* | | |
| nature of mantle rocks | harzburgites and abundant dunites down to about 10 km below Moho (1,2,3,4,5,6,7) | plagioclase lherzolites and abundant dunites within ≤ 2 km below Moho (8,10) |
| HT plastic flow structures | flat foliation (1,2,5,6,7), locally vertical with vertical lineation (2,3,4,5) | usually steep foliation and moderately plunging lineation (8,9,10,11) |
| neoblast grain size | large (~4mm)(1,2,3,4,7) | small (~0.5mm)(8,10) |
| chromite pods | present (1,2,3,4,5,6,7) | absent (8,9,10,11) |
| diabase occurrence | uncommon (1,2,3,4,7) | common in top of section (8,9,10) |
| serpenti-nization | lizardite (1,2,3,4,6,7) | lizardite and antigorite (8,9,10,11)<br>ophicalcites (9) |

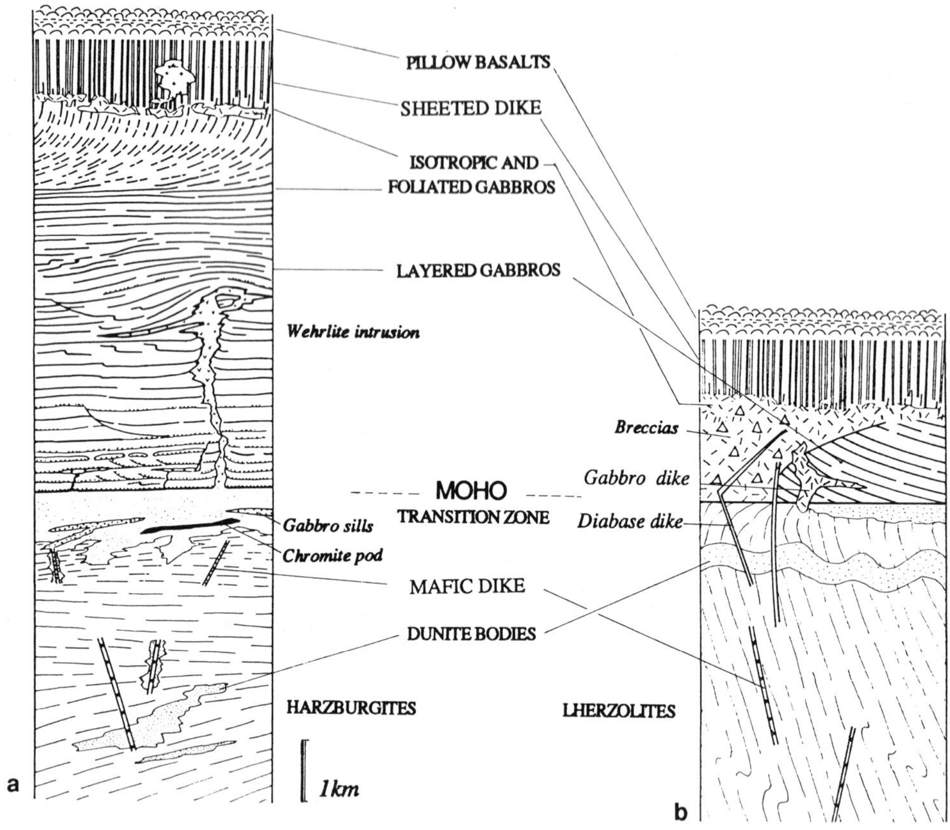


Fig. 8.1. Compared logs in a) Harzburgite (HOT) and b) Lherzolite (LOT) Ophiolites Types. The internal structures and relative thickness of the main units are approximate (redrawn from Boudier and Nicolas, 1985).

*Environmental formations* - HOT is usually covered by marine sediments, mainly umbers and cherts. LOT is overlain by more diverse sediments. The eastern margin of the Othris ophiolite is in contact with undersaturated basaltic lavas associated with fragments of ophiolites, shallow-water marine sediments and continental margin formations (Menzies, 1976). The Trinity ophiolite is covered by a breccia containing ophiolite xenoliths in a matrix of shallow marine sediments (chapter 4). The Piedmont-Ligurian lherzolite forms locally the sea floor of the Piedmont basin (§ 4.4).

Many HOT complexes rest on a sole of metamorphosed oceanic crust formations below an intraoceanic thrust (chapter 12). With the exception of the Trinity complex, which may have such a sole (§ 4.3.4), all the other LOT massifs are in direct contact with, or in the vicinity of metamorphic continental crust.

*Mafic section* - The crustal section overlying the ultramafic section is thicker in the HOT case(fig. 8.1a). In Oman, where this section is best exposed and least dismembered, the thickness ranges from 4 to 6 km in a longitudinal cross-section (§ 3.3.2). The internal composition and organization are preserved over large areas and they conform (fig. 8.1a) to the generally accepted model of ophiolite crust; in particular the gabbro sequence is well developed as compared to typical LOT. It seems that the difference in crustal thickness between the two types is mainly due to the different development of the layered gabbro unit. The importance of wehrlite intrusions is increasingly recognized in ophiolites of HOT type (Benn and Laurent, 1987 ; Benn et al., 1988 ; Juteau et al. 1988). In Oman, these intrusions can attain several hundred meters and reach the level of the sheeted dike complex (§ 3.3.2).

The LOT crustal section (fig.8.1b) is altogether thinner and less organized than the HOT one, and may even be absent locally. In the crustal sequence of the Trinity body, the layered gabbros and pyroxenites are not ubiquitous : they are apparently deposited within small and discontinuous magma chambers. There is a widespread development of magmatic breccias at the expense of these gabbros. We ascribe these breccias to hydrous anatexis related to high temperature hydrothermal circulation. The particular abundance of plagiogranite dikes may be due to this anatexis. The common occurrence of Fe-Ti rich noritic gabbros as differentiated facies at the top of the plutonic section is interpreted as indicating a magma chamber evolution in a closed system (Juteau et al., 1988). This contrasts with HOT where the exceptional occurrence of such gabbros suggests that magma chambers function mainly as open systems (steady-state magma chambers). Typically, in LOT ophiolites the layered and isotropic gabbros are strongly deformed in amphibolites to greenschist facies conditions leading, in particular, to 1-100 cm thick shear zones. Finally, the crustal section is invaded by diabase sills and dikes, the latter being prevalent in its upper part. The crustal section associated with the Piedmont-Liguria ophiolites (§ 4.4) shows an even more limited extent of layered gabbros. In several localities it is totally absent with serpentinites constituting the sea-floor.

*Nature of the ultramafic section* - The HOT massifs are dominantly composed of harzburgites interlayered with minor orthopyroxenites and transected by dunite bodies and veins. Clinopyroxene and feldspar are locally present mainly in the uppermost harzburgites, thus reconstituting lherzolites as a result of a secondary magmatic impregnation (§ 2.5.3). The dunites are observed throughout the exposed sequences, but their abundance increases upward to become dominant in the transition zone below the crustal formations. The thickness of the transition zone is highly variable ranging from a few meters to several hundred meters (Nicolas and Prinzhofer, 1983).

The LOT massifs are composed of homogeneous feldspathic lherzolites with a websteritic layering. Although the distinction is sometimes, criteria can be found (§ 2.5.3) showing the feldspar results from partial melting of the host lherzolite and not from a secondary impregnation. The only significant volumes of dunites are present in the uppermost part of the section. They form tabular bodies in the transition zone and just below it. The transition zone in Trinity does not seem to exceed some 100 meters.

Diabase dikes and sills are more abundant in the upper ultramafic section of LOT than in HOT. In Trinity and in Lanzo, dikes and lenses of pegmatoid gabbros a few meters to a few tens of meters wide also occur in the transition zone. A critical difference between the two types is that chromite pods are restricted to HOT (§ 10.5.2).

*Structures and microstructures of the ultramafic section* - Microstructures related to the asthenosphere flow (§ 2.5.5) in LOT peridotites are porphyroclastic with a neoblast size of around 0.5 mm (Boudier and Nicolas, 1980). Microstructures in HOT peridotites range from coarse grained to porphyroclastic (fig. 2.7) with a neoblast size around 3.0-5.0 mm. This records a higher flow stress in LOT than in HOT when the asthenospheric flow was frozen.

The pattern of the high-temperature plastic flow deduced from structural mapping is also distinct. When related to the paleo Moho attitude in HOT, it is either flat with a regular flow direction, or organized following a diapiric pattern (chapter 9). In the LOT massifs where the plastic flow structures have been studied (§ 4.3 and 4.4) the foliations are steep and parallel, and lineations, horizontal to moderately plunging (fig. 9.4). The shear flow records the same sense of shear over kilometers across the massifs.

*Serpentinization* - Though difficult to generalize because the studies on the serpentinization of ultramafic sequences are not exhaustive, some differences between HOT and LOT can be identified.

The harzburgite massifs are variously serpentinized. Except locally, in the vicinity of large shear zones (§ 3.4.2) or in faulted areas where they can be altered by high temperature assemblages of antigorite, talc, chlorite and tremolite, these massifs are altered in a pervasive way only by the low-temperature and post-deformational chrysotile-brucite-lizardite assemblage. This serpentinization can be ascribed to contamination by sea or ground water at low temperature (Barnes et al., 1978 ; Bonatti et al., 1984). In contrast, most lherzolitic massifs mentioned above and other described in literature (Taiwan, Ernst and Liou, 1984) show an early episode of antigoritic serpentinization (with accessory talc, chlorite, tremolite and magnetite) developed in high temperature tectonic environments and preceding the lizardite episode. The gabbro dikes are commonly rodingitized, an alteration accompanying high temperature serpentinization. They contrast with similar gabbro dikes, usually unaltered in harzburgite massifs. Interestingly, in Xigaze, it has been observed that, though the gabbros dikes are rodingitized, the diabase dikes are not (§ 4.2.6), demonstrating that the rodingitization predates the injection of diabase dikes. Early serpentinization and rodingitization recording conditions of the greenschist facies are ascribed to the hydrothermal alteration taking place close to the spreading center where the corresponding oceanic crust has been formed (§ 11.7). This indicates that at LOT spreading centers, hot sea water can penetrate into the mantle sequence in a pervasive way. Considering the respective crustal thickness estimated in LOT (3 km) and HOT (6 km) massifs, this penetration of water being limited at depth by confining pressure explains why only the LOT upper peridotites are affected in a pervasive way by this hydrothermal alteration (see § 11.3.2 and 11.3.3). The importance of hydrous recrystallization and anatexis in the LOT gabbros probably relates to the same cause.

### *8.2.2. Harzburgite and lherzolite types of ophiolites and mantle partial melting*

Several lines of evidence indicate that HOT corresponds to a larger degree of melt extraction in the mantle than LOT. They have been developed by Boudier and Nicolas (1985), Ishiwatari (1985) and Nicolas (1986 a, b) (see also § 7.2.2) and they are summarized below.

i) The harzburgites and plagioclase lherzolites are derived from pristine lherzolites through about 20% and 15% melt extraction respectively. This is deduced both from the physical analysis of melt extraction (§ 7.2.2. and fig. 7.2) and from the petrology of the residue (Mysen and Kushiro, 1977; Jaques and Green, 1980).

ii) The crustal section is thicker in HOT than in LOT (fig. 8.1). In the considered ophiolites, the crust is attached to the underlying mantle wedge and should result from its melting (see however § 4.2.6.). The crustal thickness reflects the degree of melt extraction from the mantle wedge. The compared thicknesses of HOT and LOT crusts indicate that melting has been more important in HOT. Trying to relate quantitatively the crustal thickness to the degree of melt extraction estimated in the attached peridotites is difficult because melt is extracted throughout a mantle wedge (§ 7.4), with contributions from various depths which cannot be simply estimated.

iii) The basalts directly associated with the considered ophiolites are quartz-tholeiites in HOT and they tend toward alkali-basalts in LOT, corresponding respectively to larger melting at shallower depth and to lower melting at greater depth (Ishiwatari, 1985).

iv) The occurrence of chromite pods restricted to HOT is indirect evidence of a higher degree of partial melting in HOT than in LOT (§ 10.5.2)..

In this discussion, we do not wish to use the additional evidence brought forward by Ishawatari (1985) and by Rocci et al. (1975) which is based on the sequence of crystallization in the layered gabbros. These authors propose that in the lherzolitic ophiolites the sequence of crystallization in the magma chamber is olivine-plagioclase-clinopyroxene and, in harzburgite ophiolites, olivine-orthopyroxene-clinopyroxene-plagioclase, suggesting respectively lower and higher degrees of melting in the mantle source. Our experience in ophiolites suggests that in most situations plagioclase and clinopyroxene crystallize together, following olivine. This is typically the case in Oman and in most other HOTs. As proposed by Juteau et al. (1988), whether orthopyroxene appears early or late in the magmatic sequence depends on local conditions, possibly related to the open or closed character of the magma chamber system ; considering the phase diagram and the chemistry of the magma, subtle changes may readily shift the sequence of crystallization.

### *8.2.3. Harzburgite and lherzolite types of ophiolites and oceanic environments*

The difference in the degree of melt extraction between HOT and LOT results from different thermal regimes and this suggests that they originated in different types of spreading centers. In all the HOT massifs there is evidence of intense magmatic activity in the uppermost mantle up to the Moho level while high temperature plastic flow was still occurring. As seen in chapter 10, this produced the dunites, the magmatic impregnations, the chromite pods and the wehrlitic intrusions. Thus, in the HOT case,the adiabatic conditions achieved in the rising mantle beneath the ridge were maintained all the way up to the base of the crust. A cooler regime in the LOT case is suggested by (1) the scarcity of layered gabbros in the crustal sequence, (2) the fact that the feldspathic melts produced during adiabatic ascent have crystallized during the plastic flow related to the ascent (§2.5.3) and (3) the higher mantle viscosity than in HOT deduced from the higher flow stress (see table 8.1).

In a rising asthenosphere below a spreading center, adiabatic conditions are preserved until the lithosphere is met. The subsequent cooling due to heat conduction with a thermal gradient steeper than the peridotite solidus, thus prevents further melting during ascent (fig. 7.2). The limited melt extraction than 30 km implies for LOT, a thicker lithosphere than for HOT (fig. 2.11). In this latter case, hypersolidus conditions are maintained up to the Moho. The lithosphere is accordingly reduced to the 6 km thick crust and even less just below a magma chamber. To explain the differences of lithosphere thickness between HOT and LOT, we will now consider the role of spreading rate or the local influence of a transform fault.

*Spreading rate* - In mantle diapirs, departure from adiabatic conditions is estimated to be at a depth of about 20 km, with partial melting ceasing soon after, if the spreading rate does not exceed 1 cm/yr (§ 7.2.2.). This is due to the penetration of the conductive cooling front at these depths within the mantle. For these low spreading rates, only about 10-15% melt is extracted, resulting in a lherzolitic residual mantle, a thinner crust, and possibly a more alkaline magmatism (Allegre and Bottinga, 1974). For spreading rates larger than 1cm/yr, adiabatic conditions are maintained throughout the mantle section and melting ceases at the Moho, 20% melt is extracted, resulting in a harzburgite residue, a thicker crust and in a tholeiitic magma. These theoretical estimates are supported by Reid and Jackson's (1981) seismic studies which show that below 1-2 cm/yr spreading rates, the total thickness of oceanic layers becomes thinner than 6 km.

Furthermore, in HOT the layered gabbros unit is thick and continuous suggesting the existence of a steady state magma chamber below the ridge of origin, whereas in LOT this unit is absent or discontinuous suggesting the presence of episodic magma chambers. Geophysical data on oceanic ridge structure discussed in § 11.4.1., accordingly indicate that for medium to fast spreading rates, long-lived magma chambers can be present while for slow spreading rates the magma chambers, if any, are only episodic. Gabbros recovered from slow spreading ridges are commonly sheared (Chernysheva, 1970 ; Miyashiro et al., 1971 ; Bonatti et al., 1975 ; Aumento et al., 1977 ; Helmstaedt and Allen, 1977 ; Prichard and Cann, 1982 ; Ito and Anderson, 1983 ; Karson and Dick, 1984; Mével, 1987, 1988 ; Leg 118 Shipboard Scientific Party, 1988), a feature seldom seen in HOT gabbros but very common in LOT. Eventually, exposure of lherzolitic serpentinites as sea floor in LOT (§ 4.4) matches similar discoveries made only in slow spreading oceans (§ 11.3).

From this discussion, it may be concluded that HOT are derived from oceanic spreading centers characterized by a spreading rate greater than approximately 2 cm/yr. Conditions for slower spreading rates generating LOT are also achieved in oceanic environments away from hot spots like Iceland (Jackson et al., 1982 ; Karig, 1980). However, as already proposed (Menzies, 1976 ; Nicolas, 1984) a juvenile rifting environment can be a good candidate. For example, Le Pichon and Francheteau (1978) have estimated the spreading rate in the Red Sea at 0.5-0.75 cm/yr for the last 4-5 Ma. The association of continental formations with some LOT ophiolites and the more hybrid nature of their sedimentary cover constitute evidence for a rift origin. However, studies of passive margins contrast the 'non-volcanic' and tectonically stretched type, which it is implicitly referred to here, and the 'volcanic' type with a narrow and thick igneous crust, which is transitional between continental and oceanic crusts (Roberts et al., 1985 ; Mutter et al., 1987). This new igneous crust, 15-20 km thick with 3-5 km of volcanics, evidently differs from the LOT crust.

*Transform fault environments* - Mantle isotherms at spreading centers close to a transform fault are perturbed over several kilometers along strike by the effect of the older and cooler truncating lithosphere wall (fig. 5.14). This transform fault effect increases for decreasing rates of spreading (Fox and Gallo, 1984 ; Sandwell, 1984).

The evidence reported in chapter 5 about transform structures in otherwise HOT ophiolites reveals a striking variability. The Wadi Tayin occurrence in Oman, characterized by a 20 km wide transform domain in the mantle section, with plastic flow structures typical of hypersolidus conditions, evokes a transform fault in a fast spreading environment (§ 5.4.3). The crustal section is similar to that outside the transform domain. This has been explained by accretion together transform activity. In contrast, the other ophiolites interpreted as slow spreading transform domains, Coastal Complex in Newfoundland (§ 5.3), Bogota (§ 5.2.3.) and Tiebaghi-Poum-Belep (§ 5.2.4.) in New Caledonia, or Antalya in Turkey (Reuber, 1984), display shear zones only a few kilometers wide and low temperature plastic flow structures grading to mylonites along the axis of the structure. The foliations are vertical, indicating a large thermal contrast between the two walls of the fault.

The petrological differences are equally striking. Whereas outside these transform domains all the considered ophiolites are typically HOT, this characteristic is preserved only in Wadi Tayin. In Bogota, hydrous phases, mainly amphibole, appear selectively in the sheared domain and in Tiebaghi-Poum-Belep there is a remarkable development of spinel and plagioclase lherzolites. The discussion on the origin of such facies in § 5.2.5 brings us back to the problem of the origin of plagioclase lherzolites in relation to oceanic transform activity.

Plagioclase lherzolites have been repeatedly dredged in the large oceanic transform faults (§ 11.3.1). The two possible origins discussed in § 5.2.5 and illustrated by figure 5.12 can be retained, namely either a pristine origin like in LOT massifs, or a secondary origin by magma impregnation of a harzburgite (Dick and Fisher, 1984 ; Nicolas and Dupuy, 1984). Conditions which inhibit melt extraction in the plagioclase lherzolite field (pristine origin) would be typically met for slow spreading rates when a transform fault brings a lithosphere older than 10 Ma in contact with an accreting asthenosphere (§ 5.2.5). In such conditions the conductive cooling front would reach a depth of 20-30 km, impeding further melting in a rising asthenosphere and hence leaving a plagioclase lherzolite as residual mantle.

*Origin of the LOT and spinel-plagioclase lherzolite massifs* - From this discussion, it appears that similar plagioclase lherzolites can originate in slow spreading environments and in many transform faults. Due to cooling of the melting mantle wedge, the overlying crust can be thinner in both situations. As in LOT, thin crust in transform faults can also explain the occurrence of high temperature antigorite and talc alteration in lherzolites from transform faults (Aumento and Loubat, 1971). Finally, the LOT massifs and the spinel-plagioclase lherzolite massifs have generally steep high temperature foliations and moderately plunging lineations, that might be a geometry expected in a transform region.

It is possible to distinguish between slow spreading and transform environments if the internal structures can be correlated with the geometry of the accreting system of origin. Thus, in the case of Zabargad Island (§ 4.5) a former transform fault interpretation

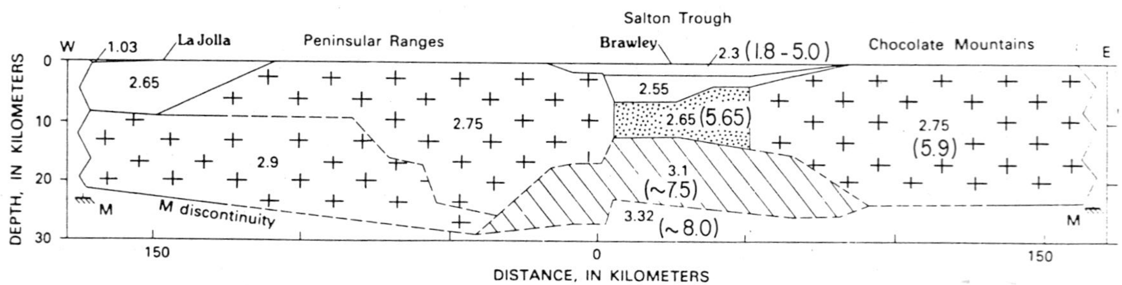


Fig. 8.2. Cross section through Salton Trough based on Fuis et al. (1984) geophysical model of continental rifting ; density estimates in plain numbers, P-waves velocities in brackets (the 7.5 velocity in the hatched domain has been recalculated from the original data). The dotted area is interpreted here as a new type of metasedimentary crust and the hatched area below as an abnormal mantle uprise ; crosses : continental basement (Nicolas, 1985a).

(Bonatti et al., 1983) had to be revised when the internal foliation induced by diapiric uprise was shown to be parallel to the Red Sea rift trend and not perpendicular to it. In addition, the associated lineation plunges at 50° NW instead of being horizontal as expected in a transform system (fig. 4.25) (Nicolas et al., 1986). The LOT massifs in Piedmont-Liguria have also been interpreted as formed in a transform environment (§ 4.4). Although the internal structures in the lherzolites have been shuffled by alpine tectonism, they trend N-S rather than E-W, as in the case for the Lanzo massif where the structures are better organized (§ 4.5), that is parallel to the expected trend of the Mesozoic Piedmont ocean (Lemoine et al., 1987). This suggests that they relate to the N-S rifting rather than to the E-W transform motion. The question is more open for the Gibraltar Arc massifs. The absence of associated ophiolitic formations suggests that they were diapirically emplaced into a continental crust (Loomis, 1975 ; Lundeen, 1978 ; Obata, 1980 ; Reuber et al., 1982 ; Frey et al., 1985 ; Tubia and Cuevas, 1987). Their deeper conditions of equilibration compared to LOT lherzolites is well in agreement with the model of diapiric intrusion into a thicker lithosphere, possibly still continental (§ 7.2.2), with, as a result, a more limited fraction of decompression melting.

Another possible criterion to distinguish slow spreading from transform environment deals with the width of the considered mantle domains. Domains reasonably ascribed to transform faults in ophiolites should not exceed 3-4 km in width, whereas in the Lanzo and Trinity massifs the observed widths are respectively 10-15 km and 50 km. A 3-4 km wide disturbance in uppermost mantle structures already requires a large fault ; a 20 km disturbance would require a transform fault as large as the Vema transform (Detrick et al., 1984). Moreover, in Lanzo or Trinity the structures are only locally mylonitic.

Finally, the common association of a few LOT with continental formations and shallow-water sediments points to a rift environment rather than an oceanic transform fault. Consequently we favor, for the lherzolite massifs considered here, the earlier interpretation proposed by Menzies (1976) ascribing, the Othris lherzolitic ophiolite with typical LOT features, to an incipient ocean forming in a rift opening at spreading rates in the range of or below 1 cm/yr. The reconstructed structure of the Trinity Complex (fig. 4.15), constitutes an example of the general structure ascribed here to an oceanic rift environment (fig. 9.4). This does not exclude however, that other LOT can also occur in oceanic lithosphere segments generated along transform faults.

Spinel-plagioclase massifs, like those from the Gibraltar arc and from Zabargad Island, reflect a smaller degree of melt extraction and are not associated with typical oceanic crust; they would correspond to a situation of important continental crust thinning or to a stage of continental rifting which has not yet evolved into an oceanic situation (Biju-Duval et al., 1979) (fig. 7.3a). These massifs are in contact with high temperature metamorphic rocks which probably represent slivers of deep continental crust upthrust together with the mantle intrusion, commonly with a large addition of basaltic material issued from the melting of the mantle diapir (magmatic underplating) (§ 4.5.4). It has been proposed that some of the metamorphic formations at their contact may also derive from sediments deposited in a trough, on top of the lherzolite diapir. They would have been heavily intruded by basalts derived from this diapir and metamorphosed up to granulite facies condition at the contact with the hot peridotites. Due to a competition between sedimentation rate and basalt discharge, turning locally in favor of the former, a 'normal' oceanic crust would not develop, but instead this mixture of sediments and basaltic intrusions would (Nicolas, 1985a). This situation of lherzolite diapirs breaking through continental crust and intruding the bottom of narrow troughs, like the Salton Trough in Southern California (Fuis et al., 1984 ; fig. 8.2), would constitute one of the possible transitions between continental and oceanic rifts.

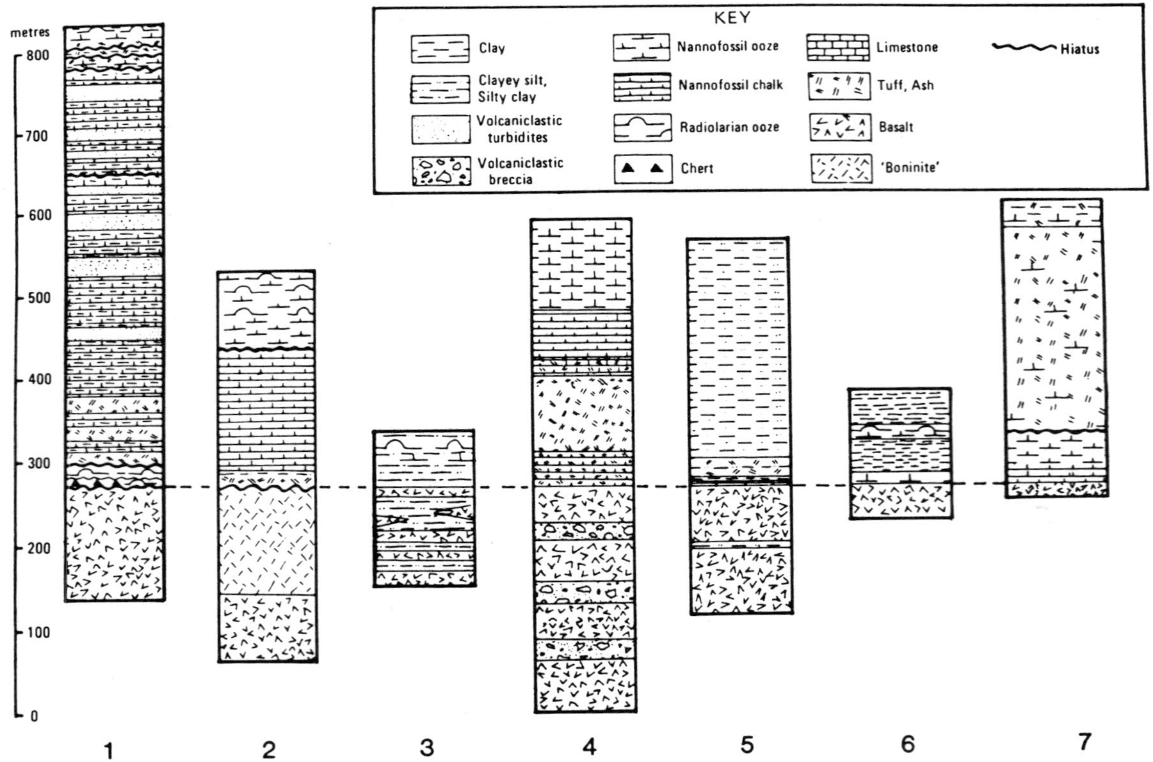


Fig. 8.3. Stratigraphic sequences overlying igneous 'basement' in modern arc elements. From DSDP sites: 1 - 459, 2 - 458, Mariana terrace ; 3 - 454 Mariana Trough ; 4 - 448, Palau-Kyushu Ridge ; 5 - 442, Shikoku Basin ; 6 - 449 Parece Vela Basin ; 7 - 205, South Fiji Basin. 1-2 : fore-arc ; 3 : active back-arc basin ; 4 : remnant arc ; 5-7 : inactive back-arc basin (after Leitch, 1984).

## 8.3. *ISLAND-ARC, BACK-ARC OR MID-OCEAN OPHIOLITES*

### 8.3.1 *Geochemical characteristics*

In the early days of plate tectonics, ophiolites were first regarded as oceanic lithosphere fragments generated at mid-oceanic spreading centers. This view was challenged by Miyashiro (1973, 1975), who proposed on the basis of major and trace element data, that ophiolites could also form in island arcs. On the basis of major elements analysis, this author recognized two sub-groups among the lower pillow lavas and sheeted dikes of the Troodos ophiolitic complex. The first presents a typical calc-alkali trend with a rapid rate of $SiO_2$ increase and a rapid decrease of FeO and $TiO_2$ with the increasing FeO/MgO ratio. The second group is tholeiitic marked by a slower rate of $SiO_2$ increase associated with a progressive enrichment of $TiO_2$ and FeO.

At the same time, Pearce and Cann (1973) and Pearce (1975) proposed elaborate variation diagrams involving the least mobile elements towards alteration processes such as Ti, Cr, Zr, Y and Rare Earth Elements for distinguishing low-K island arc tholeiites, calc-alkali basalts and ocean floor basalts from one another. Such diagrams have been further extensively used to characterize the geological setting of ophiolitic complexes. However in the Ti vs Zr diagram, the data from a given ophiolitic complex (e.g. Bay of Islands) generally overlap the three fields (island arc, calc-alkali and ocean floor basalts) defined by Pearce (1975). On the other hand, in the Ti vs Cr diagram, most of the data plot in the island arc tholeiitic field. Such diagrams suggest that most well known ophiolitic complexes contain calc-alkali components. The difficulty in reconciling these data in a simple way with an origin at major ocean ridges has lead to a quasi-general consensus among geochemists who conclude that most well studied ophiolites around the world represent the crust of small ocean basins adjacent to, or inside island arc systems, either in fore arcs, immature island arcs or back-arc basins (Cameron et al., 1980 ; Alabaster et al., 1982 ; Hawkins and Evans, 1983 ; Moores et al., 1984 ; Coleman, 1984). In this discussion, one should bear in mind that the data on incompatible elements concern only the extrusives, and that i*n many documented cases, the diversity of patterns is due to lavas in a supra-ophiolite position,* capping those associated with the cooling lithosphere, which generally have a MORB signature. Thus if a geodynamic environment can be assigned to particular element ratios, it may reflect the oceanic history of a fragment of oceanic lithosphere subsequent to its formation at a spreading center, and not necessarily correspond to the site of its creation which is recorded in the lavas directly in contact with the plutonic section of the ophiolite. The complex history of ophiolites, as recorded in their extrusive section, has been stressed by Pearce (1975, 1979), Menzies (1976), Hopson and Frano (1977) and Stern (1979).

In the mantle section of ophiolites, the most commonly studied isotopic parameter $\varepsilon_{Nd}$, varies between +7 and +12 values, indicative of depleted sub-oceanic mantle. Similar values are obtained for the associated gabbro section, suggesting parental relationships. This is the case for Bay of Islands (Jacobsen and Wasserburg, 1979), Oman (McCulloch et al., 1980 ; Richard et Allègre, 1980), and partly for Trinity (Jacobsen et al., 1984). The high values of $\varepsilon_{Nd}$ are correlated with depleted chondrite-normalized REE patterns, overdepleted in LREE as in Bay of Islands (Suen et al., 1979), Troodos (Kay and Senechal, 1976) and Trinity (Lapierre et al., 1987). In the peridotites from Oman (Pallister and Knight, 1981) and from New Caledonia (Prinzhofer and Allègre, 1985 ; Nicolas and Dupuy, 1984), depleted REE patterns are obtained, with occasionally a LREE slight enrichment (U-shaped REE pattern) which are interpreted as resulting from contamination by a magma percolating through a previously depleted mantle (Nicolas and Dupuy, 1984 ; Navon and Stolper, 1987).

More variable REE patterns and Nd isotopic data characterize the extrusives. The large set of REE data are grouped into three trends : a MORB trend, an overdepleted trend, and a LREE-enriched trend. These variable trends are present in Bay of Islands (Suen et al., 1979), Oman (Pallister and Knight, 1981 ; Lippard et al., 1986 ; Ernewein et al., 1988), Troodos (Kay and Senechal, 1976 ; Smewing and Potts, 1976 ; Cameron, 1985 ; Rautenschlein et al., 1985 ; Taylor and Nesbit, 1988), Othrys (Menzies, 1976), East Taiwan (Jahn, 1986), Sarmiento, Chile (Stern, 1979) and Trinity (Lapierre et al., 1987). In Troodos, low $\varepsilon_{Nd}$ correlate with a LREE-enriched patterns (McCulloch and Cameron, 1983).

In Oman, where the lavas stratigraphy is well depicted, the REE patterns correlate with the chronology of lavas (see § 3.3.2) : $V_1$ lavas lie in the MORB field, $V_2$ tend toward the overdepleted field, and $V_3$ are LREE-enriched (Lippard et al., 1986). In the Troodos, overdepleted REE pattern with occasionally a slight LREE enrichment characterize the upper lavas. Hence, in both cases, patterns diverging from the MORB trend characterize lavas in supra-ophiolite position (data in other areas are not precise enough to control this chronology). This has however large implications in retracing the ophiolite history and dynamics of detachment. In the Oman case, again, a genesis of the ophiolite in an arc-basin environment has been proposed on the basis of lavas geochemistry by Pearce et al., (1981), Alabaster et al. (1982) and Beurrier (1987) (§ 3.5.3). On the contrary, Boudier et al. (1988), and Ernewein et al. (1988) have proposed a mid-oceanic origin for the ophiolite and explained the island arc characteristics of the subsequent volcanism by a phase of magmatism occurring immediately after the crust generation, during the initiation of the oceanic detachment (§ 3.5.3.). This detachment would take place at the ridge itself while the ridge activity was waning (§ 3.4.3). The last melts extracted from the mantle could react with the seawater contained into the underthrust crust and/or with the products of its hydrous remelting. Dacitic products could derive directly from crust remelting.

Whether this last interpretation of the secondary magmatism in Oman is accepted or not, it has the merit to recall that ophiolites have been through a sequence of events, in particular those connected to their emplacement, which have not been recognized or identified in the various modern spreading environments to which they are referred. Before their obduction, some ophiolites seem to have been emplaced by oceanic thrusts in a fore-arc environment (§ 12.4.2, fig. 12.7b), a situation which could also account for contamination by island arc magmatism.

Related to this is the problem of the boninites in ophiolites. With reference to their occurrence in the Bonin immature island arc (Hickey and Frey, 1982), the discovery of some boninites or affiliated volcanic rocks in many ophiolites has supported the idea that such ophiolites derive from immature island-arcs. Alternatively, Casey and Dewey (1984) have suggested that possible conditions for boninitic magma generation is the initiation of subduction at a transform or along a ridge segment. We suggest here that some boninitic and picritic lavas may represent volcanic products associated with the wehrlitic magmatism, now well documented in a few ophiolites (§ 3.3.2) (table 8.I). Their magnesian character could be ascribed to the two causes discussed in § 10.3 and 10.4, i.e. their charge in olivine xenocrysts and the dissolution of orthopyroxene from the harzburgites into the melt which would also displace the original tholeiitic composition towards that of boninite.

### *8.3.2. Other criteria*

To conclude the preceding discussion, the use of geochemical discriminant diagrams to identify the oceanic environment of origin of a given ophiolite should be treated with much caution. The respective volume of the various volcanic products associated with the considered ophiolite should be accounted for, and their relation with the crust generation, in particular the timing of their injection (§ 3.4.3) better constrained. Other criteria, discussed below, should also be looked for.

The ophiolite nature and history should be more carefully taken into account. As an example, paleogeographic reconstructions suggest an island arc origin for the Canyon Mountain ophiolite (chapter 6) which has also two specific characteristics expected in an island arc environment : the dike swarm unit is not basaltic but keratophyric and a pervasive high temperature hydrous activity is recorded both in the upper mantle section (abundant amphibole-rich impregnations) and in the crustal section (hydrous metamorphism and anatexis). Finally, the overall structure, characterized by mantle diapiric intrusions penetrating high into the crustal section at high temperature, has not been reported previously in any other ophiolite.

Eventually, more attention should be paid to the nature of the sediments deposited on the ophiolitic crust as already pointed out by Moores (1982). Present day sedimentary sequences overlying the volcanic basement of arc systems (fig.8.3), are predominantly greywackes or andesitic tuffs. When similar sediments are found overlying an ophiolite crust as in the California Cordilleran ophiolites (Moores, 1982), the Betts Cove (Williams and Malpas, 1972) or Baie Verte (Kidd, 1977) ophiolites in Newfoundland, or the Karmoy ophiolite in Norway (Pedersen, written comm.), the arc-related origin proposed for these ophiolites is more convincing. Alternatively, when the sediments are only umbers or radiolarites as in Oman, suggesting an environment below the carbonate compensation level and away from sedimentary sources, a mid-oceanic ridge situation should not be dismissed, although radiolarites may also form in small bassins, transform-dominated or arc-related (Jenkyns and Winterer, 1982).

# Chapter 9

# MANTLE FLOW, LITHOSPHERIC ACCRETION AND SEGMENTATION OF OCEANIC RIDGES

## *9.1. INTRODUCTION*

Lithospheric accretion and mantle flow pattern beneath oceanic spreading centers are studied here following two approaches. The main one relies on geophysical data collected at oceanic sites and the other, on geological observations made in ophiolites.

Geophysical studies suggest two kinds of models for the asthenosphere flow and lithosphere accretion at the ridges. In the first model, the lithosphere is created in its entire thickness at the axis, by a vertical and narrow ascending asthenospheric flow, which is very rapidly frozen. This is the 'dike intrusion model' or 'plate model' (McKenzie, 1967 ; Cann, 1974 ; Kusznir, 1980). In the second model, the plate thickens away from the ridge and grows progressively at the expense of a near horizontal asthenospheric flow. This is the 'thickening plate model' or 'half-space model' (Langseth et al., 1986 ; Parker and Oldenburg, 1973 ; Forsyth, 1977). Both models provide a good explanation of the regional variations of heat flow, bathymetry and gravity away from the ridge (Sclater and Francheteau, 1970 ; Parsons and Sclater, 1977) up to ages around 70 Ma ; in older oceanic basins these geophysical parameters do not show further significant variations with seafloor age. However, these models are not adapted to explain local variations close to the ridge, for seafloor ages ≤ 5 Ma. The dike intrusion model can help to explain the presence of an axial valley for slow spreading centers (Lachenbruch, 1976), but fails to explain the structure of fast-spreading ridges (Sleep and Rosendahl, 1979), although in this case the limited area of magmatic activity at the ridge suggests a narrow upwelling flow in the asthenosphere (Bottinga and Allègre, 1978 ; Macdonald, 1982 ; Choukroune et al., 1984).

In the early 80's, it was discovered that the East Pacific Rise was segmented along its axis by overlapping spreading centers and transform faults (Macdonald and Fox, 1983 ; Lonsdale, 1983). This rapidly led to the idea that the segmentation could reflect deeper processes taking place in the mantle and, in particular, a partition of the rising asthenosphere into diapirs due to the partial melting occurring in the mantle (fig. 9.9). The punctuated character of the volcanic activity along the Red Sea also suggested that in slow spreading environments, the asthenospheric flow could be partitioned into discrete diapirs (fig. 9.10). Models of such diapiric intrusions were presented by Whitehead and his co-workers (1984).

Structural and petrological studies in ophiolites provide independent information on the local dynamics of the asthenosphere at the immediate vicinity of the ridge. In this approach, it is assumed that : 1) ophiolites can be equated with oceanic lithosphere created at spreading centers, 2) the structural frame of the spreading center can be retrieved from the ophiolite considered (§ 2.2), and 3) the plastic flow structures produced by asthenospheric flow beneath the spreading center can be unambiguously identified in the ophiolite considered and related to the spreading center structural framework (§2.5.5.). Provided these conditions are fulfilled, this source of information can be used to obtain more precise models on the asthenospheric flow and how new lithosphere is accreted.

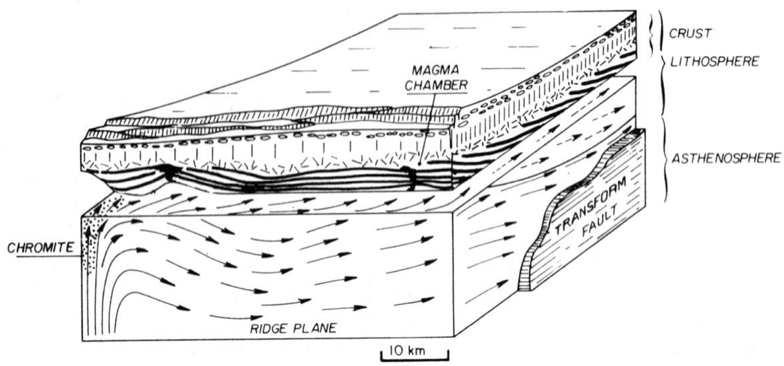


Fig. 9.1. Model of asthenosphere diapirism based on structural data in harzburgite type ophiolites. The mantle flow diverges in every direction from 10 km wide diapirs. It is progressively channelled at a right angle to the ridge by the cooling effect of transforms. Solid arrows : slip lines frozen in the lithosphere ; bold lines : layered gabbros ; vertical lines : dike swarm ; ellipses : volcanics (redrawn from Nicolas and Violette, 1982).

Although many spreading center models derived from ophiolites incorporate a picture of the flow pattern in the ultramafic tectonites, this picture is usually poorly constrained because it does not rely on structures measured in the ultramafic tectonites. The first attempts at incorporating such structural data are those of Avé Lallemant (1976 ; § 6.4) and Juteau et al. (1977). Since then, detailed structural studies conducted in over a dozen of ophiolitic massifs or districts where the conditions mentioned above apply well have led to the increasingly sophisticated models which are presented below. These models rely on the method of kinematic analysis in plastically deformed rocks presented in § 2.5.4 and on general considerations (§2.6.) of the expected asthenospheric flow patterns below spreading centers.

These detailed structural studies in ophiolites have revealed that, in extreme cases, both the 'thickening plate' and the 'dike intrusion' geophysical models apply in ophiolites with relation to their harzburgitic or lherzolitic nature respectively. Incorporating other characters, the Harzburgite Ophiolite Type (HOT) and the Lherzolite Ophiolite Type (LOT) have been presented in chapter 8 and discussed in terms of spreading rate at the ridge or rift of origin. We will see below how the thickening plate model corresponds to HOT situations and to medium to fast spreading rates and how this is illustrated by the study of the flow patterns in the Oman ophiolite. Alternatively, the dike intrusion model corresponds to LOT situations and to small spreading rates which can be illustrated by the Trinity and other lherzolitic ophiolites.

Another result of structural studies in the ultramafic tectonites of ophiolites has been the discovery of structures typical of diapiric intrusions in the Zambales massif (Philippines) and in the Troodos massif (fig. 9.1). Since then, other similar structures have been reported in the Oman ophiolite, where four diapirs are mapped and can be partly related through mantle flow structures (see below). These discoveries made in HOT may be extended to LOT as discussed below. They were used to develop models of diapiric intrusions in the upwelling and partially melting mantle, driven by plate tectonics (Rabinowicz et al., 1984, 1987).

The dimensions of the mapped diapirs prove that the asthenospheric flow structure below a ridge is much smaller than the 100 km currently proposed by geophysicists, and illustrated for the thickening plate model by figure 7.6. The size of the diapirs, in the 10-15 km range, is well constrained when they can be directly mapped as in Oman (figs. 3.17, 3.20 and 3.21) or more indirectly as in the Massif Central (France) or Eifel (West Germany) volcanic rifts, where they are identified by the nature of mantle xenoliths sampled by volcanic vents (Nicolas et al., 1987 ; Witt and Seck, 1987). Geophysical distribution of xenoliths and thermal modelling of their cooling history has led to the model in figure 7.3a for a continental rift.

## *9.2. MANTLE FLOW IN THE OMAN OPHIOLITE*

### *9.2.1. Introduction*

On the basis of the structural maps of high temperature flow in Oman peridotites (fig. 3.8), Ceuleneer et al. (1988) have distinguished typical flow patterns . Three of them (figs. 9.2, a,b,c) are ascribed to steady state conditions as defined in § 2.6, meaning that they illustrate situations in which the asthenospheric flow is progressively frozen into a lithospheric structure during drift away from the ridge. The fourth one, the diapiric structure (fig. 9.2d), would correspond to the sampling of a mantle flow situation while flow was active. This seems possible in Oman because the oceanic detachment occurred in the vicinity of the ridge (§ 3.4.3). Such diapirs do not represent necessarily ridge features; they may underlie off-axis volcanoes, propagating ridges, and other non-permanent

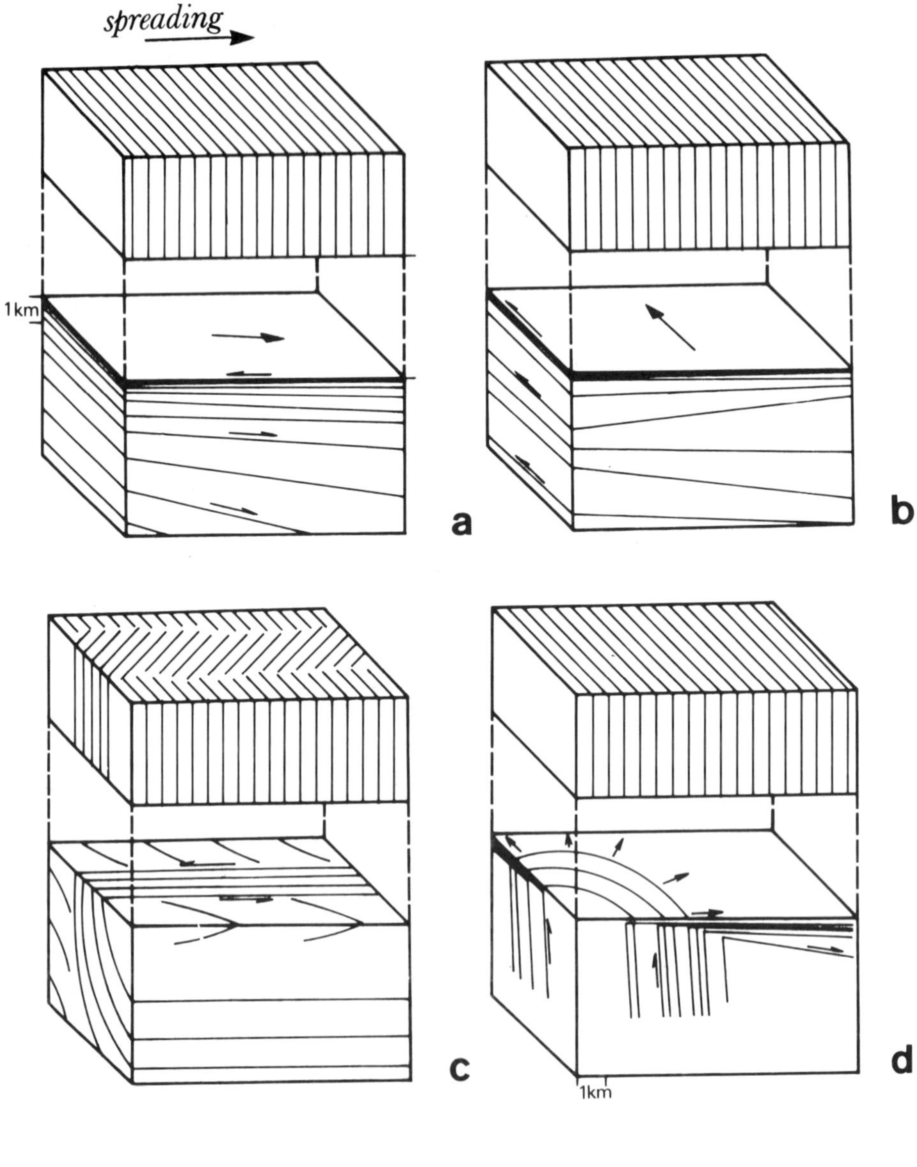


Fig. 9.2. Main mantle flow situations in the Oman ophiolite. Lower parts of blocks : deforming mantle with slip planes and arrows indicating the direction of flow and the associated shear sense ; the tightening of slip planes reflects the shear strain. Upper parts of blocks : crustal section hatched parallel to the diabase dike swarm. a) Homogeneous flow away from the ridge. b) Flow channelled parallel to the ridge. c) Flow in a transform fault. d) Flow in and around a diapir (redrawn from Ceuleneer et al., 1988).

crustal edifices. It is believed that wherever they punctuate the accreting lithosphere, mantle diapirs should be much alike because they should originate as similar, small and local gravity instabilities in the extended melting layer produced at some 50 km depth by the uprise of an asthenosphere cell or shoulder (Rabinowicz et al., 1984, 1987 ; chapter 7). Combining the information on accidental diapiric and other steady state mantle structures, it is possible to reconstitute the mantle flow pattern and the lithospheric accretion beneath a ridge.

### *9.2.2. Homogeneous mantle flow away from the ridge - Relation with seismic anisotropy*

In this configuration, the high temperature foliations in peridotites are parallel to the Moho defined by the surface of transition between peridotites and layered gabbros ; the mineral lineations are oriented at high angles to the trend of the diabase sheeted dike complex. In the referential of § 2.2 and with the kinematic assimilation of § 2.6, this corresponds to a flat-lying flow surface in the mantle below a ridge with flow lines normal to the ridge axis. This configuration, illustrated by figure 9.2a is the most common in the Oman ophiolite. It also corresponds to the flow pattern in the uppermost mantle below a fast spreading ridge as predicted by the anisotropy of seismic wave propagation.

Seismic anisotropy has been related to plastic flow directions in the peridotites through the plastically induced lattice fabrics, a theory which has recently been reviewed (Nicolas and Christensen, 1986). As a result of large plastic flow, intracrystalline slip in olivine has the effect of aligning the [100] slip direction parallel to the flow direction (§ 2.5.4) and, commonly, to orient the [010] axis, which is normal to the main slip plane, at a high angle to the flow plane. The [100] crystallographic direction happens to be the fastest direction for P waves (9.87 km/s at room pressure and temperature) and [010] the slowest direction (7.73 km/s). As a consequence, it is possible to make, from the seismic anisotropy pattern, some inference about the in situ plastic flow orientation. This plastic flow can be frozen if the anisotropy is measured in the lithosphere, or still active if the anisotropy is measured in the asthenosphere. The fact that the largest P or S wave velocity is parallel to the spreading direction indicates that the flow direction should be close to the spreading direction. The large degree of anisotropy measured in the oceanic mantle is compatible with a sub-horizontal orientation of the (010) slip plane of olivine.

The consistency between the seismic structure of present-day oceanic upper mantle and the tectonic structure of most ophiolitic peridotites leads us to attribute the monotonous flow pattern of figure 9.2a to the accretion of the lithospheric mantle under steady-state spreading conditions, after rotation of the ascending flow into a horizontal attitude at some distance from the ridge axis.

Within the top 500 m of the mantle section in Oman, an increase in shear strain and a reversal in shear sense with respect to deeper horizons have been described (§ 3.4.2) (fig. 2.2). Such a vertical evolution of the mantle flow structure had been already reported in other ophiolites where the shear sense inversion also occurs, however at a deeper level (Prinzhofer et al., 1980, fig. 5.2 ; Girardeau and Nicolas, 1981 ; Nicolas and Violette, 1982).

### *9.2.3. Channeling of mantle flow along the ridge axis*

This configuration is characterized by mineral lineations being regionally horizontal and parallel to the sheeted dike strike ; the foliations are generally parallel to the Moho, but locally in Oman, they are poorly defined and in zone around the lineation, in agreement with the dominantly linear character of the deformation (L-tectonites) (fig. 9.2b). The

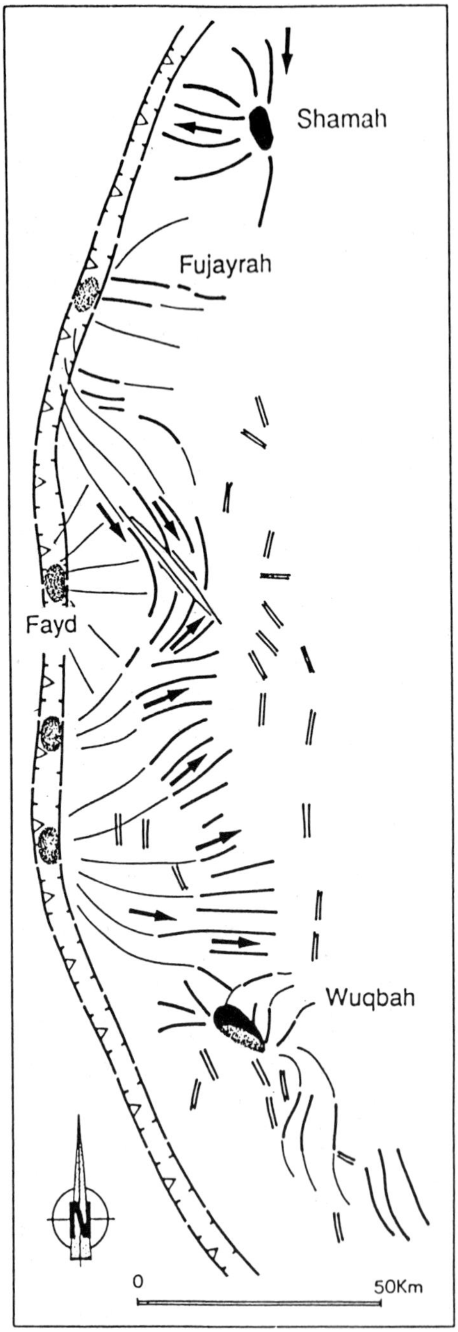


Fig. 9.3. Interpretative map of the asthenospheric flow pattern below the ridge where the northern Oman ophiolite originated (location in fig. 3.8a). Black and shaded areas : mapped and inferred diapirs respectively; bold and fine lines : measured and inferred flow trajectories respectively with arrows pointing to relative motion of upper mantle compartment ; short double lines : average orientation of sheeted dike complex ; facing barbed lines : presumed ridge of origin ; dashes with triangles : front of the nappe and presumed detachment locus (Nicolas et al., 1988).

strain is very large and no shear sense reversal has been noted in vertical sections.

### 9.2.4. *Mantle flow in transform faults*

This configuration is ascribed to a transform fault activity because the foliations are steeply dipping and bear a mineral lineation which is at a right angle to the regional azimuth of the sheeted dike complex (fig. 9.2c). It is met only in the Wadi Tayin area of Oman (§ 5.4). Compared to other transform configurations in ophiolitic peridotites (chapter 5), this transform domain is much wider, its deformation is of the high temperature type and the foliations are not vertical. The 50-60° dip of foliations observed here indicates that the lithosphere isothermal surface of the cooler wall of the tranform fault was not vertical but inclined (§ 2.6), suggesting that the cooler wall was not deep and that, therefore, the difference in age between the adjacent lithospheric segments was small (§ 5.4.3).

### 9.2.5. *Mantle flow in diapirs*

With respect to former descriptions (fig. 9.1), diapiric structures in peridotites just beneath the mafic crust are best examplified in the Oman ophiolites where four such structures have been mapped. The zone of vertical mantle flow in such diapirs can be viewed as a pipe slightly elliptical in cross section. It attains 15 km parallel to the ridge axis and, in the direction perpendicular to the axis, its width does not exceed 13 km, a value also ascribed to the basal width of the magma chamber (figs. 11.8 and 11.14). At the top of the pipe, the flow rotates into a horizontal attitude and spreads in all directions (fig. 9.2d).

The most unexpected and consequential structural feature of the Oman mantle diapirs is the *extraordinary thinness (300 m) of the zone in which the vertical flow rotates into a horizontal attitude* (fig. 3.17c). In a constant viscosity mantle, this zone might have been expected to have a vertical thickness equal to the radius of the vertical channel. The fact that it is more than ten times thinner implies that there is a major rheological discontinuity at the top of mantle diapirs (§ 7.5).

### 9.2.6. *Mantle flow patterns beneath the Oman paleo-ridge*

The four mantle flow configurations distinguished in the Oman ophiolite can be incorporated into a general pattern by analyzing the maps of figure 3.8. Ideally, it should be possible to trace the mantle flow pattern and the ridge segmentation for the entire ophiolitic belt. Unfortunately, the relation between massifs is locally obscured as a consequence of both the tectonic events which occurred during emplacement and of an insufficient structural coverage and kinematic results in a few massifs. A preliminary regional interpretation has been proposed for two areas by Nicolas et al. (1988) who discuss more fully its tenets and limits. The interpretation for the northern area (fig. 9.3) illustrates how regionally the various configurations could interact. The figure is a simplified map of the trend of the high temperature mantle flow line trajectories. Important distortions appear in areas where the foliations have been tilted. Fortunately, this is regionally the case only in the Fayd-Rajmi area. As a first approximation, the data of figure 9.3 can be, therefore, regarded as representing the mantle flow line trajectories in the asthenosphere of a spreading domain just beneath the Moho of the newly accreted lithosphere. The field documents must be improved before definitive conclusions can be drawn.

The detachment, thought to correspond to the tip of the present day ophiolite nappe, occurred in a very young lithosphere in the ridge vicinity (§ 3.4.3). The flank of the paleoridge from which a piece of ophiolite is issued, can be inferred from the sense of shear flow in the mantle section (§ 2.2). Applying this admittedly speculative rule in northern Oman, it is deduced that the ridge was located to the West of the present massifs. For the sake of simplicity, it is assumed that the detachment took place along the ridge itself. This is why the paleoridge is located in figure 9.3 along the tip of the ophiolite nappe.

A consequence of this analysis is that the Shamah and Wuqbah diapirs are not located on the main ridge. This would not be the case in the Maqsad massif, where the shear sense is reversed on each side of the diapir, indicating that it was located on the ridge itself. Thus mantle diapirs may be located either on the main ridge like in Maqsad or off-axis as in Shamah or Wuqbah, which could be discrete structures located beneath off-axis volcanoes or be part of propagating rifts. Presumed diapiric areas along this ridge have been shaded in figure 9.3. Their existence is deduced from petrological considerations (§ 3.4.2). Other interpretations are also acceptable with, for instance, all diapirs being located along a ridge itself segmented by wide spaced overlapping spreading centers (Nicolas et al., 1988).

Whichever model is preferred, the divergence of flow lines from the diapirs is confirmed with a tendency for them to be channelled parallel to the ridge axis closer to the diapirs and clearly at a high angle to the ridge axis further away below an older lithosphere, that is easterly in the ophiolite nappe.

## *9.3. MANTLE FLOW IN THE TRINITY OPHIOLITE AND LHERZOLITE MASSIFS*

The structural model built for the lherzolitic Trinity ophiolite, on the basis of the field measurements (fig. 4.15) is mainly characterized by steep foliations and horizontal to moderately inclined lineations which in azimuth are parallel to the ridge axis as defined by the sheeted dike complex. Similar conclusions are reached in the case of the spinel-plagioclase lherzolite massifs described in § 4.5.3., where the ridge referential can be identified. For example, on Zabargad island foliations are steep and strike parallel to the Red Sea axis and lineations plunge at 50°, which is the largest plunge so far recorded in this type of massif. The question of the environment of origin for ophiolites which belong to the lherzolite ophiolite type has been discussed (§ 8.2.3) and a slow spreading environment, like that of an oceanic rift, has been favored for Trinity, Zabargad, Lanzo and the other Ligurian and Corsican massifs (see also § 4.4 and 4.5.3). A continental rift origin is proposed for the Gibraltar Arc massifs.

The high-temperature structures measured in these various massifs lead to the general model sketched in figure 9.4. In spite of the fact that foliations are steep and lineations moderately plunging, they cannot be incorporated into a transform fault model, at least in massifs where the conditions of spreading are somewhat constrained. Accordingly, in Trinity, Lanzo or Zabargad the lineation trend is parallel and not perpendicular to the expected ridge direction. Caution is however necessary when dealing with plagioclase lherzolites because this peridotite facies is commonly associated with transform faults in present oceanic situations (§ 8.2.3).

The model of figure 9.4 explains the steep foliations observed in these massifs as a consequence of the steep attitude of the lithosphere-asthenosphere boundary at the very shallow depth considered here. The steep boundary is itself interpreted as a consequence of a slow spreading rate (§ 2.6, fig. 2.10), possibly following a stage of lithospheric tensional fracturing. This defines an asthenospheric channel narrowing upward along the ridge axis. The gravity driven mantle diapirs rising from the partially molten underlying mantle uprise (Rabinowicz et al., 1984, 1987 ; Whitehead et al., 1984 ; Bonatti, 1985) when penetrating this narrowing channel would diverge with *flow directions presenting a large horizontal component parallel to the ridge axis* ; this is now recorded in the

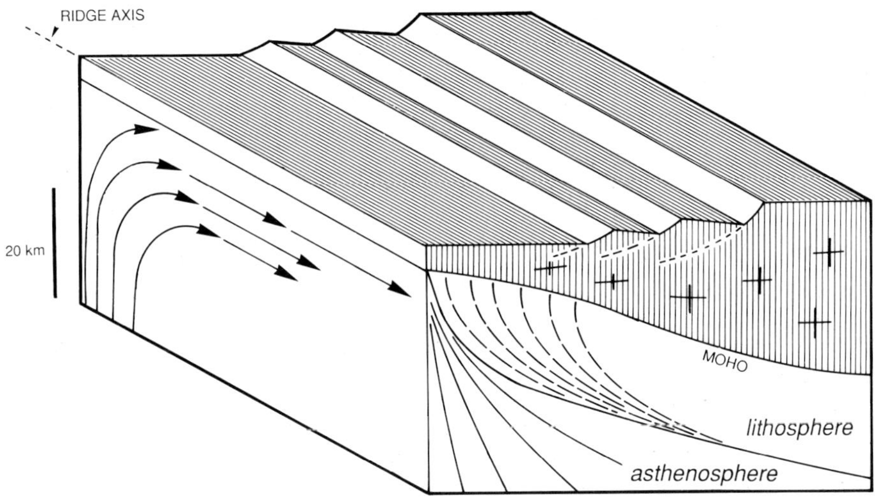


Fig. 9.4. Model of asthenospheric diapirism in solid lherzolite type ophiolites. The asthenospheric flow uprising from diapirs is channelled along the rift axis by steep lithospheric walls. Arrows : traces of the slip lines seen in the axial plane of the ridge ; solid lines : traces of the active slip plane and dashed lines : traces of fossil slip planes in the lithosphere. The crustal section is hatched parallel to the diabase dike swarm (redrawn from Le Sueur and Boudier, 1986).

moderately plunging lineations. Parmentier and Forsyth (1985) have ascribed the deepening of the median valley of slow spreading ridges toward ridge-transform intersection to the dynamic effect of this longitudinal asthenospheric flow between channelling lithospheric walls. Finally, seismic anisotropy measurements below the Rhine graben in the same depth range also show that the higher velocity in the horizontal plane, related to the flow direction (§. 9.2.2), is roughly parallel to the rift elongation (Fuchs, 1983).

## *9.4. MANTLE DIAPIRISM AND RIDGE SEGMENTATION*

### *9.4.1. Introduction*

We wish to discuss here the origin of the diapirs mapped in ophiolites and to see how they might relate to the segmentation of oceanic ridges evoked above. The diapirs identified in ophiolites are at the scale of 10 km and their spacing is about 50-100 km. Clearly, they should not be confused with mantle plumes and related hot spots which from Iceland to Azores and Tristan de Cuhna punctuate to Mid Atlantic Ridge on a scale of a few thousands of kilometers and originate deep in the mantle (Wilson, 1973 ; Schilling, 1973). We will also look for the scale of the ridge segmentation which correlates with the spatial periodicity of these mantle instabilities and address the question of the stability in time of the diapirs.

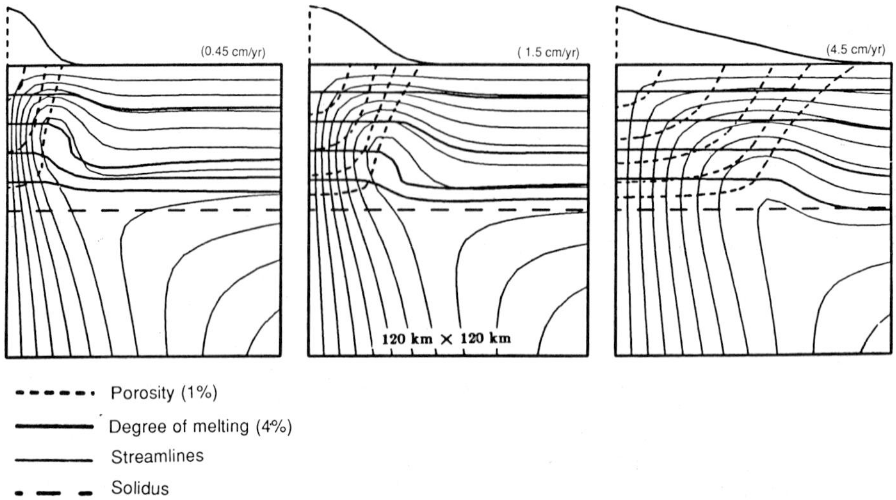


Fig. 9.6. 2D-Models of asthenosphere diapirism for increasing spreading rates. The contours of porosity (small dashed line) are at intervals of 1% ; the contours of the degree of melting (bold continuous line) are at intervals of 4% ; the contour interval used for the stream function (fine continuous line) is proportional to the spreading velocity. The curve above the mantle box represents the spatial distribution of melt in the crust (Scott and Stevenson, in press).

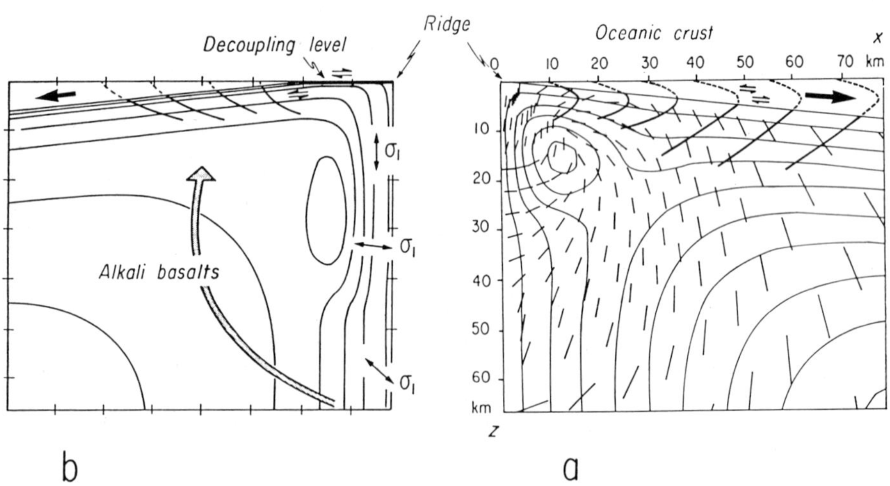


Fig. 9.5. 2D-models of asthenosphere diapirism for a 5 cm/yr spreading rate. a) First numerical model (for the physical parameters see the text and Rabinowicz et al., 1984) ; fine continuous lines : slip lines ; short lines : $\sigma_1$ stress trajectory ; bold continuous lines in the asthenosphere and short dashes when frozen in the lithosphere : shear strain profiles. b) New model based on an analytical solution for the flow pattern close to the axis taking into account a $10^4$ viscosity drop just below Moho (Rabinowicz et al., 1987).

### *9.4.2. Models of mantle diapirs*

Mantle upwelling can be induced by passive motion of the asthenosphere responding to lithosphere drifting away from spreading centers. Another possibility is that it is driven by diapirism due to the buoyancy of a partially melted mantle layer. The success met in explaining the regular spacing of island arc volcanoes by a Rayleigh-Taylor instability (Marsh, 1976) encouraged Whitehead et al. (1984) to propose a similar model to explain the ridge segmentation. Their model is based on analog experiments involving a water-glycerine mixture injected into glycerine.

Building on the structural evidence which had led to the earlier discovery of mantle diapirs in ophiolites (Nicolas and Violette, 1982), Rabinowicz et al. (1984) have proposed that the diapirism is due to a complex interaction of buoyancy-driven instability with the upwelling flow related to plate drift. Their model is based on two key-pieces of information deduced from field studies : 1) the rising peridotites are partially molten and, although melt is periodically extracted, permanently they retain enough melt (5-10 %) to develop a 1-2 % negative density anomaly with respect to unmolten peridotites, and 2) the uppermost peridotites, which are intensely strained at and above their solidus, present a shear sense inversion within 1-2 km below the Moho (§ 9.2.2) ; this is clearly indicative of a forced plastic flow diverging from the diapiric intrusion, which is, in turn, indicative of a forced, or buoyant diapir intrusion. The model presented in figure 9.5a presents a flow pattern resulting from the mixing of the buoyancy-driven local flow with the passive regional flow driven by the sliding of the lithosphere. A 6° dip angle for the lithosphere-asthenosphere boundary is assumed as well as a 1% negative density contrast and a viscosity two orders of magnitude smaller than in the mantle of the ridge corner, down to 30 km depth. The resulting flow pattern shows the development of a 10-15 km diameter diapir, explaining the equally constrained tectonic and magmatic activity at ridges.

Scott and Stevenson (in press) have elaborated on this general scheme and have developed a more comprehensive model with important implications. In particular, in the critical problem of the density contrasts, they add the effect of the degree of melt extraction to the effect of porosity. Instead of considering, as above, a fixed porosity in the diapir which is stabilized above the level of permeability by hydrofracturing extracting melt beyond a threshold of 5%, they assume a porosity depending on the upwelling velocity in the diapir, in keeping with the percolation theory. Introducing these variable parameters in their model, Scott and Stevenson find solutions similar to the preceding one (compare figs. 9.5a and 9.6) with however, as expected, a spreading rate dependence. With respect to the profiles of figure 9.6, increasing the porosity and decreasing the viscosity in the uprising mantle have the similar effect of increasing the lateral confinement. These authors also define a field of time-independent solutions for spreading velocities above 0.45 cm/yr with below, an oscillatory regime. In this latter regime, the temporary diapiric behaviour should be associated with melt delivery to the crust and periods in-between, with magmatic quiescence. This predicted behaviour explains the episodic rifting postulated for slow spreading ridges (§ 11.6).

The detailed field work conducted in Oman has revealed that rotation of the vertical flow was surprisingly sharp, occurring within 300 m below the Moho (fig. 3.17). This implies a sharp drop in viscosity, the consequences of which on processes taking place in the transition zone are discussed in § 7.5. Figure 9.5b presents a new model based on this information. Both figures 9.5a and b illustrate the complexity of the principal compressive stress trajectory, which controls the path of hydrofractures extracting the melt from the rising diapirs (§ 7.6).

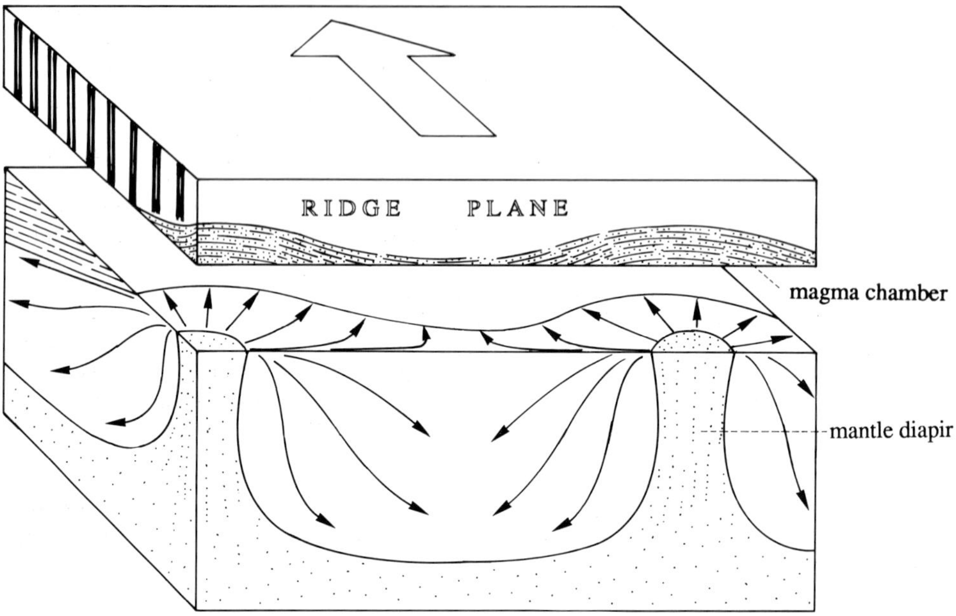


Fig. 9.7. Sketch of the possible asthenosphere diapiric structure below a ridge, illustrating the return flow in depleted harzburgites (arrows on a blank background). Dots : melt in rising lherzolites ; dashes : lithosphere; hatches : crust.

### 9.4.3. *Return flow and thickness of the buoyant layer*

Buoyant diapirism implies that the surrounding medium through which the diapir rises is denser ; for the periodic Rayleigh-Taylor instabilities evoked above to develop, a buoyant layer must be individualized within the asthenosphere uprising on a regional scale. The floor of the layer is easily identified as the horizon, at a depth of around 75 km (§ 7.2.3) where the adiabatic dry melting is initiated. For the top of the layer to become individualized, a return flow must drive down the depleted and partly dried peridotites which are expelled from the top of the diapir, as shown in figures 9.6 and 9.7. Scott and Stevenson have analyzed the interaction of the upwelling and return flows.

The deduction that in the uprising and melting asthenosphere, the horizon of partial melting where the diapirs initiate does not extend upward to the base of the lithosphere and has a definite upper limit, can be justified, and the depth of this limit can be estimated, recalling an important observation made in ophiolites. Very few diabase dikes, in contrast with the abundant gabbro dikes, are found cutting the mantle and deep crustal section of ophiolites. The critical difference between these two types of dikes is the temperature of the wall peridotites, which is below around 450°C at the time of injection of diabase dikes, and above 450°C in the case of gabbro dikes (§ 11.4.5) ; it is even above the peridotite solidus for the indigenous dikes (§ 2.5.2). This proves that the zone of melt delivery is narrowly confined to the ridge vicinity where the deep crust and uppermost mantle are still at a high temperature. It is concluded that the melting mantle from which this melt is extracted is similarly confined ; outside this zone which is evidently the top of the diapir, no or very little melt (reflected by the diabase dikes) is delivered to the crust. Hence, outside the area of the diapir, the melting layer at depth is not thick enough to promote melt extraction by hydrofracturing, meaning that it should not be shallower than around a 50 km depth (§ 7.4.1).

### 9.4.4. *Spacing of mantle diapirs and ridge segmentation*

An estimation for the spacing of diapirs can be deduced from the physical modelling of Rayleigh-Taylor instabilities. Selig (1965) has proposed the following formula for the periodic spacing of salt diapirs, provided $\eta_1 >> \eta_2$

$$\lambda = \left(\frac{2\pi h}{2.15}\right) \left(\frac{\eta_1}{\eta_2}\right)^{1/3} \qquad (1)$$

where

$\lambda$ = dominant wavelength or spacing of mantle diapirs
h = thickness of the low density layer
$\eta_1$ = viscosity of surrounding mantle
$\eta_2$ = viscosity of the low density layer

Marsh (1979) has proposed the following formula relating to melt diapirs ; the diapir diameter D is defined in terms of the parameters above :

$$D = h \left(\frac{\eta_1}{\eta_2}\right)^{1/4} \qquad (2)$$

It is possible to eliminate the viscosity ratio which is poorly constrained by combining (1) and (2) :

$$\lambda = \frac{2\pi}{2.15} (D)^{3/4} (h)^{1/4} \quad (3)$$

and to deduce λ from the estimations of h made in the preceding paragraph, and D from the measurement on diapirs in Oman. The domain of vertical flow in Oman diapirs is about 10 x 15 km ; assuming that the contours in the deeper diapir are probably slightly smaller, we take D = 10 km. The thickness h of the buoyant zone can be taken as the total thickness of partial melting layer, from its initiation at a depth of 75 km to the level of melt extraction at 50 km, that is 25 km ; alternatively, it can be somewhat reduced, say to 20 km, considering that a threshold of melt ratio should be attained before the layer can be considered as buoyant. Equation (3) is not very sensitive to this parameter ; thus, taking one value or the other, induces a change of a couple of kilometers in λ estimation. With h = 20 km we find λ = 35 km. This estimate should be taken only as indication that the scale of segmentation of oceanic ridges, thought to correspond to the spacing of mantle diapirs is probably in a comparable range.

Along the Mid-Atlantic Ridge (MAR), the most obvious segmentation is given by the spacing of transform faults, corresponding to a scale of 40-50 km. At a smaller scale of around 20-30 km, discontinuities could however exist ('en echelon relays', 'zero offset transforms',...) (H. Schouten, pers. com.). Assuming that the transform fault spacing corresponds to the significant segmentation distance suggests that, below the MAR, mantle diapirs should be equally spaced by 40-50 km (Francheteau and Ballard, 1983). This assumption is supported by the analysis of crustal thinning and deepening of sea-floor in the vicinity of transform faults which cannot be explained solely by the thermal effect of a colder wall and seems to require melt alimentation from a feeding center located between consecutive transforms as shown by figure 9.1 (see also § 9.3) (White et al., 1984 ; Forsyth and Wilson, 1984 ; Parmentier and Forsyth, 1985 ; Whitmarsh and Calvert, 1986). Direct observations with the *Alvin* submersible along the MAR in the vicinity of the Kane Fracture Zone suggest to Karson et al. (1987) the existence of a magmatic cell extending 40 km along strike. Adjacent to this area southward, another cell, probably slightly older, with apparently no present day magmatic expression (fig. 11.16) has been identified by seismic refraction experiments ; it has a comparable dimension along strike (Purdy and Detrick, 1986).

The problem of segmentation in the MAR may however be more complex because beyond the scale of the topographic roughness associated with the transform faults segmentation, there is a longer wave-length periodicity of 370 km (variance of 130 km) correlated with gravity and geochemical anomalies (Le Douaran and Francheteau, 1981). Hamelin et al. (1984) confirmed this correlation with a larger set of geochemical data and ascribed the anomalies to hot spot-related intrusions. Recently Gibert and Courtillot (1988) documented, from Seasat altimetry data in the South Atlantic, a regular pattern of geoid roughness with again a 400 km periodicity.

The segmentation along the EPR is still more complex than along the MAR. First the spacing of transform faults is variable and larger, attaining 650 km in the equatorial EPR (fig. 9.8). Next, topographic highs which coincide with the central parts of the ridge between transforms in the MAR are less periodically spaced along the EPR where their spacing attains 50-200 km (Crane, 1985). The EPR is segmented by major depth discontinuities marked by transform faults, tips of propagating rifts, and large overlapping spreading centers (the OSCs) defining long-wavelength ondulations > 200 km (Macdonald et al., 1988), and next on an average scale of 75 km by transform faults and smaller OSCs (Macdonald and Fox, 1983 ; Lonsdale, 1983 ; Macdonald et al., 1986) (fig. 9.8). These segmentation scales coincide more or less with those of topographic highs,

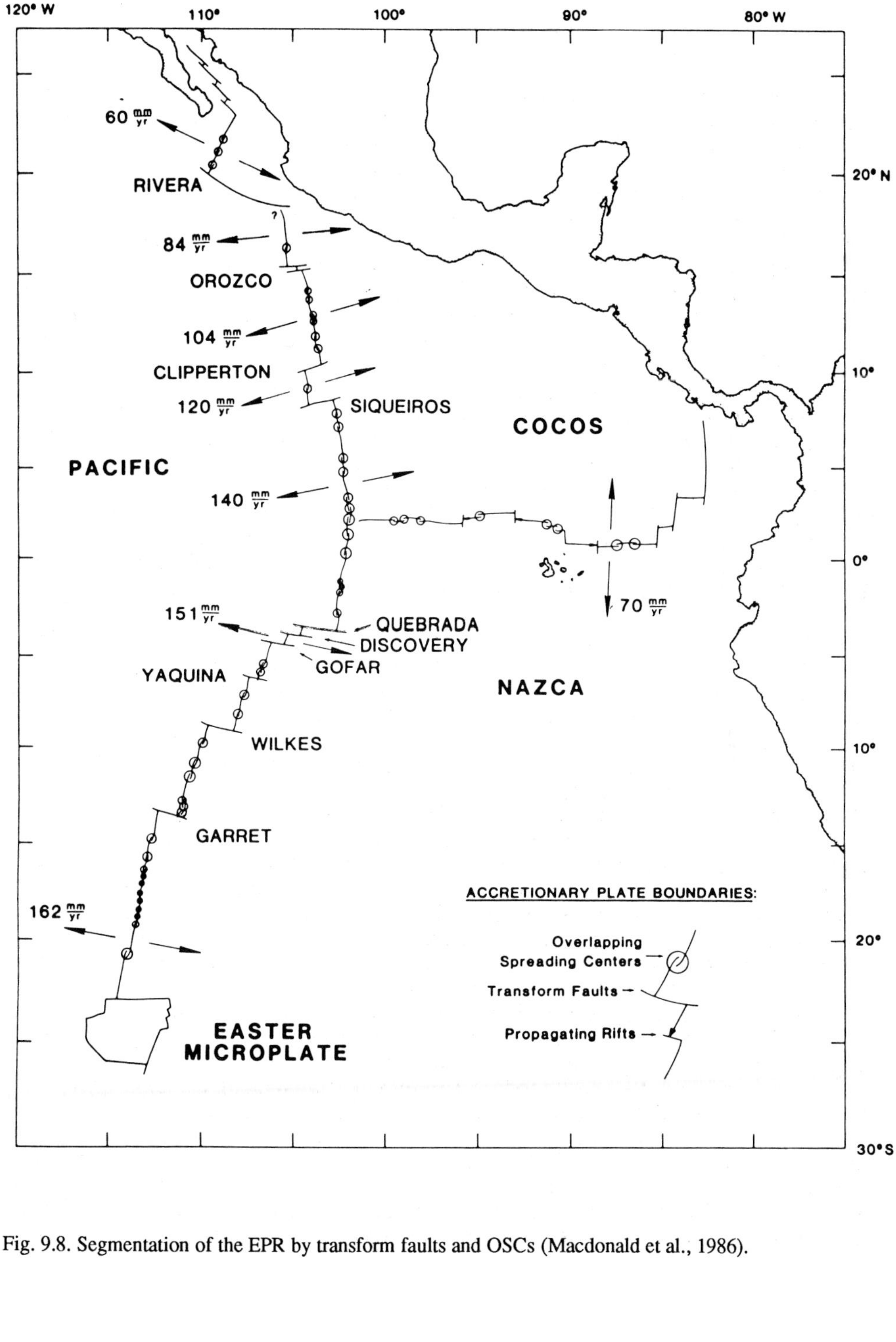


Fig. 9.8. Segmentation of the EPR by transform faults and OSCs (Macdonald et al., 1986).

depending on the differences in elevation which are taken into account (fig. 9.9). Finally, small topographic saddles between OSCs can mark deviations from the axial linearity of the spreading centers (DEVALs of Langmuir et al., 1986) and small non-overlapping offsets (SNOOs of Batiza and Margolis, 1986). DEVALs and SNOOs segment the ridge on a new scale of 10-20 km. The scars left by these various discontinuities on the drifting oceanic floor suggest that the segments on the scale of 100 km are stable over a period of 1-5 Ma and those on the 20 km scale, over only 0.1 Ma.

Francheteau and Ballard (1983) have proposed that major topographic highs on a scale of ~ 350 km are above the principal magma reservoirs feeding the ridge accretionary segment along strike, a conclusion somewhat supported by the finding of a progressive fractionation from more primitive basalts above the topographic high, to more fractionated basalts closer to OSCs (Thompson et al., 1985) . This conclusion is however disputed by Langmuir et al. (1986) who, on the basis of an extensive dredging of basalts along the EPR, conclude that the elementary geochemical segmentation scale is that of DEVALs, one order of magnitude smaller than the preceding one.

In chapter 7, it was emphasized that there should be no first order difference in the asthenosphere structure beneath fast and slow spreading ridges away from hot or cold spots, the difference being a result of the depth at which the lithosphere is met by the ascending asthenosphere flow. Thus the partially molten horizon in which the diapirs are initiated should be identical below the MAR and the EPR and consequently, the predicted spacing between mantle diapirs should be the same in both situations, presumably around 40-50 km. This is obviously a first order analysis. It ignores second order effects such as mantle flow analyzed in the preceding section on the shaping of instabilities in the melting horizon. Data on ridge segmentation and on the spacing of topographic highs (Crane, 1985) and of volcanic centers in rifts and ridges (fig. 9.10) suggest that the diapirism wavelength may increase with spreading rate and that a 100 km figure may be more appropriate for the EPR.

Results from mapping in the Oman ophiolite cast further light on mantle diapirs spacing. In the contiguous Semail and Wadi Tayin massifs, two diapirs separated by a transform fault are 70 km apart (fig. 5.22). The spacing between two other mapped diapirs in the central and northern Oman belt is 170 km, but this figure represents an upper limit because the existence of several other diapirs between them is suspected (fig. 9.3).

*These pieces of evidence encourage one to look for a figure of about 50 km in the spacing of diapirs below the MAR and of about 100 km below the EPR*, and in the latter case, to search for a correspondence with the segmentation on this scale, dominated by OSCs. Magma supply below the highs of ridge segments limited by OSCs, as proposed by Thompson et al. (1985), also suggests that mantle diapirs are located below these highs. Segmentation on the smaller scale of the DEVALs and SNOOs should rather be equated with preferred diking directions in the crust, because the extension of these structures is comparable to that of the basaltic dikes (§ 11.5.2). These structures would be controlled by the elastic properties of the lithosphere at the ridge.

The scale proposed here for mantle diapir spacing is smaller than that predicted by Crane (fig. 9.9a) or by Macdonald et al. (fig. 9.9b). These authors relate mantle diapir spacing with the major scale segmentation between transform faults and large propagating rifts and ascribe the OSCs segmentation to crustal magmatic activity (fig. 9.9b). This question of scale is obviously very open. The proposed 50-100 km scale for mantle diapir spacing is supported by the spacing of transform faults in the MAR and by ophiolite data. The 400 km scale, supported by a corresponding spacing in the MAR should probably be related to convective cells occurring within the whole upper mantle (Fleitout and Yuen, 1984).

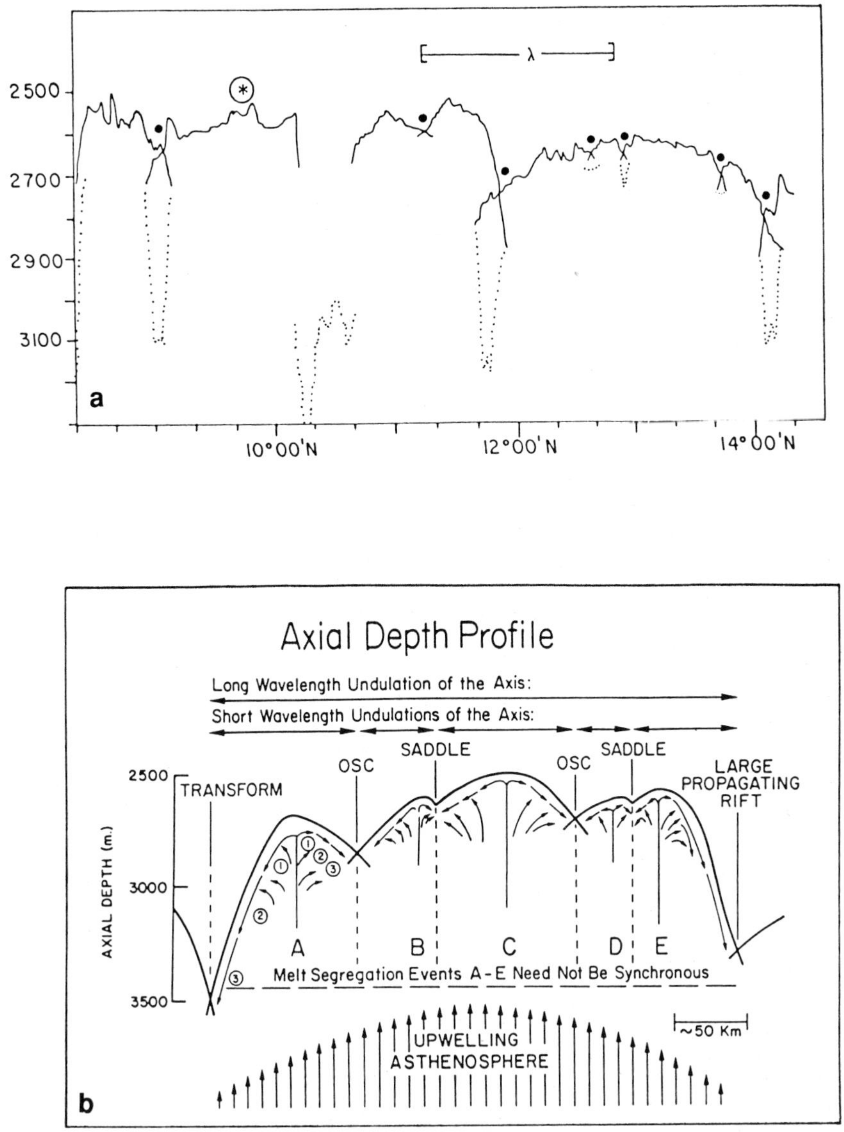


Fig. 9.9. Axial bathymetric sections of the EPR. a) Actual profile between 8°N and 14°N. Dots above the profile represent overlapping spreading centers. The encircled star represents the intersection of a seamount chain with the rise axis. Bathymetry is in meters. λ represents a typical wavelength between the intermediate-sized domes. b) Model relating the various scales of segmentation and topographic highs to mantle upwelling and melt discharge from magma chamber (a) Crane, 1985 ; b) Macdonald et al., 1988).

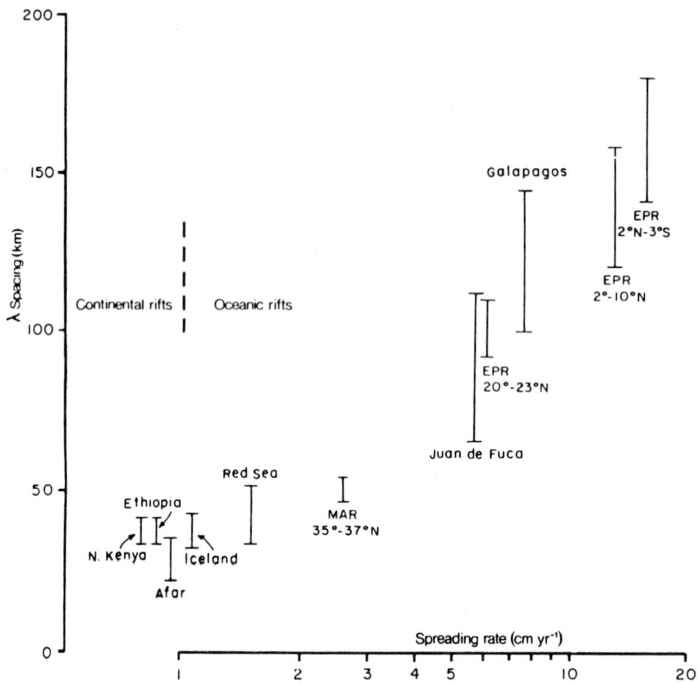


Fig. 9.10. Spacing of axial volcanism in continental, proto-oceanic and oceanic rifts, versus spreading rate for the oceanic rifts (Bonatti, 1985).

### *9.4.5. Stability of mantle diapirs*

It was proposed in § 9.4.3. that a return flow composed of depleted peridotites is necessary to define a buoyant horizon within which the diapirs originate at a depth of a few tens of kilometers in the mantle. We discuss here the time-dependent nature of these diapirs : are they individual blebs of melting mantle which force their way up or tubes, channelling a flow of this melting mantle in a steady state or quasi steady state fashion ? In their modelling of diapiric uprise, Scott and Stevenson (in press) conclude that above spreading rates of about 0.45 cm/yr, the diapiric structure is stabilized by the return flow and that below this rate, unstable solutions appear, generating an episodic diapirism, as blebs.

The geometry of return flow can also be sensitive to the shape of the lithospheric lid. It can be speculated that below a slow spreading ridge the comparatively steep slope of the lithosphere lower boundary in the spreading direction and the vertical slope along transform faults should constrain the return flow respectively in the spreading direction and normal to it. Diapirs should be spatially constrained, which does not exclude episodicity in time. In fast spreading environments, this lithospheric control should be looser and possibly inoperant because the slope in the spreading direction is smaller and the spacing of transform faults, larger.

Another possible means of approaching this question consists of estimating the duration of a mantle bleb activity, which should obviously be shorter than that of a tube, and of comparing the obtained figure with the age of possible scars left in the young oceanic lithosphere by the diapiric activity. Such scars could be topographic, like trails of volcanic highs originating on the ridge and possibly located above the mantle diapir, or the extremities of the major ridge segments, namely transform faults and overlapping spreading centers ; they could also be magmatic ; in the MARK area of the MAR, Karson et al. (1987) estimate from the pattern of magnetic anomalies that a magmatic cell has been active for at least the last 3 Ma. Along the EPR, the 100 km long segments would be stable for periods of 1-5 Ma.

The life span of a mantle bleb can be estimated using Stokes'law :

$$v = \frac{2 g D^2 \Delta\rho}{9\eta}$$

with v = ascent velocity ; g = gravity constant ; D = diapir diameter taken at 10 km ; $\Delta\rho$ = density difference between partially molten diapir with 7% melt fraction (§ 7.3) and surrounding depleted peridotites : $\Delta\rho = 1.5\%$ ; $\eta$, the viscosity of surrounding mantle, estimated at $10^{19}$ poises. With these data, v $\sim$ 10 cm/yr. Assuming that the diapir initiates at $\leq$ 50 km (§ 7.2.4) and rises to 10 km below sea surface, the life span of a mantle bleb would be $\leq$ 0.4 Ma. This is somewhat smaller than the figures mentioned above, suggesting that diapirs may rather be tubes.

The existence of off-axis volcanoes raises the question of the possible existence of off-axis mantle diapirs. Off-axis volcanoes can be fed by melts extracted from the deep part of an axis-related diapir. As proposed by Sleep (1984) and Rabinowicz et al. (1987), the maximum principal stress trajectory which is followed by the fractures extracting melt can depart significantly from the axis if the fracture originates at depth (fig. 9.5b). As expected, the resulting seamount should be composed of alkali basalts and be comparatively small because the column drained by the fracture in the melting mantle is short (§ 7.4). Large seamounts with MORB affinities would be better explained as being located above an off-axis mantle diapir. Observations in Oman (fig. 9.3) suggest this possibility.

The existence of off-axis diapirs would be a logical consequence of the existence of a buoyant layer extending in the spreading direction at a large distance as illustrated by figures 9.5 and 9.7. This certainly seems possible in fast spreading situations where the lithosphere remains thin at large distances from the ridge axis.

# Chapter 10

# MAGMATIC PROCESSES IN THE UPPERMOST MANTLE AT OCEANIC SPREADING CENTERS

## *10.1. INTRODUCTION*

The magmatic processes responsible for oceanic crust generation at spreading centers, which are considered in the next chapter, depend on the mechanisms of melt extraction from the underlying mantle diapirs. The latter process was examined in chapter 7 where was shown that, on the basis of data from ophiolites, the uppermost mantle is a critical zone, corresponding in ophiolites to the transition zone between ultramafic and mafic sections. In particular, it was proposed in § 7.5 that a dynamic pressure prevailing in this zone could retain melt inside the peridotites and locally create conditions for melt percolation in an undercompacted medium, thus making a continuous transition to the crustal magma chamber. The magmatic processes operating in the deep crust below a ridge are hence coupled with those occurring in the mantle diapir, by and through this transition zone. This justifies a detailed discussion of the origin of its typical formations, dunites, wehrlites and, occasionally, chromite pods. Descriptions of typical transition zones can be found for the two main ophiolite types in § 3.3.3 for HOT and in § 4.2.3 and 4.3.3 for LOT. LOT transition zones share many common characteristics with HOT. As seen in table 8-I, the tabular dunite bodies, rich in melt impregnation, which were described in Trinity are comparable with those of HOT transition zones. In Trinity, it has been noted that the foliation, which is systematically steep in LOT, rotates to become parallel to the Moho a few tens of meters below it (fig. 4.12), a feature common in all observed HOT situations. In contrast, table 8-I also shows that neither the wehrlite sills and intrusions nor the chromite pods, so common in HOT transition zones, have been reported in LOT. For these reasons, and because much better observations can be made in HOT, we will focus below on HOT transition zones.

## *10.2. PRINCIPAL CHARACTERISTICS OF TRANSITION ZONES*

In HOT occurrences, the transition zone is located between the top of the tectonic harzburgite unit and the base of the continuous layered gabbro unit, constituting the lower part of the crustal sequence. The transition zone thickness varies, on the horizontal scale of a few kilometers, between a few meters and several hundreds of meters. We summarize here its most remarkable features, referring mainly to the Oman case (§ 3.3.3 ; see also Nicolas and Prinzhofer, 1983).

i) It is a zone rich in dunites. Downwards, the dunites root into the harzburgites, following a vein pattern which, depending on local plastic strain, is more or less transposed into a lenticular transition. Upwards, they form a continuous screen below the layered gabbros. With the exception of local diapiric areas, dunites record a solid state deformation with very large strains.

ii) It is where nearly all chromite pods, whether discordant or concordant with respect to the foliation inherited from asthenospheric flow, are located (§ 10.5).

iii) It is a zone of intense solid-melt interaction resulting in diffuse plagioclase-clinopyroxene impregnations and injections of gabbro or websterite dikes and sills. The sills are specially abundant at the top of the transition zone, making it often difficult to identify precisely the base of the gabbro layered unit.

iv) It is where wehrlite dikes, sills and intrusive bodies commonly observed in the crustal section are rooted (§10.3).

v) It is where solid flow grades into viscous magmatic flow. Immediately above the plastically deformed dunites, layered gabbros display magmatic flow structures such as magmatic foliation or lamination, lineation and boudinaged layers. However, in other ophiolites like those of Bay of Islands, the plastic-viscous flow transition occurs within the basal gabbros unit itself (Casey and Karson, 1981 ; Girardeau and Nicolas, 1981).

vi) It is where in diapiric areas, the asthenospheric foliations and lineations rotate within a few hundreds of meters from a steep attitude with respect to the Moho surface to an attitude parallel to it (§ 7.5).

These different properties of a typical transition zone can be accounted for by magmatic accumulation of dunites and wehrlites on a floor constituted by a network of depleted harzburgites and residual dunite veins. Continued plastic flow in the tectonic peridotites and melt circulation would eventually affect the overlying ultramafic cumulates. These properties can be alternatively explained by the model of melt percolation through residual peridotites, which was presented in § 7.5 and was illustrated by data from the Maqsad area of Oman (figs. 3.17 and 3.19). It has been proposed in particular that both the melts impregnating the rising harzburgites and those injected from depth through dikes tend to be retained just below the Moho. The occurrence of gabbro or pyroxenite sills within the dunites at the top of the transition zone is ascribed to this melt retention. A crystal mush can thus be created progressively above the solid harzburgitic mantle. When the asthenosphere moves away from the ridge this mush is squeezed with the following consequences : 1) the reconstitution of a dunitic solid frame in the formations of the transition zone which had been dismembered ; 2) the possibility that the mush locally will intrude into the cooling crustal section as wehrlitic intrusions and 3) the resumption of solid state flow (fig. 3.17).

## *10.3. ORIGIN OF THE WEHRLITIC INTRUSIONS*

Wehrlitic intrusions in the crustal section of ophiolites have recently been discovered in a few ophiolites (table 3.I). The description of the Oman ophiolite has shown the volumetric importance of wehrlitic intrusions into the crustal section and their various occurrences ; early intrusions into unconsolidated gabbros appear as smaller dikes and sills (plate 3.1f, g), whereas later intrusions into consolidated but still plastic gabbros give rise to large discordant bodies, which induce large distortion of the gabbro layering at various scales (plate 3.1h). Since these wehrlitic dikes have been traced down into the dunites of the transition zone but so far have not been observed in the harzburgites below, it is concluded that wehrlites originate within the transition zone.

The origin of this magmatism has been already discussed above and in § 7.5 and 3.3.2.

The wehrlitic mush is envisaged originating from the dismembering of the peridotite solid frame in the transition zone just below the ridge. The ultramafic nature of this mush would be due to its charge in olivine and possibly chromite crystals which could be largely xenocrysts, also called 'megacrysts'. These would derive from the dismembering of the dunites. In the wehrlitic mush they could be mixed in unknown quantity with olivine phenocrysts, precipitated from the melt. As we will see in the next section, the effects of reequilibration at high temperature in the wehrlites may make it difficult to distinguish olivine xenocrysts from phenocrysts on purely geochemical grounds. The presence or absence of a dislocation substructure typical of plastic deformation (Nicolas and Poirier, 1976) may provide evidence. Unfortunately, the wehrlites are commonly serpentinized and have experienced sufficient plastic strain to induce a new substructure in olivine phenocrysts.

This wehrlitic ridge magmatism may have a bearing on the origin of certain basalts and picrites. It has been recently argued that the olivine from some basalts and picrites were xenocrysts and not cumulates. This is based on the observation in the olivine of a few picrites of a dislocation substructure typical of plastically deformed mantle olivine (Mid-Atlantic Ridge, Stakes et al., 1984 ; Réunion and Hawaii, Martin, 1987 ; Réunion, Albarède and Tamagnan, in press ; Cyprus, F. Boudier, pers. com.). Albarède and Tamagnan also show that the incompatible elements behaviour in the Réunion picrites rules out a cumulate origin of olivine. In any case, the euhedral habit of olivine in picrite does not prove its cumulate origin, as it has been seen that this habit can be produced by melt corrosion (§ 2.5.3 and fig. 2.6).

## *10.4. ORIGIN OF DUNITES*

### *10.4.1. Introduction*

The origin of dunites in ophiolite massifs has been ascribed either to magmatic accumulation (O'Hara, 1968 ; Jackson et al., 1975 ; Elthon et al., 1982 ; Komor et al., 1985 ; Furnes et al., 1988) or to reaction of a lherzolite or harzburgite with a fluid, being thus essentially residual (Boudier and Nicolas, 1972, 1977 ; Sinton, 1977 ; Dick, 1977 ; Dungan and Avé Lallemant, 1977 ; Leblanc et al., 1980 ; Cassard et al., 1981 ; Loomis and Gottschalk, 1981 ; Nicolas and Prinzhofer, 1983). This is an important question because of its implications on the nature and depth of origin of primary melts and on the thermal state in the upper mantle. If the dunites are magmatic, the primary melts have to be more picritic in order to be able to segregate that much olivine, a view advocated by many authors (O'Hara, 1968 ; Green et al., 1979 ; Elthon, 1979 ; Stolper, 1980 ; Jaques and Green, 1980). A larger degree of partial melting must be attained in order to obtain a picritic melt (over 30 % for Jaques and Green, 1980), implying that the mantle is hotter (1450°C) at the comparatively greater depth ( 60 km) where picritic melt can form (Stolper, 1980). If the dunites are essentially residual the primary melt could be more tholeiitic (Green and Ringwood, 1967 ; Kushiro, 1973 ; Presnall et al., 1979 ; Fujii and Bougault, 1983 ; Takahashi and Kushiro, 1983 ; Presnall and Hoover, 1984), generated at shallower depth and from a cooler upper mantle (1250°C at 30 km) (see however § 7.4.3.). For these reasons and also because of the bearing on the origin of the transition zone and chromite pods, it is necessary to address this problem in some detail. With our improved understanding of the processes believed to occur in the transition zone, it now appears possible to reconcile these conflicting interpretations.

### *10.4.2. Field occurrences*

Dunites in ophiolitic massifs, whether harzburgitic or lherzolitic, are found in the following settings ( see also § 2.5.2) :

i) They define, together with pyroxenite layers, the ubiquitous compositional layering of mantle peridotites. They constitute strictly parallel layers and lenses, usually a few centimeters thick, occasionally thicker than 10 cm. Their interlayering with pyroxenites is entirely at random and all attempts to define a sequential order have failed or are unconvincing.

ii) They form discordant veins cutting through this layering and the peridotite foliation.The veins are commonly a few centimeters to a few tens of centimeters thick. There is a continuous transition to the larger discordant bodies described below. In lherzolite massifs, the dunite veins constitute the walls of 'indigenous' gabbro or websterite-gabbro dikes (§ 2.5.2). In harzburgite massifs, the indigenous dikes are gabbro, websterite-gabbro or pyroxenite dikes and chromite pods, but these dikes and pods are not continuous longitudinally ; thus a dunite vein often contains only scattered patches of feldspar, pyroxenes or chromite and elsewhere it is entirely sterile.

iii) They constitute discordant bodies within the harzburgite massifs. Ranging in size from a few meters to a few tens of meters, occasionally more, these bodies have irregular contacts with the enclosing peridotites, grading into the veins described above. In areas of large plastic deformation such dunite bodies are foliated and stretched, becoming lens-shaped ; their contacts with the enclosing harzburgites are rotated toward the surrounding peridotite foliation attitude. In the Oman ophiolites, the basal banded series (§.3.3.3) is composed of 1 m to 50 m thick dunite bands interlayered with harzburgites. In this case the tectonic transposition is due to the obduction-related large plastic strain affecting the basal peridotites.

iv) In the transition zone between the tectonic mantle formations and the overlying layered gabbros in ophiolites, dunites form either tabular bodies surrounded by less depleted peridotites in lherzolite massifs such as Trinity (§ 4.3), or the top horizon of the transition zone in harzburgitic massifs. In the latter case, this dunitic zone is usually a few tens of meters thick, more rarely in the range of one meter or a few hundred meters. The lower contact of the dunites with the harzburgites is discordant, the dunites having their roots within the harzburgites in the form of the discordant veins described above. Where the high-T plastic deformation has been large, this type of contact is transposed into a dunite-harzburgite interlayering. The upper limit of the transition zone typically seen in Oman (§ 3.3.3) is marked by an increasing number of gabbroic bands and lenses within the dunites, marking a rapid transition to the overlying layered gabbros.

v) In the lower part of the layered gabbros unit, dunites associated with wehrlites and troctolites are commonly interlayered with the gabbros, ranging in thickness from a few centimeters to a few tens of centimeters.

vi) Finally, dunites also associated with troctolites may constitute parts of the wehrlitic intrusions.

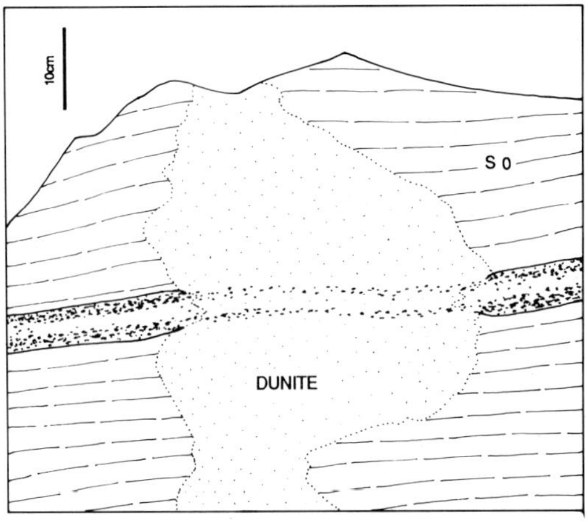


Fig. 10.1. Cross-cutting of a spinel-bearing websterite layer (So) by a dunite vein. The continuity of the layer through the dunite vein is shown by a trail of chromite grains in the dunite. This demonstrates that the dunite is not an intrusive dike (drawn after a plate in Boudier and Nicolas, 1977).

### *10.4.3. Residual/magmatic origin*

As mentioned above, several authors have proposed that the dunite veins discordant on the peridotite layering and foliation have a residual origin ; their depletion would result from the local action of some melt or fluid phase, traces of which may be still present as indigenous dikes (§.2.5.2). Direct structural evidence can be produced (figs. 10.1 and 10.5) demonstrating the residual character of a dunite vein when a chromite-bearing layer which extends in the surrounding peridotites as a spinel-bearing pyroxenite or a chromitite layer is traced throughout the dunite vein, an observation which excludes the infilling of the vein space by a magmatic dunite. Sterile dunite veins, that is enclosing no dikes, would result either from the squeezing out of the melt or fluid when the fluid pressure becomes insufficient to maintain the dike open, or from a subsequent remelting of the dike, a likely possibility considering that such dunites are usually observed in the most depleted harzburgite-dunite sections of ophiolites. However, in chromite pods where a magmatic olivine has precipitated along with chromite, early pull-apart can lead to the fragmentation of the massive chromite and injection of dunite dikelets perpendicular to the stretching lineation.

The origin of large dunite bodies which are within comparatively fertile lherzolites or which belong to the transition zone of harzburgitic ophiolites is more debatable on account of 1) their spatial extension and thickness, which can exceed several hundreds of meters, 2) their common compositional layering, defined by plagioclase, pyroxenes or chromite enrichment, and 3) in the case of harzburgitic ophiolites, their location at the base of the mafic layered section. As seen above and in § 10.5.5, for most authors large dunite bodies are entirely or largely magmatic (Quick, 1981), being formed by olivine and chromite accumulation within small magma chambers in the mantle section (lherzolitic ophiolites), or at the base of the crustal magma chamber (harzburgitic ophiolites). This interpretation has arisen with the undoubted existence of ultramafic cumulates, including dunites, at the base of stratiform intrusions like Stillwater (Jackson, 1961).

A magmatic origin of these dunites is also supported by geochemical evidence. The forsterite (fo) value of olivine in dunites is equal to, or more commonly less than that in

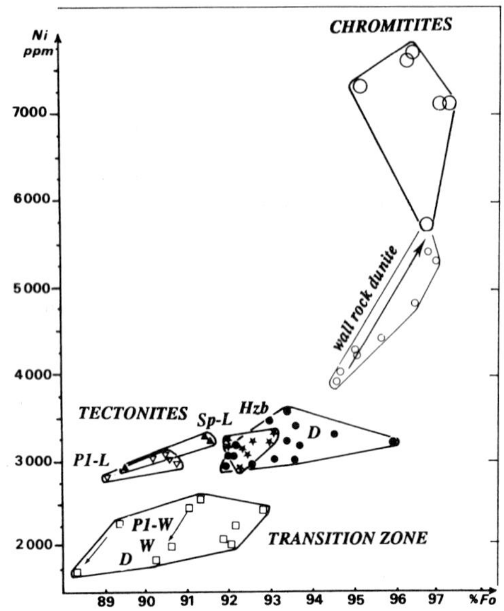


Fig. 10.2. Ni versus forsterite (Fo) values in olivine from the New Caledonia ophiolites. Sp-L : spinel lherzolite ; Pl-L : plagioclase lherzolite ; Hzb : harzburgite ; D : dunite ; W : wehrlite ; Pl-W : plagioclase wehrlite. Arrows point to specimens separated by a few centimeters, indicating a local desequilibrium (Leblanc et al., 1984).

harzburgites as observed by many workers (Sinton, 1977 ; Komor et al., 1985), although it can also be higher (Bodinier, 1988)(fig. 10.2). Barbot (1983) has systematically measured the fo value in olivine of harzburgites and dunites along a section from the basal peridotites up to the layered gabbros in two complexes of the Bay of Islands ophiolite in Newfoundland. Her results are shown in figure 10.3: fo varies from 88.5 to 90.6 in dunites and from 89.5 to 90 in harzburgites and lherzolites. Similarly, in the dunites at the base of the layered gabbro sequence in Oman, Pallister and Hopson (1981) report fo values from 89.0 to 90.5 whereas in the underlying tectonic harzburgites they range from 90.5 to 90.8 (see also fig. 10.2 for New Caledonia). If dunites represent a more depleted residue than harzburgites produced by interaction with a melt, a higher fo content should be expected. The same reasoning applies to the compared abundance of compatible elements which may be correlated with the fo composition. For example, in Barbot's measurements (fig. 10.3), the NiO values range from 3900 to 3400 ppm from the lower to the higher harzburgites and lherzolites and from 3500 to 2100 in dunites. Interestingly, a dunite from the top of the transition zone has given a 3200 ppm value and one from 100 m below, 2100 ppm, showing that the low and high values are not related to the position in the section, an observation also made for the same complex by Komor et al. (1985). The fo and NiO values fall drastically in the olivines from the overlying layered gabbros and from a magmatic wehrlite located well into the mafic sequence of the Blow-Me- Down section (fig. 10.3). On the contrary, in a wehrlite from the top of the transition zone of the Table Mountain section, values are near those recorded in the mantle. Similar conclusions are drawn from Leblanc et al.'s data (1984) in New Caledonia (fig. 10.2).

A convincing example of magmatic origin for the dunites and associated formations from

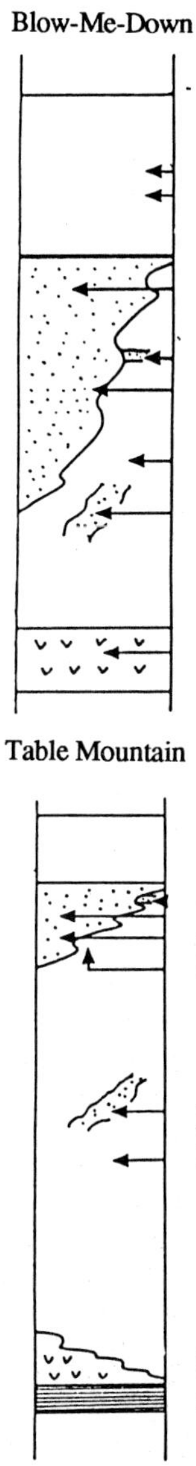


**Blow-Me-Down**

| SPECIMEN | Fo | NIO |
|---|---|---|
| Wehrlite | 80.3 | 1600 |
| Olivine gabbro | 73.2 | 400 |
| Top dunite | 89 | 2700 |
| Dunite layer in hzb | 90.6 | 3400 |
| Stratiform dunite | 88.9 | 2200 |
| Top harzburgite | 89.8 | 3400 |
| Dunite | 90.5 | 3500 |
| Basal lherzolite | 89.9 | 3700 |

**Table Mountain**

| SPECIMEN | Fo | NIO |
|---|---|---|
| Olivine gabbro | 73.5 | 800 |
| Wehrlite | 86.7 | 2600 |
| Dunite layer in Hzb | 90 | 3200 |
| Top dunite | 88.5 | 2100 |
| Basal dunite | 89 | 2900 |
| Top harzburgite | 89.5 | 3900 |
| Dunite | 89.6 | 3000 |
| Harzburgite | 90 | 3800 |
| Basal lherzolite | 89.2 | 3700 |
| Amphibolite | | |

Fig. 10.3. Forsterite (Fo) and Nickel (NiO) values in olivine from two logs through the Bay of Islands ophiolites (redrawn after Barbot, 1983).

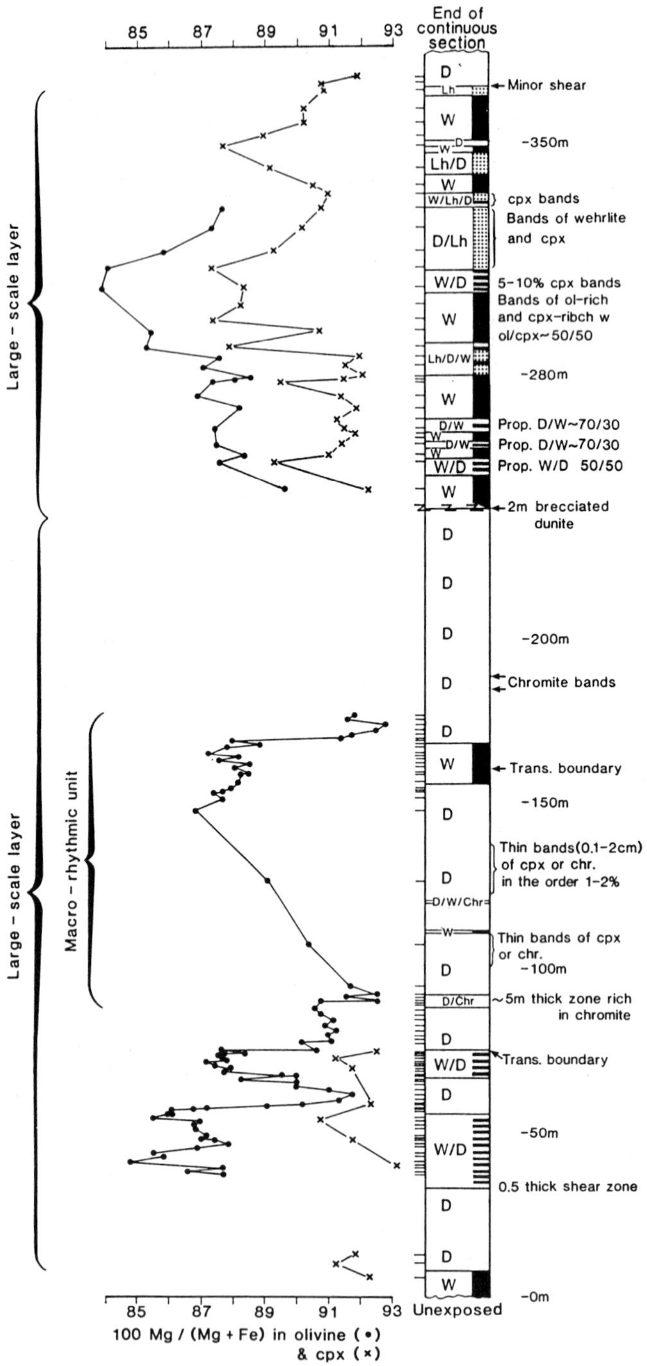


Fig. 10.4. Magnesium-iron ratios in olivine and clinopyroxene from stratified units from the transition zone in the Leka ophiolite. D : dunite ; Lh : lherzolite ; W : wehrlite ; Chr : chromite (Furnes et al., 1988).

an ophiolitic transition zone is found in the Leka ophiolite from the Caledonides of Norway (Furnes et al., 1988). There, below the layered gabbro unit and above the tectonic harzburgites into which the dunites are rooted, there is a transition zone which is now divided into blocks separated by shear zones but which probably attained several hundred meters. The dunites are interlayered on scales from hundreds of meters to a few centimeters with pyroxenites, wehrlites and chromitites. Two facts suggest a cumulate origin : 1) the remarkable longitudinal continuity, with for example a chromitite layer thought to extend over 3.5 km and 2) the large and relatively ordered changes in olivine and clinopyroxene composition inside dunite-wehrlite stratified units with sharp chemical discontinuities at the limit between such units (fig. 10.4). Such a series of observations has also been reported by Elthon et al. (1984) in some Bay of Islands ophiolites. Although the Leka series has been intensely deformed at high temperature together with the underlying harzburgites, it seems difficult to explain such a continuous layering and cryptic chemical layering by tectonic transposition, as proposed below for the more common lens-shaped layering.

The magmatic origin of the dunites of the transition zone was first questioned by Sinton (1977) who interpreted the dunites and wehrlites as residual mantle rocks more or less impregnated by a trapped melt. This conclusion was more recently supported by structural studies in the peridotites of ophiolites (Nicolas et al., 1980 ; Savelyeva et al., 1980 ; Nicolas and Prinzhofer, 1983 ; Nicolas et al., 1988). A structural evidence is the rooting of the dunites from the transition zone into the harzburgites by a network of residual dunite veins (see above) separated by screens which are relics of the enclosing peridotites. Structures within the peridotite screens are concordant with those within the host peridotites (fig. 10.5). This situation, implying reactions in the solid state, contrasts strikingly with that described by Ozawa (1983) in the Miyamori ophiolite of NE Japan. There, xenoliths of tectonic harzburgites and dunites characterized by their strong fabrics are found in random orientations within apparently cumulate dunites and wehrlites from the base of the magma chamber.

The dunites of the transition zone also share the same high-T plastic deformation as the enclosing harzburgite. Field observations and petrofabric analysis (Nicolas and Prinzhofer, 1983) show that this deformation can precede an episode of magmatic impregnation responsible for the local development of wehrlites and troctolites (§ 2.5.3 and plate 3.2g and 3.3d) and thus cannot be ascribed to a late deformation superimposed on a dunite cumulate. Finally, the structural evidence (cumulate textures and dunite-chromitite- harzburgite layering) in favor of the magmatic interpretation has been discussed in more detail by Nicolas and Prinzhofer (1983) and Nicolas (1985b) who show 1) that textures identical to cumulate ones can be produced by corrosion and magmatic impregnation in a solid dunite medium (fig. 2.6), and 2) that the layered aspect can result from plastic deformation. Such a layering at various scales in dunites as well as in chromitite bands (§ 10.5.3) can be produced by tectonic transposition of dunite or chromitite veins and bodies. This view is supported by field observations and petrofabric studies showing that the uppermost peridotites of HOT ophiolite complexes have suffered a remarkably large plastic deformation at hyper - to subsolidus temperatures (§.2.5.5 and § 3.4.2). This fact is generally overlooked in dunites because in the field little is to be seen within these rocks and because their microstructures resemble adcumulates, except for the very strong lattice fabric. Analysis of deformation in chromite pods which originate in this environment confirm this interpretation by revealing that there is a direct correlation between their degree of tectonic transposition and their internal deformation (§ 10.5.3).

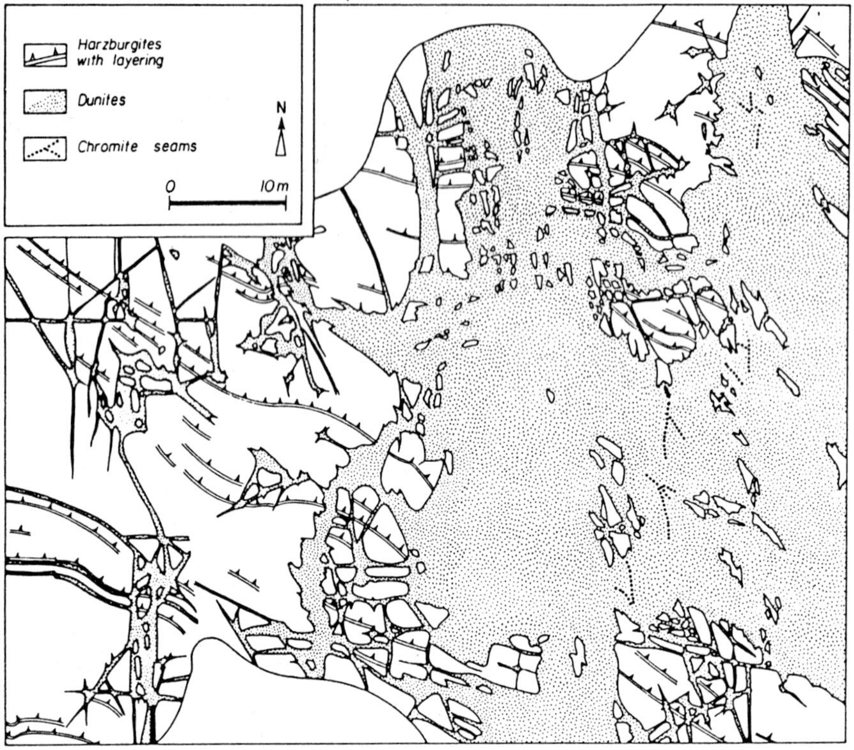


Fig. 10.5. Map of the relations between harzburgites and dunites in Polar Urals. The pyroxene layering is not displaced from one harzburgite block to the next through the dunites demonstrating the residual and not magmatic character of the dunites (redrawn after Savelyeva et al., 1980).

Fig. 10.6. Resorption of orthopyroxene and recrystallization of spinel into chromite across a transition in the field over a few meters, from harzburgite (left) to a dunite vein (right) (Leblanc et al., 1979).

### *10.4.4. Mechanism of formation of residual dunites.*

In the interpretation of an essentially residual origin for the dunite veins, the formation of the dunite is ascribed to a reaction of the peridotite with an introduced fluid. This could be either a water-rich solution which leaches the wall rocks (Dungan and Avé Lallemant, 1977) and precipitates pyroxene- hornblende along its channel (Loomis and Gottschalk, 1981), or a magmatic melt which reacts with a peridotite heated above its solidus (Boudier and Nicolas, 1972, 1977 ; Sinton, 1977 ; Berger and Vannier,1984). Except when amphiboles are present in the indigenous dikes, the origin of the dunite should be explained by reaction with a melt rather than with an hydrous fluid.

Considering a melt circulating in a dike, the composition of which is basaltic as suggested by the mineralogy of indigenous gabbro dikes, two factors, overheating and/or chemical action, can explain the melt reactivity with the enclosing peridotite. To produce the melting-out of orthopyroxene by overheating alone requires a melt circulating at temperatures in excess of 1365°C (Green et al., 1979). This is totally incompatible with models of melting by adiabatic decompression (chapter 7) and would necessitate the injection of foreign super-heated liquids. This seems unrealistic.

Sinton (1977), Dick (1977), Quick (1981) and Berger and Vannier (1984) have proposed an interpretation which satisfies the available observations. They explain the origin of the residual dunites by the incongruent melting of orthopyroxene due to the interaction of an olivine-tholeiite magma with a peridotite located in the stability field of quartz-tholeiites. The olivine-tholeiite, thought to have formed at depths greater than 30 km (Jaques and Green, 1980 ; Presnall and Hoover, 1984) and circulating through the dikes would react with the peridotite at pressures lower than 0.6 GPa, which is about the lower limit of the quartz tholeiite field (Jaques and Green, 1980) and thus also the lower limit of incongruent melting of orthopyroxene. Experimentally, this reaction takes place at 1200°C at room pressure, yielding a high-magnesium andesite (Fisk, 1986). Such a resorption of orthopyroxene, along with recrystallization and chemical re-equilibration of chromite-spinel as subeuhedral crystals has been described in harzburgite xenoliths partly reequilibrated at low pressure with the host basalt (Berger, 1985) and can be observed in ophiolite massifs at the contact of dunite veins with their enclosing harzburgites (fig. 10.6).In this interpretation the network of residual dunites found in harzburgites and some feldspar lherzolite massifs represents the channels followed by melts circulating in a peridotite close to or above its solidus at depths shallower than 15-20 km (fig. 10.7). The depth of formation of dunite veins and bodies is indeed within the feldspar lherzolite field (< 1.0 GPa). This is indicated by the facts that the associated dikes are usually feldspar-bearing and that dunitic walls along dikes are observed only in the feldspar lherzolites stability field. In the spinel lherzolites stability field, the only dunites observed belong to the layering, large dunite bodies have not been described and dunite screens along websterite dikes are quite exceptional.

It should be clear that in this interpretation the residual dunite is in fact composed of olivine having two or three distinct origins : truly residual crystals, olivine crystals produced by the incongruent melting of orthopyroxene, and possibly also magmatic olivine fractionating from the melt (fig. 10.11). However, except locally in transition zones (see § 10.4.6 and 7.5), at no time is the solid framework destroyed since both the preferred orientations and the trace of spinel-bearing layering are preserved throughout the dunite veins. Physically this indicates that the second threshold of percolation has not been passed and that therefore the molten fraction present at a given time in the rock never exceeded some 35%-40% (Wickham, 1987). The fact remains that these dunites should have been pervasively penetrated by the reacting melt in order to achieve the complete melting-out of orthopyroxene. Bodinier (1988), on the basis of a trace element study of the Lanzo plagioclase lherzolites and associated dunites, also concludes that dunites are

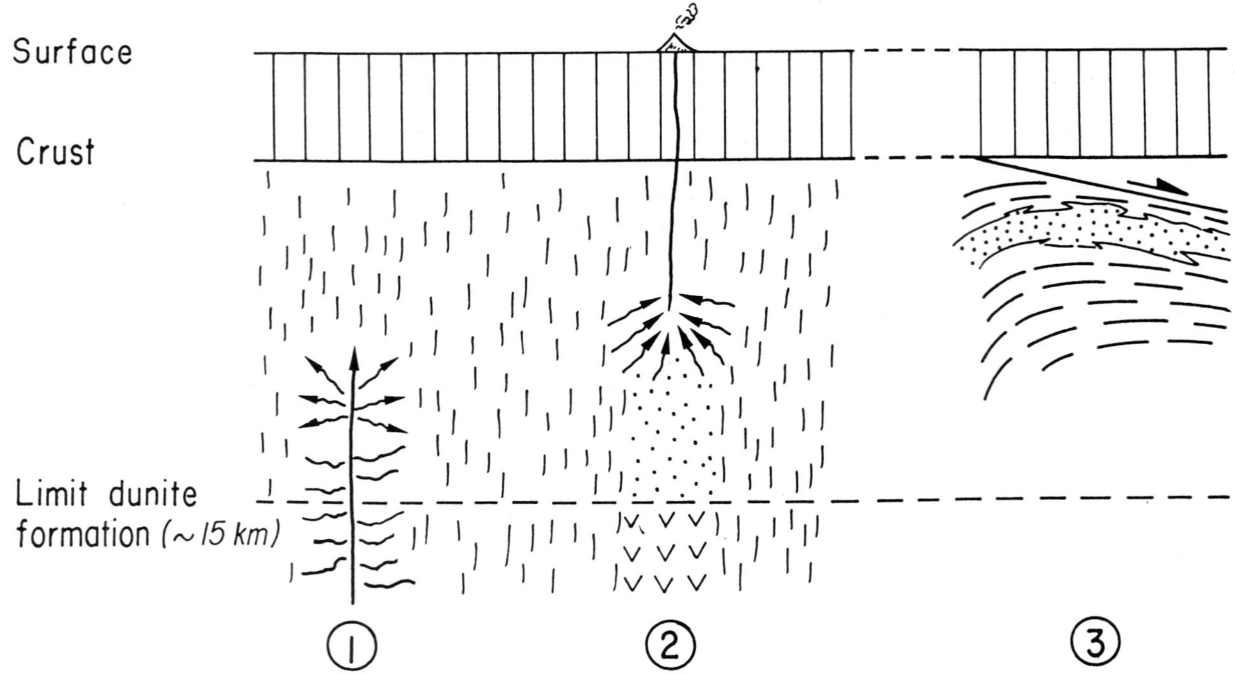


Fig. 10.7. Sketch of residual dunite formation. 1- Incoming in the opx-incongruent melting field of a melt conduit, opening its way by hydrofracturing and whose melt impregnates the local connected melt network. 2- Melt drainage in the zone of melt impregnation, leaving harzburgites and dunites (respectively, vs and dots) as residue. 3- Drifting away from the ridge of a tectonically rotating dunite lens. The vertical orientation of the melt conduit is arbitrary ; it should follow the $\sigma_1$ trajectory which may be inclined ; the dunite bodies are not to scale (Nicolas, 1986a).

residues intensely percolated by a melt.This seems feasible with the model of melt extraction proposed in chapter 7.

In this model, two stages are necessary to create the residual dunites in relation with the opening, within a melting horizon, of a hydrofracture carrying melt from depth (fig. 10.7). Propelled by the overpressure at the head of the fracture, this melt first filters through the permeable peridotites and reacts with them. Next it is drained back to the fracture by the depression created at the base of the fracture once communication with the surface is achieved. The fluid pressure should be high enough to impregnate and then drain the dike margins over a maximum distance of approximately 50 m on each side in the case of the largest dunite bodies.

### *10.4.5. Geochemical reequilibration*

For dunites originating at their maximum possible depth around 20 km, the drainage pressure is still large (P = 75 MPa) and an efficient melt extraction is possible. This is impossible for the dunites originating in the transition zone just below the Moho because the drainage pressure falls to almost nothing as a result of the small density difference between the overlying melt column and the enclosing crust. However, in this particular zone the presence of a pressure gradient has been envisaged and the compaction-percolation mechanism seems to be able to operate (§ 7.5).

In the interpretation of the dunites as a residue, we must take into account and explain the composition of olivines, which, in the dunites, are close to magmatic compositions, rather than being more refractory than olivines in the harzburgites (fig. 10.3). It should be recalled that in the proposed model, the olivines in the dunites have been in contact at 1250° C, with a melt with which they were in desequilibrium. This desequilibrium is shown on a scale of a few centimeters by variable olivine compositions in the transition zone and at the margins of chromite pods in New Caledonia (fig. 10.2). The exchange with an ephemeral melt circulating through the peridotites is thought to have promoted a ion-exchange, resulting in an increase in Fe-content of these olivines. This interpretation relies on two assumptions, 1) that the desequilibrium with the percolating liquid is such that it tends to decrease the magnesium and the nickel values in the peridotite olivine and 2) that the diffusion coefficients for these exchanges are sufficiently high, considering the temperature and time parameters. The first assumption is obviously correct because just above the dunites of the transition zone the same melt precipitates olivine with still lower forsteritic content and nickel values (fig. 10.3) ; thus the direction of the exchange reaction is correct. The fo value is also found to decrease in olivine from harzburgite and dunite xenoliths upon reaction with their host basalt (Berger and Vannier, 1984 ; Berger, 1985).

Let us now consider the second assumption. That the fo and NiO values remain higher in dunites than in mafic cumulates may largely be a question of kinetics. In the model of chapter 7, the temperature is 1250°C and the duration of the contact of the peridotite with the melt carried by a hydrofracture has been estimated at a few weeks (§ 7.4.2, see also § 10.5.5). Considering the available diffusion coefficients and data (Freer, 1981 ; Lehmann, 1983 ; Jurewicz and Watson, 1988) it appears that, above 1200°C, Ni-Mg and Fe-Mg interdiffusions are very fast in olivine in contact with an exchanging solid or liquid phase. The NiO diffusion coefficient seems to be the smallest. Applying this coefficient to an olivine crystal with an initial NiO content of 3700 ppm, which is 5mm in diameter and in contact with a melt at 1250°C, it would necessitate a period of 20 years to produce a zoning over 0.4 mm with a NiO content lowered to 500 ppm, the value measured in magmatic olivines. Afterwards, within a few hundred years this olivine crystal would be homogenized by subsolidus intracrystalline diffusion to a 2000 ppm value, which corresponds to an average for the dunites. This latter time lapse is very short compared to the cooling time of the lithosphere drifting away from the ridge. The 20 year period of contact between melt and dunites would correspond to the duration of the process of compaction and percolation affecting the transition zone (§.7.5). It is substantially longer than the duration of melt flood through the hydrofractures in the mantle formations (a few weeks). If this latter process is envisaged as providing a lower boundary for the exchange reaction, during a time lapse of one month and with the same data as above, a depletion halo in the olivine crystal 0.03 mm thick of NiO = 500 ppm and 0.1-0.2 mm of fo = 0.80 is created. After homogenization, the average NiO value would be 3584 ppm and the forsterite ratio fo = 0.89. Although these results are somewhat high, mainly regarding the nickel, they are still compatible with the composition of some dunites in figure 10.3. The erratic but altogether decreasing content in nickel going upsection in the peridotites could reflect a variable but increasing duration of contact with melt, although the effect of other parameters such as oxygen or sulfur fugacity cannot be ruled out (Barbot, 1983 ; Jurewicz and Watson, 1988). Indeed, Lorand's (1987) mineralogical study of sulfur in peridotites from the same ophiolites shows that a sulfur saturated basaltic melt percolated through the dunites of the transition zone during and after their plastic deformation, and Talkington's et al. (1984) report of phlogopite and pargasite inclusions in chromite points to the possible role of a fluid phase.

### *10.4.6. Conclusions as to the origin of dunites*

Structural evidence has been given in favor of an essentially residual origin for dunite veins enclosed in less depleted peridotites, and convergent observations allow to extend this conclusion to the dunites of many transition zones in ophiolites. As discussed above, the geochemical evidence is equivocal. If the petrological process (incongruent melting of orthopyroxene) is the same in dunite veins and in dunite bodies from the transition zone, the physical processes controlling melt circulation are probably different : hydrofracturing in the case of dunite veins and compaction- percolation in the case of dunites from the transition zone. As discussed above (§ 7.5, and 10.2), the intervention of percolation processes in the transition zone sheds new light on the discussion of the origin of dunites and affords the possibility of reconciling the two points of view.

Following these interpretations, the question of cumulate/residual dunites in the transition zone becomes a question of melt/solid ratio in this zone (see plate 3.3). Suppose that the melt/solid ratio never exceeds 35-40 % at a given time, a fraction below which the solid frame is not destroyed (Van der Molen and Paterson, 1979 ; Wickham, 1987), the dunites may be considered residual, althougth a few tens of percent olivine derive from the incongruent melting of orthopyroxene and an unknown fraction derives from interstitial precipitation from the percolating melt. The pyroxenitic and gabbroic layers derive from sills and impregnation pockets, well identified in Oman (§ 3.3.3), and the chromite layers from pods, all these formations being transposed into lenses and layers by the large plastic flow. Suppose on the other hand that the melt fraction exceeds the 35-40 % threshold over large areas. Then, the dismembered peridotites, mainly dunites, provide dunite fragments and olivine xenocrysts which can settle together with newly formed crystals when the system cools. The origin of the interlayered dunites, chromitite, pyroxenites and gabbros is now cumulative. The tectonic peridotite blocks described in a dunite matrix by Ozawa (1983) could have originated in this way.

There is, however, a critical difference between this interpretation of a cumulative transition zone and the more classical one evoked in § 10.4.1. In this interpretation, the rising melt can be tholeiitic because a large fraction of the olivine and possibly of chromite may be xenocrysts, inherited from the dismembering of the peridotites. On the contrary, in the previous interpretation the melt must be picritic and these minerals considered entirely magmatic.

## *10.5. STRUCTURE AND ORIGIN OF THE CHROMITE DEPOSITS*

### *10.5.1. Introduction*

Following the classification of Thayer (1960), two distinct types of chromite deposits have been distinguished. The first type lies within stratified igneous complexes, such as the Bushveld and Stillwater. Here, the chromite forms layers, up to 1.5 m thick and extending over a few tens of kilometers, which are parallel to the igneous layering and display clear sedimentary magmatic structures. Each chromite band corresponds to the lower part of a cumulative stratigraphic unit. The layers are not deformed, are composed mainly of small euhedral chromite crystals, and do not exhibit nodular or orbicular textures. In the Bushveld complex, much of the chromite is associated with olivine and pyroxene (bronzite) (Kuschke, 1940) ; however, in the Stillwater complex, the chromite is always associated with olivine (Jackson, 1961). The second type of chromite deposits is located within ophiolite complexes, more specifically HOT (Nicolas, 1986b), and are very different from the stratiform type. Ophiolite chromite deposits are usually irregular in form and have a lateral extension one to three orders of magnitude smaller than in the great complexes. The walls of the chromite band are dunitic in composition, and may be either concordant or discordant with the structures in the host peridotite. The chromite itself may exhibit orbicular or nodular textures of the type described by Johnston (1936). Frequently, the chromite shows deformation structures such as pull-apart and strong lineations. Crystal grains are generally xenomorphic and variable in size. Orebodies with these typical characteristics have been called 'podiform' by Thayer (1964a).

Exceptionally, stratiform-type chromite may also be present within ophiolites. A careful consideration of the structure and setting is necessary (Ceuleneer and Nicolas, 1985) to distinguish authentic stratiform chromite bodies deposited at the bottom of crustal magma chamber from initially irregular pods transposed into parallel streaks (concordant orebodies) by a large deformation. We will briefly present below the main structural types of chromite bodies in ophiolites as they have been classified by Cassard et al. (1981), after the pioneering work of Thayer (1964a, b). The last section discusses the origin of chromite bodies in HOT.

### *10.5.2. Setting of chromite deposits*

Chromite bodies in ophiolites are remarkably restricted in their location, and appear to obey the following rules :

i) Chromite bodies thicker than about one meter are restricted to HOT (table 8-I). In LOT and more generally in lherzolite massifs the chromite concentrations do not exceed a few clots within small dunite veins. To the author's knowledge, the only exception is represented by the Beni Bousera, Ronda and Collo lherzolite massifs of western Mediterranean (fig. 4.16) where meter-thick chromite pods have been reported (Leblanc, 1984) ; the pyroxenite matrix between chromite and the association with arsenide-chromite and sulfide-graphite-chromite mineralization suggest a specific origin. The Othrys ophiolite in Greece where plagioclase lherzolites and large chromite deposits are present does not invalidate this rule because, to our knowledge, the deposits are exclusively located within the harzburgites and diopside-harzburgites which are associated with the sterile plagioclase lherzolites. The chromite-HOT association can be matched with the higher chromium content of spinel in harzburgites than in lherzolites (Reid and Woods, 1978 ; Jaques and Green, 1980 ; Dick and Bullen, 1984). The higher activity of chromium in the harzburgite situation with respect to the lherzolite may be explained by the fact that the melts are enriched in chromium as a result of the extensive melting of the chrome-bearing clinopyroxene (Potvin, 1983 ; Nicolas, 1986a).

A puzzling problem is, however, raised by the comparative scarcity of chromite deposits in certain HOT ophiolites such as Oman and Bay-of-Islands. We have ascribed a fast spreading rate to the Oman ophiolite (§ 3.5.2). It is speculated that for fast spreading rates, temperatures at Moho level below the ridge may have been too high to allow large chromite fractionation within the melt-bearing conduits (see below). Large chromite deposits (over the million of tons) would be restricted to comparatively slow-spreading HOT massifs.

ii) Chromite bodies are mainly located within the transition zone, occasionally above the Moho or to depths exceeding 1 km below it. For example, in Oman where the depth with respect to the Moho is best constrained, out of over 100 pods visited by the author or reported by other workers we know only a few ones, which are located well inside the harzburgites, and most of these deep pods are in the Wadi Tayin transform structure (§ 5.4), where the thermal conditions believed to be necessary to precipitate chromite (see below) could have been met at depth. Similarly, only one stratiform body located above the Moho, has been reported in Oman (Ceuleneer and Nicolas, 1985).

iii) Although this may not be a strict rule, there is a correlation between the abundance and the size of chromite bodies and their location in a mantle diapiric structure. This relation, first proposed by Nicolas and Violette (1982) on the base of studies in the Philippines and Cyprus, is remarkably illustrated by the Maqsad area in Oman (§ 3.4.2, fig. 3.17) which is both a diapir and the richest chromite district in the country. This association can be explained by speculating that, as it is usually the first phase to crystallize from a basaltic melt, chromite should be selectively precipitated in diapiric areas where, expectedly, the most pristine melts have been circulating. It seems, however, difficult to use this correlation as a strategic guide in chromite prospecting because diapiric structures are only exceptionally preserved. Rajmi, which is the second largest chromite district in Oman, could have been initially a diapiric structure, subsequently transposed into a flat-lying attitude during plastic flow away from the ridge of origin (see discussion in § 9.2.5 about the possible use of petrological criteria to locate former diapiric areas). The near absence of chromite in the Shamah diapir (fig. 3.21) is ascribed to the level of erosion, which in this area is clearly below the transition zone where chromite concentrates.

### *10.5.3. Structure of chromite deposits*

All chromite bodies visited by us in various ophiolites or described with sufficient detail in the literature fall into Cassard et al.'s (1981) structural classification, initially established for the pods of New Caledonia. The only exceptions are a few stratiform deposits, the three following to our knowledge : Kizilyuksek Tepe in the Pozanti-Karsanti ophiolite of Turkey (Rahgoshay et al., 1981), Leka in Norway (Furnes et al., 1988) and one body in the Maqsad district of Oman (Ceuleneer and Nicolas, 1985).

*Stratiform chromite deposits* - With respect to the chromite pods presented just below, the stratiform deposits are characterized by : 1) their location within the crustal section or just below within dunites and 2) an exceptional vertical and horizontal extension of the ore layers. Though already restrictive, these criteria are not entirely satisfactory. For instance, Engin et al. (1987) describe in the Guleman chromite district of Turkey a layered chromite deposit with six major layered horizons, followed for up to 1600 m along strike. This

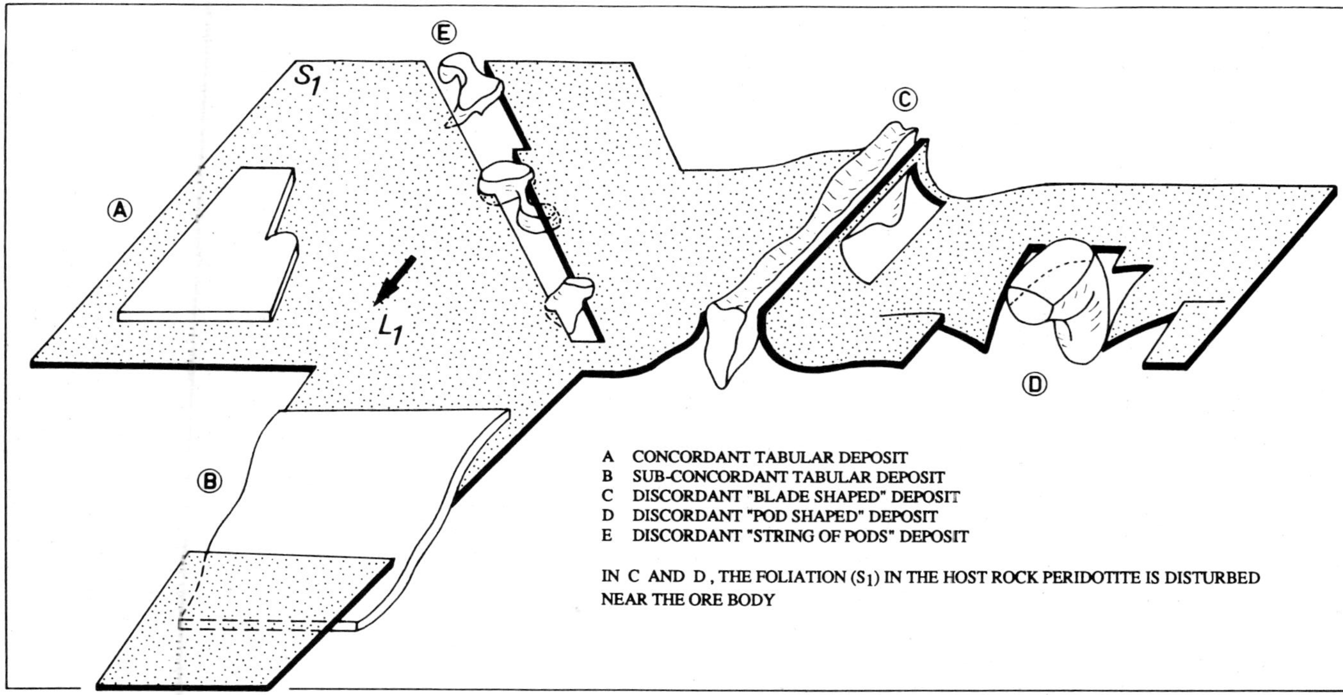


Fig. 10.8. Diagram showing the form and relation of podiform chromite deposits to the foliation and lineation in the host peridotite (Cassard et al., 1981).

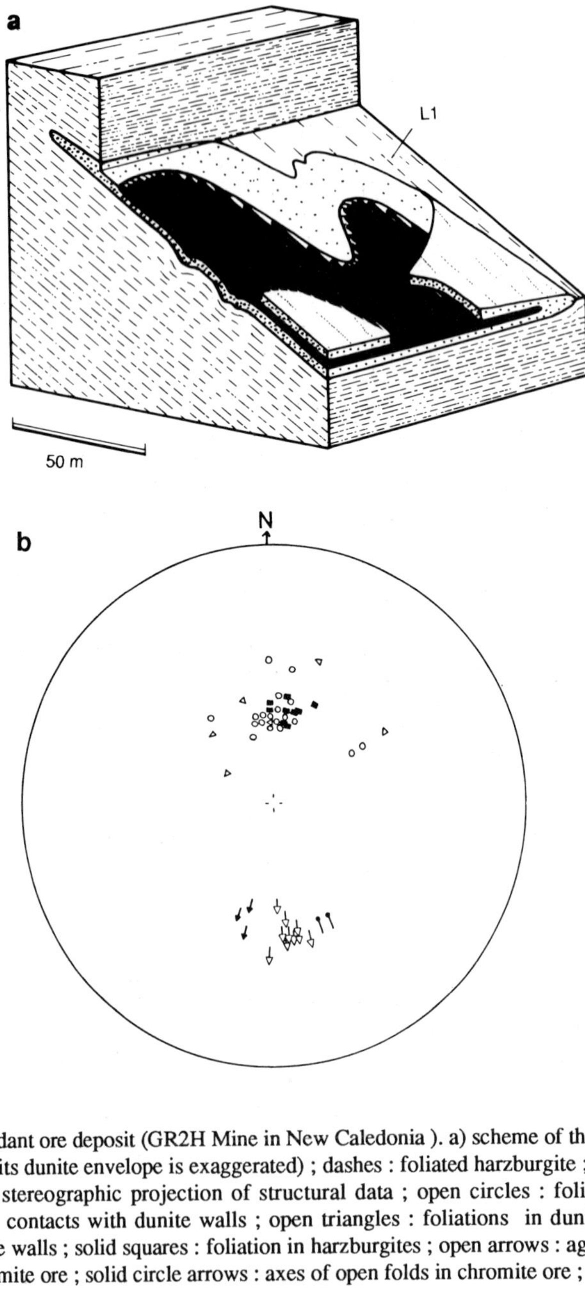


Fig. 10.9. Concordant ore deposit (GR2H Mine in New Caledonia ). a) scheme of the deposit (thickness of the ore sheet and its dunite envelope is exaggerated) ; dashes : foliated harzburgite ; dots : dunite ; black : chromite ore. b) stereographic projection of structural data ; open circles : foliation and layering in chromite ore and contacts with dunite walls ; open triangles : foliations in dunites ; solid triangles : dunite-harzburgite walls ; solid squares : foliation in harzburgites ; open arrows : aggregate and pull-apart lineations in chromite ore ; solid circle arrows : axes of open folds in chromite ore ; solid arrows : mineral lineations in harzburgites (Cassard et al., 1981).

extension is more in the range of stratiform deposits than that of concordant pods ; however, their location inside the harzburgite mantle excludes the former interpretation. Other criteria are even more difficult to use. As seen in § 10.4.3, demonstrating that the enclosing dunites are cumulates and not modified restites is not straightforward unless a sequential evolution can be traced (fig. 10.4) ; on the other hand, the superimposition of plastic deformation does not exclude a stratiform cumulate origin. Examination of the three chromite deposits identified above as stratiform illustrates these difficulties. They have a common setting within a thick dunite sequence, which attains 2 km in Pozanti-Karsanti and 3 km in Leka, with probably an important thickening in the latter massif by faulting and folding (Furnes et al., 1988) ; these dunites are themselves in a lower contact with harzburgites and grade upsection into wehrlites, pyroxenites and gabbros of the plutonic section. The situation is somewhat different in Maqsad where the stratiform chromite body, surrounded by a dunite lens, a couple hundreds meters thick, is inserted into the layered gabbro section, some 200 m above the Moho. In the Maqsad area, the dunites of the underlying transition zone are comparatively thick, attaining 100-200 m. The Maqsad deposit is not affected by the plastic deformation of the transition zone and, as shown by Ceuleneer and Nicolas (1985), its textures and fabrics match closely those of the stratiform chromite-dunite layers in Stillwater. The Pozanti-Karsanti body is also undeformed by plastic flow and it has preserved viscous flow structures. The horizontal (4 km) and vertical (1500 m) extension of chromite layers also surpasses that of chromite pods. Finally, the Leka occurrence, though it is plastically deformed, should be regarded as primarily stratiform for the reasons discussed in § 10.4.3 (see also fig. 10.4).

*Chromite pods* - In their structural classification of chromite pods, Cassard et al. (1981) distinguish between concordant, subconcordant and discordant pods, depending on their attitude with respect to the foliation-lineation reference frame in the surrounding plastically deformed peridotites (fig. 10.8). These bodies only occasionally exceed a few hundreds meters in their largest dimension.

*Concordant deposits* - Concordant deposits are the most common, making up about 50 percent of the chromite occurrences in New Caledonia and an even higher proportion in Oman. Most are tabular in shape, but they may have a pencil form. When such a deposit consists of several sheets, these sheets or plates have a dunite wall and are always separated from each other by harzburgite. The ore lens and its internal foliation and lineation are parallel to those in the host-rock peridotite (fig. 10.9). Contact between the dunitic wall rock of the orebody and the harzburgite is made by a series of fingerlike projections that parallel the harzburgite foliation. Lineations control the form of the mineralized zone and systematically indicate the direction of extension of the orebody.

The concordant pods are commonly cut by dunite veins, with or without associated gabbroic or websteritic segregations. These veins and dikes are normal to the lineation and they are interpreted as the result of the infilling of pull-apart structures developed at the scale of the entire pod. Finally, concordant pods may be folded in early tight isoclinal folds with attenuated or boudinaged limbs and axes parallel to the lineation and in late folds, commonly more open and whose axes can be at a large angle with respect to the mineral lineation (fig. 10.10).

*Subconcordant deposits* - Subconcordant deposits are less well represented than concordant deposits forming about 25 % of the chromite deposits in New Caledonia. They can be defined as follows : the orebody and its internal structures make a slight angle ($\leq$ 25°) in strike and/or dip with structures in the dunite wall and surrounding peridotites (figs. 10.8 and 10.10) ; locally, however, such structures may be parallel. Lineations in

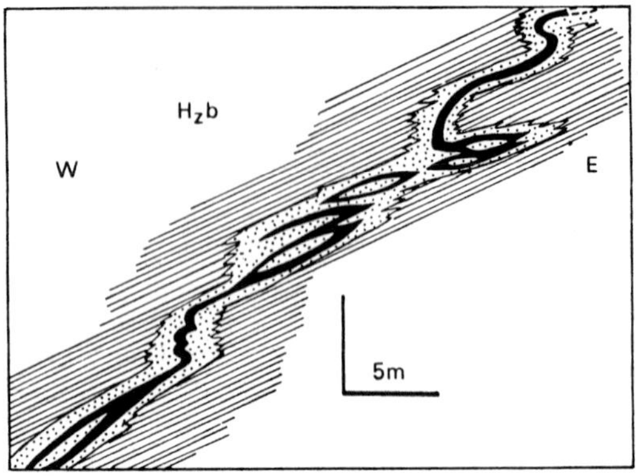


Fig. 1Q.10. Subconcordant pod (Marais Kiki in New Caledonia). Cross section perpendicular to the harzburgite lineation and foliation (hatches) ; dots : dunites ; black : chromite ore (Leblanc, 1987).

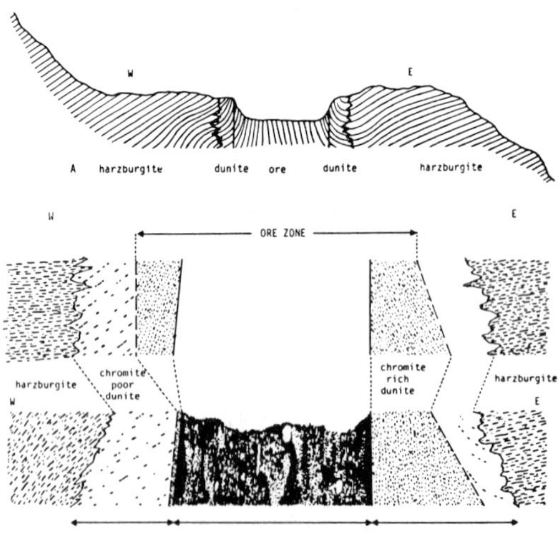

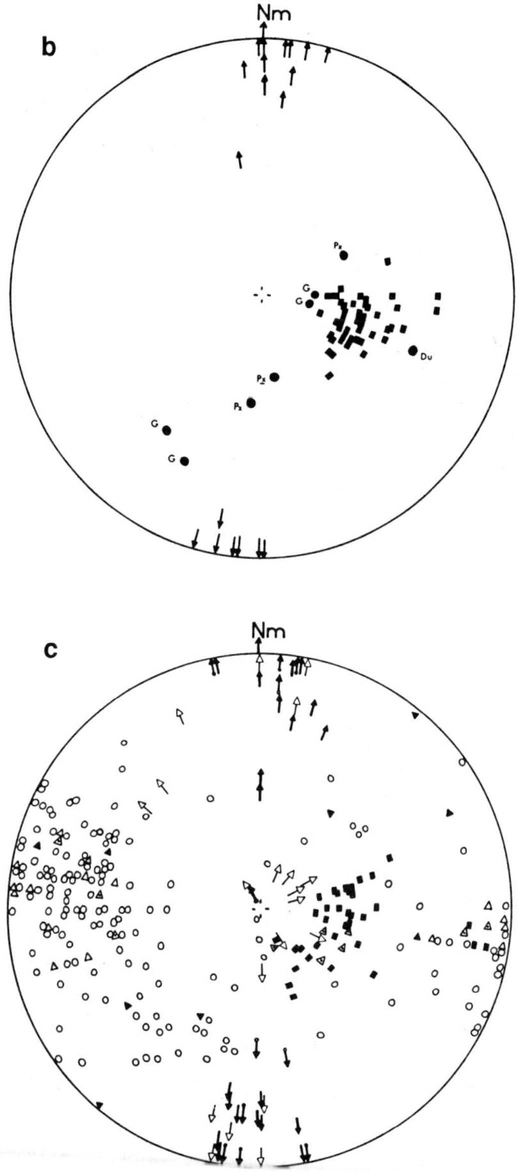


Fig. 10.11. Discordant pod (Anna Madeleine in New Caledonia). a) Cross section showing, above, the rotation of the foliation in the vicinity of the pod and, below, how this occurs in the dunite wall. The chromite-poor dunite wall in structural continuity with the harzburgite could have a residual origin (melting-out of orthopyroxene) and the chromite-rich dunite wall, a magmatic origin like the olivine clots inside the pod. b) and c) Stererographic projections of structural data, b) in country peridotites and c) within the pod and in this vicinity. Same symbols as in figure 10.9 with in addition ; solid circles : dikes (Du : dunites ; G : gabbros ; Px : pyroxenites) ; dotted triangles : foliation in chromite-poor dunite wall (Cassard et al., 1981).

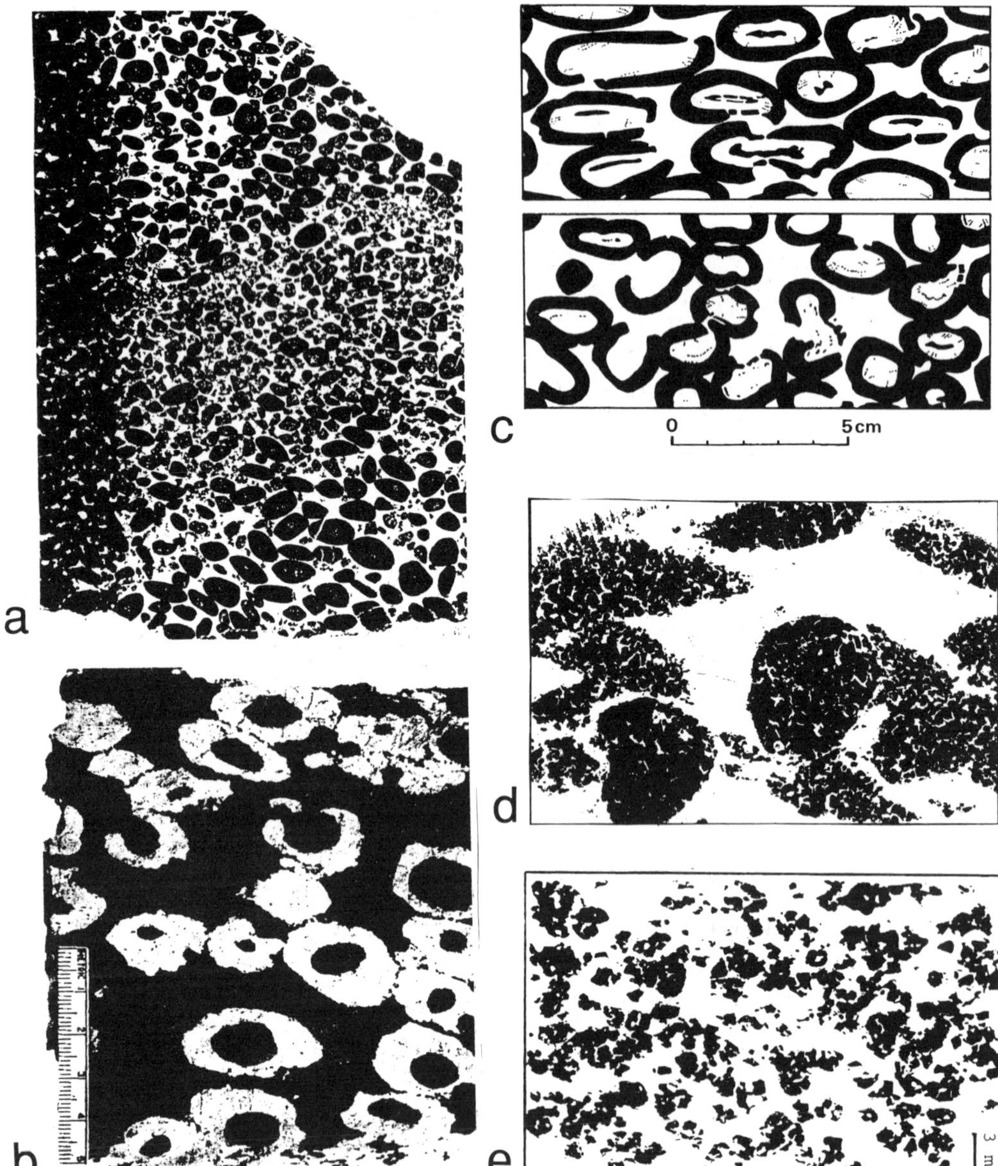
a
b
c
0
5cm
d
e
3 mm

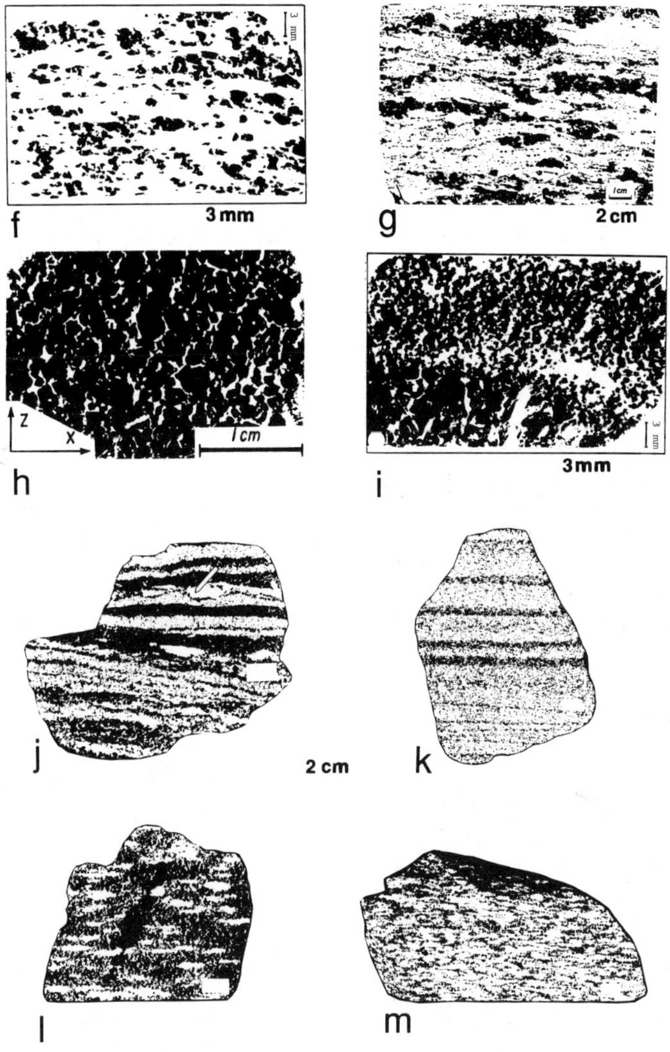


Fig. 10.12. Typical textures in chromite pods. In all figures, the magmatic or plastic stretching lineation is E-W and the foliation, perpendicular to figure. Chromite : black ; serpentinized olivine and accessory plagioclase or clinopyroxene : blank. a), b), c) and j) Textures of *discordant pods* : a) nodular, suggesting sedimentary bedding with size grading ; b) and c) orbicular ; j) banded with oblique stratification. d) and e) Textures of *subconcordant pods* : d) dismembered nodular ; e) disseminated. f), g), h), i), k),l) and m) Textures of *concordant pods* : f) and g) disseminated (g) taken in a schlieren) ; h) and i) massive, with pull apart ; k) banded and disseminated ; l) and m) antinodular. (a) Camaguey, Cuba (Thayer, 1969) ; b) West Pakistan (Thayer, 1969) ; c) Fethiye, Turkey (Leblanc et al., 1981); h) Maqsad, Oman (Ceuleneer and Nicolas, 1985) ; all the others, New Caledonia (Cassard, 1980).

the surrounding rocks indicate the elongation direction of the deposit, whereas lineations within the ore can be used to follow local changes in the dip of the ore sheet. These deposits are typically tabular in form and the ore is usually less deformed than ore in the concordant deposits.

*Discordant deposits* - Discordant-type deposits are about as abundant as the subconcordant deposit type in New Caledonia. In Oman, they are less common. Whereas concordant and subconcordant deposits share a number of common characteristics, discordant-type deposits are quite dissimilar. This is a result of their differing structural relations with the surrounding rocks and their varying form and ore types. The main features of this class of deposits may be summarized as follows.

Generally, structures in the rocks immediately surrounding the deposit are disturbed with respect to their regional attitude. The orebody shape and internal structures are clearly discordant with respect to regional structures in the surrounding rocks. Lineations in the ore are not always present because the ore is sometimes only slightly deformed. They may be quite oblique to those in surrounding rocks and follow the orebody shape. Lineations in the surrounding peridotite may give the general direction of extension of the orebody.

Two examples from New Caledonia are presented to illustrate the differences which can exist within this class of deposits. One, the Dyne Mine, is formed of irregularly-shaped pods, a few meters to a few tens of meters across imbedded in a large dunite zone (type E in figure 10.8). The other, the Anna Madeleine Mine, (type C in figure 10.8) is an elongated body parallel to the regional lineation, some 200 m long in exposure and several meters wide. As shown by figure 10.11, this latter body is clearly discordant with respect to the surrounding foliation. It has an internal foliation, possibly entirely magmatic, parallel to its walls and the associated lineation, which is generally flat, steepens locally, possibly in relation with local rooting of the ore-blade.

| Discordant deposits | Subconcordant deposits | Concordant deposits |
|---|---|---|
| Massive ore in foam textures | Slightly deformed massive chromite ore | Massive ore with pull-apart textures |
| Massive, disseminated, and antinodular (occluded) ore, slightly to moderately deformed. Leopard-type (nodular ore). Orbicular ore | Slightly to moderately deformed disseminated ore, foliated and lens-shaped relicts of leopard-type structures. | Antinodular ore, very deformed and stretched-Disseminated chromite ore strongly foliated |
| Breccia-type chrome ore Banded ore with sedimentary features | | Flattened lenses and layers of chromite-ore |
| strain → | | |

Table 10-I. Ore textures met in the three structural types of chromite deposits and relation with increasing plastic strain (modified from Cassard et al., 1981).

The typical ore textures related to these three structural types of chromite deposits are summarized in table 10-1 and illustrated in figure 10.12. In the matrix, olivine may be associated with clinopyroxene, amphibole, phlogopite or plagioclase.

The order in table 10-1 and figure 10.8, from discordant to concordant bodies, reflects an increasing plastic deformation of magmatic settling textures at very high temperature. The spectacular nodular and orbicular texture, with rounded to flattened chromite balls attaining 2 cm and 4 cm respectively, are produced by growth in a dynamic magma as shown by their internal concentric structure (Thayer, 1969). They settled while they were still soft with mutual impression at their contacts (Pavlov et al., 1977 ; Leblanc et al., 1981 ; fig. 10.12c). Their progressive dismembering and scattering into disseminated ore can be followed in subconcordant and concordant pods (fig. 10.12d). Here we disagree with Hock and Friedrich (1985), who regard the nodules as being elongated by plastic strain and thus typical of plastically deformed bodies. Massive chromite is very difficult to deform (Sécher, 1982 ; Doukhan et al., 1984 ; Christiansen, 1986) and yields by fracturing (pull-apart). Nevertheless, we agree with Hock and Friedrich in considering the flattened antinodular texture as typical of plastic strain ; it is associated with pull-apart texture in more massive chromite, and is itself a result of the straining of weaker olivine inclusions in a chromite-rich matrix.

This structural classification is justified by the fact that the peridotites belonging to the transition zone, where most chromite pods are located have suffered a large plastic flow at near or above solidus temperatures (§ 9.2). If they are emplaced early during solid-state flow (Doukhan et al., 1984 ; Christiansen, 1986), chromite pods, whatever their primitive orientation, are tectonically reoriented parallel to the peridotite foliation and elongated parallel to the lineation. Pods emplaced late, during or after plastic deformation would be respectively subconcordant and discordant, with regard to the foliation. The genetic model proposed below predicts that the chromite pods are formed by magmatic accumulation along the conduits channeling the magma extracted from the rising mantle diapir toward the accreting oceanic crust. Consequently, pods would be originally discordant and, depending on the time of their formation, would remain so or be internally deformed and reoriented toward the foliation (fig. 10.16). These views are supported by considering the chromite-ore textures (table 10-1). Only the discordant pods preserve the most delicate textures typical of the accumulation process (nodular, orbicular, chromite-net, silicate-occluded, etc.). The concordant pods are composed of these disrupted textures. A positive test of this interpretation is that undeformed magmatic textures are not observed in concordant pods. We know only two exceptions ; one is in Poum Island (New Caledonia), where the Poum 30 deposit (Sécher, 1981) is a concordant pod with some nodular ore ; however, its well layered structure suggests that it could have been directly emplaced parallel to the regional foliation ; the other is the Voidolakkos Mine in the Vourinos ophiolite (Greece) which is formed of concordant pods in a mylonitic zone, preserving some nodular ore in their central parts. Borchert (1964) had already observed that nodular and orbicular chromite ore were restricted to pods with steep walls (our discordant pods). He explained this observation, and that of orbicules accumulating downsection, by a dynamic sedimentation controlled by a steep floor.

### *10.5.4. Composition of chromite deposits*

As shown by figure 10.13a, the field of chromite composition in ophiolites is wide compared to those occurring in stratiform complexes and in komatiites. This field shows the same Cr/Cr + Al range in the chromite of ore deposits as in accessory Cr-spinel of the

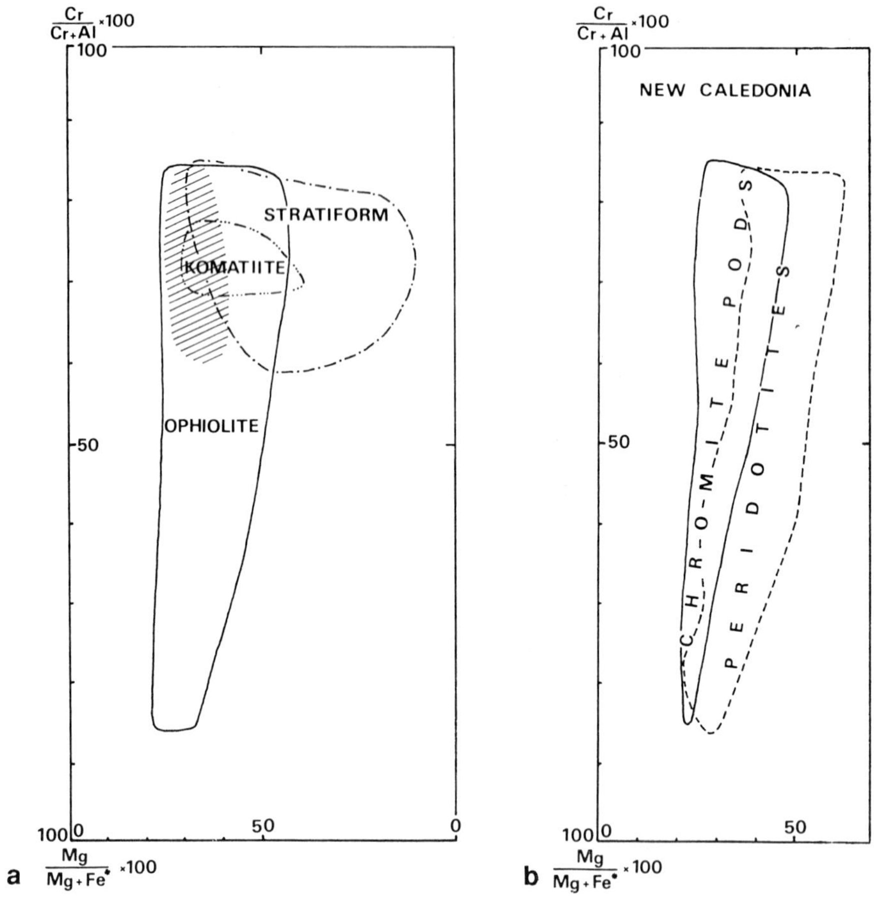


Fig. 10.13. 100 Cr/(Cr + Al) versus 100 Mg/(Mg + Fe) diagrams showing areas of the chromite pods and of the accessory chromite in peridotites. a) Fields of chromites from different types of ultramafic complexes ; hatches : dominant composition in pods, detailed in fig. 10.14. b) Compared composition of chromite in pods and Cr-spinel in peridotites from New Caledonia (Leblanc, 1987).

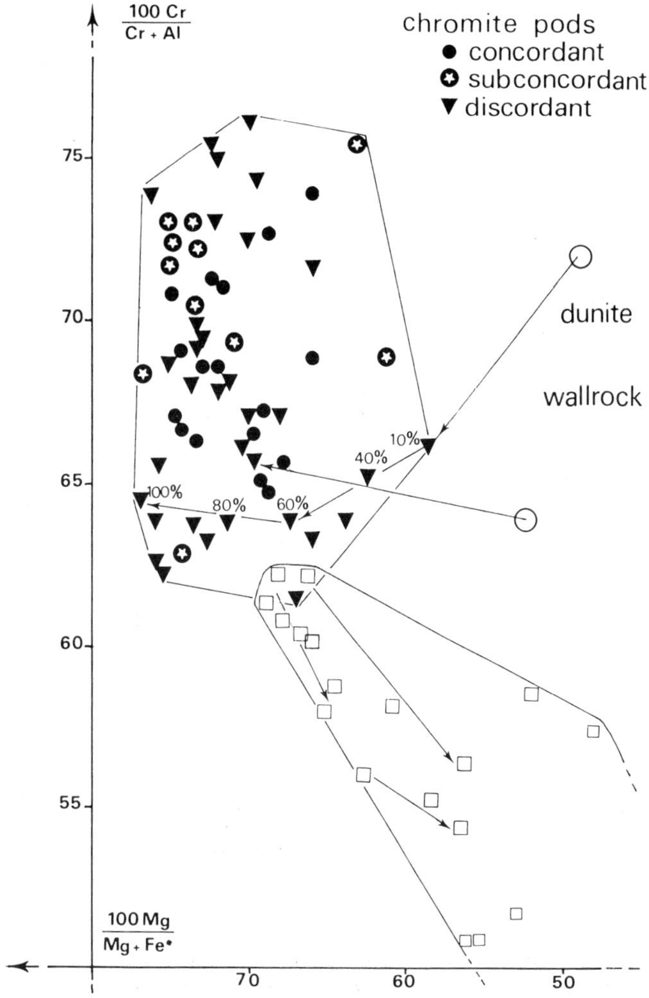


Fig. 10.14. Plot, in the hatched area of fig. 10.13, of the composition of some chromite pods (black) and some chromite concentrations in the layered gabbros (empty squares) from the Massif du Sud (New Caledonia). The various pod types lie in the same compositional field. Connected samples are joined by tie lines ; the chromite in the ore is more magnesian than the accessory chromite of the dunite wall rock, and it displays a greater magnesium enrichment from chromite-poor to chromite-rich ore than the massive chromite ore. In the layered gabbros pile, chromite shows an upward Al and Fe enrichment (arrows) corresponding to a fractional crystallisation trend (Leblanc, 1987).

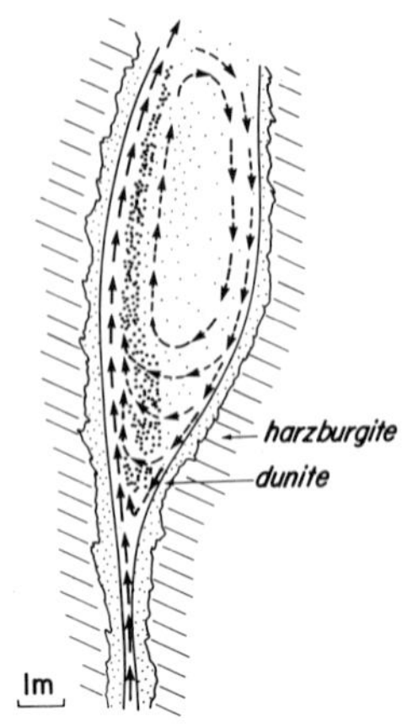


Fig. 10.15. Schematic model of chromite pod formation in a cavity along magma dikes in the tectonic harzburgite (location as dotted lenses in fig. 10.6) (Lago et al., 1982).

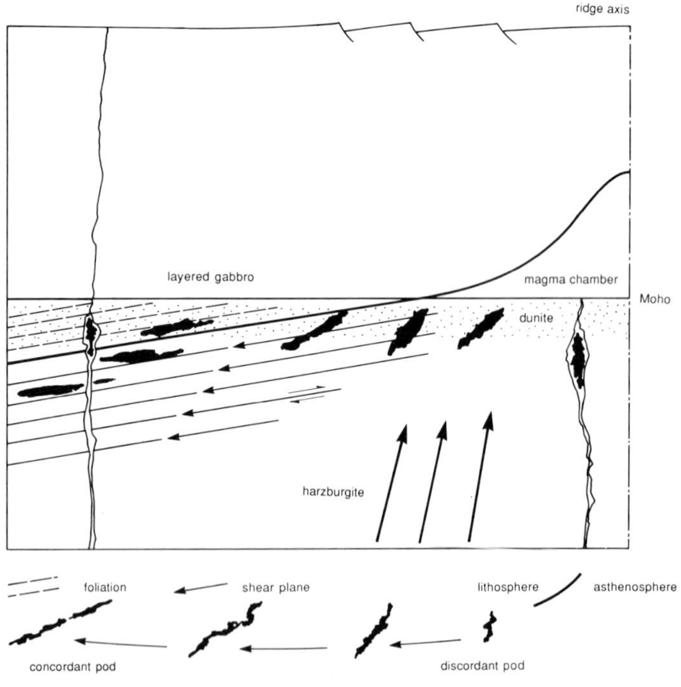


Fig. 10.16. Sketch of the genesis and evolution of the chromite pods in uppermost oceanic mantle beneath an active spreading ridge (Lago et al., 1982).

associated peridotites, with a shift of the ore-chromite toward more magnesian compositions (fig. 10.13b). Inside the ore-chromite field, most deposits concentrate between 0.6 and 0.8 Cr/ Cr + Al (fig. 10.13a), the tail toward lower ratios corresponding to particular districts, for instance Poum-Yandé-Belep as opposed to all other districts in New Caledonia (Leblanc, 1987) (§ 5.2) or White Hills as opposed to Bay of Islands in Newfoundland (Laurent and Kacira, 1987). These districts are characterized by a more lherzolitic composition, confirming the link between chromite in the ore deposits and Cr-spinel in associated peridotites. However, in other ophiolites like Zambales in the Philippines, this relationship falls down, with strongly differing Cr contents in chromites from Acoje and Coto, which are both harzburgitic, and an inverse relationship between chromite in the ore deposits and Cr-spinel in the associated peridotites (Leblanc and Violette, 1983). Focusing on the hatched area of figure 10.13, enlarged in figure 10.14, which represents most of the New Caledonia pods, the following general comments can be made :

- there is no obvious chemical difference between discordant, subconcordant and concordant pods ;
- there is a clear enrichment in Mg ratio with increasing chromite-olivine ratio, from wall-rock dunites to massive ore (fig. 10.2). This Mg variation is related to subsolidus reequilibration resulting in Fe with respect to Mg enrichment of chromite when olivine is abundant (Irvine, 1967 ; Lehmann, 1983) ;
- More Cr-rich pods tend to be deeper and Al-, Fe-rich pods more superficial, an observation also made in other ophiolites (Golding and Johnston, 1971 ; Brown, 1980 ; Leblanc and Violette, 1983 ; Engin et al., 1987 ; Laurent and Kacira, 1987).

### *10.5.5. Origin of chromite deposits*

Like the dunites to which they are clearly linked, from the beginning chromite pods in ophiolites have been considered as magmatic cumulates settled at the base of the magma chamber, below the layered gabbros of the plutonic unit. This interpretation was again inspired by the analogies between chromite deposits in ophiolites and the stratiform ore bodies in great complexes. This interpretation, which is still largely accepted (Burgath and Weiser, 1980 ; Augé and Roberts, 1982 ; Christiansen, 1982 ; Hock and Friedrich, 1985), is evidently valid for the exceptional stratiform chromite bodies described above. However, it fails to explain the general location of pods below the Moho, often clearly sited within the harzburgites, and the discordant shapes of somes pods. Authors aware of such difficulties have proposed various interpretations which start from the deposition by gravity settling, either in mantle 'mini-chambers' (Brown, 1980) or at the base of the main magma chamber. The latter interpretation solves the problem of having a large enough reservoir to account for the volume of chromite in the pods. Various explanations have been offered to explain the existence of discordant pods : a magmatic accumulation inside trenches in the chamber floor for Borchert (1964) ; a tectonic insertion inside the peridotites, either related to tight folding for Greenbaum (1977) and Burgeth and Weiser (1980), or related to gravity sinking for Dickey (1975). Thayer (1969) proposes that chromite precipitates as layers in stratiform complexes located in the upper mantle, which are subsequently emplaced at solidus to hypersolidus temperatures at its present upper level. Lago et al. (1982) have discussed these interpretations and noted that none of them was compatible with the structures described by Cassard et al. (1981) in the chromite pods and in their wall peridotites. A new genetic interpretation has been developed by these authors.

Lago et al. (1982) assumed that chromite accumulated from the rising basaltic melt, inside dikes crossing through the mantle diapir below the ridge. These dikes are normally narrow (5-50 cm thick) and the active upstream magmatic flow prevents the settling of any chromite, even the biggest nodules. A discordant pod may be created if there is a local enlargement of the dike to the dimensions of the future orebody, say 100-200 m high and 2 to 5 m thick. Rising within a peridotite which is already above its solidus, these dikes belong to the category of the indigenous dikes (§ 2.5.2), meaning that they develop a dunite wall. Indeed, such a dunite wall is always observed around chromite pods. Various factors can induce chromite and olivine crystallization from the rising basalt, including pressure release in the dike, drop in temperature or increase in oxygen fugacity (Brown, 1982 ; Maurel and Maurel, 1982). Cooling of the melt by the conduit walls becomes very effective when the mantle diapir has crossed the limit between adiabatic and conductive heat transfer. Below a ridge or at a distance of a few tens of kilometers, this should happen at the Moho level or just below, where indeed most pods are now located. Common inclusions of hydrous phases in chromite, mainly amphibole and phlogopite, suggest that chromite fractionation could have been facilitated by the presence of some water which has an important effect on oxygen fugacity. If this water permeated from above locally in the transition zone, it would cool the temperature and modify the oxygen fugacity, possibly favoring chromite fractionation from the basaltic melt.

In order to explain nodular and orbicular textures in the discordant pods a strong convection must be envisaged in the cavity. This cannot be achieved by the upstream flowing magma in a meter-sized cavity, unless unrealistic velocities of meters per second inside the decimeter-sized dike feeding this cavity are envisaged (see § 7.4.2). Thermal convection instead seems necessary. This would be promoted by a thermal contrast between the flowing magma and the colder walls of the conduit. Figure 10-15 illustrates the pattern of the circulation that can be expected according to the preceding scenario. This scenario has been modelled numerically by Lago et al. (1982). With a 15° C difference between the temperature of the magma and that of the inner wall and with the dimensions chosen above, the convection velocity and the upward velocity of magma necessary to prevent the biggest nodules from obstructing the feeding dikes, are both found to be around 1cm/s ; the model is thus physically consistant. The prediction of mixing particles of different sizes is also verified. Depending on the choice of the initial thermal contrast, the time span necessary to create a pod is remarkably short, in the range of one month, in keeping with estimates for the duration of melt injection through the dike (§7.4.2). After this time, the convection may become too sluggish to sustain the biggest nodules, which fall and obstruct the feeding dike ; alternatively, the source of magma may start drying up and the melt pressure falling, resulting in collapse of the pod. The elastic closure of the cavity induces the first deformation in the soft pod (fig. 10.12c) and expels the melt. Trapped melt crystallizes as poikilitic diopside and amphibole or plagioclase. If this pod, belonging to the discordant type, is formed close to the ridge axis, it is swept by plastic flow and becomes progressively transposed into a concordant pod during drift away from the ridge axis (fig. 10.16). To be preserved as a discordant pod, it should have been emplaced at a sufficient distance from the axis in order to escape plastic deformation (§ 10.5.3).

# Chapter 11

# GENERATION OF OCEANIC CRUST

## *11.1. INTRODUCTION*

Significant differences between ophiolites (table 8.I) led us to distinguish a harzburgite ophiolite type (HOT) and a lherzolite ophiolite type (LOT) which were ascribed respectively to fast and slow spreading situations (chapter 8). Similarly, significant differences between fast and slow spreading ridges encourage comparisons with these two types of ophiolites. In this chapter we will therefore discuss how HOT and LOT crustal sections are generated at oceanic spreading centers, and will hopefully increase our understanding of the processes of crustal generation in fast and slow spreading environments respectively.

At fast spreading ridges, geological observations at sea introduce constraints mainly on the processes of ridge volcanism and hydrothermalism. For processes occurring deeper than a few hundred meters, HOT ophiolites constitute the only direct source of information. However, the ridge models which can be proposed on the basis of observations in such ophiolites must be confronted with the increasingly sophisticated models derived from indirect evidences such as ridge topography or the multiple geophysical signals that can be measured there. In slow spreading environments, the balance, as a source of information, is clearly more in favor of oceanic data rather than data from ophiolites. This is so because, on one hand, higher reliefs and more active superficial tectonics at these ridges provide more direct information on deep crustal processes than in the case of fast spreading ridges and, on the other hand, LOT ophiolites are altogether less common, more dismembered and thus more difficult to integrate in ridge models than HOT ophiolites. In order to combine information on both ophiolites and oceanic crust, we must first assess the correspondence between the well known seismic layering of oceanic lithosphere and the main lithological units of ophiolites.

From the data in ophiolites summarized in table 8-I, it seems that HOT and fully developed LOT have comparable sheeted dikes and volcanics sections, and that their major differences stem from the processes occurring deeper in the crust at the spreading center of origin. The homogeneity of the HOT crustal structure with, in particular, a thick and continuous section of layered gabbros suggests that HOT crustal accretion responds to more continuous and general processes than LOT, where marked crustal heterogeneities have been described (see chapter 4), and where the crust may even be absent (§ 4.4), or of a new type (§ 8.2.3). Drilling by the DSDP-ODP in the slow spreading Mid Atlantic Ridge has revealed a complex crustal structure even at the scale of adjacent holes (Hall and Robinson, 1979 ; Juteau et al., in press). Consequently, when considering HOT, we will address more specifically the problem of the structure and functioning of long-lived magma chambers and, dealing with LOT and marine data from slow spreading ridges, the problem of the nature of the deep crust, and the episodicity of spreading processes at slow spreading ridges.

## *11.2. LITHOLOGY OF OPHIOLITES AND SEISMIC STRUCTURE OF THE OCEANIC CRUST*

Oceanic crustal structure was first defined by seismic refraction studies and more recently by multichannel seismic reflection profiling. The first studies have identified a layered structure (Raitt, 1963 ; fig. 11.1a) which was readily compared with the grossly layered mafic sections of ophiolites (Coleman, 1971 ; Moores and Jackson, 1974 ; Christensen and Salisbury, 1975). Combining seismic information with petrological observations on dredged and drilled specimens, the following general stratigraphic section is still largely accepted. The top layer, called the 'sedimentary layer' or 'Layer 1', consists of sediments in various states of consolidation that are transgressive onto the igneous crust. Their thickness varies with the age of the crust and source of the sediments. Average seismic compressional velocities (Vp) are between 2.0 and 2.8 km/s. Below, upper part of the basement layer, also called 'Layer 2', is composed of basaltic pillowed or massive flows which in the DSDP hole 504 B are 500 m thick, grading down over 150 m into a sheeted dike complex extending from a depth of 1000 m to the bottom of the hole at 1562 m (Becker et al., 1988). Layer 2 is on average 1.6 km thick and it has an average Vp velocity of 5.0 km/s. Oceanic layer, also called 'Layer 3', has an average thickness of 5 km and an average Vp velocity of 6.73 km/s, compatible with a gabbroic composition. The base of Layer 3 is the *original Mohorovicic discontinuity which, strictly speaking, is defined by a sharp increase in compression wave velocity* to an average of 8.15 km/s. This high velocity, combined with a large velocity anisotropy, points to the presence of tectonic peridotites below the discontinuity. Anomalous mantle characterized by VP in the range of 7.2 - 7.7 km/s has been identified in oceanic mantle younger than 15 Ma.

The 'standard crustal section' (fig. 11.1a) has been refined using air gun-sonobuoy refraction techniques (Maynard, 1970 ; Sutton et al., 1971 ; Hussong, 1972 figs. 11.1b and c). In figure 11.1b, Layer 3 is divided into a thin sublayer 3A, with a Vp of 6.4 km/s and a thick sublayer 3B, with a Vp of 7.1. In figure 11.1c, Layer 3 is composed of a 6.8 km/s horizon overlying a basal layer with a velocity varying from 7.2 to 7.6 km/s. This basal layer was identified in lithosphere older than 15 Ma and thus cannot be attributed to anomalous mantle.

In 1975, at the time of Christensen and Salisbury's synthesis, the analogy between oceanic crustal models based on seismic refraction data and ophiolites was largely accepted at the expense of an earlier model which had a Layer 3 composed of serpentinite, below a basaltic Layer 2 (Hess, 1962). The ophiolitic model was, however, hampered by the evidence that many ophiolites seemed to have a thinner crust (Moores and Jackson, 1974) and were lacking rock facies corresponding to a 3 km thick basal layer with a Vp of 7.2 to 7.6 km/s.

Concerning the problem of comparing crustal thicknesses in ophiolites and oceanic crust two remarks should be made. Firstly, we have seen in § 3.3.2 that, unless careful structural studies are conducted in non dismembered ophiolites, it is very difficult to estimate accurately their crustal thickness. Many published data need to be reevaluated. Secondly, early seismic refraction studies showed that the seismic structure of the oceanic crust changes slightly and its thickness increases with age ; in particular Layer 3 thickens by 2 km within 40 Ma (Le Pichon, 1969), possibly by means of mafic intrusions in the uppermost mantle or more probably, by partial serpentinization of this horizon, (§ 11.3.3). This point raises the problem of extending the definition of the Moho with respect to the original seismic definition, as being now the limit between mafic and ultramafic formations. A further extension has been the concept of a 'petrological Moho' which is located between the tectonic and the 'cumulate' ultramafics, assuming that this distinction has a general sense (chapter 10). For seismologists, serpentinized peridotites

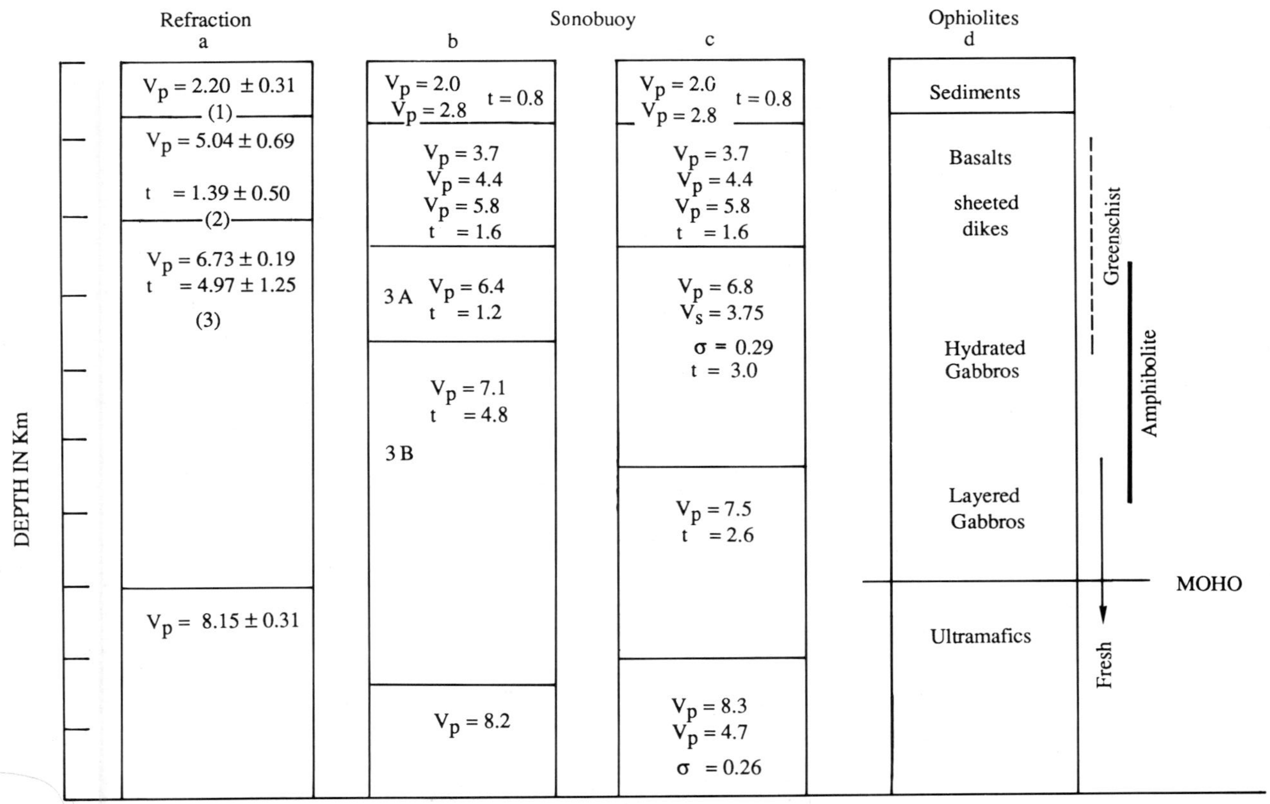


Fig. 11.1. Comparison of oceanic crustal structure (a,b,c) and ophiolitic section (d). a) Standard oceanic structure deduced from refraction data. b) and c) Oceanic structure deduced from sonobuoy data. d) Composite ophiolite model integrating new data from Oman (§ 3.3.2) to previous models (see text) (metamorphism log from De Wit and Stern, 1976). In logs a, b, c, ; VP and Vs : compressional and shear velocities in km/s ; t : thickness in km ; σ : Poisson's ratio.

with $V_p$s below 7.2 km/s are now above the Moho and are thus incorporated in Layer 3, whereas for the ophiolite community they still represent mantle. Christensen and Salisbury (1975) rightly observed that most ophiolites represent young lithosphere (see also § 12.4.2) and should rather be compared with young oceanic crust, thus avoiding the problems of crustal aging. This has been successfully achieved by Kempner and Gettrust (1982a) (fig. 11.2).

During the last decade, oceanic crust models have greatly improved and the analogy with ophiolites have been futher confirmed (fig. 11.2). New experimental measurements of Vp and Vs have been conducted under higher confining pressures on typical ophiolite specimens (Christensen, 1978) and incorporated into ophiolite sections (Christensen and Smewing, 1981). Synthetic seismograms obtained from such data on the Bay of Islands and the Oman ophiolites have been compared interactively with high quality seismic refraction measurements in oceanic lithosphere (Spudich et al., 1978 ; Spudich and Orcutt, 1980 ; Nicholls et al., 1980; Kempner and Gettrust, 1982 a and b). Multichannel seismic reflection profiles from the North Atlantic revealed a complex structure at Moho depth on a scale which matches that obtained on synthetic seismograms constructed from the Bay of Islands ophiolite (Brocher et al., 1985 ; Collins et al., 1986).

Kempner and Gettrust (1982b) reinterpreted the base of Layer 3 as a 1 km thick layer of partly serpentinized ultramafics at the top of Moho, overlain by a 2 km thick layer of layered gabbros with an average Vp of 7.1 km/s. In this model, Layer 3A would be composed of hydrated massive gabbros. These authors locate the Layer 3-Layer 2 boundary within the sheeted dike unit, at the limit between the underlying amphibolite facies and the overlying greenschist facies resulting from hydrothermal alteration near the ridge. For Clague and Staley (1977) and Christensen and Smewing (1982) the Layer 3-Layer 2 boundary is located at the base of the sheeted dike unit. Figure 11.1d is an hybrid log from ophiolites, trying to relate the reevaluated thickness of the Oman ophiolites (§ 3.3.2) to the seismic structure of oceanic crust.

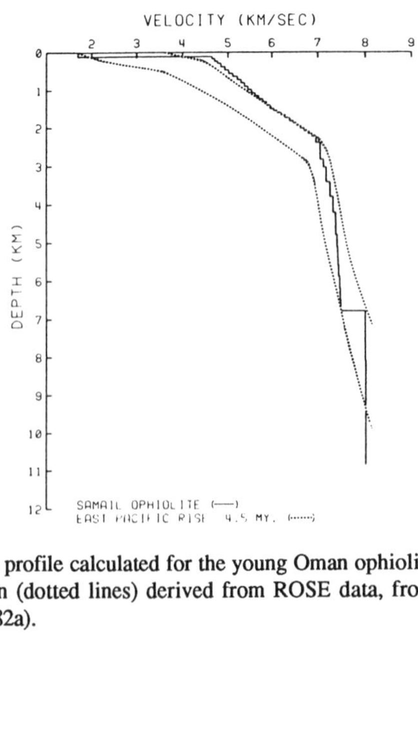


Fig. 11.2. P velocity-depth profile calculated for the young Oman ophiolitic crust (full line), bounded by the velocity-depth function (dotted lines) derived from ROSE data, from the EPR of comparable age (Kempner and Gettrust, 1982a).

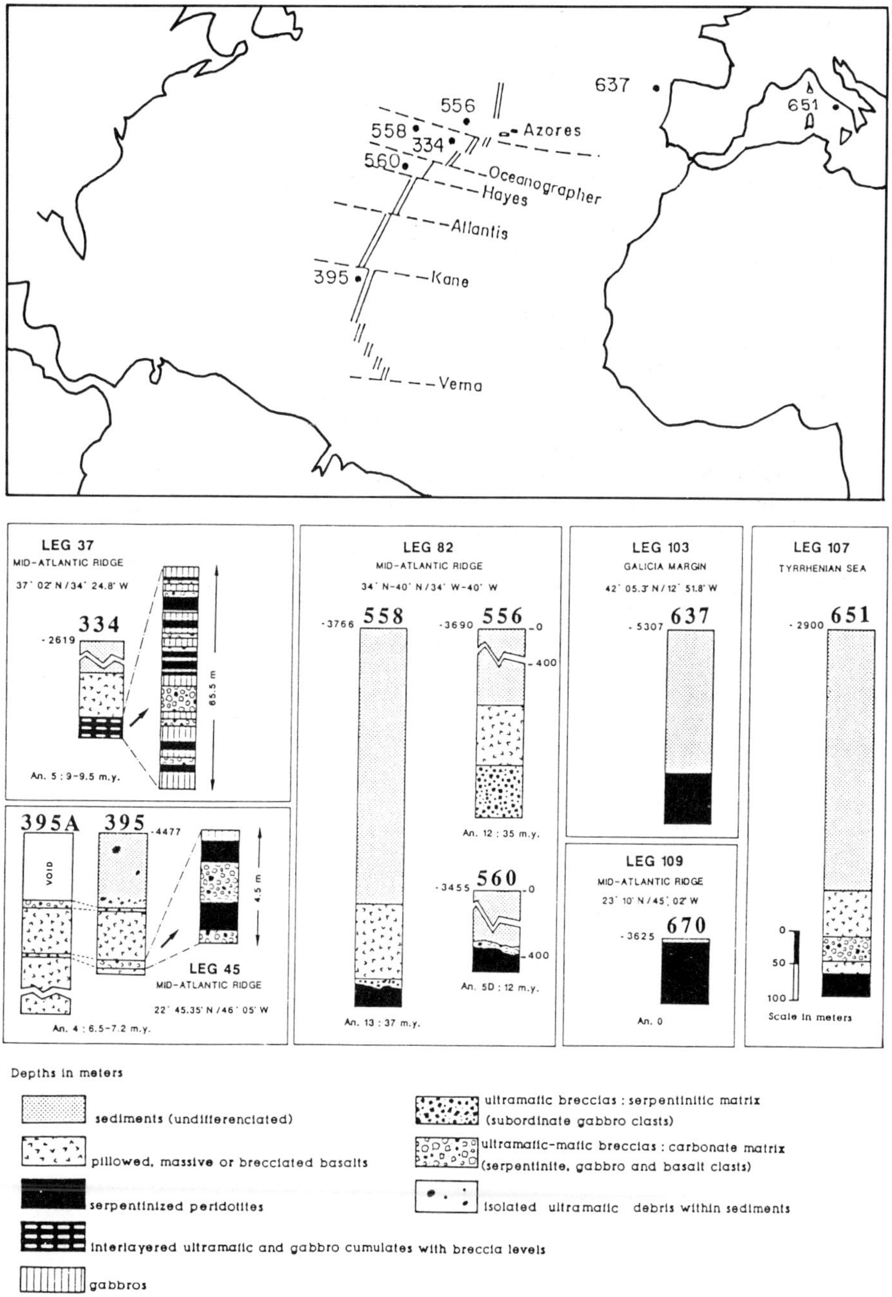


Fig. 11.3. Compilation of DSDP-ODP holes where, serpentinites and peridotites have been drilled (Juteau et al., in press).

## *11.3. SERPENTINITE SEAFLOOR IN SLOW SPREADING ENVIRONMENTS AND LOT*

In spite of the successful analogy of many oceanic and ophiolitic crustal sections, it is necessary to reevaluate Hess's(1962) interpretation advocating serpentinized peridotites as forming Layer 3 of the oceanic crust below a basaltic Layer 2. This is imposed by the repeated recovery of serpentinized peridotites in the oceans, and by recent discoveries in ophiolites of the western Alps, showing that in these LOT ophiolites the crust is largely reduced and may locally be totally missing, with sediments or lavas resting directly on serpentinites (§ 4.4 ; fig. 4.17).

### *11.3.1. Abyssal and ophiolitic peridotites*

With the few exceptions of trenches and transform faults offsetting the fast spreading East Pacific Rise (see Hebert et al., 1983), the abyssal serpentinized peridotites recovered so far come from slow spreading ocean floor and mainly from their transform faults. Since the reviews of Cann and Simkin (1971), Christensen and Salisbury (1975) and Hekinian (1982), recent petrological and structural studies on these rocks include those of Hamlyn and Bonatti (1980), Nicolas et al. (1980), Michael and Bonatti (1984), Dick and Bullen (1984), Dick et al. (1984), Fisher et al. (1984), Evans and Girardeau (in press), Cannat et al. (in press) and Juteau et al. (in press). These authors describe, as parent facies to the abyssal serpentinites, peridotites ranging from harzburgites to spinel and plagioclase lherzolites, with microstructures similar to those in ophiolitic peridotites. It is generally admitted that, with respect to peridotites from ophiolites, the abyssal rocks are richer in diopside and plagioclase. The microstructures show that this is due both to magmatic impregnation of a depleted peridotite, and to a more moderate depletion of the source lherzolite (Nicolas et al., 1980 ; Dick et al., 1984 ; Cannat et al., in press) (§ 2.5.3). The moderate depletion of many abyssal peridotites is confirmed by an extensive study of the chromium and aluminum ratios in spinels (Dick and Bullen, 1984). The more refractory character of many ophiolitic peridotites, compared to oceanic ones, led Dick and Bullen to suggest that ophiolites did not represent 'normal' oceans and were formed from a larger melting of the mantle in presence of water, like in island arcs. We observe that the most depleted ophiolitic peridotites (Dick and Bullen's type III), which are typically HOT, are rare in the oceans, and that the least depleted ones (these authors' type I), which are typically LOT, are dominant in the oceans. *We ascribe the difference between oceanic and ophiolitic peridotites to the oceanic sampling occurring preferentially in slow spreading oceans and along fracture zones* ; this supports our contention of a predominantly LOT mantle in slow spreading oceans and in fracture zones (§ 8.2.3).

### *11.3.2. Serpentinized peridotites as seafloor*

Fault reliefs, uplift motion and the well documented crustal thinning close to transform faults (Cormier et al., 1984 ; White et al., 1984 ; Fox and Gallo, 1984 ; Mutter et al., 1984 ; Whitmarsh and Calvert, 1986) help us to understand the preferential recovery of mantle specimens in this environment. More intriguing is the increasing recovery of mantle specimens in 'normal' oceanic crust, mainly close to the ridge or along passive margins. Figure 11.3 summarizes the discoveries of the DSDP-ODP drilling programs in the Atlantic Ocean. They point to the presence of serpentinites directly beneath sea-floor sediments or separated from them by a veneer of basalts 100 thick or less. Sediment-serpentinite or gabbro-serpentinite breccias can be met at these contacts. These serpentinites are derived from harzburgites and from plagioclase or spinel lherzolites.

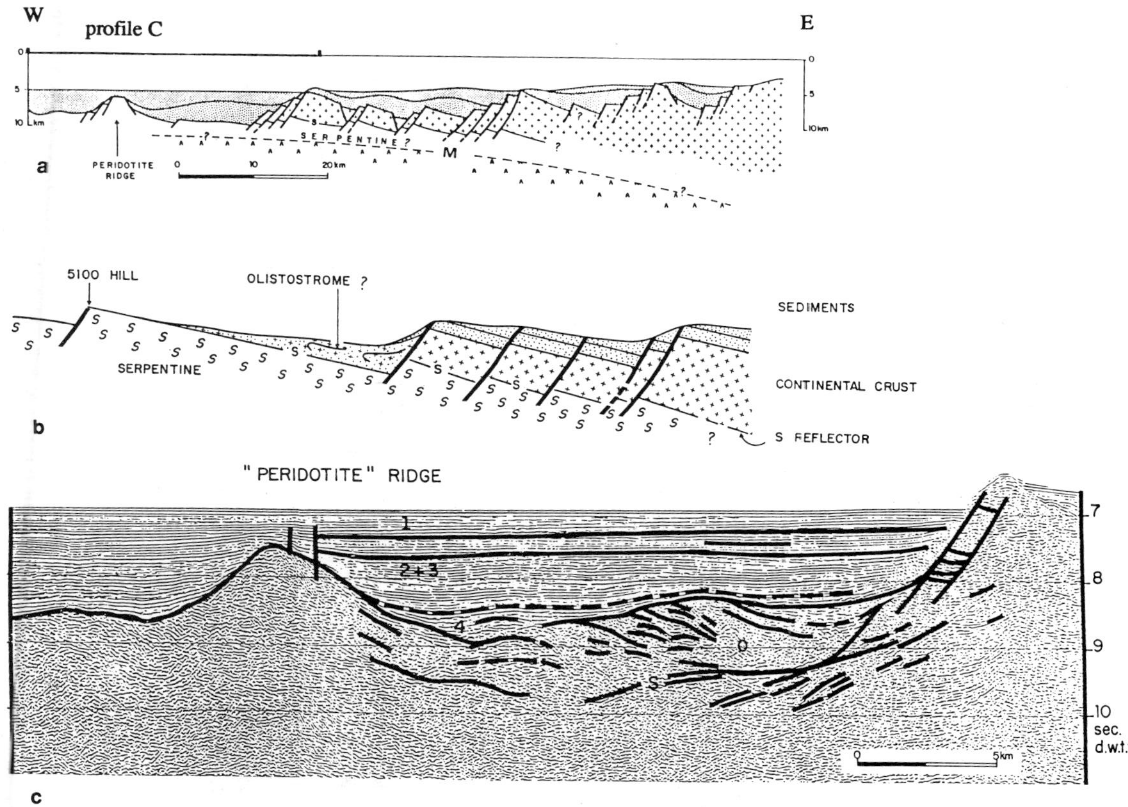


Fig. 11.4. Profiles along the Galicia passive margin. a) General setting. b) Interpreted structure of the 5100 Hill or 'peridotite Ridge' from the seismic reflection profile c) ; S : 'S-reflector' assimilated to a serpentinitic Moho ; O : olistostrome ; 1-3 : post-rifting sediments ; 4 : syn-rifting sediments. a) and b) Boillot et al., in press ; c) Mauffret and Montadert, 1987.

Recent ODP leg 107 records a similar situation at a spreading center, now inactive, of the Tyrrhenian Sea (Kastens et al., 1986). Examples of lherzolites constituting the seafloor along passive margins are found in the southern Indian Ocean at the SW margin of Australia and in the Atlantic along the Galicia margin of Spain (fig. 11.3), as well as in the Zabargad Island in the Red Sea (§ 4.5). In the foremost occurrence, only spinel lherzolites have been recovered, whereas in the two last mentioned plagioclase lherzolites are present. The lherzolite ridge of Galicia margin (fig. 11.4) has been extensively studied by dredging, drilling (ODP Leg 103) and by diving with the *Nautilus* submersible (Boillot et al., in press). It extends N-S over more than one hundred kilometers at the boundary between the stretched continental crust and normal oceanic crust. The plagioclase lherzolites bear evidence both of partial melting during high-T plastic flow and of subsequent mylonitic deformation (Girardeau and Evans, in press), as do those of the Zabargad Island.

From this brief review, it is clear that, possibly restricted to local areas of oceans characterized by a small spreading rate, Hess' interpretation of serpentinite as Layer 3 needed to be rehabilitated (Lewis, 1983). New models of discontinuous magmatic activity in slow spreading environments can account for such situations (§ 11.6) ; however, they raise a perplexing new question discussed below : why does the Moho depth remain at a fairly constant 6 km depth below the top of the oceanic crust, in spite of its contrasted nature and origin.

### *11.3.3. Nature of the Moho*

It has been proposed (Clague et Straley, 1977 ; Lewis and Snydsman, 1979 ; Lewis, 1983) that the Moho depth at around 6 km reflects the deepest penetration of seawater into the newly formed lithosphere and that the Moho may therefore be the limit between serpentinized and unaltered peridotites. In older lithosphere, the progressive thickening of Layer 3 with age is best explained by serpentinization progressing downward. The nature of the serpentinization at the base of LOT normal crustal sections shows that hot hydrothermal water could have penetrated in a pervasive way through the uppermost mantle peridotites, in contrast with HOT sections, where this alteration occurs only in faulted areas (§ 8.2.1). The Xigaze ophiolite is demonstrative in this regard (§ 4.2) ; here the mafic crust is obviously too thin compared to oceanic crust (fig. 4.4), unless one includes the top 2 or 3 km of the mantle section, which was serpentinized at the ridge (§ 4.2.3). The penetration of seawater should be limited downwards by the closure of cracks due to the increasingly confining pressure. The figure of 100-200 MPa for this limit, quoted in rock mechanics (Paterson, 1978), is attained for depths in the range of 3-7 km, bracketing a Moho depth of 6 km. Laboratory measurements of acoustic velocities in peridotite specimens at increasing temperatures and pressures have shown that at 275°C, the velocity dependence on confining pressure after a sharp decrease becomes a linear function beyond 150 MPa, suggesting that beyond this confining pressure, the cracks were closed (Peselnick et al., 1974, 1978). These data, particularly relevant to the situation considered here, predict the correct 6 km depth for the Moho below the top of the crust.

From this analysis, it may be tempting to conclude that the 6 km depth for the Moho in the oceans reflects, in HOT cases, the thickness of the mafic crust. In LOT, where this thickness may be locally smaller, it represents the depth of serpentinite penetration into the mantle peridotites. A step further would be to conclude that slow spreading ridges like the Mid Atlantic Ridge to which we have ascribed a crust with LOT affinities, have a Moho marked by the serpentinite-peridotite boundary. However, generalizing this interpretation would lead us to the prediction that the Moho depth in the Atlantic should everywhere be 6 km below the top of the crust. Data show clearly that this is not the case. In particular, the Moho depth becomes shallower in the vicinity of transform faults (Fox and Gallo, 1984 ; White et al., 1984 ; Whitemarsh and Calvert, 1986 ; Potts et al., 1986) and at slow spreading ridges (Reid and Jackson, 1981). Hence, the systematic interpretation of the Moho as a serpentinite-peridotite boundary seems questionable. It should be recalled here that serpentinization of a peridotite is accompanied by a volume increase of up to 30 % (Hostetler et al., 1966 ; Nicolas, 1969). Serpentinization along the walls of a fracture would thus have a sealing effect on this fracture. This may preclude serpentinization penetrating to greater depths in the lithosphere unless the transform is actively deforming. Indeed within fracture zones like the Tydeman fracture zone, where mantle velocities of 7.2-7.5 km /s have been recorded, the sustained deformation may be responsible for the deep penetration of sea water (Calvert and Potts, 1985).

In conclusion, it is suggested that the 6 km depth of Moho beneath the top of crust generally reflects the limit of the mafic crust, in particular in HOT ophiolites and fast spreading oceans. Typical LOT would correspond to oceanic environments where this limit and thus the Moho are shallower, as in the Arctic Ocean (Jackson et al., 1982). However, a LOT thin crust may overlie an uppermost mantle serpentinized at the ridge. Consequently, in this case the 6 km-deep seismic Moho is a serpentinite-peridotite boundary.

## *11.4. THE PLUTONIC SECTION AND THE PROBLEM OF MAGMA CHAMBERS*

### *11.4.1. Introduction*

The problem of magma chambers beneath oceanic accretion centers is a problem pendent between the marine geophysical and the ophiolitic communities, as it is between the marine geophysical and petrological communities (Nisbet and Fowler, 1978 ; Stakes et al., 1984). As recalled in Macdonald's (1982) and Orcutt's (1987) reviews, active magma chambers have not been found by seismic methods along the Mid Atlantic Ridge except in the Reykjanes hot spot, whereas along the East Pacific Rise they are either absent (McClain and Lewis, 1982) or, when identified, they display a narrow roof between 2 and 6 km in width (Hale et al., 1982 ; McClain et al., 1985 ; Detrick et al., 1987). Thermal modelling predicts that for spreading rates lower than 1 cm/yr, magma chambers are either absent (Lewis, 1983) or discontinuous and intermittent (Kuzsnir, 1980), and that for larger spreading rates they should also be narrow at the roof, adopting a dome or tent-shaped profile because their walls should be parallel to some isothermal surface close to the magma solidus (fig. 11.10).

Most ophiolites have a continuous and thick layered gabbro section implying the existence of a magma chamber. Petrological evidence suggests that these magma chambers operated as open systems and were thus relatively long-lived (Jackson et al., 1975 ; Juteau and Whitechurch, 1980 ; Pallister and Hopson, 1981 ; Smewing, 1981). A popular model (fig. 11.5) claims that the roof width is in the range of 20 km in width. The existence of such a large roof is in conflict with the marine data mentioned above. The heat loss would also be so large that they could not have a lifetime of more than 3 x $10^4$ years (Lister, 1983). However, this apparent contradiction may be less severe because some ophiolites, like Xigaze (§ 4.2) and Trinity (§ 4.3), have a crustal section with no or minor layered gabbros and thus do not require the existence of a large and permanent magma chamber ; the Antalya ophiolite in Turkey (Juteau and Whitechurch, 1980) and Karmoy ophiolite in Norway (Pedersen, 1986) may also reflect conditions of non-permanent magma chambers. Alternative models attempting to reconcile oceanic spreading center and ophiolite data are discussed below.

### *11.4.2. Origin of the layering in the plutonic gabbro sequence*

The descriptions of magma chamber models in the next paragraph will make it clear that models depend critically on how the origin of the magmatic layering is interpreted. Firstly, it should be recalled that there is a variety of planar structures in the plutonic sequence of ophiolites (§ 2.4.1), suggesting that more than one process may have been operative. Secondly, the vast literature on this question addresses directly or refers to the case of stratiform layered complexes or intrusions (for recent and comprehensive studies on this subject, see Irvine, 1980 ; Rice, 1981 ; McBirney et al., 1985 ; Turner and Campbell, 1986).

To summarize our current knowledge of stratiform intrusions, we can state that crystals accumulate or grow and form layers either in the static medium of a boundary layer separating a crystallizing front from a convecting magmatic medium, or by deposition from density currents (Irvine, 1980). Mechanisms for differentiation inside a boundary layer include gravity settling (Pallister and Hopson, 1981), oscillatory crystallization due to either heterogeneous nucleation (Campbell, 1978 ; Brandeis et al., 1984) or to double diffusion (thermal and chemical), which also leads to separating the magma into double diffusive convection layers (Turner and Chen, 1974 ; McBirney and Noyes, 1979). Density currents sweeping the floor of a magma chamber can deposit crystals which have been sorted by the dynamic effects and gravity field acting within the current. Such currents can originate from the slumping to the base of the chamber of a mush crystallizing along the walls. Irvine (1980) proposes that the modally graded layering (§ 2.4.1) and in particular the graded bedding can be explained by density currents and that the uniform or isomodal layering (§ 2.4.1) can be explained by in situ crystallization in a boundary layer.

Fig. 11.5. Wing-shaped magma chamber model. Note that the approximate limit of the ultramafic cumulates (black lines) defines a major lithologic subdivision which is parallel to the floor of the chamber and oblique with respect to the cumulate gabbros layering (dashed lines) ; hatches : high-level gabbros ; vertical hatches : sheeted dikes (Pallister and Hopson, 1981).

In the case of ophiolites, structural studies in Oman (§ 3.3.2) have shown that the magmatic layering, whatever its origin, *has been transposed by a large viscous flow*, following a process well documented in plastically deformed metamorphic rocks (§ 2.4.2). Accordingly, the layering would result from the stretching and rotation of preexisting heterogeneities into parallel lens-shaped layers by the viscous flow affecting the magma. Some of these heterogeneities may be inherited from a mixture of a gabbroic melt and a wehrlitic crystal mush. This specific transposition would produce the well contrasted isomodal layering (Benn et al., 1988). On the other hand, the modally graded layering mainly observed in the basal part of the layered gabbro unit of Oman may result from crystal settling. The gravity settling could occur statically within sills of fresh magma, 100- 200 m thick (Browning, 1984), injected into the solidifying chamber (§ 11.4.5), or dynamically inside density currents sweeping the floor of the main chamber. The observation of dynamic sedimentation casts in these graded layers does not necessarily militate in favor of density currents because most likely these layers have also been affected by the viscous flow.

In view of the forthcoming discussion of magma chamber models it should be noted that layering resulting solely from gravity settling or deposition by density currents is necessarily formed on the floor of the chamber. A layering due to oscillatory crystallization in a boundary layer is formed along both the floor and the walls of the chamber. Eventually, layering, however how it originated, which has been transposed by a large magmatic shear flow, tends to become oriented parallel to the shear flow surfaces. In a long-lived magma chamber, one would expect that this layering should be frozen with an orientation parallel to the walls of the magma chamber (see below).

### *11.4.3. Magma chamber models*

*The wing-shaped magma chamber* - This model proposed by Greenbaum (1972) Parot and Ricou, (1976) and Pallister and Hopson (1981) for ophiolites, and by Cann (1974) for ocean ridges (the 'infinite onion') remains popular (fig. 11.5). These authors make the premise that the magmatic accumulation takes place on the floor of the magma chamber. The cumulate pile grows from nothing at the precise axis of the ridge, where new floor is created, to a maximum thickness of 1-3 km, which is attained at the lateral edge of the chamber. There the cumulate floor growing upwards due to increasing exposition time to crystal deposition meets with the isotropic or plated gabbro roof growing downwards due to cooling from above. Olivine and chromite which are the first minerals crystallizing from the rising melt at the ridge axis, preferentially settle near the axis, thus contributing to the ultramafic and chromite-rich nature of the lowermost cumulates. On the other hand, plagioclase, which appears later and is less dense, preferentially enriches the highest cumulates. Thus a gross vertical stratification in the chamber is explained with ultramafic cumulates at the base and anorthositic gabbros at the top. Pallister and Hopson (1981), point out the settling surface which joins the floor of the chamber at the axis to the top of the cumulates at the edge of the chamber should not coincide with the grossly horizontal limits of the compositional stratification, a prediction also made by Casey and Karson (1981) (fig. 11.5). The inward slope of the floor is limited by stability in the gravity field of the newly settled layers. The stability slope is a poorly constrained parameter, with slope estimates derived from dips in stratiform complexes varying between 15° and 50° (Fergusson and Palvertaft, 1963 ; Goode, 1976). It is also questionable whether this depositional surface is planar, as would be inferred from a well convecting chamber smoothing the variations in melt composition and viscosity, or progressively incurved upsection (Smewing, 1981), implying that the melt viscosity increases toward the chamber roof or that oscillatory crystallization occurs along the walls. These questions are

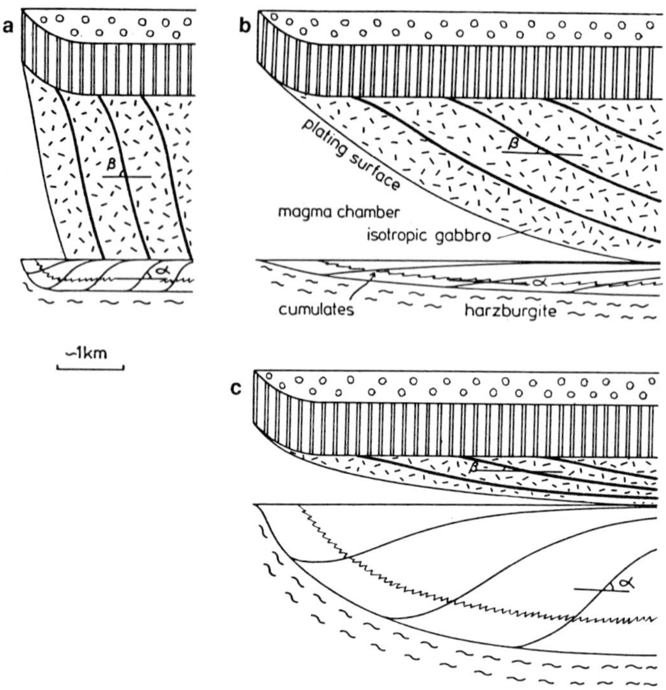


Fig. 11.6. Subsiding magma chamber model, showing the dependence of thickness of plated gabbros and layered gabbros on geometry of magma chamber and height of axial lherzolite welt. Solid lines : layering of gabbros ; zigzag lines : major lithologic sudivisions (Dewey and Kidd, 1977).

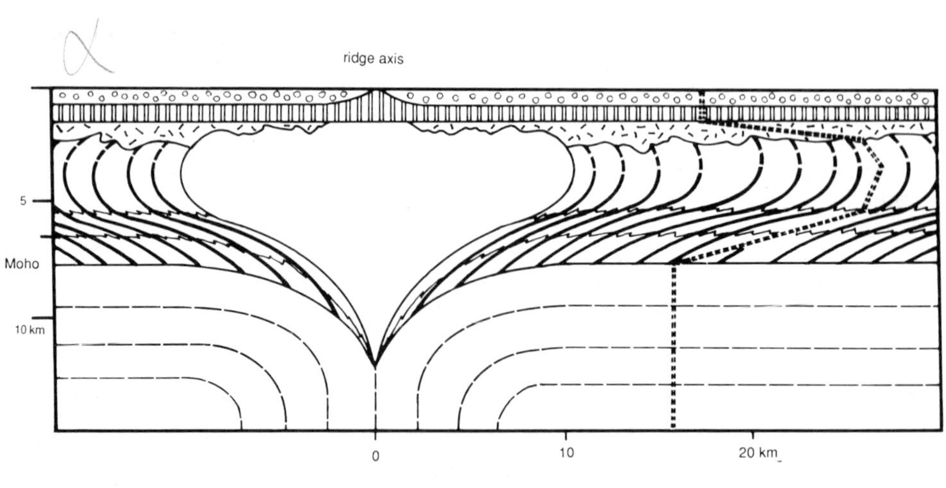


Fig. 11.7. Axial trough magma chamber. This model is proposed for a fast spreading ridge. Dashed lines indicate 1.25 x $10^5$ yr isochron ; bold lines above Moho : layering in gabbros and foliation when dashed ; zigzag lines : major lithologic subdivisions ; fine dashed lines below Moho : flow line in mantle (Casey and Karson, 1981).

critical because the slope angle controls the transverse width of the roof for a given thickness of layered gabbros. Thus, in the model of figure 11.5, the half width is 15 km ; this width, derived from a thickness of 3 km of layered gabbros, is reduced to a 7 km half width for 1 km of layered gabbros. Varying the slope angle from zero at the axis to 70° at the chamber margin, Smewing (1981) proposes a half width of 10 km. We have already mentioned the incompatibility of such large roofs at 3 km below the ridge with the present day marine data (§ 11.4.1).

Regarding the Oman ophiolite, which has inspired developments of the wing-shaped magma chamber model (Pallister and Hopson, 1981 ; Smewing, 1981), three structural observations need to be considered. Firstly, the layering plane in the basal gabbros is always strictly parallel to the limit with the ultramafics and to the foliation in the uppermost peridotites, except in faulted areas (§ 3.3.2 and fig. 3.9), in contradiction with the predictions of figure 11.5. Secondly, the dip of the layering increases in the uppermost layered gabbros. These two observations lend more support to Smewing's model than to Pallister and Hopson's one. Finally, the most fundamental assumption of this magma chamber model, that is the origin of the layering is a direct magmatic structure, either due to gravitational crystal settling, to oscillatory crystallization in a boundary layer (Browning, 1984) or to deposition by density currents, is in contradiction, at least in Oman, with the data presented in § 2.3 and 11.4.2, which show that the layering has been transposed by magmatic flow.

*The subsiding magma chamber* - This model has been proposed by Sleep (1975, 1978), Bryan and Moore (1977) and Dewey and Kidd (1977). As shown in figure 11.6, it is assumed here that the asthenosphere rises along the chamber axis, making a narrow welt which recedes laterally, thus producing subsidence of the cumulates. It is thus possible to obtain a wide and flat floor on which the cumulates can be deposited. After subsidence, layering surface dips inwards with variable angles reflecting the accumulation rate at each point and, on average, the ratio between cumulate thickness and the width of the magma chamber floor (fig. 11.6). As in the preceding case, the magmatic stratification, which is supposed to reflect the proximity to the axis where melt replenishement occurs, would be at an angle to the layering. In Sleep's model the roof remains comparatively large and the molten part of the magma chamber may be very thin. In Dewey and Kidd's model, the overhanging walls of the magma chamber which are cooled by hydrothermal circulation are plated with isotropic gabbros, also called 'plated gabbros' by these authors. The depth of the magma chamber compared to that in the wing-shaped model is reduced and can be equated to the thickness of the plated gabbros. The subsidence angle and the ratio of plated to cumulate gabbros control the shape of the magma chamber : a large angle and a large ratio determine a small chamber (fig. 11.6a), whereas a small angle and a small ratio would determine a very wide magma chamber (fig. 11.6c). A common situation in ophiolites corresponds to a few hundred meters of isotropic gabbros for 1-3 km of cumulates, and thus a ratio in the range of 1/5 to 1/10. This would favor the model of figure 11.6c.

This model is indeed compatible with the available data on the shape of magma chambers below oceanic ridges (fig. 11.9). It accounts partly for the increased dip of the layering upsection, although the flattening at the contact with the isotropic gabbros has not been reported in ophiolites (Casey and Karson, 1981 ; Nicolas et al., 1988). However, two discrepancies with respect to field data in Oman should be noted. This model predicts the existence of an angle between the lowest cumulates and the tectonic peridotites, which is not observed, and the frozen mantle diapirs mapped so far (fig. 3.17) do not show the peridotite welt, but a systematic flattening of the tectonic foliation at a few hundred meters below the layered gabbros.

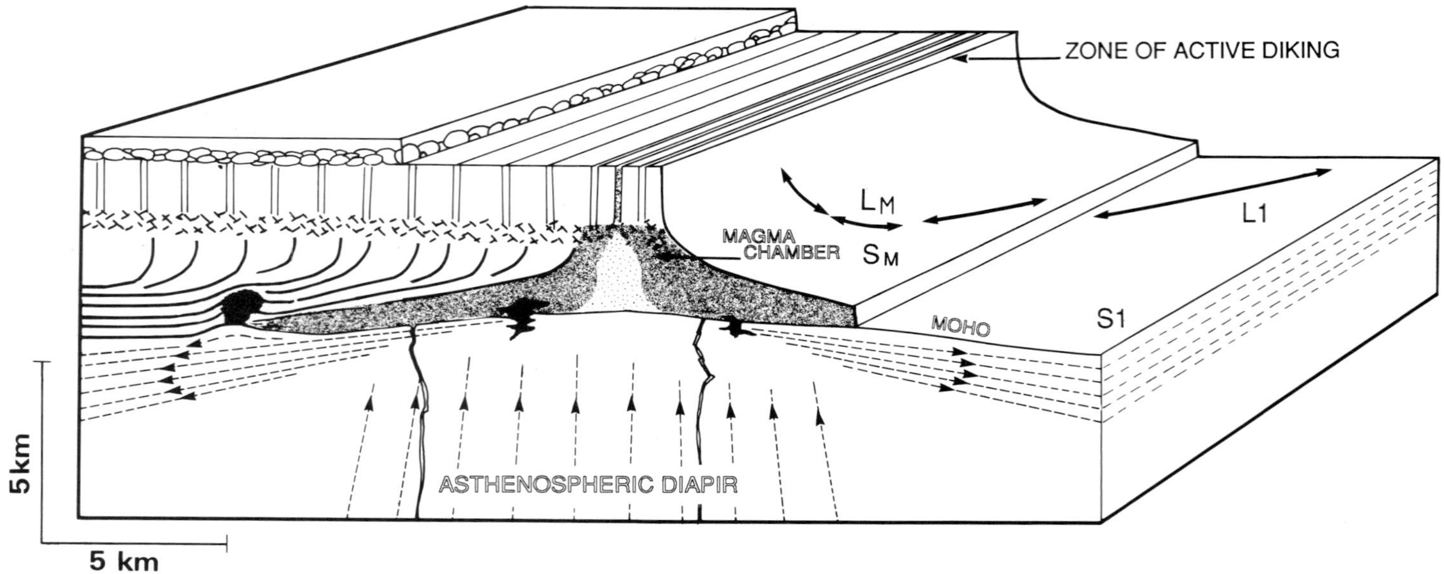


Fig. 11.8. Tent-shaped magma chamber model. Heavily dotted area : zone of thick crystal mush and lightly dotted area : zone of thinner magma (see text). Solid lines : layering and magmatic foliation in gabbros ; hatches : isotropic gabbros ; black decoration : wehrlites. The magma chamber is shown sitting on a mantle diapir whose flow structure is represented by fine dashed lines. Note that the magmatic layering Sm and lineation Lm in gabbros become parallel to the plastic foliation S1 and lineation L1 in peridotites close to the Moho (Nicolas et al., 1988b).

*The axial trough model* - Casey and Karson (1981) and Elthon et al. (1984) have presented a model (fig. 11.7) which relies on two assumptions, 1) the divergence of the asthenosphere below the ridge creates an axial trough and 2) the layered gabbros are crystallized in a stagnant boundary layer along the walls and the floor of the magma chamber. The layering can thus be at any angle with respect to the horizontal. The large scale magmatic stratification remains horizontal for the same reason as in the preceeding models. In Elthon et al.'s model the rooting of the ultramafic cumulates in the diverging trough is as deep as 40 km.

This model accounts for the large rotation of the layering plane in gabbros from bottom to top of the section which the authors observed in the Bay of Islands ophiolites. The inward curvature of this layering is deduced from one-way chilling statistics in the sheeted dike unit and from the overall ridge structure integrating the Coastal Complex transform fault (fig. 5.19). In the absence of any constraint on the initial attitude of the layering, the shape of thc walls of the magma chamber could be adapted to fit any magma chamber model imposed by marine data.

On the other hand, there is neither geophysical (Lewis, 1983 ; Detrick et al., 1987) nor geological (fig. 3.17) evidence to support the trough model in the diverging mantle. This assumption relies solely on the hypothesis of a constant viscosity in the uppermost mantle, which is doubtful (§ 7.5). Finally, the in situ crystallization requires a rather static medium, which is not in agreement with the dynamic conditions called for in § 2.4.3.

*The tent-shaped magma chamber* - Structural studies (§ 3.3.2) in the Oman ophiolites have revealed three important features concerning the crustal unit which lead us to propose a new model of magma chamber (Nicolas et al., 1988b) :

i) There is a striking *structural continuity between the tectonic peridotites and the layered gabbros*. As seen in figures 3.8, 3.9 and 3.12, the plastic foliations and lineations in the peridotites are parallel to the layering and magmatic foliation plane and to the lineation within the gabbros respectively.

ii) In most places, *the layered gabbros have been affected by a large magmatic flow* (§ 2.4.3). As discussed below and in § 11.4.2, as a result of the large rotations which can be imposed by this flow, the orientation of gabbro structures now reflects the magmatic flow field and no longer an attitude related to an 'in situ' crystallization process.

iii) *In the upper gabbros, foliations rotate progressively toward parallelism with respect to the dikes of the sheeted complex*, (figs. 3.9, 3.12 and 3.22). The mineral lineations depart from the general trend and become steep. These rotations are often difficult to observe because the magmatic deformation is less intense than down section and is overprinted by an hydrous recrystallization into isotropic amphibole gabbros.

Inspired by these discoveries, a new model of magma chamber (fig. 11.8) displays a flat floor, a narrow roof and a layering parallel to the solidifying walls of the magma chamber (Nicolas et al., 1988b). Noting the parallelism of structures in adjacent peridotites and gabbros, the viscous flow responsible for the formation of the magmatic layering, foliation and lineation in the gabbros is ascribed to mechanical coupling with the plastic flow in peridotites and not to gravity currents. This coupling has also been documented in Table Mountain ophiolite in Newfoundland (Girardeau and Nicolas, 1981). These uppermost peridotites are expected to be drifting at a velocity much greater than the ridge spreading velocity (Rabinowicz et al., 1987) and in a direction which does not necessarily coincide with the spreading direction, thus creating a drag on the overlying crystallizing gabbros. This suggests that at least close to its walls and its floor away from the ridge axis, the magma chamber is filled with a crystal mush sufficiently viscous to be able to transmit the shear stress induced by the drift of the underlying asthenosphere. There may be in the inner part of the magma chamber a narrow zone of free melt, convecting thermally (fig. 11.8), but its flow structure is no longer visible. By a reasoning similar to that held for the asthenospheric flow trajectories in the uppermost peridotites which are regarded as being parallel to the surface separating lithosphere from asthenosphere (§ 2.6.), it is proposed that the magmatic flow surfaces now observed have been frozen during flowage parallel to the colder overhanging walls of the magma chamber. The walls coincide with the isothermal surface for which a basaltic melt solidifies. The magma-solid transition occurs at around 60-65 % solid/melt fraction, corresponding for a dry basalt to a temperature around 1185° C (Usselman and Hodge, 1978). This temperature should be lowered by a few hundred degrees for the hydrous magma present at the top of the chamber. Thus the shape of the magma chamber walls below a ridge can be deduced from the shape of the 1185° C isothermal surface (Kusznir, 1980 ; Morton and Sleep, 1985) in the main part of the chamber and from an isothermal surface closer to 1000°C at the top of the chamber. In an ophiolite, the magma chamber shape is deduced from the pattern of the magmatic foliation surface as shown in figure 3.22. In the model of figure 11.8, the adopted profile inspired by field results (fig. 3.12) fits fairly well the 1185°C isotherm profile calculated by Morton and Sleep (1985) for the East Pacific Rise (fig. 11.9). Flaring of the walls upward is expected from their parallelism with cooler isothermal surface. The width of the roof is modelled to fit the width of the seismic reflectors observed in the East Pacific Rise (Detrick et al., 1987). The width at Moho level is equated with that of the mantle diapirs, i.e. around 10 km. This is supported by the idea that as the tholeiitic melt is being delivered equally throughout the diapir, there should be enough input of matter and energy to maintain a molten horizon above the diapir. A similar magma chamber width has been estimated by Sleep and Rosendahl (1979) from thermal modelling of a 5 cm/yr ridge. This does not rule out the possibility of a magma chamber with a thin tail extending beyond the boundary of the diapir. Alongstrike, the chamber would extend at least over the length of the diapir (15 km) as proposed in figure 11.14.

Two types of walls can be distinguished in the magma chamber model of figure 11.8. The outer wall coincides with the plastic-viscous boundary ; though deforming viscously, the formations close to this wall can sustain a shear stress. An inner wall could separate this mush from a central cavity of free melt (melt/crystals ratio above 35%). The lozenge-shaped transverse profile of chamber modelled by McClain et al. (1985) from seismic refraction experiments on the EPR is compatible with our expectations both in shape and size for this internal domain.

### *11.4.4. Conclusions about magma chamber models*

With the exception of the last model, the previous models refer, at least on the origin of the layering (§ 11.4.2), to the stratiform layered complexes. *There are, however, striking differences between these static and cold-floored intrusions, and the dynamic and hot-floored magma chambers of ophiolites.* In a hot-floored magma chamber, it is understood neither 1) how crystallizing minerals should accumulate on a floor composed of peridotites at such a temperature that these minerals could remelt, nor 2) how magmatic settling structures could be preserved during the motion of this asthenospheric floor with respect to the overlying crust. Hence, interpretations depending too much on the stratiform complex analogy meet with these two severe difficulties ; both are avoided by the last

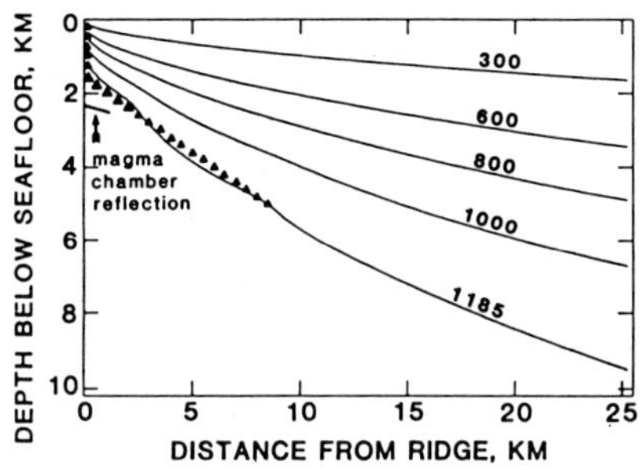


Fig. 11.9. Temperature profile in degrees Celsius for a ridge spreading at 3 cm/yr half rate with the latent heat (solid triangles) distributed along the top and sides of the magma chamber. Bold line shows top of magma chamber determined by seismic reflection (Morton and Sleep, 1985).

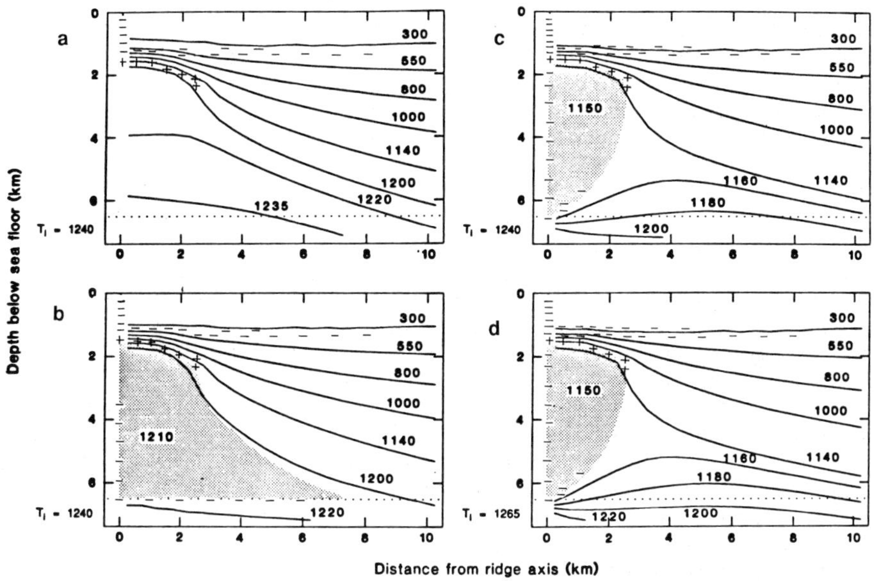


Fig. 11.10. Isotherms from thermal models, with the upper part of the magma chamber constrained to match the seismic reflection data of Detrick et al. (1987). Plus and minus symbols indicate discrete sources and sinks of heat used as input to the models. a) Model without magma convection. The temperature gradients above 1200°C in the crust (above dotted line at 6.5 km) are unphysical. b) Model for a hot (1210°C) magma chamber with minimal convection. The convecting region is shaded. This chamber is much too wide to match the seismic data. c) Model for a cold (1150°C) magma chamber with substantial heat convection. This geometry is consistent with the seismic reflection data, as well as with cross-ridge seismic refraction data (McClain et al., 1985). d) Model modified from c) by increasing the intrusion temperature at all depths by 25° and increasing convection to maintain the 1150°C chamber temperature. The greater heat content of the upper mantle results in ~10° greater reheating of the lower crust (Wilson et al., in press).

model in which the layering reflects viscous flow and is frozen against the overhanging walls of the chamber. This model is so far supported by data only from the Oman ophiolite and it is possible that it might not apply to other ophiolites where a magma chamber might have formed in more static conditions.

In order to compare the merits of the various magma chamber models in the light of the available data on oceanic ridge magma chambers, a brief comment is necessary on how a magma chamber is defined. Seismological techniques give reflectors corresponding to seismic velocity contrasts and average velocities through certain wave paths. From this, magma chamber contours and contents in terms of velocity can be deduced. Recent seismological models (R.S. Detrick, pers. com.) seem to distinguish a central molten domain based on the lowest measured velocities and a surrounding low velocity zone ascribed to cooling gabbros still trapping some melt. Thermal models taking into account the possibility of thermal convection in the magma chamber (Wilson et al., in press) also distinguish an internal domain where the medium is essentially liquid, the low viscosity allowing for thermal convection, and a peripheral domain cooled by this convection where the medium is near, or at crystallizing temperatures (fig. 11.10).

In all ophiolitic models except the last one of figure 11.8, the magma chamber margin is viewed as the internal limit as defined above, i.e. the limit between a liquid with crystals in suspension, possibly affected by a thermal convection, and a non convecting crystal-melt mixture. In the tent-shaped model, the limit of the 'magma chamber' is the limit between an inner viscous domain where stress-induced, or forced convection, is possible and an outer domain where the medium becomes solid. In this model, the magma chamber limit is somewhere within the seismic low velocity zone. A more precise location would require better seismic velocities data in basaltic crystal mushes, as the transition from solid to magma (second percolation threshold) is itself somewhat constrained about 35% melt .

Given the present state of knowledge, it is difficult to reconcile any one of the ophiolite-inspired models of magma chambers with all geophysical data. Problems of size of the inner, thermally convecting, domain tend to disqualify the wing-shaped model. The subsiding floor model zone and the axial trough model predict respectively a melt and a keel which are not identified at the base of frozen magma chambers in ophiolites. Eventually, the tent-shaped model meets with two intrinsic difficulties. First, it implies that the low velocity zone identified above the Moho along the East Pacific Rise (Rosendahl et al., 1976 ; Detrick et al., 1987) is composed of a crystal-liquid mixture with 35 % or more liquid, a ratio which may be too high to produce the comparatively high velocities measured in this zone. Second, in this model the layering in gabbros dips away from the ridge axis, in opposition with the preceding models. Tests from ophiolites on the sense of layering dip support this model in Oman, but in other ophiolites they are equivocal (§ 2.2.2.) ; in the North Atlantic, seismic reflection lines have identified strong reflectors in the lower oceanic crust which moderately dip toward the present day ridge axis (McCarthy et al., 1988). They are interpreted either as possible reflections on the gabbro layers or as fractures and shear bands in gabbros. These are certainly abundant in the gabbros dredged in the Atlantic (§ 8.2.3) and if they are related to listric faults they should also dip preferentially toward the ridge axis.

### *11.4.5. Plating of gabbros and diking at the roof of magma chambers*

The formations between the highest layered gabbros and the overlying sheeted dike unit constitute a complex zone, with structural and petrological characteristics varying from one ophiolite to another and even within the same ophiolite (§ 3.3.2, 4.2.3, 4.3.3 and 6.3; see also descriptions by Dewey and Kidd, 1977, Rosencrantz, 1983, or Pedersen, 1986). It is a zone of highly variable thickness (from a few tens of meters to a few hundred meters) comprising : 1) the transition between foliated and isotropic gabbros recrystallized in an hydrous environment and 2) the transition between these gabbros, with few diabase dikes and the sheeted dike unit composed uniquely of dikes.

Descriptions of the transition between layered or foliated gabbros and isotropic gabbros in various ophiolites show that it is difficult to draw a line between them. The layered gabbros progressively lose their layering and grade into poorly foliated gabbros, where the pyroxenes tend to disappear in favor of magmatic amphiboles. The foliated gabbros in this transition zone are highly recrystallized by high temperature hydrothermal circulation. This circulation generates the isotropic gabbros which are characterized by a great heterogeneity in grain size and by successive generations of amphiboles, and induces wet anatexis at the expense of all gabbros (Payne and Strong, 1979 ; Gerlach et al., 1981). The products of this anatexis are leucocratic tonalites and plagiogranites which invade the gabbros as a dike or vein-like network and locally dismember them into a magmatic breccia. Distinguishing plagiogranites produced by this anatexis from others ascribed to the crystallization of highly evolved melt at the top of the chamber (Coleman and Peterman, 1975 ; Tilton et al., 1981) or to melting below the ophiolite nappe (Boudier et al., 1988) relies on chemical criteria which are not discussed here (see Pedersen and Malpas, 1985).

The diabase sheeted dikes complex also roots into this zone, and is emplaced at the same time as the plagiogranite dikes, as demonstrated by their mutual cross-cutting relationships. The intense tectono-magmatic activity of this zone is also emphasized by the common occurrence of diabase breccias with a dioritic matrix and of numerous shear bands also related locally to this leucocratic anatexis. Such flat-lying shear bands transforming isotropic gabbros into flaser gabbros have been reported from submersible observations in the root zone of diabase dikes of the Gorringe Bank in the Atlantic (Auzende et al., 1978).

How the diabase dikes relate to the gabbros is a major question. Does the basaltic melt circulating in the dikes represent differentiated products of the magma chamber partial crystallization or primitive melt feeding the chamber with little contamination during its ascent through the chamber ? Among the structural observations made in the root zones of the sheeted dike unit of the Oman ophiolite (§3.3.2), three observations may help to answer this question : 1) the magmatic foliation in the gabbro unit steepens upsection and tends to become parallel to the diabase dikes (with generally steep mineral lineations), although it is locally much disordered in the root zone itself, 2) when this foliation is not destroyed by recrystallization to isotropic amphibole gabbro in the root zone, it is commonly underlined by ribbons of finer-grained and darker material which derives from the dismembering and straining of diabase dikes, and 3) a number of these diabase dikes have no chilled margins and comparatively coarse textures.

These observations may be explained by considering the situation created at the top of a magma chamber. It seems necessary to postulate the existence of a thermal boundary layer separating the magma chamber affected by a magmatic convection, from the sheeted dike complex in which a hydrothermal convection is now well documented (§ 11.7). Through this critical layer, the temperature drops from about 1000°C at the top of the magma chamber (§ 11.4.3), to 400°C ± 50°C, a temperature fixed by the pervasive hydrothermal circulation at the base of the sheeted dikes complex (§ 11.7). At this low temperature, the channels carrying basaltic melt to the surface are easily identified as diabase dikes with chilled margins. At the higher temperatures encountered on the lower inside boundary layer, the melt channels may be more difficult to identify. Below the sheeted dike unit, one may first meet diabase dikes without chilled margins and then, diffuse zones of melt transport as indicated by the steep magmatic foliation. At this deeper level, proto-dikes developing a finer grain size along their walls may be partly digested by the continuing

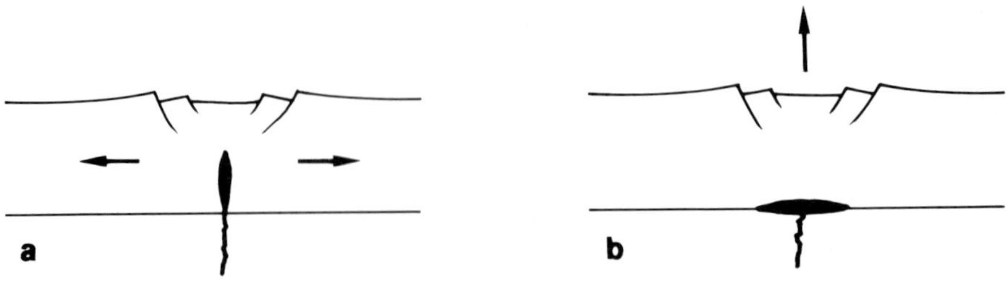


Fig. 11.11. Sketches of possible ways to initiate a new magma chamber. a) Dilational fissuring of a vertical dike. b) Swelling of a horizontal sill. Dilation directions are indicated by arrows.

melt flow with, as a result, the inclusion of darker and finer-grained material into the gabbroic foliation. The hydrous recrystallization and anatexis affecting this critical zone suggest that heat was not entirely transported by conduction, and that hydrothermal convection may also have contributed (11.7).

The absence of rooting evidence for the dikes within this critical zone or below within the foliated gabbros, and the fact that it is a zone of melt migration and not the top of a magma chamber, as commonly believed, do not support the idea of the melt being issued from the differentiation at the top of the magma chamber. However, these observations and the proposed interpretation may apply only to specific areas (top of magma chamber) of ophiolites which, as the Oman ophiolite, have well developed magma chambers. Other situations exist which may lead to other conclusions.

### *11.4.6. Initiation of a new magma chamber*

In the case of a ridge characterized by non permanent magma chambers, the development of a new magma chamber might be caused by dilational vertical fissuring and progressive opening, giving rise to the 'infinite leek' of Nisbet and Fowler, (1978), or by the progressive swelling of a sill emplaced at the base of the crust (Gudmundsson, 1986, 1987) (fig. 11.11). Both models face a space problem, because in the oceanic crust, which is a system cooled by hydrothermal circulation, it is quite unlikely that magma chambers form by melting of host rocks. Vertical fissures due to elastic deformation of the lithosphere are probably limited to a few hundred meters (see below) and further opening depends on permanent spreading strain ; swelling of a sill requires a change in topography which is also limited by gravity force. During the magmatic phase of the supposed cyclic activity affecting a slow spreading ridge (§ 11.6), it seems possible that the slow spreading ridge morphology with a 2 km deep central rift may swell into a fast spreading one with no rift. Otherwise, stretching related to permanent spreading is required to open a lens-shaped magma chamber.

In the model of magma chambers opening by vertical fracturing, the only direct evidence on the width of opening which can be ascribed to elastic deformation is provided by the thickness of diabase dikes. These dikes have been filled by a single batch of melt, as there is a general absence of internal structures which would be indicative of sequential intrusion separated by partial cooling. Their average thickness is 1 m in the sheeted dike unit and it does not exceed 10 m for individual dikes intruding the peridotite and the

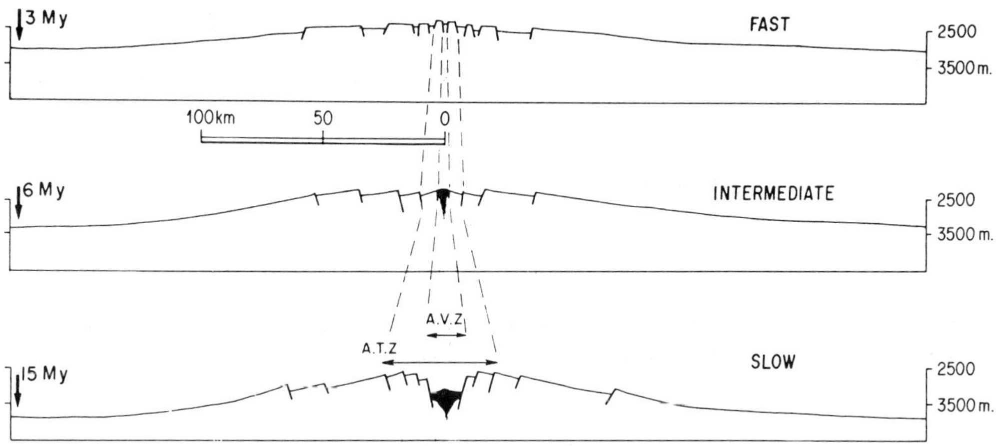


Fig. 11.12. Dependence, on the spreading rate, of the width of active volcanic (AVZ) and tectonic zones (ATZ) at a ridge. Note also the variation in topography and mode of surface deformation which is through fissuring in the case of a fast ridge and through inward-facing normal faulting in the case of slow and intermediate ridges (Choukroune et al., 1984).

gabbro units (§3.3.2). Assuming that elastic deformation is caused by magma pressure, and adopting Lister's (1983) estimations for the magmatic pressure and its distribution, the length of lithosphere in the spreading direction deformed elastically by these intrusions is 7.5 km for a 1 m dike and 75 km for a 10 m dike. Opening of a magma chamber 2 km wide would require compressing nearly half the extension of the Earth's lithosphere and therefore necessitate a global synchronization of the spreading activity (Lister, 1983). This 2 km figure is obviously a maximum. To account for the larger magma chambers expected from ophiolite studies, it is necessary to imagine that, following the initial elastic opening, there is a continuous opening caused by the spreading- related tensional strain. The morphology would have to evolve in order to reduce heat loss to the surface through the narrow sides of this magma chamber.

The cooling problem is less crucial in the case of a sill injected at the base of the crust, because the hydrothermal circulation, which is the main cooling agent, hardly penetrates to this deep level. In Iceland, Gudmundsson (1986) explains the generation of sills and lens-shaped magma chambers as a result of the existence of stress barriers. In a rifting context, where the principal compressive stress is vertical, favoring the opening of vertical conduits for the basaltic melt, the occurrence of mechanically contrasted layers can induce a shift of the stresses at the contact between mechanically weak and strong rocks, such that the principal compressive stress locally becomes horizontal. This creates a stress barrier for the uprising melt which would open sills at the contact between layers of thick and weak volcanic breccias and layers of thin and strong volcanic flows. Another way to stop the ascent of melt and to generate a lens-shaped magma chamber is to build up a dynamic pressure barrier as proposed in § 7.5. At the top of the transition zone in a mantle diapir, a sharp dynamic pressure gradient would be created as a consequence of an equally sharp decrease in viscosity. This would induce by solid compaction a segregation of melt at Moho depth and thus initiate a new magma chamber.

## *11.5. SHEETED DIKES AND VOLCANIC UNITS*

### *11.5.1. Introduction*

Ophiolite studies show that whether the base of the crustal sequence is composed of a distinct unit of layered and isotropic gabbros or not, the upper section is always composed of a diabase sheeted dike or sheeted sill unit covered by volcanic formations, generally pillowed. The occurrence of a dominant sheeted sill unit seems to be associated with a series of other characters defining some LOT ophiolites (chapter 8). Together with the overlying volcanic formations, the sheeted dike unit constitutes the frozen carapace of a magma chamber. It is partly accessible to direct studies in rifts as in Iceland and in oceanic ridges. Consequently, the following discussion on the structure and physical aspects of the formation and evolution of these extrusive and intrusive units is able to incorporate data from present day rifts and ridges as well as from ophiolites.

### *11.5.2. Generation at rifts and ridges*

*Extension of the zone of crustal accretion* - Marine studies point to the remarkably restricted extent of volcanic and tectonic activity on either side of a ridge, and show in particular that the faster the spreading rate is the narrower appear to be these zones (fig. 11.12). Two observations in ophiolites also point to the same conclusion and may help to explain it. Firstly, many diabase dikes of the sheeted dike complex are split by younger dikes intruding inside them and parallel to them ; this is commonly repeated with, as a result, swarms of dikes with a single chilled margin (§ 2.2.2). This indicates that for a time the opening of new dikes followed the same weak zone. Changes in facing direction for the chilled margin, however, suggest that the opening direction can also jump an unknown distance. Secondly, the observation of diabase dikes cutting the uppermost peridotites and layered gabbros units is quite exceptional, although more common in LOT than in HOT situations (table 8.I) ; gabbro dikes cutting the plutonic section are also rare in HOT. This means that under normal circumstances the melt feeding the crust, 1) crossed the uppermost peridotites whilst they were still above 400-450°C (§ 11.4.5) to permit, within dikes, the slower crystallization as gabbro rather than the faster one as diabase, and that the melt 2) reached the crust inside the magma chamber limits and exceptionally outside. The magma chambers wider in HOT than in LOT explain why diabase dikes are more common in the second situation.

As exposed in § 7.6, the model by which small mantle diapirs channel the asthenosphere flow from where the melt is extracted, just below a magma chamber (fig. 11.8) accounts well for the focusing of the ridge activity within a similar width. For the volcanic activity itself, the focusing would be increased even further by the fact that, in the magma chamber model depicted in this figure, the fracturing responsible for the dike injection would be expected to occur at the thinner top of the chamber.

*Attitude and shape of the diabase dikes* - As already mentioned in § 2.4, the data from present day rifts and the admittedly limited data from oceanic ridges indicate that the sheeted dike azimuth is parallel to the local rift or ridge trend, implying that they result from identical tensile stress fields. Observations in rifts (Olafsson, 1977 ; Gudmundsson, 1983 ; Helgason and Zentilli, 1985) show that these dikes are nearly vertical, a conclusion verified in the most favorable sections of ophiolites, where it is observed that the attitude of sheeted dike unit is perpendicular to that of the overlying lava flows and the deposition plane of umbers and radiolarites.

Studies conducted in Iceland on the dynamics of magma transport at shallow depths cast a light on the shape of active dikes, thought to be equivalent to the dikes of the sheeted

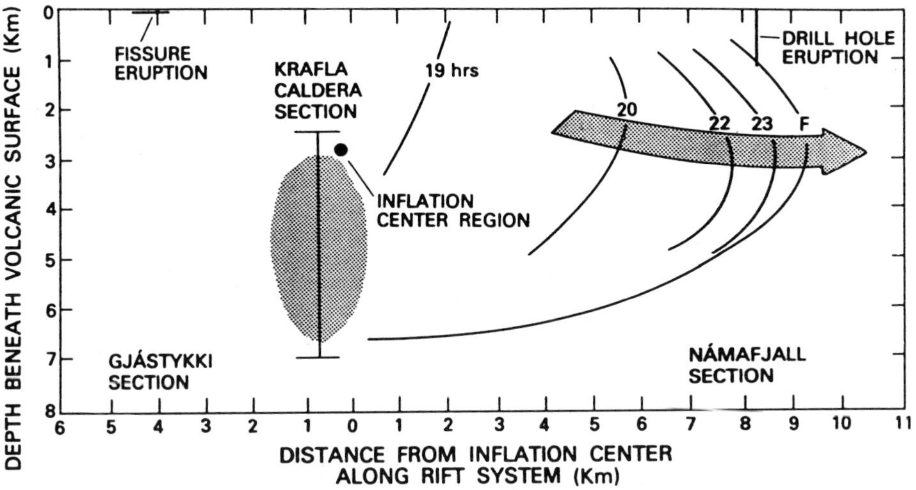


Fig. 11.13. Cross sectional profiles of the sequential position of the magmatic fracture front during rift zone intrusion from the Krafla central volcano, northeast Iceland. Numbers on the profiles are the times (in hours) following the onset of the intrusive phase. 'F' denotes the final position of the magmatic fracture front. The large shaded arrow delimits the time-depth-distance pathway followed by the parabolic apex of the intrusion as it moved along the inferred horizon of neutral buoyancy. The vertical bar that penetrates the aseismic region (shaded) beneath the Krafla caldera, delineates the maximum depth range of subcaldera magma storage. Parabolic intrusion profiles have been inferred by the progression of seismic hypocenter shifts within the rift zone. The inflation center is in essential coincidence with the inferred point of neutral buoyancy. The intrusion sequence culminated in a geothermal borehole eruption on 8 September 1977 (Ryan, 1987).

dike unit. It is shown that magma issued from magma chambers can be transported laterally as far as 30 km in a direction parallel to that of the rift through fissures a few meters across and extending to depths of about 4 km, although some dikes may reach deeper (Einersson and Brandsdottir, 1980 ; Bjornsson, 1985). In a study based on the comparative densities of basaltic melts and crustal rocks, Ryan (1987) estimates that there is a neutral buoyancy level at a depth of about 3 km. As a result, the melt below this level is lighter than the surrounding rocks and tends to ascend whilst the melt above, being heavier, tends to descend. The neutral buoyancy level would coincide with the parabolic nose of the expanding magmatic fracture front, moving along this horizon of neutral buoyancy (fig. 11.13). Structural (Baer and Reches, 1987) and magnetic (Knight and Walker, 1988) studies in comparable dikes point to magmatic flow lines predominantly horizontal, in keeping with the similarly extended dike shape proposed by the preceding authors.

Applying this model to the sheeted dikes of ophiolites, it is predicted that in a sagittal section, the diabase dikes should also be blade-shaped and extend horizontally over a few tens of kilometers ; downward, their trace is lost at the top of the active magma chamber ; their vertical extension follows the top of the magma chamber along the ridge strike (fig. 11.14). Consequently, variations in the thickness of the sheeted dike section along strike

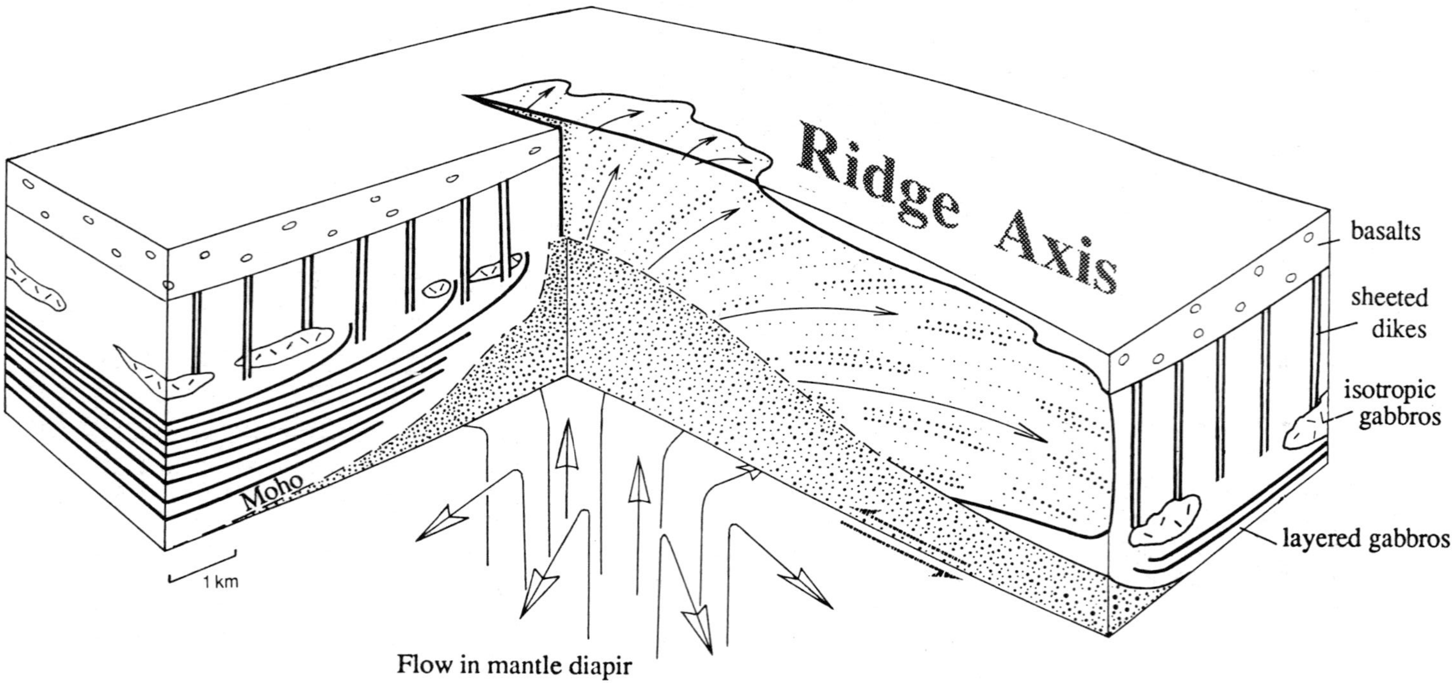


Fig. 11.14. Sketch of an along-strike section of a ridge magma chamber (heavily dotted area), illustrating the extent of crust which can be fed in this direction by a single dike (aligned dots following assumed flow lines in the dike).

at the expense of layered gabbros could be used to determine the pattern of the segmentation (§ 9.4.4) of the ridge of origin in a given ophiolite. In this regard, it is noteworthy that the wavelength of ridge segmentation, which is in the range of 50-100km, as determined by the average spacing of transform faults in slow spreading ridges and of overlapping spreading centers in fast spreading ridges, is compatible with the wavelength illustrated by the ridge model of figure 11.14. Mantle diapirs, modelled after those mapped in Oman (figs. 3.17, 3.20 and 3.21) would extend over 15 km along strike. To accrete crust continuously, they would need to be spaced such that melt issued from two adjacent diapirs, and carried laterally by feeder dikes 10-30 km along strike, join or overlap. If a dike is assumed to feed the crust to a distance of 20 km along strike, the typical segmentation length would be 15 km + 2 x 20 km = 55 km. .

### *11.5.3. Structural evolution of the volcanic-hypovolcanic units*

The sheeted dikes emplaced vertically at the rift or ridge site and the lava flows erupted initially horizontally may have been tilted during drifting away from the spreading axis. Outward tilting is commonly 5-15° along the flanks of the Mid-Atlantic Ridge, where it contributes to the stretching of the newly formed lithosphere (Tapponnier and Francheteau, 1978) but can attain 30° (Macdonald and Luyendyk, 1977 ; Atwater, 1979) and tilting of as much as 70° has been reported from submersible observations (Karson, 1987) and from structural and paleomagnetic measurements on DSDP cores (Flower, 1977 ; Hall and Robinson, 1979 ; Verosub and Moores, 1981).In eastern Iceland the dip of lavas toward the rift remains modest, 6-9° at sea level, but seems to increase with depth (Gudmundsson, 1983), attaining 20° dip at about 200 m below sea level in a drill core (Gibson, 1979). In ophiolites, evidence for early tilting has been reported in a few occurrences. In order to exclude the possibility that the rotation was induced by subsequent tectonic events, the physical and chemical conditions of cataclastic and ductile flow in the faults must be carefully analyzed. From field and paleomagnetic evidence, Varga and Moores (1985) and Allerton and Vine (1987) estimate that rotations of up to 78° have been suffered by the dike complex of Cyprus. In the same ophiolite, Boyle and Robertson (1984) in an analysis of the volcano-sedimentary filling of grabens, documented rotations up to 30°. Harper (1982) in a comprehensive study of the angular relationships between gabbro layering, diabase dikes and sedimentary bedding in the Josephine ophiolite (Oregon) calculates that a 50° rotation occurred prior to the Nevadan folding. Locally in Oman, we have documented a 45° rotation of the complete crustal section (§ 3.4.2). However, it must be stressed that outside such areas, there is no indication of early tilting. On the contrary, the best sections (fig. 3.15) suggest that the volcanic flow and associated sediments are parallel to the Moho and the sheeted dike complex normal to both.

Tilting of oceanic crust at a spreading center around an horizontal axis, presumably parallel to the ridge direction, can be accomplished either by subsidence due to increasing volcanic load while the crust is drifting away from the axis (Moore et al., 1974 ; Cann, 1974 ; Rosencrantz, 1982), or by block faulting (Verosub and Moores, 1981)(fig. 11.15). The subsidence model requires that the lava flows dip toward the ridge axis, whereas the sheeted dikes may dip away from the axis or remain vertical, depending on the mechanism involved in subsidence (see Rosencrantz, 1982). The block faulting model, in contrast, involves tilting of the lava flows away from and tilting of the dikes towards the ridge. In Iceland, tilting occurs by subsidence, and in all the ophiolite occurrences mentioned above, by block faulting. Tilting by block faulting is well documented in rifts (Morton and Black, 1975) and in some typically sediment-starved and non volcanic passive margins (Ginzburg et al., 1985). Interestingly, other passive margins like the Outer Voring Plateau

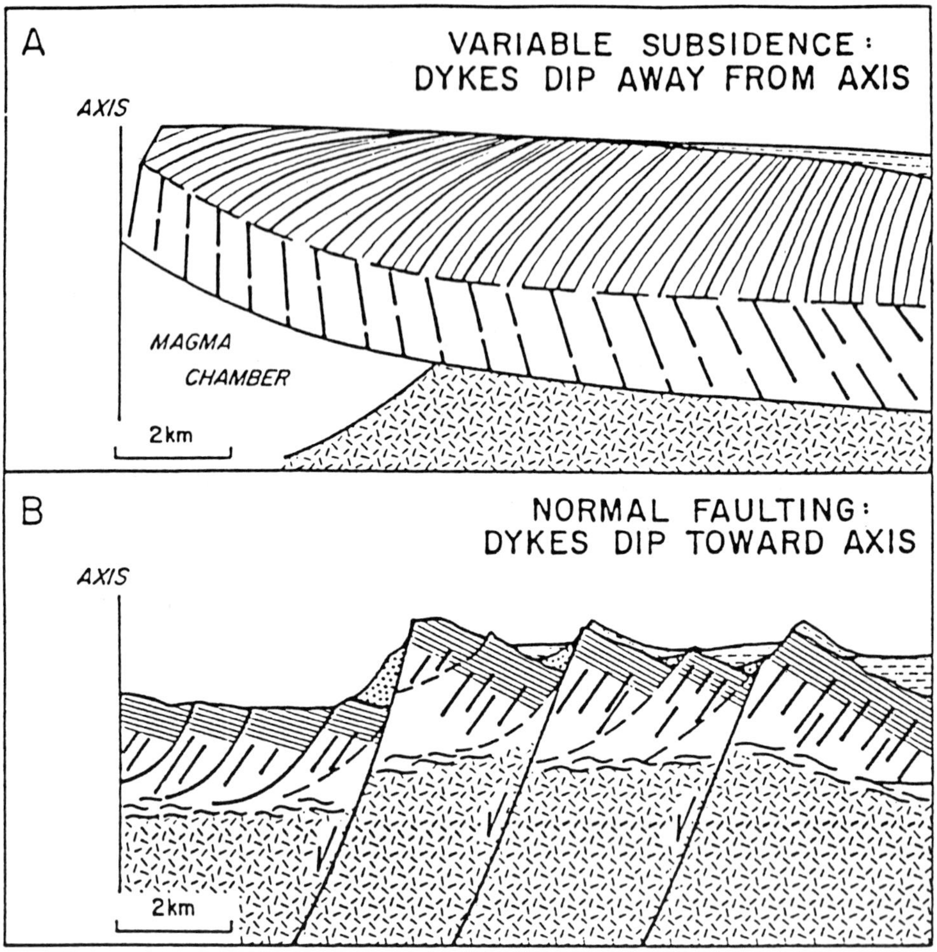


Fig. 11.15. Tilting of dikes near spreading centers. a) Subsidence : dikes (bold, dashed lines) are initially formed in a vertical orientation above a magma chamber. Overlying lava-flow units (fine lines) are thick near their source and subside rapidly near the spreading axis, but more slowly in older crust so that dikes dip away from the axis and lavas dip steeply toward the axis. b) Block faulting : initially vertical dikes and overlying lavas and sediments (fine dashed lines = chalks, stippled pattern = talus) are tilted by movement along listric normal faults that flatten with depth into a subhorizontal zone of plastic deformation in the upper portion of the gabbro unit, where crustal extension is accomplished without dike intrusion. Away from the axis, steeper faults (solid lines), which may be listric at depth, penetrate all units and cut inactive faults (bold broken lines) to create rift valley walls. Dikes dip toward the axis whereas sediments and lavas dip away (Karson, 1987).

in the North Sea characterized by a thick volcanic pile, present seaward dipping seismic reflectors indicative of subsidence (Mutter and Buck, 1986 ; White et al., 1987). Block faulting implies the existence of listric faults, while subsidence implies various amounts of vertical faulting, depending on the operating mechanism.

In the Josephine, Oman and Cyprus ophiolites, these listric faults merge into flat detachment shear zones, respectively, just below the Moho, at the Moho itself, and within the plutonic section. Harper (1985) reports that fault-plane solutions for near-axis earthquakes along the Atlantic Ocean Ridge indicate normal faulting on faults which are parallel to the ridge and have moderate to possibly flat dips at depths in the range of 5-8 km beneath ocean floor. These seismically-resolved faults could match the listric faults of ophiolites.

Tilting by subsidence or block faulting may depend on a balance between tectonic stretching and magmatic outflow at spreading centers as proposed by Karson (1987). Early tilting in the oceanic crust has been documented more often at slow spreading than fast spreading ridges. If the magmatic activity of slow spreading ridges is episodic (§ 11.6), it can be speculated that between magmatic pulses, the stretching, much like that of a rift, would give rise to large rotations by block faulting, whereas a magmatic episode with copious lava extrusion could lead to subsidence as in Iceland. In the case of fast spreading ridges, the modest reliefs limit the observation in the vertical direction. DSDP hole 504 B in the East Pacific Rise has shown that chilled margins of dikes seen in cores were dipping at 30°-40° (Anderson et al., 1982). However, the dikes are brecciated and the direction of rotation with respect to the ridge strike is not known ; this evidence for tilting seems too slender to permit any definite conclusion. Obviously, this subject deserves more study, particularly in ophiolites. The evidence available so far suggests that tilting in the mafic part of ophiolites is accomplished by block faulting and not by subsidence. Many sections in Oman also suggest that there is no significant tilting in the mafic sequence.

## *11.6. CRUSTAL DISCONTINUITIES IN LHERZOLITE TYPE OF OPHIOLITE AND EPISODIC OCEANIC SPREADING*

The continuity of the layered gabbro unit in the field, which is one of the main characteristics of HOT, points to the idea of a continuous basaltic melt delivery through time at the spreading centers that generated this ophiolite type. In contrast, it has been shown in the LOT case that the layered gabbro unit is altogether less well-developed and discontinuous ; locally the crustal section is totally absent (§ 4.4 and 11.3.2). This can be explained by a variable or discontinuous basaltic melt influx either in space, in time, or both.

### *11.6.1. Variable basalt delivery along ridge-strike*

Evidence given in chapter 9 showed that both in HOT and LOT situations, creation of a new lithosphere results from discrete mantle diapirs rising beneath the ridge or rift axis with a spacing probably comparable to the length of the ridge segments, that is several tens of kilometers. Variations in layered gabbro thickness in HOT may be related to an along-strike variation in the delivery of basaltic melt, itself related to the distance from the diapir which is expected to be the main source of melt. In LOT situations, where less melt is extracted from the mantle, it can be assumed that at some distance from a diapir along the strike of the ridge, the delivered melt and heat are insufficient to maintain a magma chamber ; this would preclude the formation of a layered gabbro unit. Along the Mid Atlantic Ridge, it is now well documented that the crust becomes thinner in the vicinity of

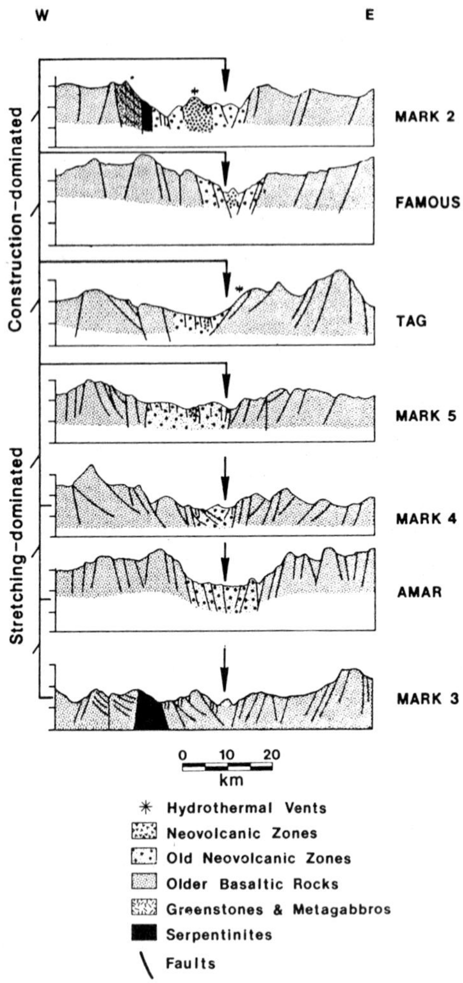


Fig. 11.16. Flow chart for median valley evolution along the Mid-Atlantic Ridge. Geometry of faults at depth is schematic. Vertical exaggeration x 3. Cross-sections based on data from FAMOUS/AMAR, TAG and MARK (Karson et al., 1987).

transform faults, an observation which has been ascribed to the cooling effect of the older lithosphere (Fox et al., 1980 ; Stroup and Fox, 1981 ; Fox and Gallo, 1984) and may also be the effect of the distance from the magma feeding center (§ 9.4.4).

### *11.6.2. Episodic basalt delivery in time*

LOT discontinuous structure may also be due to episodic spreading or rifting with successive episodes of magmatic and tectonic activity, although the idea of cyclic magmatism superimposed on steady-state tectonic extension is favored (see below). An episode of magmatic activity would last long enough to permit the building up of a new magma chamber, probably by the mechanism of sill-swelling (§ 11.4.5). It would be followed by an episode of magmatic quiescence, during which the tectonic stretching is more or less continuous (Karson and Winters, 1987 ; Norrell and Harper, 1988). This is illustrated by figure 11.16, showing the contrasted nature of sections through the MAR, organized following this presumably cyclic process. The common occurrence of sheared gabbros in the layered gabbro unit (§ 8.2.3) suggests that the magma chamber is totally or largely crystallized between magmatic episodes. Between two such episodes, tectonic extension can be large enough to allow a denudation down to the upper mantle ; this explains the exposure of serpentinites along the MAR, preferentially in the vicinity of transform faults where the crust was already thinner (Karson and Dick, 1984 ; Karson and Winters, 1987).

This episodic activity is clearly on a time scale much larger than the episodic rifting on the scale of 13 years envisaged in Iceland (Bjornsson, 1985). With an average spreading rate of around 1 cm/yr, 13 years represents the time span necessary to create, by fracturing following elastic extension of the lithosphere, the space for a single dike in an ophiolitic dike swarm. In Bjornsson's interpretation the cause for this episodicity is tectonic : fracturing of the thin lithosphere above a magma reservoir releases the elastic extensional forces which have built up progressively in response to spreading. Alternatively, dike intrusion is a response to the overpressure created in the crust by melt discharge from the mantle (chapter 7).

The time necessary to build up and to cool a magma chamber ($10^4$ to $10^5$ yr for Stakes et al., 1984), and to denude the upper oceanic crust by tectonic extension, are in the range estimated by Courtillot et al. (1984) for episodic spreading in the Asal propagating rift or by Gente (1987) and Kappel and Ryan (1986) in the MAR. Courtillot and his co-workers conclude that a magmatic episode, culminating in the creation of a magma chamber, would last $10^6$ yr and a propagating episode, $10^5$ yr. A complete cycle would last $10^6$ yr for Gente and 2.5 x $10^5$ yr for Kappel and Ryan. Van Andel and Ballard (1979) and Gente (1987) envisage similar cycles for the fast spreading EPR with time constants of 1 to 5 x$10^4$ yr. In this latter case, the episodicity of spreading would not alter the continuity of magma chamber activity.

Although the magmatic and tectonic activities are in the long term coupled, it is still equivocal whether the episodicity results from an instability in the magmatic or in the tectonic activity. In the case of a tectonic discontinuity, the episodicity origin would have to result from pulsations in spreading rates with a similar period. Because of the amplitudes of the displacements (a few kilometers to a few tens of kilometers), they would have to be globally coupled. Such pulsations have not been yet demonstrated. For this reason the magmatic origin of episodicity is favored. The source of the discontinuity would then be located in the mantle diapirism, caused by episodicity in the uprise of fresh peridotite through a mantle conduit, as envisaged by Scott and Stevenson (§ 9.4.2) or by uprise of independent mantle blebs, whose ascent duration could be indeed in the $10^5$ - $10^6$ yr range (§ 9.4.5).

## *11.7. EARLY METAMORPHISM IN OPHIOLITES AND HYDROTHERMAL ACTIVITY AT OCEANIC RIDGES*

### *11.7.1. Introduction*

Ophiolites commonly bear the imprints of several metamorphic events. For instance, in the ophiolites of the western Alps, a greenschist facies metamorphic event which can be ascribed to a late alpine thermal event because of its regional imprint and of its alpine age, is easily recognizable. This metamorphism is superimposed on a blue schist-eclogite metamorphism also ascribed to the alpine subduction by the same reasoning. However, in the least deformed and metamorphosed areas like Mongenèvre, it has been shown that a greenschist-amphibolite facies metamorphic event, preferentially displayed in flasergabbros, is related to a ridge activity because the flasergabbros were cut by diabase dikes akin to the ophiolite volcanics (Mével et al., 1978 ; Steen et al., 1980).

This type of early metamorphism in ophiolites was soon ascribed to seawater circulation within cooling crust adjacent to oceanic ridges (Williams and Malpas, 1972 ; Gass and Smewing, 1973), on the basis of the similarity of metamorphic facies with rocks dredged from the oceans (Melson and Van Andel, 1968 ; Cann, 1969 ; Aumento and Loubat, 1971 ; Hekinian, 1982). The scientific interest in this metamorphism and the accompanying sulfide mineralization sharply increased when the importance of the seawater circulation at oceanic ridges was realized (Lister, 1972 ; Williams et al., 1974 ; Wolery and Sleep, 1976) and when its most spectacular manifestation as 'black smokers' was observed directly by manned submersible (Corliss et al., 1979 ; Rise, 1980 ; Rona, 1984).

It is beyond the scope of this book to describe and compare the petrology and geochemistry of hydrothermal metamorphism in ophiolites and in the various oceanic spreading centers where it has been identified. Instead, we recall below the main features of this metamorphism and briefly address some physical aspects of water circulation through the ridge crust, in the light of the structural prints left in the crustal formations of ophiolites.

### *11.7.2. Metamorphic zonation in ophiolites*

The existence of a generally static hydrous metamorphism, grading downwards from zeolite facies in the extrusives to greenschist facies in the sheeted dikes and thence amphibolite facies in the high-level gabbros, seems to be an ubiquitous feature of ophiolites. The metamorphic zonation can be illustrated by figure 3.10 (see also § 3.3.2) which was established for the Oman ophiolite. Gillis and Robinson (1988) observed that both in ophiolites and oceanic drill holes, the transition between metamorphic facies is sharp and irregular, being controlled mainly by permeability contrasts, themselves related to rock facies. As a consequence, the relative stratigraphic levels of metamorphic zonation shown in figure 3.10 should not be taken too strictly.

Descriptions of this metamorphism have insisted either on its petrological zonation (Stern and Elthon, 1979 ; Liou and Ernst, 1979 ; Evarts and Schiffman, 1983 ; Lippard et al., 1986 ; Harper et al., 1988) or its geochemical budget in terms of major and trace elements (Coish, 1977) and correlated stable isotope zonation (Gregory and Taylor, 1981; Cocker et al., 1982 ; Elthon et al., 1984 ; Agrinier, 1987 ; Schiffman et al., 1987), or eventually the physico-chemical parameters of the hydrothermal fluids, as determined by fluid inclusion studies (Spooner and Bray, 1977 ; Richardson et al., 1987 ; Schiffman et al., 1987 ; Nehlig and Juteau, 1988).

The $\delta^{18}O$ profile through the crust of ophiolites (fig. 11.17) reflects the exchanges with seawater at the ridge ; $\delta^{18}O$ depletion or enrichment with respect to the standard 5.7 ‰ mantle value depends on the temperature and volume of the exchange and thus can help in calibrating these parameters throughout the exchanging section, as exemplified by Harper et al.'s (1988) work in the Josephine ophiolite. The microthermometric studies of fluid inclusions showed that the massive sulphide mineralizations have formed from fluids with a seawater salinity. After penetrating to the base of the sheeted dikes complex and being heated to 400°C, they ascended adiabatically through this complex, and were mixed with

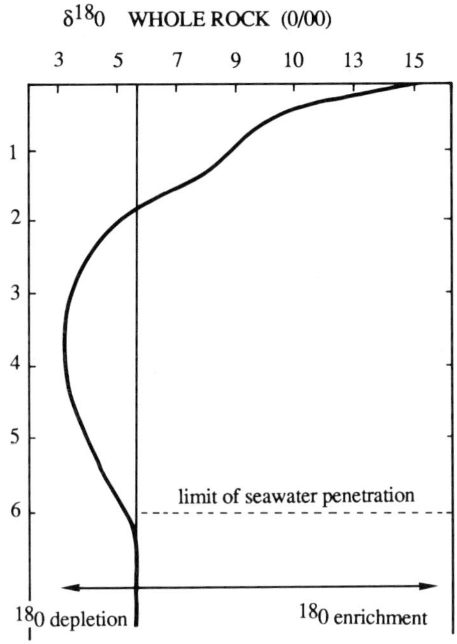


Fig. 11.17. $\delta^{18}O$ profile due to high temperature hydrothermal circulation in an ophiolitic crust (after Gregory and Taylor (1981) and Cocker et al. (1982), in Agrinier, 1987).

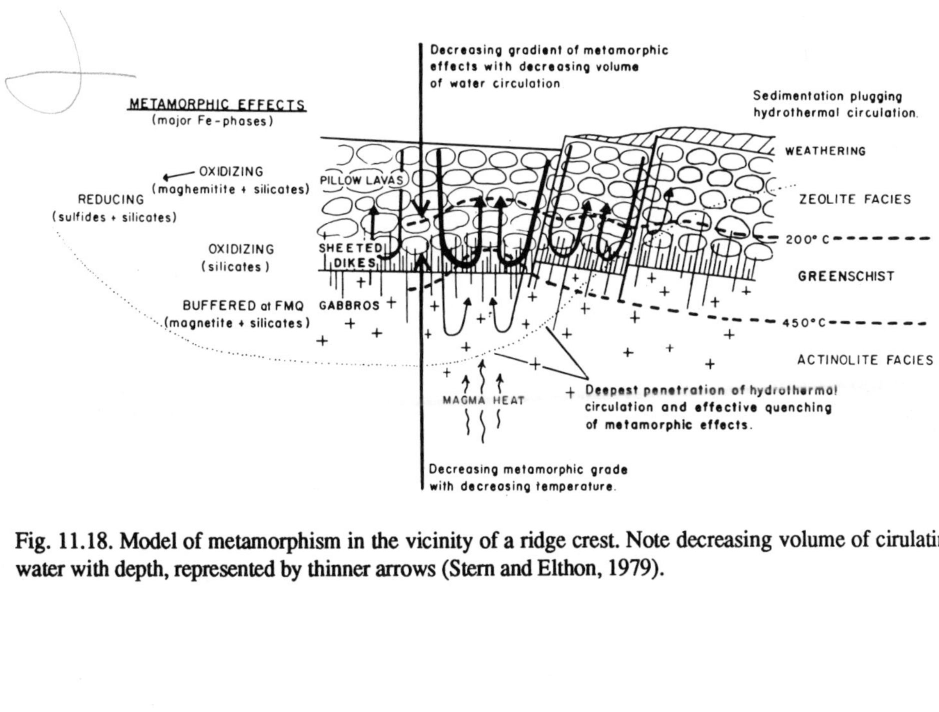


Fig. 11.18. Model of metamorphism in the vicinity of a ridge crest. Note decreasing volume of cirulating water with depth, represented by thinner arrows (Stern and Elthon, 1979).

colder seawater in the volcanic unit (stockwork) before erupting on the seafloor at temperatures of 300-400°C typical of oceanic 'black-smokers'.

Figure 11.18 is a sketch locating the metamorphic zones with respect to presumed hydrothermal circuits at an oceanic spreading center. It illustrates a few important points concerning the structure of the hydrothermal system, which are deduced from studies in ophiolites. The massive alteration occurs down to the base of the sheeted dike unit. As the volume of circulating seawater is large, this metamorphism reflects an oxidizing environment. The products of this metamorphism are observed to a depth estimated at 6 km (§ 11.3.3), where the effects of confining pressure and temperature close all open fractures. Above this depth, the metamorphic zonation would be grossly controlled by rock facies, via permeability contrasts. Thus, below the base of the fractured sheeted dikes, the metamorphism becomes less pervasive, as there is a less dense fracture network; the conditions also become more reducing, being buffered by the FMQ buffer of the igneous rocks. The high-level gabbros locally recrystallize into amphibole-pegmatitic gabbros in amphibolite facies conditions. Wet anatexis of these gabbros produces some plagiogranitic melts (§ 11.4.4).

In their study of the Oman ophiolite, Nehlig and Juteau (1988) and Nehlig (1989) note a sharp decrease of the porosity in the root zone of the sheeted dike unit. From a study of fluid inclusions, Nehlig (1989) also concludes that, at this level, the descending seawater is at 35 MPa, which is a hydrostatic and not a lithostatic pressure and suggests that it could have reached its critical point. These results could explain the difference in the hydrothermal alteration above and below the base of the sheeted dike unit. Nehlig determines the temperature of this boundary at 405°-410°C, from the location of the critical point of seawater in its P, T phase diagram and the above-mentioned P determination. The structural analysis of the orientations of the hydrothermal conduits (Nehlig and Juteau, 1988 ; see also § 3.3.2) reveals that in the sheeted dike unit there is, on average for every dike, one quartz-epidote vein corresponding to a fossil channel of seawater circulation. These veins are parallel to the dikes and thus to the presumed direction of the oceanic ridge of origin. This orientation for hydrothermal circulation at ridges was predicted (Cann, 1979) on the basis of the observed fracture pattern on ridge seafloor. Seismic anisotropy in oceanic crust also points to the existence of fractures in upper Layer 2, which would be aligned parallel to the ridge (Stephen, 1985). The pervasive alteration of the dikes proceeds from this dense network of fissures.

Below, in the upper gabbros where the diabase dikes are rooted, the water conduits, marked by the patchy recrystallization of isotropic amphibole gabbros or diorites, become more dispersed and grossly horizontal. This suggests a closure of the hydrothermal circuit at this level. However, greenschist and amphibolite facies metamorphism locally affects the lower layered gabbros and antigorite-talc-tremolite recrystallization in peridotites (§ 2.5.6) can be traced locally along fractured and sheared areas (Girardeau and Mével, 1982). Amphibolite facies metamorphism along shear zones producing flasergabbros, is much more conspicuous in LOT than in a typical HOT such as Oman. This point was made earlier in § 8.2.1, where it was also emphasized that in LOT the high temperature hydrothermal circulation commonly attains the top of the mantle section, producing an antigoritic serpentinization.

### *11.7.3. Relationship with the sequence of hydrothermal alteration in oceanic crust*

Summarizing our present knowledge of hydrothermal alteration in oceanic crust, it can be said that the high temperature hydrothermal metamorphism is produced in oceanic crust very close to the ridge, as shown by the location of hydrothermal vents within 1 km from the ridge axis. Water outpouring at 350°C from these vents precipitates massive sulfides, forming copper deposits of the Troodos type (Macdonald, 1982). The temperature related to this ridge hydrothermalism attains 400°C in the sheeted dike unit, inducing greenschist facies metamorphism as shown by the analysis of hole 504B of the DSDP-ODP Program (Laverne, 1987). In somewhat older crust, hydrothermal vents at 10°C precipitate manganese and iron oxides and smectite. A brownstone facies alteration in the basalts is induced by cold and oxidizing waters (Cann, 1979), the mineralogy of which has also been reviewed by Honnorez (1981). This low-temperature circulation depends on fractures in a crust which is still tectonically active and, thus, must occur at distances from the ridge which do not exceed a few tens of kilometers. Eventually, when the crust older than a few Ma becomes blanketed by a sedimentary cover, 150-200 m thick which has a sealing effect, the temperature can rise again, depending on the heat flow and on the nature and thickness of this sedimentary cover (Anderson and Hobart, 1976). This alteration occurs, in reducing conditions, at temperatures which do not exceed 200°C (Lister, 1982). Its mineralogy has been reviewed by Honnorez (1981). Its effects are superimposed upon those of the higher temperature ridge metamorphism, giving rise to a retrogressive zeolite to low greenschist facies metamorphism. This is also observed in ophiolites (fig. 3.10), where these retrograde effects decrease downsection and become insignificant in the gabbros (Stern and Elthon, 1979).

Deeper effects of the high temperature oceanic alteration have been observed, mainly in the Atlantic ocean, by manned submersibles and on drilled specimens. The most spectacular evidence is the dynamic metamorphism responsible for the common recovery of flasergabbros equilibrated in the amphibolite facies (§ 8.2.3). As recalled above, such flasergabbros are very common in LOT. From LOT evidence and from seismic data (§ 11.3.3), it can be speculated that hot seawater could have penetrated in the Atlantic down to the top of the mantle section, inducing the antigoritic serpentinization.

Finally, it should be recalled that the seismic layering of the oceanic crust has been related by some authors to the metamorphic zonation (§ 11.2). In particular, it has been proposed that the limit between Layers 2 and 3 is located at the greenschist to amphibolite facies boundary which may itself coincide with the lower limit of pervasive hydrothermal circulation, at the base of the sheeted dike unit.

# PART IV

# EMPLACEMENT OF OPHIOLITES TROUGH SPACE AND TIME

# Chapter 12

# OPHIOLITES EMPLACEMENT

## *12.1. INTRODUCTION*

Starting with the general papers of Dewey and Bird's (1971) and Coleman's (1971), the problem of the emplacement of ophiolites in the frame of plate tectonics has been considered by many workers (Christensen and Salisbury, 1976 ; Smith and Woodcock, 1976 ; Brookfield, 1977 ; Parrot and Whitechurch, 1978 ; Gealey, 1980 ; Moores, 1982 ; Michard et al., 1985). An extensive review of the possible emplacement models has been presented by Dewey (1976). At some stage, all models imply subduction of oceanic lithosphere. As proposed by Moores (1982) and Coleman (1984), the models can be divided into two families depending on whether the ophiolites were incorporated into continental crust at a passive margin (the *Tethyan ophiolites*) or an active margin (the *Cordilleran ophiolites*). In the former case, they rest as giant thrust sheets upon a continental substrate ; in the latter case they are incorporated, often dismembered or as mélange, into accretion terranes. Confronted with the complexity of certain natural situations and the profusion of models, for the sake of simplicity, we will contrast below these two types, namely ophiolite nappes with associated high temperature metamorphic aureoles which have been emplaced on passive margins and ophiolite bodies affected by high pressure metamorphism which have been accreted at active margins.

In the models dealing with emplacement of tethyan ophiolite nappes on a passive margin, the sequence comprises three stages ; there is first a phase of *oceanic detachment and thrusting*, followed by the *obduction* onto the passive margin (fig. 12.1a) ; the obduction may be eventually followed by squeezing into a *collision* belt. 'Ophiolites' found in an oceanic environment like those of Yap, Macquarie Island or even Gorringe Bank have not evolved beyond the oceanic detachment stage, whereas the Oman and the southwestern Pacific ophiolites have stopped at the second stage. The Tibetan ophiolites (§ 4.2) squeezed between the Eurasian and the Indian continents, have been through the complete sequence. In the second series of models corresponding to cordilleran ophiolites, these ophiolites occur rather as blocks of various sizes in blueschist mélanges. Whether they derive from the subducted plate or from the overlying oceanic plate, they have been accreted to a continental active margin (fig. 12.1b), before being eventually squeezed into a collision belt.

The oceanic thrusting of a piece of oceanic lithosphere onto a passive margin, which is the best documented and apparently most common emplacement process (Davies, 1971, 1980 ; Aubouin et al., 1977 ; Mattauer et al., 1980 ; Gealey, 1980 ; Silver et al., 1983), raises several difficult questions however. How can such a huge nappe of oceanic lithosphere which, in Oman or New Guinea, was originally 10-15 km thick, be emplaced onto a continental passive margin ? How did it travel for hundreds and possibly thousands of kilometers in the ocean of origin as suggested by some emplacement models ? Is there in modern oceans evidence of such an imposing event ? How can the double metamorphic belts commonly observed at the base of the ophiolite nappes be explained ? In particular, how were the high pressures responsible for blueschist and eclogite metamorphism, estimated at 1.2 GPa (35 km) in Oman (§3.3.5), produced at the base of these nappes ?

The occurrence of high pressure metamorphism in the ophiolitic fragments and bodies met in accretion prisms of active margins is more easily understood by assuming that they were first buried within the subduction zone (Hsü, 1971). There remain, however, several

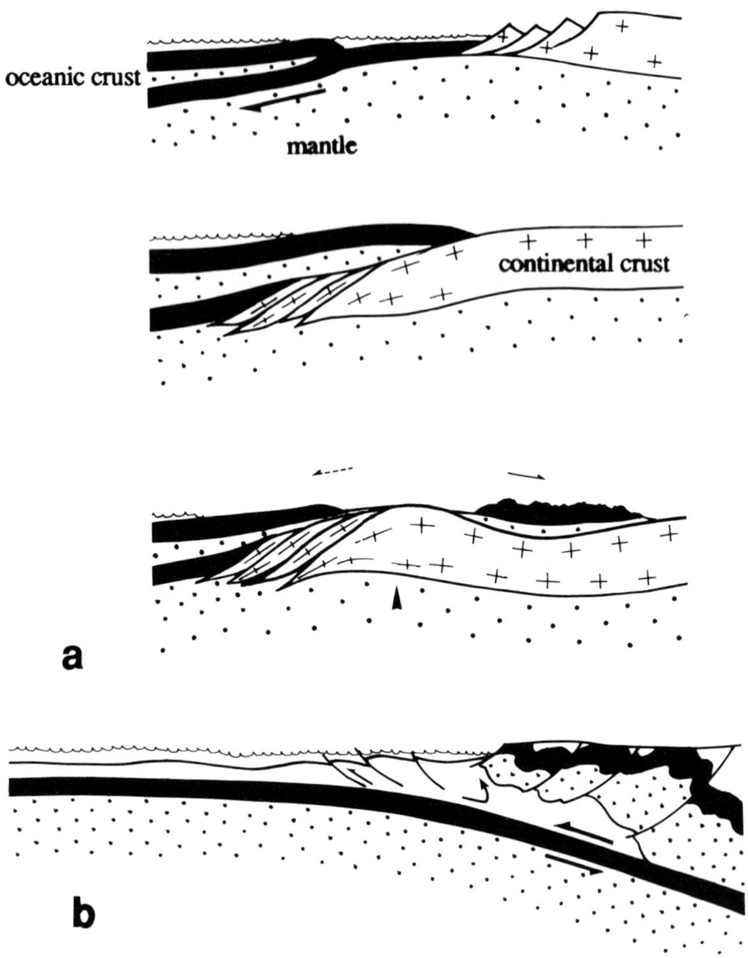


Fig. 12.1. Schemes of the two end-member models of ophiolite emplacement. a- Successive stages of emplacement of the Tethyan ophiolites. b- Cordilleran ophiolites with possible derivation from either the subducting plate or the overlying one (arrows).

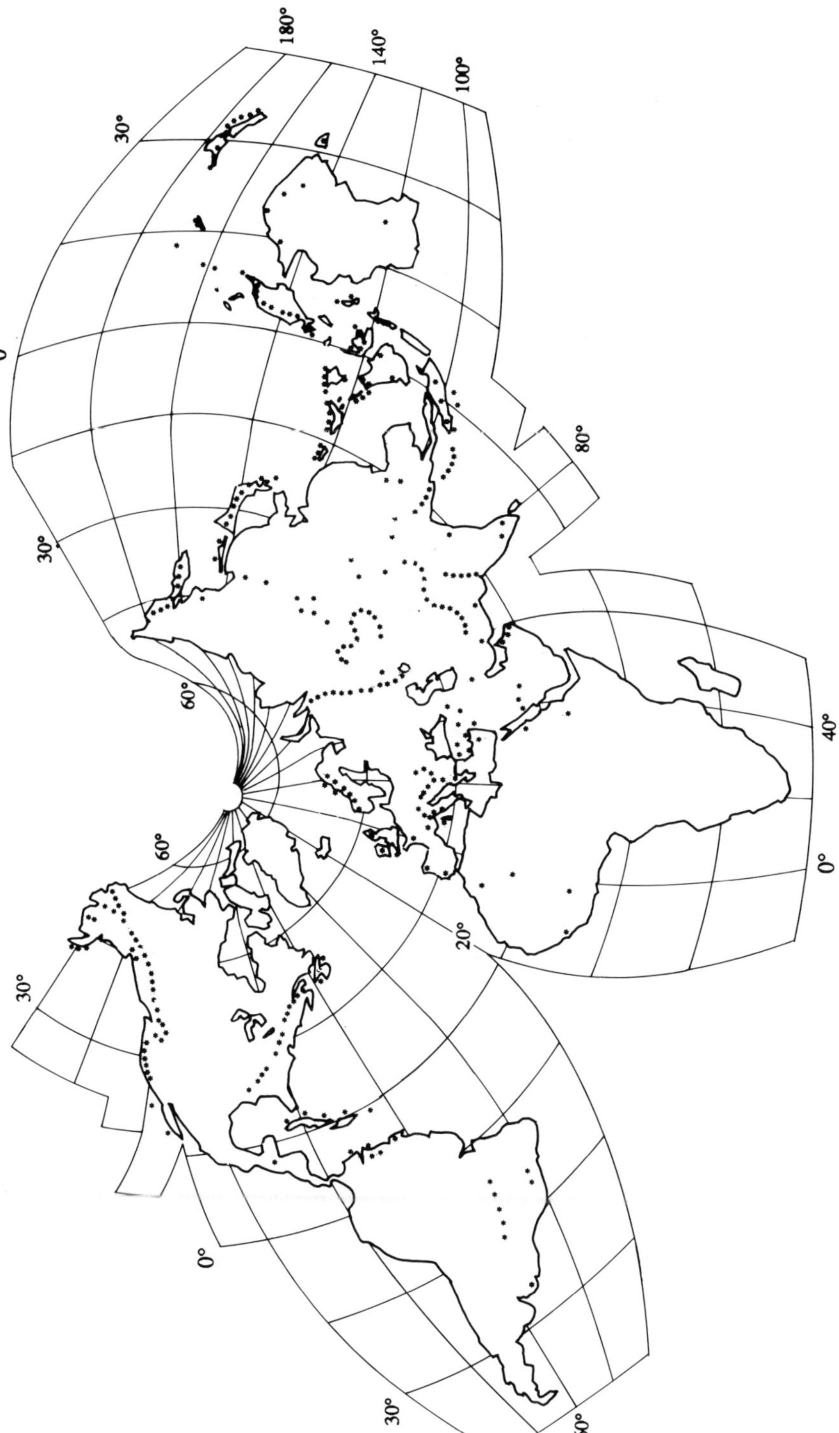


Fig. 12.2. North Polar projection showing the principal ophiolite belts of the world (after Irwing and Coleman, 1972).

questions pertaining to ophiolites in this environment. Do they represent slivers derived from the subducting plate or from the oceanic lithosphere located above the subduction zone (fig. 12.1b)? How were they upheaved into their present location ?

Some of these questions will be addressed in the section discussing the proposed models for ophiolite emplacement, after two more factual sections, one on the geological environments where ophiolites are actually found, and the other on the main emplacement-related features found in ophiolites. Some of these features are more fully described in the Oman ophiolite (§ 3.3.4).

## *12.2. OPHIOLITE BELTS*

Ophiolites are not randomly distributed on the Earth's surface but confined within belts (fig. 12.2). Although they mutually overlap, three main types of ophiolitic belts can be distinguished. The first two types reflect the two modes of emplacement, namely along continental passive margins and along continental active margins, and the last type, their common ultimate fate, continental collision.

### *12.2.1. Passive margins of continents*

This mode of ophiolite occurrence is best exemplified by the Western Pacific ophiolites, including New Caledonia (§ 5.2) (fig. 12.3a) and Papua-New Guinea (fig. 12.3b) (Davies, 1971 ; Aubouin et al., 1977). The Oman ophiolite can be considered as the most western extension of this belt. As shown in figure 12.3, these ophiolites represent a thrust clearly extending seaward into an oceanic lithosphere. Their obduction onto a passive margin has not been followed by a collision, explaining why such ophiolites display the most complete and best preserved sections. The chances that a passive margin will be involved in a continent or island-arc collision increase with time, thus explaining why such ophiolites are generally young (Cenozoic in the Western Pacific, Upper Cretaceous in Oman). The example of the Oman ophiolite is demonstrative in this respect. The Arabic margin is here facing the Makran subduction zone (fig. 3.3.) ; considering that the oceanic crust in the Gulf of Oman extends over no more than 300 km and, if the present-day kinematic frame and subduction rate of 5 cm/yr are maintained, a collision between Oman and Iran can be predicted within the next 6 Ma. In this respect, the Bay of Islands ophiolite which belongs to this group and is nearly 500 Ma old represents a noteworthy exception.

### *12.2.2. Active margins of continents*

Ophiolites are associated with the active continental margins of the circum-Pacific belt. In these margins, many of which have been active since at least the Mesozoic, ophiolites of decreasing age and commonly decreasing metamorphic grade oceanward reflect the successive stages of continental accretion (Ernst, 1975). As mentioned above, these ophiolites occur typically as glaucophanitic to eclogitic blocks within a High Pressure (HP) metamorphic mélange. This is illustrated by the eclogitic 'knockers' which are inserted into the Franciscan mélange of California. In their review of ophiolite occurrences in western Cordilleras of North America, Irwin and Jones (1981) distinguish other settings : in particular ophiolite massifs overlain by very thick (~4-10 km) sequences of volcano-clastic and/or carbonate formations ; these massifs are thrust over rocks which are not part of the craton. These categories obviously correspond to ophiolites with distinct origins and modes of emplacement (§ 12.4) ; they also testify to the complexity of continental accretion at active margins. Some of these ophiolites belonging to the

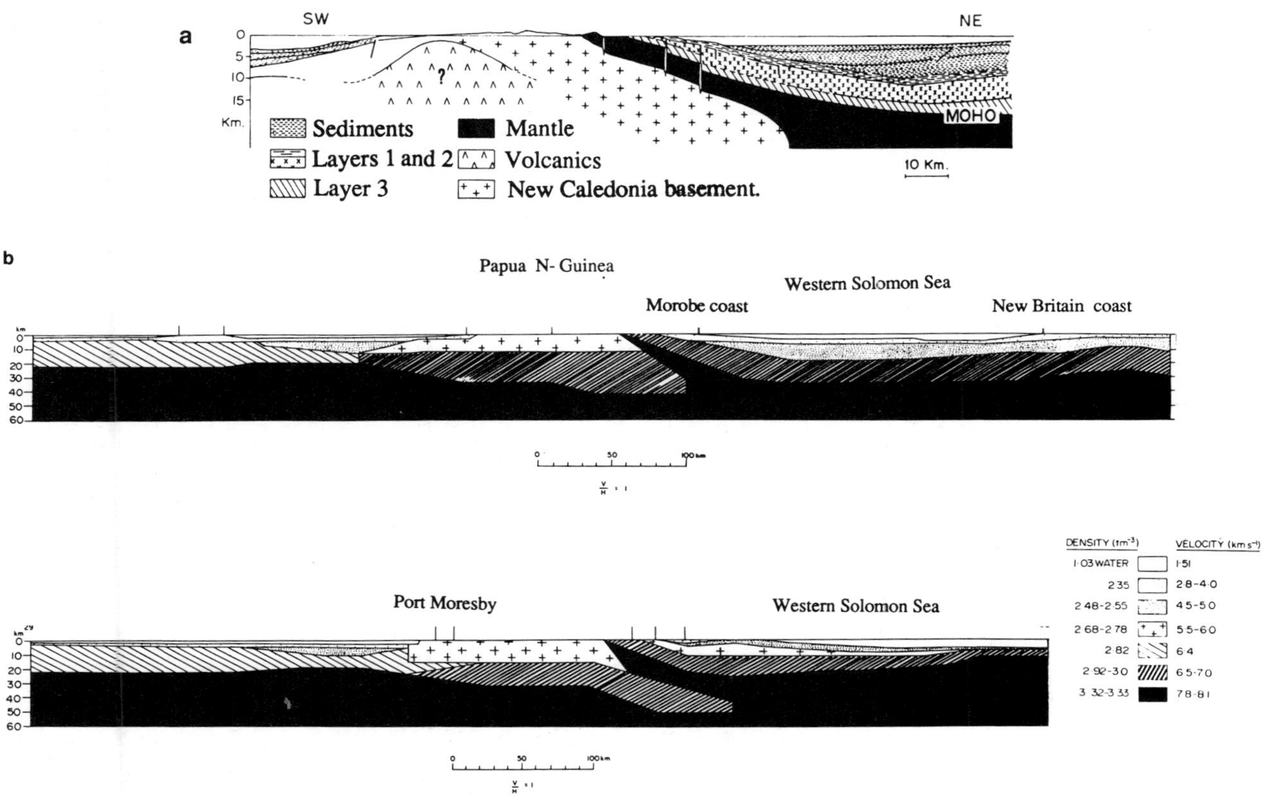


Fig. 12.3. Cross sections based on geological and geophysical data showing the continuity between oceanic lithosphere and the New Caledonia (a) and the Papua-New Guinea (b) ophiolite nappes emplaced on passive margins (a) Collot et al., 1987 ; b) Davies, 1980).

Phanerozoic accretion belt of the Cordillera result from collision between arc-systems and continent. They will be considered under the next heading.

If ophiolites are commonly emplaced along active margins (fig. 12.2), it is still not understood why, as noted by Gansser (1974), ophiolites or ophiolitic mélanges are missing along 5000 km of Andean subduction zone.

### *12.2.3. Collision belts*

Most exposed ophiolites are now in this situation (fig. 12.2), underlining the oceanic suture zones within continents or between continents and arc systems. Depending on the level of continental erosion, which is roughly a function of the age of the collision, they are variously preserved. They are generally most complete in the youngest alpine belt extending from Southeast Asia to Gibraltar, with a remarkable development in the Mediterranean belt. This rule is obviously very crude as the Cenozoic suture of Tibet (fig. 4.1), in which the Xigaze and neighbouring ophiolite massifs (§4.2) are inserted, is also locally reduced to slivers of sheared serpentinites a few tens of meters thick. On the other hand, in the Hercynian belt of the Polar Urals (Ivanov et al., 1975 ; Dobretsov, 1978) and the Caledonian belt of western Norway (Furnes et al., 1980), large and comparatively little dismembered ophiolite massifs have been preserved. They contrast with the highly dispersed and commonly highly metamorphic ophiolitic fragments which mark the Hercynian and Caledonian sutures in central and southern Europe (Girardeau et al., 1986 ; Dubuisson et al., 1988). In the Massif Armoricain and Massif Central of France, it is difficult to trace these sutures which over large distances are only marked by boudinaged eclogites (Bard et al., 1980).

## *12.3. EMPLACEMENT-RELATED FEATURES IN OPHIOLITES*

### *12.3.1. Introduction*

The emplacement-related events are best recorded by the deformation and associated metamorphism which they induce both within the ophiolite and in its surrounding formations. Two main metamorphic suites can be distinguished : the High Temperature-Low Pressure suite (HT suite) which is recorded in metamorphic aureoles of a number of ophiolites, and the High Pressure suite (HP suite) which is mainly developed in ophiolitic mélanges accreted to active margins. However, HP metamorphic belts and mélange nappes are also commonly found below the HT metamorphic aureoles of the former ophiolites (table 12.I), and ophiolitic mélanges are not necessarily associated with HP metamorphism. The following description shows that HP metamorphism associated with the HT suite is restricted to the continental basement of the ophiolitic nappes. Consequently, in this HT suite the *ophiolitic mélanges should not be affected by HP metamorphism,* suggesting as a criterion for identifying an HP suite would be the fact that the *HP metamorphism affects the ophiolitic bodies and mélanges* as well as the surrounding formations. Non-metamorphic ophiolitic mélanges of the HT suite along the Tethyan belt have been reviewed by Gansser (1974), who stresses that they derive from a sedimentary mixture (olistostromes) with superimposition of tectonic deformation. Ophiolitic mélanges of the HP suite may be produced by a tectonic mixture, as demonstrated by the occurrence of blocks with different metamorphic histories (§ 12.4.3).

### *12.3.2. Ophiolite nappes and high temperature aureoles*

The existence of metamorphic aureoles beneath ophiolite nappes has been first described

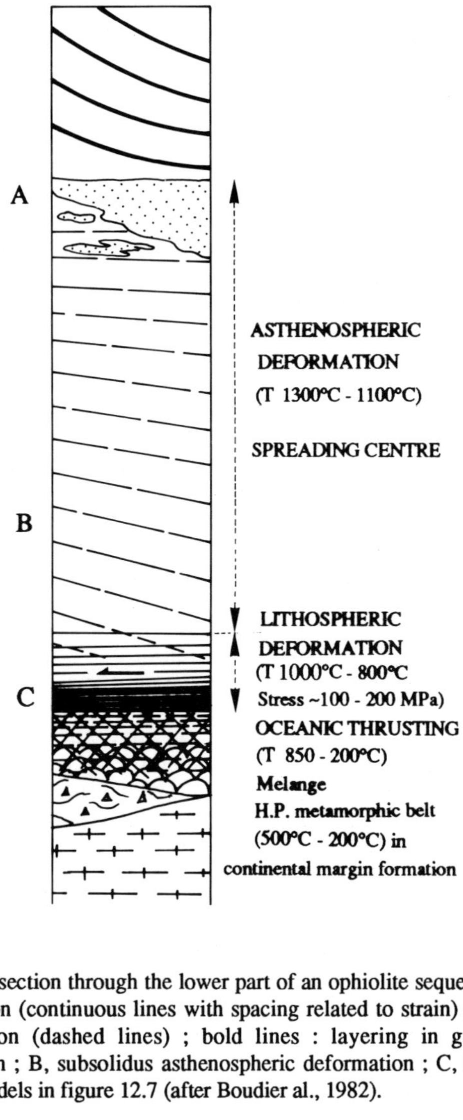


Fig. 12.4. Schematic cross-section through the lower part of an ophiolite sequence showing structures of the lithospheric deformation (continuous lines with spacing related to strain) superimposed on that the asthenospheric deformation (dashed lines) ; bold lines : layering in gabbros. A, hypersolidus asthenospheric deformation ; B, subsolidus asthenospheric deformation ; C, lithospheric deformation. These zones refer to the models in figure 12.7 (after Boudier al., 1982).

by Church and Stevens (1971), and Williams and Smyth (1973) in the Bay of Islands ophiolite. Typical aureoles have been described in this ophiolite (Dewey, 1976 ; Malpas, 1979 ; Jamieson, 1981) and in the Oman ophiolite (§ 3.3.4). In both sites as in many other ophiolites (Boudier et al., 1982), the metamorphic aureole is overlain by basal mylonitic peridotites. The main features of these aureoles, recently reviewed by Spray (1984), can be summarized as follows (fig. 12.4).

i) In the overlying peridotites within 1-2 km from the basal contact, a high stress-low temperature deformation appears superimposed on the low stress- high temperature deformation ascribed to asthenospheric flow (§ 2.5.5). The strain rapidly increases down section and becomes mylonitic in the last hundred meters above the basal contact. Blocking temperatures of 900-950°C have been recorded by the pyroxene of these mylonites (Jamieson, 1981), and an initial temperature of 1000°C is retained (Malpas, 1979 ; Boudier et al., 1988). Spray (1984) reports examples where the usual harzburgitic mantle becomes more lherzolitic in these mylonitic peridotites and ascribes this to the tectonic dispersal of mafic segregations constituting the ubiquitous layers and dikes in high temperature peridotites (§ 2.5.2).

ii) A metamorphic sequence is developed below the basal thrust within basalts and cherts of oceanic origin. It is characterized by an inverse thermal gradient, which is particularly strong (around 1-2°C/m), leading from pyroxene hornfels, granulites or garnet-amphibolites at the peridotite contact with local evidence of partial melting (peak temperature of 900°C, Jamieson, 1986), to greenschists, some 200 m below. The estimated pressures are usually moderate (0.2-0.5 GPa) (Ghent and Stout, 1981) with a maximum of 1 GPa (Jamieson, 1986). The metamorphic sequence can be discontinuous and broken, as nearly all have been involved in tectonic transport.

iii) There is complete continuity between the basal peridotites and the adjacent formations of the metamorphic aureole, in terms of both metamorphic conditions of equilibration and deformation (mineral lineations and foliations, shear sense). Strain moves downwards during thrusting (Jamieson, 1981 ; Pavlis, 1986) and tectonic discontinuities are common (Malpas, 1979), however, in the documented case of Oman a coherent kinematic path is maintained throughout the strained domain (fig. 3.29).

A few other characteristics have been reported in several of these ophiolites, suggesting that they may have some general meaning. An interesting one is the occurrence of an High Pressure Belt (HPB ophiolites in table 12-I) below the HT aureole of the ophiolite ; in this belt the metamorphic grade increases toward the ophiolite from pumpellyite-lawsonite schists, to glaucophane schists and eventually eclogites. Interestingly, this metamorphism is imprinted *within the underlying continental margin formations* and, typically, it does not affect the overlying ophiolites (Parrot and Whitechurch, 1978). The continental connection of HP belts is also apparent in the SW Pacific, where Parrot and Dugas (1980) note that the HP metamorphism occurs only when subductions involve continental crust ; when they occur in an intra-oceanic environment only HT metamorphism is found under the ophiolites. In Oman, the associated deformation and the age (within the brackets of table 3.1) are comparable to those recorded in the greenschist formations of the lower part of the metamorphic aureole, suggesting that this HP tectonic-metamorphic event affected the continental margin in response to the burden of the overlying and progressing unmetamorphosed ophiolite thrust sheet(s), already largely cooled (Goffé et al., 1988). In addition to the examples referred to in table 12-I, other examples of HP belt are reported by Gealey (1980) below the Brooks Range ophiolite (Alaska), and Mattauer et al. (1980)

for the Cap Corse ophiolite (Corsica).

Another feature common to several ophiolites of this type is their intrusion by granitic to tonalitic bodies. These intrusions vary in nature, size and age of emplacement. Small plagiogranite and locally potassic granite dikes and stocks like those described in Oman (§ 3.3.2) are commonly found crosscutting various levels of the ophiolite pile. They are commonly ascribed to a limited anatexis of the metamorphic aureole (MacKenzie, 1960 ; Challis, 1965 ; Pamic, 1977 ; Parrot and Whitechurch, 1978 ; Jamieson, 1980 ; Searle and Malpas, 1982 ; Pedersen and Malpas, 1985 ; Boudier et al., 1988). We have accepted this interpretation in the case of the Oman ophiolite, but anatexis of the underthrust continental crust has also been envisaged (Lippard et al., 1986 ; see § 3.3.4). This latter explanation has been proposed for the origin of granitic to tonalitic massifs by Abbotts (1978) in the Masirah ophiolite (Oman), and Gealey (1980) in the Brooks Range ophiolite (Alaska), and it can be envisaged in the Urals ophiolites (Ivanov et al., 1975), in the New Caledonia ophiolite (Prinzhofer, 1981) and in the Domingo Belt ophiolites of Cuba (Wadge et al., 1984). These granite intrusions, several kilometers across, are emplaced after the obduction of the ophiolite on the continental margin. Assuming they are produced by a crustal anatexis induced by a still hot ophiolite thrust sheet, such granites should be emplaced during or soon after the obduction. In this way, it should be possible to distinguish them from clearly subsequent and independent intrusions, as for instance the Nevadan granodiorites intruding the Ordovician Trinity ophiolite (§ 4.3) or the Eocene tonalites intruding the Cretaceous Papua-New Guinea ophiolite (Jaques and Chappell, 1980).

Eventually, as pointed out by Moores (1982), several of these ophiolitic sheets, including Oman (§3.2.2), have been emerged during their emplacement, developing laterites at the expense of exposed peridotites ; the eroded ophiolite is overlain by shallow water or continental deposits.

It should be realized that a large number, if not a majority, of ophiolites possesses a HT metamorphic aureole. It is not always easy to ascertain the existence of this feature from published descriptions ; hence, the following list of 62 ophiolites throughout the world certainly represents a minimum number (table 12.I).

Table 12-I. Compilation of ophiolites in which a HT metamorphic aureole and/or an overlying high-stress deformation band in peridotites have been described. Question marks refer to those occurrences where either the evidence is insufficient or has been disputed. HPB, for High Pressure Belt, points to the ophiolites characterized by paired HT, HP, belts. This compilation is partly based on compilations, geographically more restricted, but more detailed than ours, which have been published by Parrot and Whitechurch (1978), Wadge et al. (1984), Spray (1984), Ogawa and Naka (1984) and Jamieson (1986).

| Environment | Ophiolite locality | References |
|---|---|---|
| | SE of Bonin (26°25'N,142°55'E) | Naka and Uehara (1983) |
| | E of Mariana (11°18'N, 147°E) | Evans and Hawkins (1979) |
| | E of Mariana (11°51'N, 144°30'E) | Dietrich et al. (1978)<br>Crawford et al. (1981) |
| | E of Yap (8°11'N, 137°30'E, and 9°40'N, 138°25'E) | Crawford et al. (1981) |
| | Yap Island | Shiraki (1971) |
| Oceanic trenches and island arcs | Central Timor Island | Sopaheluwakan et al. (in press) |
| | Tonga | Fisher and Engel (1969) |
| | New Hebrides | Coleman (1970) ; Parrot and Dugas (1980) |
| | Solomon Islands | Coleman (1966), Parrot and Dugas (1980) |
| | Palawan Island (Philippines) *HPB* | Violette (1980) ;<br>Raschka et al. (1985) |
| | Middle America Trench (Guatemala) (leg 84 DSDP) | Bourgois et al. (1984) |
| | New Caledonia *HPB* | Prinzhofer et al. (1980) |
| | Papua-New Guinea *HPB* | Davies (1980) ;<br>Davies and Jaques (1984) |
| Passive margins | Oman *HPB* | (§3.3.4, 3.3.5, 3.4.3).<br>Williams and Smyth (1973) |
| | Bay of Islands (White Hills) and St Anthony (Canada) | Jamieson (1981, 1986)<br>Girardeau and Nicolas (1981)<br>Girardeau (1982) ; McCaig (1983) |
| | Trinity (USA) | § 4.3 |
| | Thetford (Canada) | Clague et al. (1981) ;<br>Laurent (1980) ;<br>Feininger (1981) |
| | Mont Albert (Canada) | MacGregor and Basu (1979)<br>Gagnon and Jamieson (1985) |
| Collision belts | Ballantrae (UK) *HPB* | Church and Gayer (1973)<br>Bluck et al. (1980) ;<br>Spray and Williams (1980) ;<br>Treloar et al. (1980) |
| | Lizard (UK) | Green (1964a,b) ; Barnes and Andrews (1984) |
| | Shetland (UK) | Prichard (1985) |
| | Limousin (France) *HPB* | Girardeau et al. (1986) |
| | Urals (Platinum belt, USSR) | Ivanov et al. (1975) |
| | Polar Urals (USSR) *HPB* | Dobretsov (1978) |
| | Zapadnyy Sayan (Central Asia, USSR) *HPB* | Dobretsov (1978) |
| | Seiad (USA) | Medaris (1966) ; Cannat (1985)<br>Cannat and Boudier (1985) |
| | Preston Peak (USA) | Snoke (1977) ; Cannat |

| | | | |
|---|---|---|---|
| | | | and Boudier (1985) |
| | Josephine (USA) | | Garcia (1982) ; Cannat and Boudier (1985) |
| | Sierra Nevada (USA) | | Saleeby (1982) |
| | Red Hills (New Zealand) | | Challis (1965) ; Walcott (1969) |
| | Dun Mountain (New Zealand) | | Coombs et al. (1976) |
| | Tinaquillo (Venezuela) | | MacKenzie (1960) |
| | Las Palmas (Puerto Rico) | | Tobisch (1968) |
| | N Guatemala | | Rosenfeld (1981) ; Wadge et al. (1984) |
| | S Guatemala | *HPB* | Muller (1979) ; Roper (1978); |
| | Domingo Belt (Cuba) | | Gealey (1980) Wadge et al. (1984) |
| Collision belts | Purial (Cuba) | *HPB* | Hernandez (1979);Wadge et al. (1984) ; Boiteau and Saliot (1982) |
| | Blue Mts (Jamaica) | *HPB* | Wadge et al. (1982) |
| | North Coast (Hispaniola) | *HPB* | Nagle (1974); Wadge et al. (1984) |
| | Western Cordillera (Columbia) | *HPB* | Bourgois et al. (1982, 1984) |
| | Xigaze, Dagzhuka (Tibet) | | § 4.2. |
| | Spontang (Ladakh) | | Reuber (1986) |
| | Muslim Bagh (Pakistan) | | Moores et al. (1980) ; Ahmad and Bilgrami (1987) |
| | Neyriz (Iran) | *HPB* | Ricou (1976); Pamic and Abib (1982) |
| | Esfandagheh (Iran) | *HPB* | Sabzehei (1974); Pamic et al.(1979) |
| | Kermanshah (Iran) | *HPB* | Pamic et al. (1979) |
| | Hadji Abad (Iran) | *HPB* | Pamic et al. (1979) |
| | Baer Bassit (Syria) | | Whitechurch and Parrot (1974, 1978) |
| | Mamonia (Cyprus) | | Lapierre (1972) |
| | Pozanti-Karsanti (Turkey) | | Cakir (1978) |
| | Antalya (Turkey) | | Juteau (1974) |
| | Lycian Nappes (Turkey) | | De Graciansky (1972) ; Sarp (1976) |
| | Crete (Greece) | *HPB* | Bonneau (1976) ; Koepke et al. (1985) |
| | Othrys-Pelion (Greece) | | Ferrière (1985) |
| | Euboea (Greece) | | Parrot and Guernet (1972) Spray and Roddick (1980) |
| | Vourinos (Greece) | | Zimmermann (1969) ; Moores (1969) ; Pichon and Brunn (1977) Spray and Roddick (1980) |
| | Pindos (Greece) | | Parrot (1967) ; Spray and Roddick (1980) |
| Collision belts | Zlatibor (Yugoslavia) | *HPB* | Pamic (1977) |
| | Krivaja (Yugoslavia) | | Pamic (1977) |
| | Borje (Yugoslavia) | | Pamic (1977) |

### *12.3.3. Ophiolitic mélanges and high pressure metamorphism*

HP metamorphic assemblages grading from prehnite-pumpellyite schists, to lawsonite and epidote-glaucophane schists and to garnet-jadeite eclogites are widely associated with ophiolites. The corresponding P-T conditions of equilibration are estimated to range from a few hundreds degrees and a few fractions of gigapascals to 500°C and 1.3 GPa (Ernst, 1981). HP metamorphism occurrences have been classified in § 12.3.1 into two categories depending on whether it has a regional extension, affecting the ophiolites and their environment together (the HP suite), or whether it is restricted to the continental substrate beneath ophiolitic nappes and their HT metamorphic aureole (the HT suite). We will consider here only the first case since the second has been described above.

Ernst (1975) has shown that both in active margins and collision belts, examplified in figure 12.5 by the cases of Japan and the western Alps, there is a HP metamorphic gradient positive in the subduction sense, that is, in the western Alps, from the external to the internal zones, and in active margins from the trench toward the fore-arc. Large metamorphic gaps separate the tectonic units mélanges are common within these units. As discussed by Hsü (1968) and recalled by Cowan (1985), the term 'mélange' covers many different rock associations with different origins, starting with disruption and fragmentation of incompletely consolidated sediments to mechanical disruption in fault zones. We recall now the main characters of ophiolitic mélanges affected by HP metamorphism (type III in Cowan, 1985), which are regarded as symptomatic of ophiolite accretion at active margins.

i) They are composed of ophiolite elements in a scaly serpentinitic or pelitic and clastic matrix. The ophiolite elements of various nature and disordered orientation have any size from small clasts to massifs in the kilometer range.

ii) These weak formations are intensely deformed and it is generally difficult to distinguish debris flow and gravity slides carrying large elements, the olistostromes, from tectonically induced deformation.

iii) They may be affected by several tectono-metamorphic episodes in the blueschist to greenschist facies. Ophiolite elements may have recorded more severe metamorphic conditions than their matrix (§ 12.4.3).

## *12.4. MECHANISMS OF OPHIOLITE EMPLACEMENT*

### *12.4.1. Introduction*

As mentioned in the first section, there is a wide number of possible models of ophiolite emplacement, reflecting the large variety of plate situations in the vicinity of a subduction environment and their conceivable evolutions (Dewey, 1976). Based on the review of ophiolites mentioned in table 12-I, we have identified two main mechanisms : thrusting of an oceanic lithosphere slab upon a passive margin (fig. 12.1a), and upheaval of lithospheric fragments within the accretionary prism of an active margin (fig. 12.1b). They can apply to different plate situations, interfere and be altered by collision events, but can still be traced in the Tethyan and Cordilleran ophiolites respectively (Moores, 1982) (§ 12.3.2 and 12.3.3).

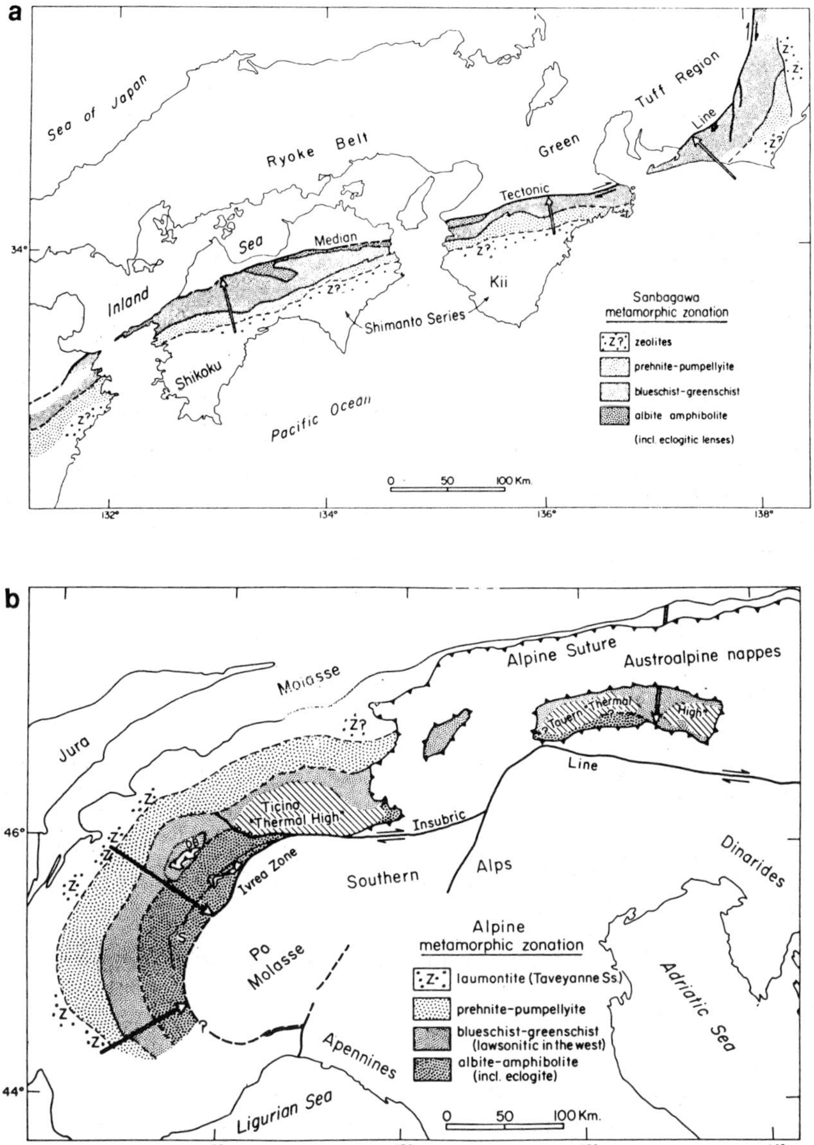


Fig. 12.5. HP metamorphic zonation in a) Japan active belt and b) western Alps collision belt. Arrows point to the direction of presumed subduction ; grade and age of metamorphism increase accordingly (Ernst, 1975).

### *12.4.2. Thrusting on passive continental margins*

*Introduction* - Ophiolitic obduction on passive continental margins is certainly facilitated if, in a compressive regime and prior to meeting with a passive margin and obduction, the oceanic lithosphere was anomalously elevated in its oceanic realm of origin. This should facilitate the margin underthrusting and ophiolite emplacement by gravity sliding. An oceanic environment characterized by a ridge or a very young lithosphere submitted to compression is an obvious candidate for ophiolite detachment. The impressive number of thin ophiolite nappes with a hot metamorphic aureole (table 12.I) points to such a young oceanic or back-arc environment of ophiolite origin, as being common, if not dominant. Other oceanic situations meet, however, with the elevation criterion : a fore-arc oceanic lithosphere above a young subduction zone, like some present active margins facing the East Pacific, or the uplifted wall of a transform fault or an anomalously thick oceanic crust, like Iceland or possibly some oceanic plateaus.

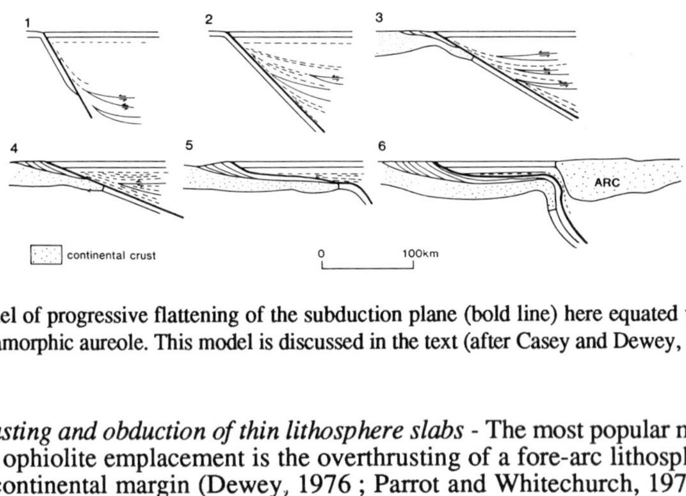


Fig. 12.6. Model of progressive flattening of the subduction plane (bold line) here equated with the basal thrust and metamorphic aureole. This model is discussed in the text (after Casey and Dewey, 1984).

*Oceanic thrusting and obduction of thin lithosphere slabs* - The most popular model in the literature on ophiolite emplacement is the overthrusting of a fore-arc lithosphere upon a subducting continental margin (Dewey, 1976 ; Parrot and Whitechurch, 1978 ; Gealey, 1980 ; Moores, 1982; Casey and Dewey, 1984). This model satisfies many observations ; in particular it provides a good explanation for a possible chemical contamination of the ophiolite by island arc magmatism. It would also account for pressures in the 1 GPa range, estimated in a few metamorphic aureoles of ophiolites on the basis of phase equilibria in amphibolites (Jamieson, 1986) ; however, these estimates are imprecise, and for other ophiolites they do not exceed 0.5 GPa. The model has an explicit (fig. 12.6) or hidden assumption, which is that the thrust plane is the subduction surface. Such an assumption is incompatible with 1) the fact that so many ophiolite nappes are thin (≤ 10 km of mantle section), and do not show in their mantle section (see for instance, § 3.4.3 and fig. 3.9) any structural discontinuity indicative of the considerable thinning of the mantle wedge during emplacement which is required by this model and 2) the hot character of the basal thrust (800-1000°C in the peridotites) with extraordinarily narrow heat and strain gradients. In a shallow (≤ 35 km) subduction environment envisaged by the fore-arc model, the movement zone is expectedly much colder (≤ 600°C, Earle, 1980) and consequently should be deformed more diffusely and in a more brittle manner. For these reasons, the fore-arc model as such should be abandoned.

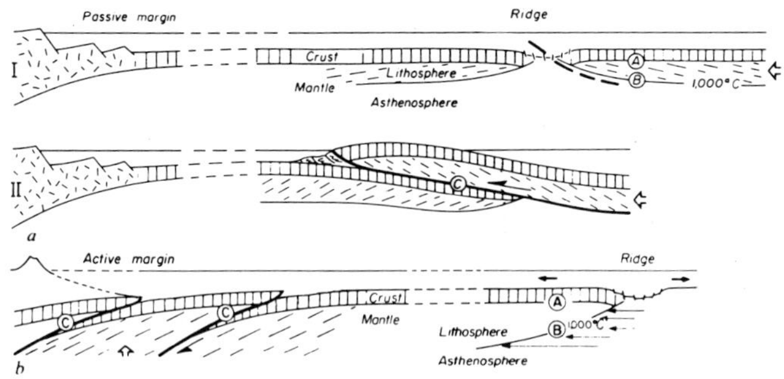


Fig. 12.7. Possible geodynamic locations of oceanic thrusting : a) at a spreading axis, along the 1000°C isotherm corresponding to the lithosphere/asthenosphere interface. A flat isothermal surface corresponding to fast spreading will favor the expected situation : b) in a subduction environment at the time when newly created lithosphere is involved in the subduction process. A, B, C indicate zones where asthenospheric and lithospheric deformations are imprinted in the peridotite ; it refers to levels A, B, C, in the ophiolite sequence of figure 12.4 (after Boudier et al., 1982).

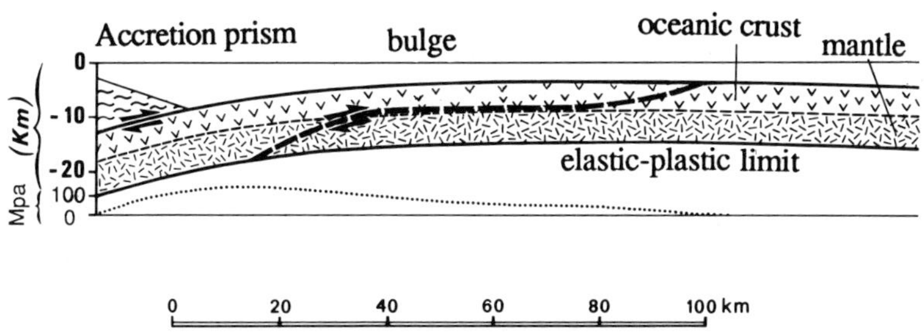


Fig. 12.8. Deformation of a thin elastic plate over a plastic lithosphere, 10 Ma old. Dotted line: the computed compressive stress at the base of the elastic plate; dashed line : possible geometry of a shear fracture (modified from Nicolas and Le Pichon, 1980).

The age of these thin and hot, or heated, oceanic lithosphere nappes is generally very young. This point, stressed by many authors (Christensen and Salisbury, 1975 ; Dewey, 1976 ; Brookfield, 1977 ; Nicolas and Le Pichon, 1980; Coleman, 1981 ; Spray, 1984 ; Abbate et al., 1985) is based on comparing the radiometric ages of the metamorphism related to the thrusting with those of the genesis of the ophiolite at a ridge. The example of Oman (table 3-I) shows that both events are in the same age bracket. In many ophiolites, the thrusting occurs within less than 20 Ma after the formation of the lithosphere.

With reference to compared subductions of oceanic lithosphere of various ages showing that the dip of the subduction plane is dependent on the age of the subducting lithosphere (Vlaar and Wortel, 1976 ; Molnar and Atwater, 1978), the generally very young age of the oceanic thrust helps to understand why the decollement plane remains so shallow over large distances and why temperatures are so high during the detachment phase. The resistance to subducting a young lithosphere (the 'Chilean' type of subduction of Uyeda and Kanamori, 1979) also explains why high deviatoric stresses (§ 3.3.3 and fig. 12.4) are recorded in the sheared peridotites of basal thrust.

*Detachment stage* - Two geodynamic environments have been proposed to explain the origin of these oceanic thrusts (fig. 12.7). The first one, illustrated by the Oman case (fig. 3.2.3), is the ridge of origin itself (fig. 12.7a) (Dewey, 1976 ; Boudier and Coleman, 1981 ; Boudier et al., 1982 ; Spray, 1984 ; Mitchell, 1985). If a shift from oceanic expansion to compression occurs within 1-2 Ma, the lithosphere-asthenosphere boundary remains very shallow, provided that the spreading rate is large enough (fig. 3.23) (2-3° slope in Oman, § 3.4.3) and constitutes mechanically a remarkable decoupling surface for oceanic thrusting. The HT metamorphism in the underthrust oceanic crust is easily explained by the temperature of the overthrust mantle still being at around 1000°C at the detachment site (Boudier et al., 1988).

Another model, first proposed by Dewey (1976), has been developed by Nicolas and Le Pichon (1980) (fig. 12.7b). It starts from the assumption that the bending of oceanic lithosphere entering a subduction zone develops large stresses within the elastic part of the lithosphere which are tensile above the neutral surface and compressive, below. At the limit between elastic and plastic lithosphere, elastic compressive stresses are in excess of 100 MPa, even in young lithosphere (fig. 12.8). This can induce a shear fracture which propagates seaward and emerges beyond the dynamic bulge of the subducting plate. To account for the thickness of the mantle section which in most ophiolites does not exceed 10 km, the limit between elastic and plastic lithosphere must be at the same depth below the Moho, implying that the corresponding lithosphere is not older than 10-20 Ma. In such a young subduction zone, compressive stress, estimated from the dynamics of this environment (Chamot-Rooke and Le Pichon, in press) and from structural piezometers (Nicolas et al., 1980), seems to be high enough (100-200 MPa) to promote thrusting along the newly created fracture. Since the thrust is generated at the limit beween elastic and plastic lithosphere (where the compressive stress is maximum), then the mantle is at a temperature which should not exceed 600°C (Watts et al., 1980). In order to explain metamorphism with higher temperatures in the aureoles, the temperature should be elevated by shear heating. Considering that the initial temperature is far from the domain of a weak rheology in olivine (~ 1200°C), that the stress level is elevated and that the movement zone is only a few hundred meters thick, shear heating seems quite possible (Fleitout and Froidevaux, 1980). Pavlis (1986) has shown, using the non-linear experimental flow laws for the rock types in the considered situation, that strain heating buffers temperatures at around 900-1000°C in deforming peridotites above an overriden plate. This buffering effect could explain why, whatever the specific situations, metamorphic aureoles in ophiolites look so similar.

A recent example of a possibly comparable oceanic thrust has been documented by Lallemant et al. (in press) and Chamot-Rooke and Le Pichon (in press) in the Shikoku Basin (fig. 12.9). Several points satisfy the theoretical model of Nicolas and Le Pichon (1980), including the age of the implied lithosphere which is not in excess of 20 Ma. However, the estimated displacement on the Zenisu thrust (some 10 km) remains too small with respect to the models of figure 12.7.

Akin to this model is the flake tectonics model (Oxburgh, 1972) which predicts that the slab is detached and thrust in the opposite direction, onto the fore-arc. This model discussed by Mattauer et al. (1980) is not supported, to our knowledge, by geological or geophysical observations.

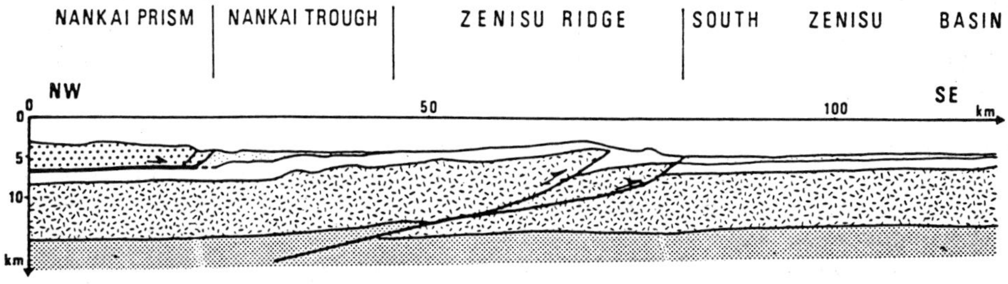


Fig. 12.9. Cross section of the Zenisu thrust in the Shikoku basin (Japan) based on geological and geophysical data. Tight dots : mantle ; dashes : oceanic crust ; blank : sedimentary cover of the Zenisu lithosphere ; spaced dots : sediments from Nankai prism (Lallemant et al., in press).

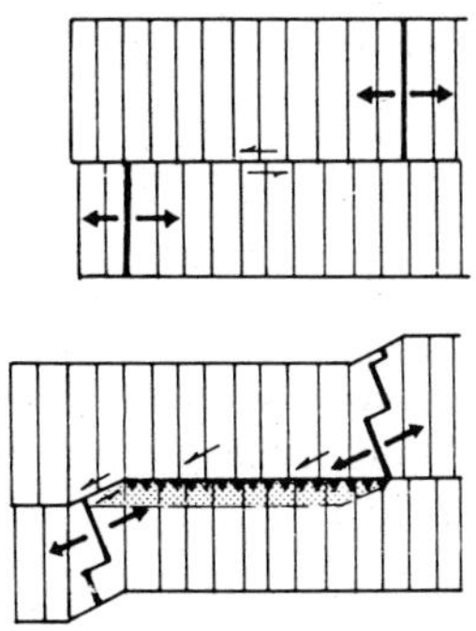

Fig. 12.10. Schematic plate model for the initiation of a subduction zone along a transform system in the case of a transform offsetting two ridge segments as the result of a change in spreading direction. Fine lines, magnetic anomalies ; bold black lines, spreading ridges ; medium straight lines, transform-fault ; triangles pointing in the direction of underthrusting of a subduction zone ; shaded areas, lithospheric overlap (Casey and Dewey, 1984).

*Oceanic thrusting and obduction* - After initial detachment, the thin lithospheric slab is thrust over oceanic crust. In the case of a ridge detachment, the nappe can either progress until it meets a passive margin upon which it is next obducted, or stay within the ocean and become an oceanic plateau, owing its elevation above surrounding sea-floor to the fact that it is composed of a doubled oceanic crust. Such a plateau might collide with an active margin during the course of subsequent plate wandering and become accreted to this margin. In the case of a continuous drift toward a passive continent margin, as illustrated by the Oman ophiolite emplacement (fig. 3.28), obducting the initially 10-15 km thick slab upon this passive margin may be facilitated by the facts that the crustal thickness of the passive margin was probably much reduced as a consequence of earlier rifting, and that erosion or gravity sliding may have removed part of the nappe thickness (in Oman, the shallow-water Maestrichtian deposits just postdating the emplacement locally rest directly on harzburgites). Again in Oman, it has been speculated that the subducting of continental margin could have blocked the system (fig. 3.28) ; as a consequence, the subduction could have jumped seaward to the still active Makran subduction zone, while the isostatic rebound of the relaxed Oman margin induced gravity slide of the nappes to their present location (fig. 12.1a ; § 3.4.3).

The HP metamorphic belt developed within the continental margin formations below many ophiolitic nappes (§ 12.3.2, table 12.I) can be explained by a transient subduction of the continental margin, possibly under a stack of oceanic lithosphere slabs thrust one over the other in response to the difficulty of further oceanic thrusting. This is again illustrated by the Oman case (§ 3.5.3 and fig. 3.28) and by the remarkably similar situation described in Papua-New Guinea by Davies (1980).

Two questions still unanswered by these scenarios are 1) How far can an oceanic slab be forcefully thrust over oceanic crust ? and 2) Why is this presumably spectacular and common situation in the past not reported in present-day oceans ? We will address the latter question in chapter 13. Dealing with the former, we find it puzzling that the applied stress, estimated above at 100-200 MPa may be able to propel the slab over distances of several hundred kilometers, perhaps even over 1000 km as envisaged in Oman (§ 3.2.2). In this respect, if some oceanic plateaus represent oceanic lithosphere doubled by thrusting, they may illustrate a situation where the applied force failed to propel the slab further, with resultant abandoning of this flat-lying subduction and jumping to a more favorable site.

In the case of detachment in front of a subduction zone (fig. 12.7b), it has been proposed, based on the geometry of the system and in analogy with physically comparable continental thrusts in the Himalaya, that the slab is thrust over around 100 km before it is relayed by a new shear fracture and thrust located seaward (Nicolas and Le Pichon, 1980). By this process the lithosphere slab is transferred to the fore-arc system and it can become the basement of islands like Yap where the typical high stress peridotites and HT metamorphics are observed. Another illustration, derived from seismic reflection evidence, could be the doubled oceanic lithosphere proposed by Green et al. (1986) at the base of the Vancouver Island accretionary structure. Final emplacement of the thrust lithosphere slab would occur if the subduction zone becomes the site of a collision.

*Obduction of transform structures* - Because they correspond to a major structural discontinuity responsible for a high relief difference, and because some mélange below ophiolite massifs could have been formed as relief screes (Lagabrielle et al. 1984), fracture zones have been considered by many authors as potential sites for oceanic detachment, at the origin of ophiolite obduction. In particular, this has been proposed by Saleeby (1978) for the Kings River ophiolite (California), Karson and Dewey (1978) for the Bay of

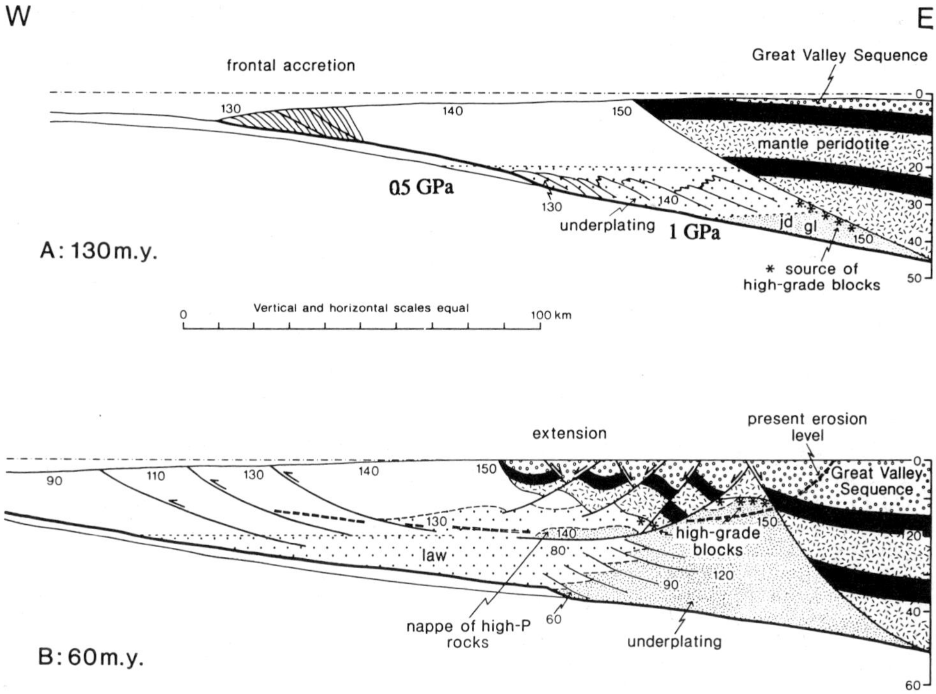


Fig. 12.11. Tectonic evolution of the Franciscan Complex, California. a) Early Cretaceous HP-LT metamorphism took place in sediments subducted beneath the leading edge of North America (represented by the Coast Range ophiolite and the underlying mantle wedge). Amphibolite, eclogite, and high-grade blueschist formed a metamorphic sole beneath the peridotite ; this provided the source for Franciscan high-grade blocks. b) Early Tertiary. Underplating and resultant extension have stretched the ophiolite and fore-arc basin sediments (Great Valley Sequence). As a result, most of the contacts between ophiolite and Franciscan are low-angle normal faults. The HP rocks have risen within reach of subsequent erosion and have been transported laterally over younger, lower-P rocks. High-grade blocks were dispersed by extensional faults. Stippled areas : HP metamorphic domains (tight dots > 1 GPa). Approximate ages of sediment in different parts of the complex are shown in millions of years before the present in order to illustrate the pattern of material circulation. Note that the present-day erosion level is 10-20 km below the original upper surface of the prism (After Platt, 1986 ; modified by assuming that the ophiolitic basement of the Great Valley is made of doubled lithospheric slabs, see text).

Islands Complex, Girardeau et al., (1985b) for the Indus-Tsang-Po ophiolites, Suppé et al. (1981) for the Taiwan ophiolite, Brookfield (1977) for the Andaman and Macquarie Islands ophiolite and Ogawa and Naka (1984) for the Setogawa and Mineoka belts in Japan. A modern illustration could be the Gorringe Bank, where the northern wall of the Azores-Gibraltar fracture zone is elevated to only 40 m below sea level as a result of a minor compressive motion : should Africa collide with this fracture zone, it would certainly scrape off the 'ophiolitic sequence' recognized by submarine along the elevated wall (Auzende et al., 1978 ; 1983). The model in fig. 12.10 illustrates this potential situation.

This mode of emplacement should be considered in the case of ophiolites obducted on passive margins which are devoid of basal metamorphic aureoles and record transform fault activity (see chapter 5).

### *12.4.3. Upheaval in the accretionary prism of active margins*

Structural and petrological analysis of high-grade blueschist knockers in the Franciscan mélange demonstrates a complex history with possibly two or more burial and upheaval stages in an accretionary prism (Cowan,1978; Moore, 1984). Fragments scraped off of the subducting lithosphere would have been sheared and metamorphosed at depths of about 30 km (0.9 - 1.2 GPa, 500°C) before and/or during upward transportation to elevated parts of an accretion prism. Next, they were deposited as olistostromes within the Franciscan mélange which was, in turn, buried to some 20 km (0.7 GPa, 160-200 °C), before the final uplift. Platt (1986) proposes a different origin, in which the knockers derive from the HP metamorphic sole of the hanging-wall lithosphere slab representing the basement of the Great Valley fore-arc (fig. 12.11).

The process bringing material from a few tens of kilometers to above sea-level is not understood. Basically, two main mechanisms can be proposed, one relying on gravity and the other on tectonic forces.

Lockwood (1972) and Moore (1984) have proposed a serpentinite diapir to carry upward deep-seated ophiolitic materials. In these deep-seated formations, in spite of a small thermal gradient, the ambient temperature should develop an antigoritic serpentinite whose density of 2.65 - 2.70 g/cm$^3$ seems inappropriate, unless the medium is heavily permeated by water. The same problem exists with other accretion prism materials. Their mobility in the gravity field would depend on their degree of undercompaction. Liquefaction of sediments by pressurized water can lead to mud diapirism (Suppé, 1973 ; Williams et al., 1984).

Assuming that the accretionary wedge has the simple geometry of figure 12.12 and is filled with weak sediments, tectonic models imply either a return flow as proposed by Cowan and Silling (1978) and Cloos (1982) or a pure shear deformation ; this latter deformation can be achieved by progressive tilting of imbricate thrusts with as a result the upthrusting of the most internal thrust slices (fig. 12.13). Return flow is generated in the unconsolidated trench sediments by their mechanical coupling with the downward motion of the subducting slab (underplating) and by the geometrical corner effect of the fore-arc basement. Different models have been presented (fig. 12.14). In Platt's model (figs. 12.14c and 12.11), extension in the rear of the accretionary wedge is promoted by continuated underplating which creates a relief in the rear of the wedge ; for a critical elevation, the body force due to the weight of the wedge formations exceeds their mechanical strength and they yield by extensional tectonics. Deep formations are exposed by the combined action of extension and erosion. Recent field evidence supports extensional tectonics in the Franciscan complex (Jayko et al., 1987).

Coming back to the overall metamorphism and age distribution in active margins, it

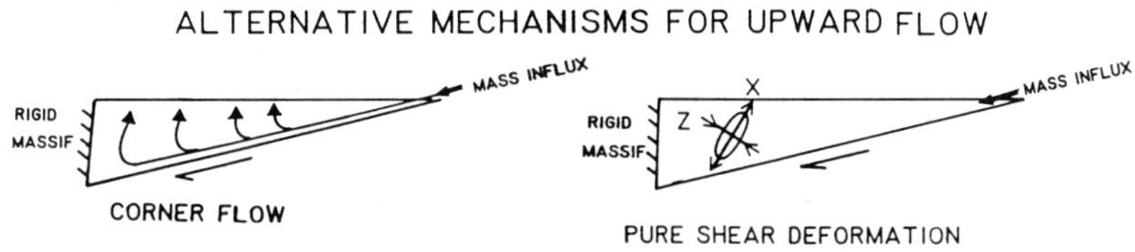


Fig. 12.12. 'End member' models which might account for uplift patterns observed in large fore-arc systems. On the right-hand side, X = max extension direction and Z = max shortening direction with X and Z assumed to be coaxial to a simple secondary stress field set up by the dynamics of underthrusting, i.e. maximum principal stress oriented 45° from the kinematic boundary (Pavlis and Bruhn, 1983).

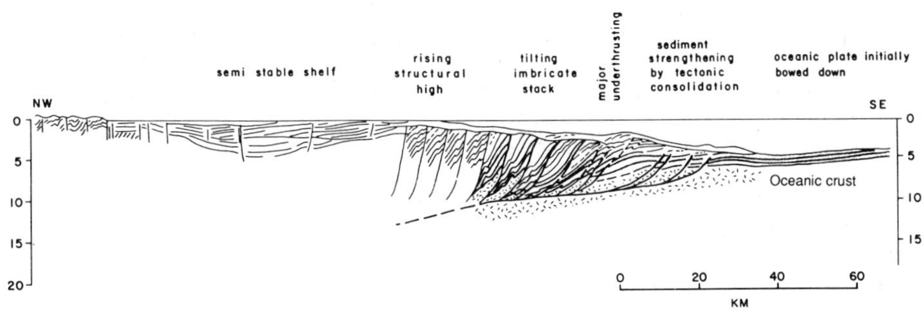


Fig. 12.13. Imbricate thrusting, progressive tilting and uprising of imbricate stack illustrated by the Kodiak (Alaska) subduction zone (After Von Huene, 1978).

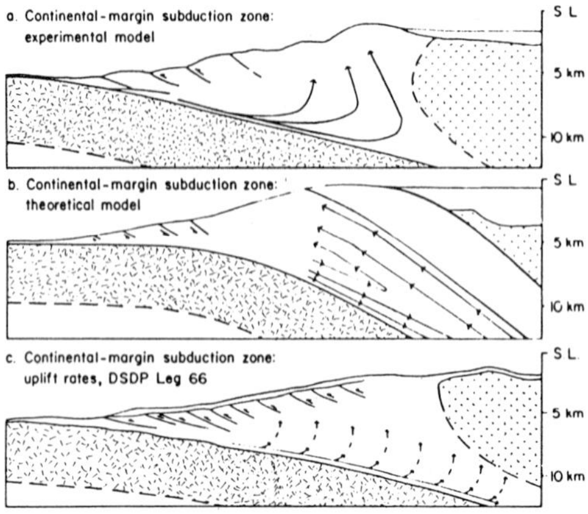


Fig. 12.14. Models of return flow. a) Dynamic, scaled clay model of a subduction zone by Cowan and Silling (1978). b) Theoretical modeling of the Franciscan accretionary wedge as a newtonian fluid by Cloos (1980). c) Model of an extension in the rear of the accretionary wedge Platt, 1986). Horizontal scale equals vertical scale ; double-barbed arrows depict particle paths.

should be noted that the observation made in a few belts (Ernst, 1975 ; fig. 12.5) of an increasing age of the HP metamorphism with increasing grade, suggests that during the course of a continuing subduction, at increasing distances from the trench or the suture, deeper terranes are progressively upheaved, and finally exposed. For forearc systems, Pavlis and Bruhn (1983) report uplift rates of 200-1000 m/Ma. The most internal and deepest formations can be uplifted of the necessary 20-30 km with a concomitant erosion, if such rates are maintained for several tens of million years as proposed by Ernst (1975).

## *12.5. SUMMARY AND CONCLUDING REMARKS*

Ophiolite emplacement is a vast topic covered by an abundant literature and a profusion of models, often poorly supported by facts. Inspired by Moores (1982) and Coleman (1984), we have tried to reduce these to two end-members models : the Tethyan and the Cordilleran models.

In *the Tethyan model*, an ophiolite nappe, 10-15 km thick, is obducted upon a passive continental margin. This process is initiated by oceanic detachment of a thin lithosphere slab, followed by intra-oceanic thrusting. Two environments are favored, an oceanic ridge where there is a rapid shift from expansion to compression, and the front of a subduction zone where a young oceanic plate is subducting (fig. 12.7). In the first case, the slab is either continuously thrust, through the ocean or basin of origin as in Oman (fig. 3.28) until it is obducted upon a passive margin, or is stopped and 'stored' in the ocean as an oceanic plateau, eventually to be emplaced when this plateau enters a subduction zone. In the second case, an ocean-directed thrust initiated in the subducting lithosphere may become the new subduction surface, thus transferring the overthrust slab to the fore-arc domain ; emplacement would occur if the fore-arc is engaged in a collision as illustrated in Tibet (fig. 4.10), or if its basement is uplifted at the rear of an accretionary wedge (fig. 12.11).

These Tethyan ophiolites have a high temperature dynamometamorphic aureole developed during the detachment phase. The contribution of shear heating has been probably underestimated, and its thermal buffer effect (Pavlis, 1986) may account for the similarity of thermal aureoles in the various ophiolites. During oceanic thrusting, deformation and lower grade metamorphism migrate downwards into the overthrust oceanic formations. An HP metamorphic belt is commonly developed below the HT belt once the nappe has attained the continent margin, probably as the result of incipient margin subducting below the stacking thrusts of ophiolites.

*In the Cordilleran model,* ophiolite massifs and blocks scattered in HP metamorphic mélanges have been uplifted in the rear of accretionary wedge of active continental margins. They could have been scraped off from the subducting plate or derived from the oceanic lithosphere constituting the basement of the fore-arc. Uplifting of formations metamorphosed at 30 km or more is still poorly understood, but return flow in the weak formations of the wedge seems to play a major role. It would be promoted by underplating of soft marine sediments ; superficial extension tectonics could largely contribute to progressively uncover the deepest and oldest formations which are observed at the rear of accretionary wedges.

These are end-member models. The complexity of most natural situations arises from their interference and from the common fate of passive and active continent margins, not to mention island arcs, to become squeezed with their ophiolites in collision belts. Some of these countless scenarios have been described by Dewey (1976). In this way, the choice of the adjective 'cordilleran' to identify one of our end-member types, is not totally satisfactory because, as shown by Irwin and Jones (§ 12.2.2) in the western Cordilleras of North America, there are ophiolites akin to the Tethyan type which have been obducted upon continental or arc terranes before their collision with the North American craton.

Finally, one is impressed by the wide, if not predominant number of ophiolites possessing an HT metamorphic aureole (table 12.I). This sets severe limitations on the oceanic environments and on the possible modes of emplacement, as seen above. Oceanic thrusting of hot and thin lithosphere slabs suggests a highly compressive environment in a young oceanic lithosphere. Detachment at a ridge along a flat lithosphere-asthenosphere boundary also suggests a ridge spreading at a fairly large rate before a rapid shift from expansion to compression. Neither this environment nor thrusting in front of a subduction zone seem common in the present-day oceans. We will return to this in chapter 13.

# CHAPTER 13

# OPHIOLITE BELTS THROUGH TIME

## *13.1. INTRODUCTION : A REAPPRAISAL OF OPHIOLITES AND THEIR OCEANIC ENVIRONMENTS*

Recalling the first question asked in the introduction of this book, how do ophiolites compare with oceanic lithosphere, we are tempted to answer that the best evidence that they do is their wide structural and petrological variety (chapter 8). This necessarily reflects a variety of oceanic situations and it excludes the possibility that a particular oceanic environment has been preferentially sampled to become an ophiolite.

However, equating ophiolites with oceanic lithosphere and in particular with mid-oceanic ridges meets with two main difficulties. The first, raised by many petrologists dealing with ophiolites, is that only a few ophiolites have a MORB affinity. As a consequence, there is a tendency to believe that most ophiolites derive from island arc-back arc environments. This question has been dealt with in § 8.3, where some caution was expressed about possibly hasty conclusions. The complexity of ridge and ridge-vicinity volcanism in Oman shows that scenarios other than the arc ones may account for the geochemical spectrum of lavas (§ 3.5.3). To emphasize the limits of present knowledge, Coleman (1984) also recalls that MORB chemistry is based on samples from the vicinity of active ridges, that very little is known in older oceanic crust, and that exposed Phanerozoic ophiolites represent probably less that 0.001 % of the corresponding oceanic crust. This question of the environment of origin is still open, and has captured a disproportionate amount of attention and diverted it from more tractable issues.

The second difficulty in comparing ophiolites with oceanic lithosphere deals with the emplacement of ophiolites as hot oceanic thrusts (§ 12.4.2). This is probably the most common process by which ophiolites are obducted onto continents. Paradoxically, it is not documented in present-day oceans and is thus considered with reluctance by the marine community. An oceanic event with the geotectonic importance of the assumed ophiolitic thrusts which would move a slab having a 10 to 15 km thickness over hundreds of kilometers in the oceans should have attracted attention. Its absence raises the problem of major changes in plate configuration through time. This point is addressed in this conjectural chapter, starting from the hint that ophiolite emplacement seems to occur episodically through geological times, as already reported by a few authors (Rona and Richardson, 1978 ; Moores, 1982 ; Abbate et al., 1985 ; Sheridan, 1987).

## *13.2. OPHIOLITE GENERATION AND EMPLACEMENT THROUGH TIME*

It should be recalled here that the age of an ophiolite is an ambiguous concept : there is first the age of crustal creation at a spreading center ; next, for many if not most ophiolites, there is the age of detachment and oceanic thrusting of a lithospheric slab, and finally there is the age of obduction onto, or of collision with, a passive continental margin. We consider here only the two first ages because they reflect the most significant geodynamic event in ophiolite generation, that is the shifting from oceanic expansion to compression.

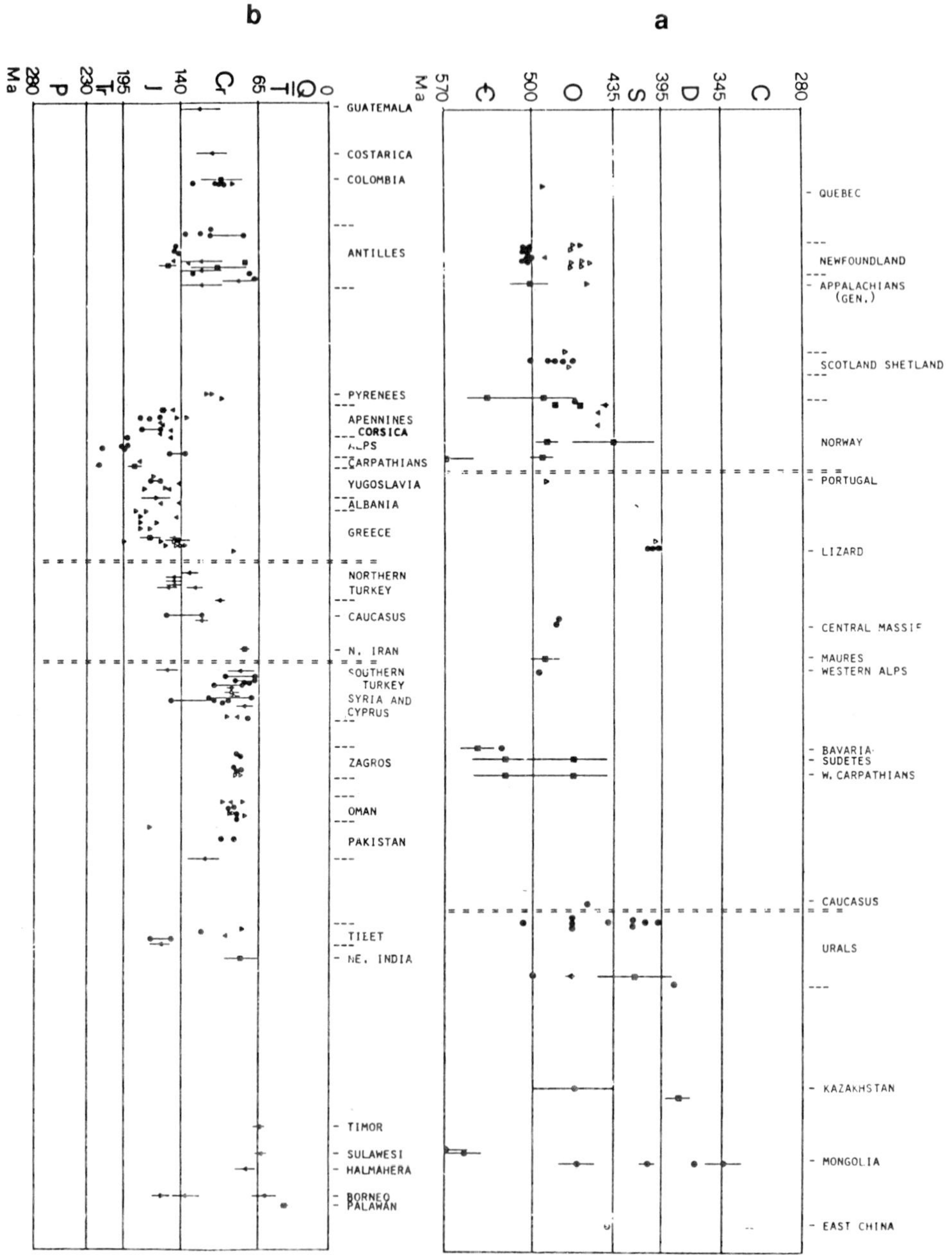


Fig. 13.1. Dating of the creation of ophiolites at oceanic spreading center. a) Caledonian belts. b) Tethyan belts (Abbate et al., 1985). Dots : radiometric igneous ages ; squares : estimations, with error bars, based on regional geology ; solid triangles : biostratigraphical ages of first sedimentary cover of ophiolite ; open triangles : oceanic metamorphism.

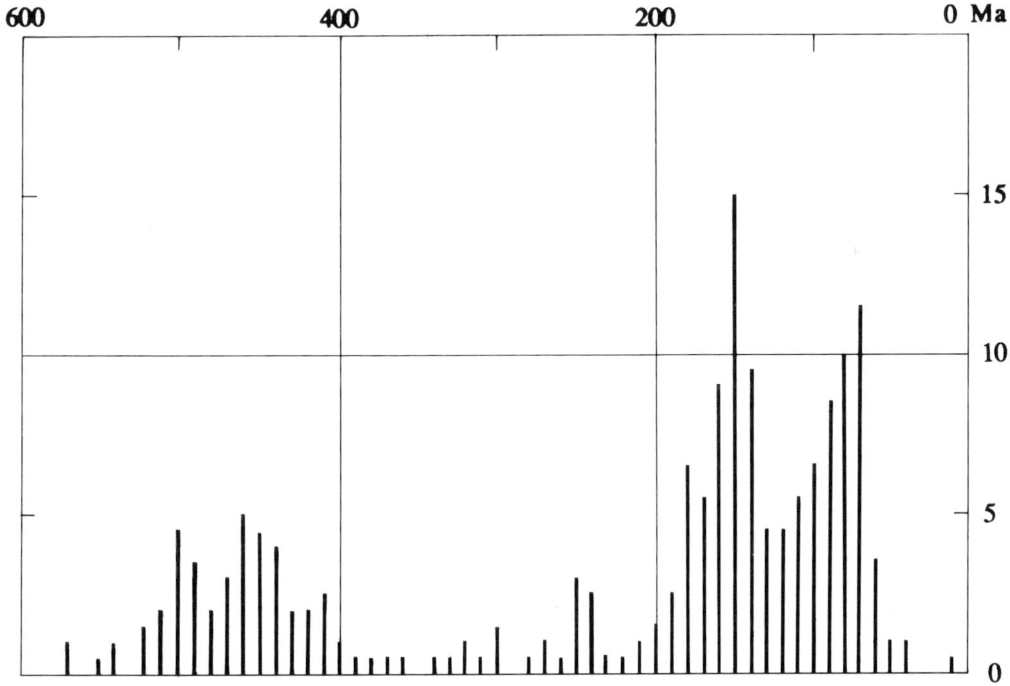


Fig. 13.2. Histogram of ages of ophiolite generation compiled from Abbate et al.'s (1985) data pertaining to the Caledonian and Tethyan belts (fig. 13.1) and to the Circumpacific belt. The relative elevations of the peaks are not necessarily significant, as they depend on the available data ; for instance, the highest Jurassic peak is an artefact due to the abundance of chronological data on the western Mediterranean ophiolites.

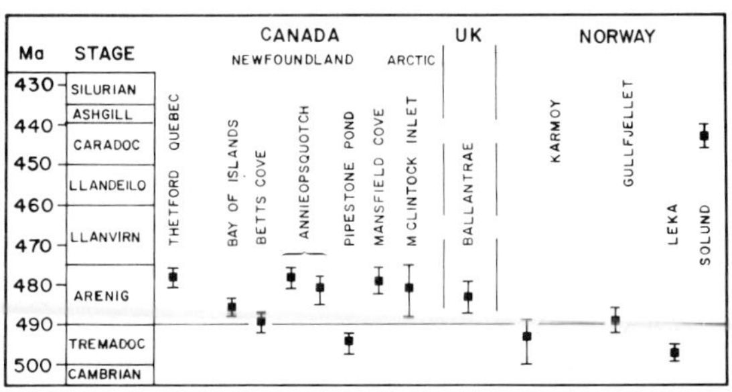


Fig. 13.3. Summary of U/Pb zircon ages for ophiolites of the Appalachian-Caledonian belt (Dunning and Pedersen, 1988).

As seen in § 12.4.2, this shift generally occurs soon after creation of the future ophiolite at a spreading center as emphasized by the hot character of the detached slab. Once the slab has been thrust it is normally bound to be emplaced as an ophiolite. Obduction or collision does not necessarily follow during the same tectonic event and can happen much later. Therefore, this age is not necessarily related to the oceanic thrusting event. In the most complete compilation of ophiolite ages made by Abbate and his co-workers (see below), this is shown by the larger scattering of obduction or collision ages compared to those of creation at a spreading center and detachment, or by the wide age dispersal of the high pressure metamorphism in ophiolitic belts of Asia (Dobretsov et al., 1987).

Abbate et al. (1985) show that many ophiolites cluster along the Caledonian and the Tethyan belts, being generated within a comparatively short time-span, mainly between 500 and 440 Ma in the Caledonian belt (fig. 13.1a) and between 200 and 65 Ma in the Tethyan belt (fig. 13.1b). This duality in time remains clear, even when these data are plotted together with, the ages compiled by the same authors for the ophiolites belonging to the circumpacific composite belts (fig. 13.2). The image emerging from these raw data becomes sharper after a certain amount of screening. Thus, considering only U/Pb zircon ages of ophiolites generation in the Appalachian-Caledonian belt which, extended originally over about 4000 ? km, the time-span narrows down to 500-470 Ma (fig. 13.3), with a single younger ophiolite, the Solund complex at 440 Ma. The Tethyan belt presents two peaks in age, a Jurassic one (200-140 Ma) predominant in the western Mediterranean ophiolites, and a Cretaceous one (140-65 Ma) predominant in the rest of the belt, with an Upper Cretaceous clustering for the eastern Mediterranean and Middle-East ophiolites. These peaks may correlate with distinct ophiolite belts, mainly LOT for the western Mediterranean, and HOT for the rest of the Tethyan domain (§ 8.2.2) with a limit between the two belts located in Greece (Koepke et al., 1985). The Jurassic belt is probably related to the Atlantic rifting, and the Cretaceous belt to the Neo-Tethys activity (Dercourt et al., 1986) (fig. 3.5).

It seems possible to conclude that ophiolites form mainly along definite belts and during restricted time periods, which from figure 13.2 could be separated by 300-400 Ma. Abbate et al. (1985) also observe that the Mesozoic ophiolites have been clearly emplaced along the periphery of the pre-Jurassic Pangea (fig. 13.4).

## *13.3. OPHIOLITES AS WITNESS OF PANGEAN CYCLES*

Those authors mentioned in § 13.1 who propose that ophiolite generation is episodic have looked for major plate reorganizations to explain this phenomenon. Rona and Richardson (1978) note that during Early Cenozoic (55-40 Ma) there is a major reorientation of relative plate motions with large N-S components rotating into large E-W components and an increase in collisional plate boundaries from about 2500 km to 28,500 km. They relate the obduction of ophiolites in the Pacific and Midle East to this reorganization, but they do not investigate the episodicity and the possible cause of the reorganization.

Sheridan (1987) relates periods of high spreading rates associated with a magnetic quiet field, alternating with periods of slow rifting and magnetic reversals, to interactions between deep mantle and core. He deduces a period of 60-100 Ma. Being smaller than the 300-400 Ma time-span between peaks of ophiolite generation, this periodicity is not relevant here ; it could, however, be superimposed on a 300-400 Ma periodicity. Indeed, Sheridan points out that the breakup of Pangea occurred during two fast spreading pulses, one during mid-Jurassic and the other during mid-Cretaceous, which correspond in figure 13-2, to two peaks of ophiolite generation. Relating ophiolite generation to episodes of fast spreading rates is supported by the volumetric dominance of HOT over LOT (§ 8.2.3). Sea level changes record three major orders of cycles (Vail et al., 1977), of which

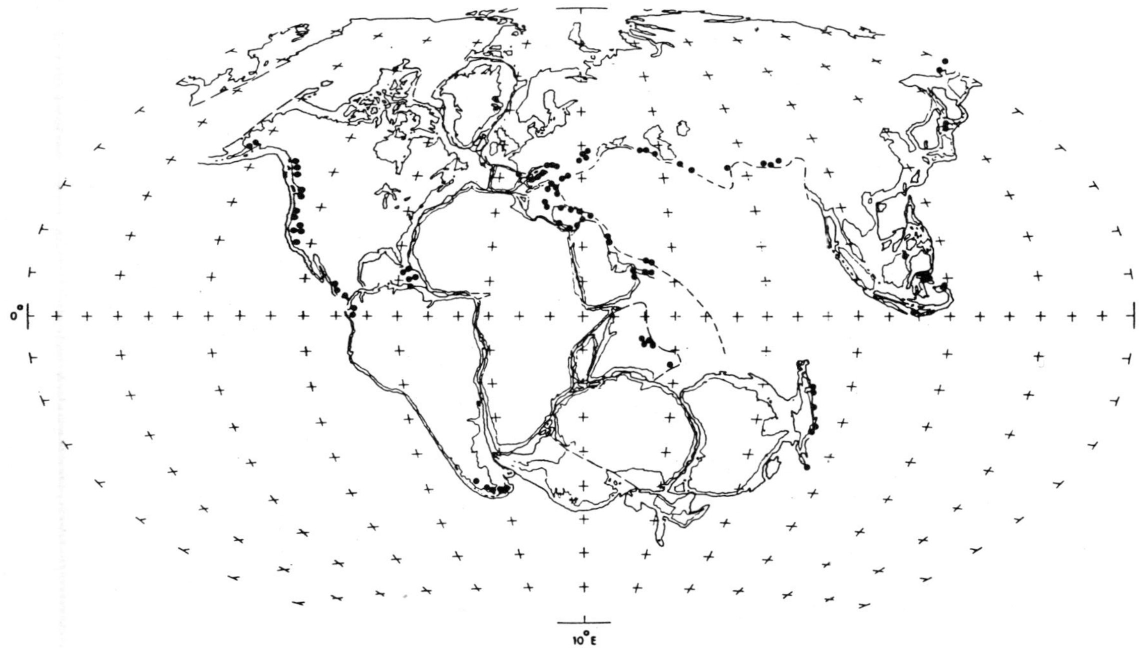


Fig. 13.4. Areas of Mesozoic ophiolites plotted on the pre-Jurassic Pangea as reconstructed by Owen (1983). The figure shows the peri-Pangea position of the ophiolites (with the exception of the Caribbean and Western Tethys regions) (Abbate et al., 1985).

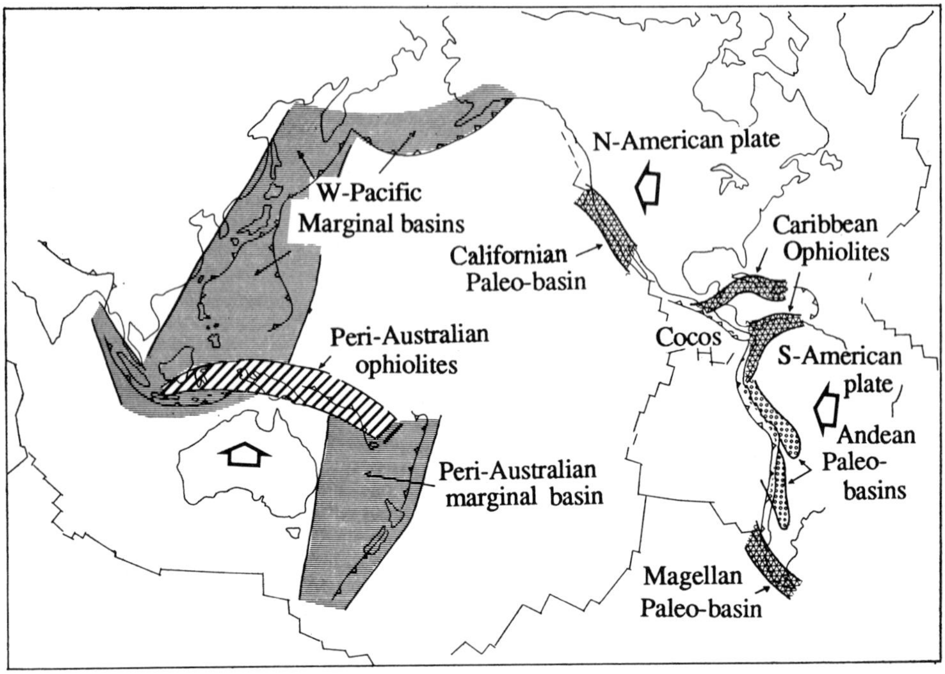


Fig. 13.5. Interpretation of the West America ophiolites as being formed in marginal basins comparable to the present-day marginal basins of Western Pacific ; obduction of these ophiolites occurs in response to a Westward motion of the two American plates, overriding the marginal basins. A similar situation is assumed in SW Pacific, related to a Northward motion of Australia. Cross patterns : West American paleo-basins with ophiolites ; dots : the same without ophiolites ; horizontal hatches : present-day marginal basins ; oblique hatches : SW Pacific Tertiary ophiolites (Chotin, 1981).

the two first (200-300 Ma and 10-80 Ma) could correspond to the periodic events envisaged above.

Abbate et al. (1985) propose that we correlate the two periods of ophiolite generation separated by 300-400 Ma to the dispersion phase of past Pangeas which are supposed to reconvene episodically, defining a pangean or Wilson cycle. This is compatible with the peripheral location of ophiolites with respect to the Jurassic Pangea of figure 13.4 and with the periodicity expected for such cycles. Paleomagnetic data suggest the existence of a paleozoic supercontinent (Cowie, 1971 ; Morel and Irving, 1978 ; Bond et al., 1984) which started to dismember between 600 and 500 Ma, some 300-400 Ma before the next Jurassic dispersal, pointing to a periodicity of 300-400 Ma. Geodynamic models, implying thermal exchange through a lithosphere with a time constant of 100 Ma and phases of dispersal and reassembly taking each around 150 Ma, come to comparable results for the entire cycle : 350 Ma for Anderson (1982), 400 Ma for Le Pichon and Huchon (1984) and 400-450 Ma for Nance et al. (1986) ; considerations on the periodicity of Phanerozoic first-order changes in sea level and stable isotope trends and, since the Archean, in orogenic and rifting events on the global scale, also suggest to the last authors a periodicity of 400-500 Ma.

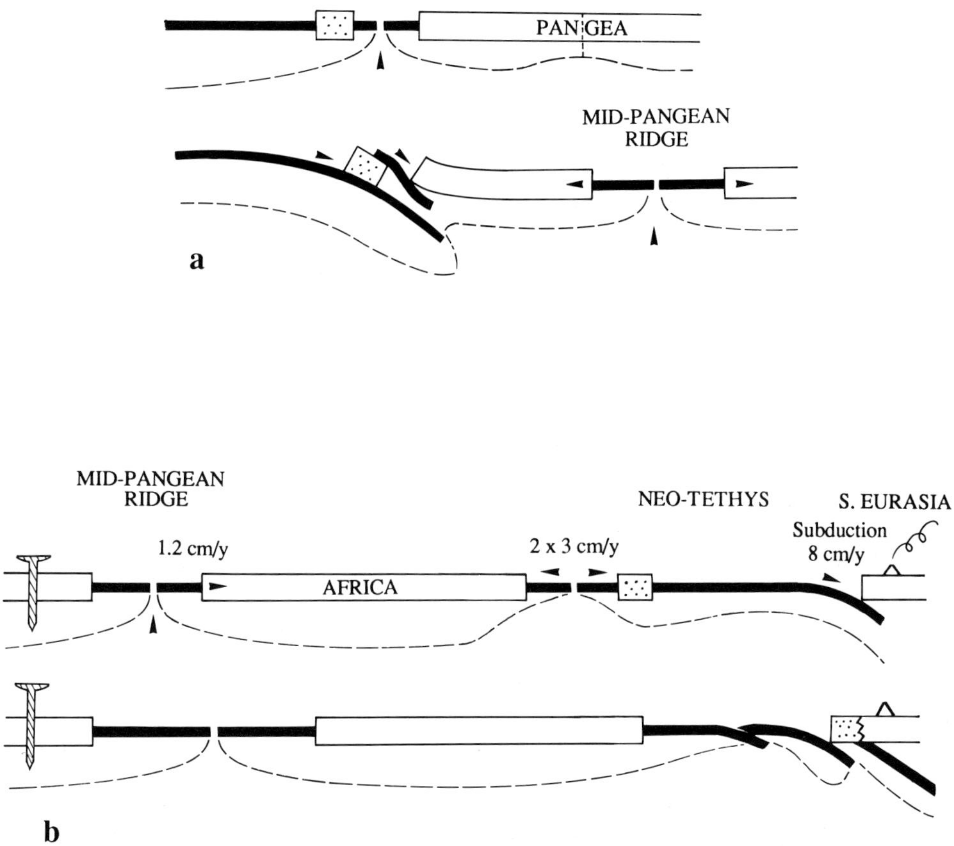


Fig. 13.6. Sketches depicting two possible models of peri-Pangea ophiolites generation and emplacement. a) Early opening of an oceanic basin in response of lithospheric heating and thinning below Pangea, followed by closure induced by mid-Pangea oceanic spreading and peripheral subduction (modified from Abbate et al., 1985). b) Opening of an oceanic basin to accomodate the fast motion of subduction below the Eurasian margin (fixed end on the left continent of the former Pangea is arbitrary ; velocities should be considered as indicative) ; the lower cartoon depicts one possible scenario responsible for ocean thrusting within the South Neo-Tethys, here the temporary choking of subduction by collision with the block detached from Gondwana by the South Neo-Tethys opening. Double arrows are intended to represent the main driving forces.

The peripheral generation of ophiolites with respect to Pangea implies an episode of sea floor spreading, rapidly followed by oceanic detachment and eventually by emplacement of the newly created oceanic lithosphere onto the supercontinent margins. This is examplified by opening of the Neo-Tethys, detaching small continental blocks from the main landmass of Gondwana (fig. 3.5) and generating the Tethyan ophiolites (Senghör, 1987). In Abbate and his co-workers' interpretation, this event preceding the Pangea dispersal is generated by the same cause, the thermal erosion of the subpangean lithosphere, following Anderson's model (1982). The subsequent opening of the major oceans closes these peripheral basins and induces ophiolite formation (fig. 13.6a). We do not see a systematic record of the initial spreading taking place preferentially at the periphery of Pangea, and the physical rationale for it is dubious. This chronological sequence exists, however, for some ophiolites emplaced along the Pacific margin of North and South America. Chotin (1981) has suggested a comparable model, although for this author the nature of the peripangean basins is different. Chotin proposed that during the Triassic, the eastern Pacific margin was bordered by a series of subduction-related marginal basins like in the present-day western Pacific. These Triassic basins were closed, with ophiolite emplacement as a consequence, by the Westward drift of North and South America in response to the Atlantic opening (fig. 13.5). He explains the recent emplacement of the SW Pacific ophiolites by a similar process, namely the Northward motion of Australia overriding former marginal basins. Present-day oceanic thrustings should be sought in these areas.

The peripheral emplacement of ophiolites along the supercontinent margins can also be explained by reference to Patriat et al.'s (1982) kinematic analysis of Gondwana-Eurasia relative motion. These authors note a major difference between the slow spreading (1-2 cm/yr) Atlantic and Indian Ocean ridges parting the Pangea supercontinent and the fast spreading Neo-Tethys ridge, parting small continental blocks on the northern edge of Gondwana (fig. 3.5). The slow spreading rates of the former ridges is related to the large continental plates which they part. Assuming that the long-lived and active (8 cm/yr of average convergence rate based on the motion of the Tibetan block) subduction zone at the southern margin of Eurasia plays an active role, it is possible that opening of the Neo-Tethys was imposed to adjust the Atlantic and Indian Oceans slow spreading motion to the fast convergence motion along this subduction zone. Thus are created oceanic basins in a highly dynamic environment. The newly created oceanic lithosphere can be locally submitted to compression and be duplicated by oceanic thrusting, preparing its emplacement as the Cretaceous ophiolites of the Middle East. The Jurassic rift, created in the western Mediterranean as an extension of the Atlantic rift (Lemoine, 1980) and rapidly closed by Africa drift, is another favorable site for ophiolite generation.

In contrast, the present situation with a large dispersal of continental massifs is not favorable for ophiolite generation. A possible exception could be the western Pacific, where complex interactions between subduction zones and various oceanic lithospheres may create the required conditions. Indeed, the youngest known ophiolites are located there.

## *BIBLIOGRAPHIE*


Abbate, E., Bortolotti, V. and Principi, G., 1980. Apennine ophiolites : a peculiar oceanic crust. *Ofioliti, Spec. issue*, 1, 59-96.

Abbate, E., Bortolotti, V., Passerini, P. and Principi, G., 1985. The rhythm of phanerozoic ophiolites. *Ofioliti*, 10, 109-138.

Abbotts, I.L., 1978. High-potassium granites in the Masirah ophiolite of Oman. *Geol. Mag.*, 115, 415-425.

Abrams, M.J., Rothery, D.A. and Pontual, A., 1988. Mapping in the Oman ophiolite using enhanced Landsat Thematic Mapper images. *Tectonophysics*, 151, 387-401.

Agrinier, P., 1987. L'évolution de la croûte océanique vue par les rapports isotopiques de l'oxygène, du carbone et de l'hydrogène. *Thèse Doc. Univ. Paris*, 100 p.

Agrinier, P., Javoy, M. and Girardeau, J., in press. Hydrothermal activity in a peculiar oceanic ridge : oxygen and hydrogen isotope evidence in the Xigaze ophiolites (Tibet, China). *Chem. Geol.*, 70.

Ahmad, Z. and Bilgrami, S.A., 1987. Chromite deposits and ophiolites of Pakistan. In : *Evolution of chromium fields*, C.S. Stowe ed., Van Nostrand Reinhold Co., New-York, 239-264.

Aki, K. and Koyanagi, R., 1981. Deep volcanic tremor and magma ascent mechanism under Kilauea, Hawai. *J. geophys. Res.*, 86, 7095-7109.

Alabaster, T., Pearce, J.A., and Malpas, J., 1982. The volcanic stratigraphy and petrogenesis of the Oman ophiolite complex. *Contr. Mineral. Petrol.*, 81, 168-13.

Alabaster, T. and Pearce, J.A., 1985. The interrelationship between magmatic and arc-forming hydrothermal processes in the Oman ophiolite. *Econ. Geol.*, 80, 1-16.

Albarède, F. and Tamagnan, V., 1988. Modelling the recent geochemical evolution of the Piton de la Fournaise volcano, Réunion Island, 1931-1986. *J. Petrol.*, 29, in press.

Allard, B. and Benn, K., in press. Shape preferred orientation analysis using digitized images on a microcomputer. *Computers and Geosciences.*

Allègre, C.J., Montigny, R. and Bottinga Y., 1973. Cortège ophiolitique et cortège océanique, géochimie comparée et mode de genèse. *Bull. Soc. Géol. France*, 7, 5-6.

Allègre, C.J. and Bottinga, Y., 1974. Tholeiite, alkali basalt and ascent velocity. *Nature*, 252, 31-32.

Allègre, C.J. *et al.*, 1984. Structure and evolution of the Himalaya-Tibet orogenic belt. *Nature*, 307, 17-22.

Allègre, C.J. and Turcotte, D.L., 1986. Implications of a two-component marble-cake mantle. *Nature*, 323, 123-127.

Allemann, F. and Peters, T., 1972. The ophiolite-radiolarite belt of the North Oman mountains. *Eclogae geol. Helv.*, 65, 657-698.

Allerton, S. and Vine, F.J., 1987. Spreading structure of the Troodos ophiolite, Cyprus : some paleomagnetic constraints. *Geology*, 15, 593-597.

Amstutz, G.C., 1980. The early history of the term ophiolites and its evolution until 1945. In : *Proc. Inter. Ophiolite Symp.*, Geol. Surv. Dpt. Cyprus, 149-152.

Anderson, R.N. and Hobart, M.A., 1976. The relation between heat flow, sediment thickness, and age in the eastern Pacific. *J. geophys. Res.*, 81, 2968-2989.

Anderson, R.N. *et al.*, 1982. DSDP Hole 504B, the first reference section over 1 km through layer 3 of the oceanic crust. *Nature*, 300, 589-594.

Andreatta, C., 1934. Analisi strutturali di rocce metamorfiche, V. Oliviniti. *Period. Mineral.*, 5, 237-253.

Andrews-Speed, C.P. and Johns, C.C., 1985. Basement diapirism associated with the emplacement of major ophiolite nappes : some constraints. *Tectonophysics*, 118, 43-59.

Anonymous, 1972. Penrose field conference on ophiolites. *Geotimes*, 17, 24-25

Anonymous, 1982. Deep drilling on the Costa Rica Rift, *Joides J.*, 8, 10-18.

Artyushkov, E.V. and Sobolev, S.V., 1984. Physics of the kimberlite magmatism. In : *Petrology*, Elsevier, Amsterdam,11B, 309-322.

Atwater, T.M., 1979. Constraints from the Famous area concerning the structure of the oceanic section. In

: *Deep drilling results in the Atlantic ocean : oceanic crust*. Maurice Ewing Ser., 2, AGU Washington, 33-42.

Aubouin, J., Mattauer, M. and Allègre, C.J., 1977. La couronne ophiolitique périaustralienne : un charriage océanique représentatif des stades précoces de l'évolution alpine. *C.R. Acad. Sci. Paris*, 285, 953-956.

Augé, T. and Roberts, S., 1982. Petrology and geochemistry of some chromitiferous bodies within the Oman ophiolite. *Ofioliti*, 2/3, 133-154.

Aumento, F. and Loubat, H., 1971. The Mid-Atlantic Ridge near 45°N. Serpentinized ultramafic intrusions. *Canad. J. Earth Sci.*, 8, 631-663.

Aumento, F., Melson, W.G. *et al.*, 1977. Initial reports of the Deep Sea Drilling Project. *Proc. ODP. Init. Repts*, 37, 1008.

Auzende, J.M., Olivet, J.L., Charvet, J., Le Lann, A., Le Pichon, X., Monteiro, J., Nicolas, A., and Ribeiro, A., (Groupe CYAGOR), 1978. Sampling and observations of oceanic mantle crust on Gorringe Bank. *Nature*, 273, 45-49.

Auzende, J.M., Polino, R., Lagabrielle, Y. and Olivet, J.L., 1983. Considérations sur l'origine et la mise en place des ophiolites des Alpes occidentales : apport de la connaissance des structures océaniques. *C.R. Acad. Sci. Paris*, 296, 1527-1532.

Auzende, J.M., Bideau, D., Bonatti, E., Cannat, M., Honnorez, J., Lagabrielle, Y., Malavieille, J., Mamaloukas-Frangoulis, V. and Mével, C., submitted. Complete section of slow spreading oceanic crust from the Vema fracture zone in the Atlantic. *Nature*.

Avé Lallemant, H.G., 1976. Structure of the Canyon Mountain (Oregon) ophiolite complex and its implication for sea floor spreading. *Geol. Soc. Am. Sp. Pap.*, 173, 49p.

Avé Lallemant, H.G., 1984. Speculations on the origin of the ophiolites of the northeastern Oregon (U.S.A). *Geol. en Mijnbouw*, 63, 151-158.

Avé Lallemant, H.G. and Carter, N.L., 1970. Syntectonic recrystallization of olivine and modes of flow in the upper mantle. *Geol. Soc. Am. Bull.*, 81, 2203-2220.

Baer, G., and Reches, Z., 1987. Flow patterns of magma in dikes, Makhtesh Ramon, Israel. *Geology*, 15, 569-572.

Bailey, E.B. and McCallien, W.J., 1953. Serpentine lavas, the Ankara melange and the Anatolian thrust. *Trans. r. Soc. Edinburgh*, 62, 403-442.

Bally, A.W. *et al.*, 1980. Notes on the geology of Tibet and adjacent areas-Report of the American Plate Tectonics delegation to the People's Republic of China. *U. S. Geol. Surv. Open file*, 80-501, 99p.

Barbot, J., 1983. Le nickel dans les olivines et les spinelles des roches basiques et ultrabasiques. *Thèse Doc. 3° cycle, Brest*, 204 p.

Bard, J.P., Burg, J.P., Matte, P., Ribeiro, A. 1980. La chaine hercynienne d'Europe occidentale en termes de tectonique des plaques. 26ème C.G.I. Paris. *Mém. BRGM*, 108, 233-246.

Barnes, J., O'Neil, J.R., and Trescases, J.J., 1978. Present-day serpentinization in New-Caledonia, Oman and Yougoslavia. *Geochim. Cosmochim. Acta*, 42, 144-145.

Barnes, R.P. and Andrews, J.R., 1984. Hot or cold emplacement of the Lizard Complex ?. *J. Geol. Soc. London*, 141, 37-39.

Batiza, R. and Margolis, S.H., 1986. Small non-overlapping offsets of the East Pacific Rise. *Nature*, 320, 439-441.

Bayer, R., Courtillot, V., Daignières, M. and Tapponnier, P., 1973. Dorsales médio-océaniques : un modèle évolutif de la zone axiale. *C.R. Acad. Sci. Paris*, 276, 2765-2768.

Beccaluva, L., Ohnenstetter, D. and Ohnenstetter, M., 1979. Geochemical discrimination between ocean-floor and island-arc tholeiites-Application to some ophiolites. *Canad. J. Earth Sci.*, 16, 1874-1882.

Becker, K., Sakai, H. et al., 1988. Site 504 : Costa Rica rift. *Proc. ODP, Init. Repts.*, 111, 35-252.

Benn, K., Laurent, R., 1987. Intrusive suite documented in the Troodos ophiolite plutonic complex, Cyprus. *Geology*, 15, 821-824.

Benn, K., Nicolas, A., and Reuber, I, 1988. Mantle-crust transition zone and origin of wehrlitic magmas :

evidence from the Oman ophiolite. *Tectonophysics*, 151, 75-85.

Benson, W.N., 1926. The tectonic conditions accompanying the intrusion of basic and ultrabasic igneous rocks. *Nat. Acad. Sci. U.S.A.*, 19, 1-90.

Berberian F. and Berberian, M. (1981). Tectono-plutonic episodes in Iran. *A.G.U., Geodynamics Series*, 3, 5-32.

Berger, E.T. and Vannier, M., 1984. Les dunites en enclaves dans les basaltes alcalins des iles océaniques : approche pétrologique. *Bull. Minéral.*, 107, 649-663.

Berger, E.T., 1985. Le concept de dunites résiduelles et la signification pétrologique de certains magmas picritiques. *Bull. Minéral.*, 108, 727-731.

Berger, E.T., 1985. Hétérogéneités pétrographiques du manteau Sud Pacifique sous l'archipel des Australes : mise en evidence et interprétation par l'étude des enclaves ultramafiques. *Bull. Soc. Géol. France*, 8, 57-66.

Bernoulli, D., 1982. Preliminary notes on geological reconnaissance field work in thé Oman Mountains, December 1981. *Unpubl. Report, Earth Science and Research Institute*, Swansea, Great Britain.

Bernoulli D. and Weissert H., 1987. The upper Hawasina nappes in the central Oman Mountains : stratigraphy, palinspastics and sequence of nappe emplacement. *Geodinamica Acta*, 1, 47-58.

Beurrier, M., 1987. Géologie de la nappe ophiolitique de Samail dans les parties orientale et centrale de l'Oman. *Thèse Doc. Etat, Paris 6*, 406p.

Bezzi, A. and Piccardo, G.B., 1971. Structural features of the Ligurian ophiolites : petrologic evidence for the oceanic floor of the Northern Apennines geosyncline ; a contribution to the problem of the alpine type gabbro-peridotite associations. *Mem. Soc. geol. italiana*, X, 53-63.

Biju-Duval, B., Letouzey, J. and Montadert, L., 1979. Variety of margins and deep basins in the Mediterranean. *Mem. Amer. Assoc. Petrol. Geol.*, 29, 293-317.

Bjornsson, A., 1985. Dynamics of crustal rifting in NE Iceland. *J. geophys. Res.*, 90, 151-162.

Blake, M.C., Brothers, R.N. and Lanphere, M.A., 1977. Radiometric ages of blueschits in New Caledonia. *Symp. Int. Geodyn. SW Pacifique, Noumea. Technip Publ., Paris*, 279-282.

Bluck, B.J., Halliday, A.N., Aftalion, M. and Macintyre, R.M., 1980. Age and origins of the Ballantrae ophiolite and its significance to the Caledonian orogeny and Ordovician time scale. *Geology*, 8, 492-495.

Bodinier, J.L., Guiraud, M., Dupuy, C., and Dostal, J., 1986. Geochemistry of basic dikes in the Lanzo massif (Western Alps) : petrogenetic and geodynamic implications. *Tectonophysics*, 128, 77-95.

Bodinier, J.L., 1988. Geochemistry and petrogenesis of the Lanzo peridotite body, Western Alps. *Tectonophysics*, 149, 67-88.

Boillot, G., Girardeau, J. and Kornprobst, J., in press. Rifting of the West Galicia continental margin : a review. *Bull. Soc. géol. France.*

Boillot, G., Recq, M., Winterer, E.L., Meyer, A.W., Applegate, J., Baltuck, M., Bergen, J.A., Comas, M.C., Davies, T.A., Dunham, K., Evans, C.A., Girardeau, J., Goldberg, G., Haggerty, J., Jansa, L.F., Johnson, J.A., Kasahara, J., Loreau, J.P., Luna-Sierra, E., Moullade, M., Ogg, J., Sarti, M., Thurow, J. and Williamson, M., 1987. Tectonic denudation of the upper mantle along passive margins : a model based on drilling results (ODP leg 103, western Galicia margin, Spain). *Tectonophysics*, 132, 335-342.

Boiteau, A. and Saliot, 1972. Métamorphisme de haute pression dans le complexe ophiolitique du Purial (Oriente, Cuba). *C.R. Acad. Sci. Paris*, 274, 2137-2140.

Bonatti, E., 1978. Vertical tectonism in oceanic fracture zones. *Earth and Planet. Sci. Lett.*, 37, 369-379.

Bonatti, E., 1985. Punctiform initiation of seafloor spreading in the Red Sea during transition from a continental to an oceanic rift. *Nature*, 316, 33-37.

Bonatti, E., Honnorez, J., Kirst, P. and Radicati, F., 1975. Meta-gabbros from the Mid-Atlantic Ridge at 06°N : contact hydrothermal dynamic metamorphism beneath the axial valley. *J. Geol.*, 83, 61-78.

Bonatti, E. and Honorez, J., 1976. Sections of the earth's crust in the equatorial Atlantic. *J. geophys. Res.*, 81, 4104-4116.

Bonatti, E., *et al.*, 1983. Zabargad (St John) Island : an uplifted fragment of sub-Red Sea lithosphere. *J.*

*Geol. Soc. (London)*, 140, 677-690.

Bonatti, E. and Seyler, M., 1987. Crustal underplating and evolution in the Red Sea rift : uplifted gabbro/gneiss crustal complexes on Zabargad and Brothers islands. *J. geophys. Res.*, 92, 12803-12821.

Bonhommet, N., Roperch, P. and Calza, F., 1988. Paleomagnetic arguments for block rotations along the Arakapas fault (Cyprus). *Geology*, 16, 422-425.

Bonneau, M., 1976. Esquisse structurale de la Crète alpine. *Bull. Soc. Géol. France*, 2, 351-353.

Bortolotti, V. and Gianelli, G., 1976. Le rocce gabbriche dell'Appennino settentrionale : I. Dati recenti su rapporti primari, posizione stratigrafica ed evoluzione tettonica. *Ofioliti*, 2, 99-105.

Bottinga, Y. and Allègre, C.J., 1978. Partial melting under spreading ridges. *Trans. r. Soc. London*, 288, 501-525.

Boucot, A.J., Dunkle, D.H., Potter, A., Savage, N.M. and Rohr, D., 1974. Middle Devonian orogeny in western North America : a fish and other fossils. *J. Geol.*, 82, 691-708.

Boudier, F., 1978. Structure and petrology of the Lanzo peridotite massif (Piedmont Alps). *Geol. Soc. Amer. Bull.*, 89,1574-1591.

Boudier, F. and Nicolas, A., 1972. Fusion partielle gabbroïque dans la lherzolite de Lanzo (Alpes piémontaises). *Bull. Suisse Minéral. Pétrol.*, 52, 39-56.

Boudier, F. and Nicolas, A., 1977. Structural controls on partial melting in the Lanzo peridotites. *Oregon Dept. Geol. Mineral. Ind.*, H.J.B. Dick ed., 96, 63-78.

Boudier, F. and Nicolas, A., 1980. Stress and strain estimates in the Lanzo peridotite massif (Western Alps). *Coll. Inter. CNRS*, 272, 221-228.

Boudier, F. and Coleman, R.G., 1981. Cross section through the peridotites in the Samail ophiolite, Southeastern Oman. *J. geophys. Res.*, 86, 2573-2592.

Boudier, F. and Michard, A., 1981. Oman ophiolites - The quiet obduction of oceanic crust. *Terra Cognita*, 1, 109-118..

Boudier, F., Nicolas, A. and Bouchez, J.L., 1982. Kinematics of oceanic thrusting and subduction from basal sections of ophiolites. *Nature*, 296, 825-828.

Boudier, F., Nicolas, A., Bouchez, J.L., Crambert, S., Dahl, R. and Juteau, T., 1983. Les ophiolites des nappes de Semail (Oman) : structures internes des massifs de Nakhl et de Rustaq. *Sci. Geol. Strasbourg*, 36, 17-33.

Boudier, F., Bouchez, J.L., Nicolas, A., Cannat, M., Ceuleneer, G., Misseri, M. and Montigny, R., 1985. Kinematics of oceanic thrusting in the Oman ophiolite : model of plate convergence. *Earth and Planet. Sci. Lett.*, 75, 215-221.

Boudier, F. and Nicolas, A., 1985. Harzburgite and lherzolite subtypes in ophiolitic and oceanic environments. *Earth and Planet. Sci. Lett.*, 76, 84-92.

Boudier, F., Ceuleneer, G. and Nicolas, A., 1988. Shear zones, thrusts and related magmatism in the Oman ophiolite : initiation of thrusting on an oceanic ridge. *Tectonophysics*, 151, 275-296.

Boudier, F., Le Sueur, E. and Nicolas, A., 1989. Structures in the Trinity Complex (Eastern Klamath Mountains). an atypical ophiolite. *Geol. Soc. Amer. Bull.*, in press.

Bourgois, J., Azema, J., Tournon, J., Bellon, H, Calle, B., Parra, E., Toussaint, J-F., Glaçon, G., Feinberg, H., De Wever, P., and Origlia, I., 1982. Ages et structures des complexes basiques et ultrabasiques de la façade pacifique entre 3°N et 12°N (Colombie, Panama et Costa-Rica). *Bull. Soc. Géol. France*, 3, 545-554.

Bourgois, J., Calle, B., Tournon, J. and Toussaint J-F., 1982. The Andean ophiolitic megastructures on the Buga-Buenaventura transverse (Western Cordillera - Valle Colombia). *Tectonophysics*, 82, 207-229.

Bourgois, J., Desmet, A., Tournon, J. and Aubouin, J., 1984. Petrology and geochemistry of mafic and ultramafic rocks drilled during DSDP Leg 84 (landward slope of the middle America trench off Guatemala). *Ofioliti*, 9, 27-42.

Bourgois, J., Azema, J., Baumgartner, P.O., Tournon, J., Desmet, A. and Aubouin, J., 1984. The geologic history of the Caribbean-Cocos plate boundary with special reference to the Nicoya ophiolite complex (Costa Rica) and D.S.D.P. results (legs 67 and 84 off Guatemala) : a synthesis.

*Tectonophysics*, 108, 1-32.
Boyd, F.R., 1954. Amphiboles. *Carn. Inst. Washington Dir. Geophys. Lab. Ann. Rep.*, 53, 93-104.
Boyle, J.F. and Robertson, A.H.F., 1984. Evolving metallogenesis at the Troodos spreading axis. *Geol. Soc. London Sp. Pub.*, 13, 169-181.
Bowen, N.L., 1927. The origin of ultrabasic and related rocks. *Amer. J. Sci.*, 14, 89-108.
Brandeis, G., Jaupart, C., and Allègre, C., 1984. Nucleation, crystal growth and the thermal regime of cooling magmas. *J. geophys. Res.*, 89, 161-177.
Brocher, T.M., Karson, J.A. and Collins, J.A., 1985. Seismic stratigraphy of the oceanic Moho based on ophiolite models. *Geology*, 13, 62-65.
Brookfield, M.E., 1977. The emplacement of giant ophiolite nappes I. Mesozoic-Cenozoic examples. *Tectonophysics*, 37, 247-303.
Brongniart, A., 1813. Essai d'une classification minéralogique des roches mélangées. *J. des Mines, Paris*, 199, 5-48.
Brongniart, A., 1821. Sur le gisement ou position relative des ophiolites, euphotides, jaspes, etc. dans quelques parties des Apennins. *Ann. des Mines*, 6, 177-238.
Brongniart, A., 1827. *Classification et caractères minéralogiques des roches homogènes et hétérogènes*. F.G. Levrault éd., Paris, 144 p.
Brothers, R.N., 1974. High pressure schists in Northern New Caledonia. *Contr. Mineral. Petrol.*, 46, 109-127.
Brouxel, M., and Lapierre, H., 1988. Geochemical study of an early paleozoic island-arc-back-arc basin system. Part. 1 : The Trinity ophiolite (Northern California). *Geol. Soc. Amer. Bull.*, 100, 1111-1119.
Brown, G.M., 1956. The layered ultrabasic rocks of Rhum, Inner Hebrides. *Phil. Trans. r. Soc. London*, 240, 1-53.
Browning, P., 1982. The petrology, geochemistry and structure of the plutonic rocks of the Oman ophiolite. *Ph.D., The Open University, Milton Keynes (U.K.)*, 404 p.
Browning, P., 1984. Cryptic variation within the cumulate sequence of the Oman ophiolite : magma chamber depth and petrological implications. *Geol. Soc. London Sp. Pub.* 13, 71-82.
Brunn, I.H., 1956. Contribution à l'étude géologique du Pinde septentrional et d'une partie de la Macédoine occidentale. *Ann. géol. Pays hélléniques*, 7, 358 p.
Brunn, I.H., 1959. La dorsale médio-atlantique et les épanchements ophiolitiques. *C.R. Soc. géol. France*, 8, 234-236.
Bryan, W.B. and Moore, J.G., 1977. Compositional variations of young basalts in the Mid-Atlantic Ridge rift valley near lat 36°40'N. *Geol. Soc. Amer. Bull.*, 88, 556-570.
Bulau, J.R. and Waff, H.S., 1979. Mechanical and thermodynamic constraints on fluid distribution in partial melts. *J. geophys. Res.*, 84, 6102-6108.
Burg, J.P., 1983. Tectogenèse comparée de deux segments de chaines de collision : le Sud Tibet (suture du Zangbo) et la chaine hercynienne en Europe (Massif Central français). *Thèse Doc. Etat, Univ. Montpellier*, 300 p.
Burg, J.P. and Chen, G.M., 1984. Tectonics and structural zonation of Southern Tibet, China. *Nature*, 311, 219-223.
Cabanes, N., and Briqueu, L., 1987. Hydration of an active shear zone : interactions between deformation, metasomatism and magmatism-The spinel-lherzolites from the Montferrier (Southern France) Oligocene basalts. *Earth Planet. Sci. Lett.*, 81, 233-244.
Cakir, U., 1978. Pétrologie du massif ophiolitique de Pozanti-Karsanti (Taurus Cilicien, Turquie) : étude de la partie centrale : *Thèse Doc. Ing., Strasbourg*, 151 p.
Calmant, S., 1987. The elastic thickness of the lithosphere in the Pacific Ocean. *Earth Planet. Sc. Lett.*, 85, 277-288.
Calmant, S. and Cazenave, A., 1986. The effective elastic lithosphere under the Cook-Austral and Society islands. *Earth and Planet. Sci. Lett.*, 77, 187-202.
Calvert, A.J. and Potts, C.G., 1985. Seismic evidence for hydrothermally altered mantle beneath old crust

in the Tydeman fracture zone. *Earth and Planet. Sci. Lett.*, 75, 439-449.
Cameron, W.E., 1985. Petrology and origin of primitive lavas from the Troodos ophiolite, Cyprus. *Contr. Mineral. Petrol.*, 89, 239-255.
Cameron, W.E., Nisbet, V.G. and Dietrich, V.J., 1979. Boninites, komatiites and ophiolites. *Nature*, 280, 550-553.
Campbell, I.M., 1978. Some problems with the cumulus theory. *Lithos*, 11, 311-323.
Cann, J.R., 1969. Spilites from the Carlberg Ridge, Indian Ocean. *J. Petrol.*, 10, 1-19.
Cann, J.R., 1971. Major element variations in ocean-floor basalts. *Phil. Trans. r. Soc. London* , 268, 495-505.
Cann, J.R., 1974. A model for oceanic crustal structure developed. *Geophys. J. r. Astron. Soc.*, 39, 169-187.
Cann, J.R., 1979. Metamorphism in the ocean crust. In : *Deep drilling results in the Atlantic ocean : ocean crust.* Maurice Ewing Ser. 2, AGU, Washington, 230-238.
Cann, J.R. and Simkin, T., 1971. A bibliography of ocean-floor rocks. *Phil. Trans. r. Soc. London*, 268, 737-743.
Cann, J.R. and Strens, M.R., 1982. Black somokers fuelled by freezing Magma. *Nature*, 298, 147-149.
Cannat, M., 1983. Cinematique de charriages oceaniques (Klameth, Semail, Groix) et convergence océanique. *Thèse Doc. 3° cycle, Univ. Nantes*, 145 p.
Cannat, M., 1985. Tectonics of the Seiad massif, northern Klamath Mountains, California. *Geol. Soc. Amer. Bull.*, 96, 15-26.
Cannat, M. and Boudier, F., 1985. Structural study of intra-oceanic thrusting in the Klamath Mountains, northern California : implications on accretion geometry. *Tectonics*, 4, 435-452.
Cannat, M., Juteau, T. and Berger, E., 1988. Petrostructural analysis of the Leg 109 peridotites (hole 670A). *Proc. ODP, Init. Repts*, 106-109.
Carter, N.L. and Avé Lallemant, H.G., 1970. High temperature flow of dunite and peridotite. *Geol. Soc. Amer. Bull.*, 81, 2181-2202.
Casey, J.F. and Karson, J.A., 1981. Magma chamber profiles from the Bay of Islands ophiolite complex. *Nature*, 298, 295-301.
Casey, J.F., Dewey, J.F., Fox, P.J., Karson, J.A. and Rosencrantz, E., 1981. Heterogeneous nature of oceanic crust and upper mantle : a perspective from the Bay of Islands Ophiolite Complex. In : *The Oceanic lithosphere, 7,* C. Emiliani ed., J. Wiley Publ., New York, 305-338.
Casey, J.F., Karson, J.A., Elthon, D., Rosencrantz, E. and Titus, M., 1983. Reconstruction of the geometry of accretion during formation of the Bay of Islands Ophiolite Complex. *Tectonics*, 2, 509-528.
Casey, J.F. and Dewey, J.F., 1984. Initiation of subduction zones along transform and accreting plate boundaries, triple-junction evolution and fore-arc spreading centres-implications for ophiolitic geology and obduction. *Geol. Soc. London Spec. Publ.*, 13, 269-290.
Casey, J.F., Elthon, D.L., Siroky, F.X., Karson, J.A. and Sullivan, J., 1985. Geochemical and geological evidence bearing on the origin of the Bay of Islands and Coastal Complex ophiolites of Western Newfoundland. *Tectonophysics*, 116, 1-40.
Cashman, S.M., 1980. Devonian metamorphic event in the Northeastern Klamath Mountains, California. *Geol. Soc. Amer. Bull.*, 91, 453-459.
Cassard, D., 1980. Structure et origine des gisements de chromite du Massif du Sud (ophiolite de Nouvelle Calédonie). Guides de prospection. *Thèse Doc. 3° Cycle, Univ. Nantes* , 152p.
Cassard, D., Nicolas, A., Rabinowicz, M., Moutte, M., Leblanc, M. and Prinzhofer, A., 1981. Structural classification of chromite pods in southern New Caledonia. *Econ. Geol.*, 76, 805-831.
Ceuleneer, G., 1986. Structure des ophiolites d'Oman : flux mantellaire sous un centre d'expansion océanique et charriage à la dorsale. *Thèse Doc. Univ. Nantes*, 217 p.
Ceuleneer, G. and Nicolas, A., 1985. Structures in podiform chromite from the Maqsad district (Sumail ophiolite, Oman). *Mineralium Depos.*, 20, 177-185.
Ceuleneer, G., Nicolas, A. and Boudier, F., 1988 . Mantle flow patterns at an oceanic spreading centre :

the Oman peridotites record. *Tectonophysics*, 151, 1-26.

Challis, G.A., 1965. High-temperature contact metamorphism at the Red Hills ultramafic intrusion - Wairau Valley - New Zealand. *J. Petrol.*, 6, 395-419.

Challis, G.A. and Guillon, J.H., 1971. Etude comparative à la microsonde électronique du clinopyroxene des basaltes et des péridotites de Nouvelle Calédonie. Possibilité d'une origine commune de ces roches. *Bull. B.R.G.M.*, 4, 39-45.

Chamot-Rooke, N., 1988. Le bassin de Shikoku, de sa formation à l'écaillage intra-océanique : évolution tectonique et mécanique. *Thèse Doc., Univ. Paris*, 220 p.

Chamot-Rooke, N. and Le Pichon, X., 1989. Zenisu Ridge : mechanical model of formation. *Tectonophysics*, in press.

Chang Chengfa, Zheng Xialand and Pan Yasin, 1977. The geological history, tectonic zonation and origin of uplifting of Himalayas. *Inst. Geol. Acad. Sinica, Beijing*, 1-17.

Cherchi, A. and Schroeder, R., 1980. *Palorbitolinoides hedini* n. gen.n.sp., grands foraminifères du Crétacé inférieur du Tibet méridional. *C.R. Acad. Sci. Paris*, 291, 385-388.

Chernysheva, V.I. and Rudnick, G.B., 1970. Serpentinized varieties of plagioclase lherzolite from the rift zone of the Western Indian Ocean Ridge. *Akad. Nauk. SSSR*, 194, 191-193.

Choukroune, P., Francheteau, J. and Hekinian, R., 1984. Tectonics of the East Pacific Rise near 12°50'N : a submersible study. *Earth and Planet. Sci. Lett.*, 68, 115-127.

Christensen, N.I., 1978. Ophiolites, Seismic velocities and oceanic crustal structure. *Tectonophysics*, 47, 131-157.

Christensen, N.I. and Salisbury, M.H., 1972. Sea floor spreading, progressive alteration of layer 2 basalts, and associated changes in seismic velocities. *Earth and Planet. Sci. Lett.*, 15, 367-375.

Christensen, N.I. and Salisbury, M.H., 1975. Structure and composition of the lower oceanic crust. *Rev. Geophys. Space Phys.*, 13, 57-86.

Christensen, N.I. and Smewing, J.D., 1981. Geology and seismic structure of the Northern section of the Oman ophiolite. *J. geophys. Res.*, 86, 2545-2555.

Christiansen, F.G., 1986. Structures of ophiolitic chromite deposits. *Thesis, Aarhus University, Denmark*, 84 p.

Church, W.R. and Gayer, R.A., 1973. The Ballantrae ophiolite. *Geol. Mag.*, 110, 497-592.

Church, W.R. and Riccio, L., 1977. Fractionation trends in the Bay of Islands ophiolite of Newfoundland : polycylic cumulate sequences in ophiolites and their classification. *Canad. J. Earth Sci.*, 14, 1156-1165.

Church, W.R. and Stevens, R.K., 1971. Early Paleozoic complexes of the Newfoundland Appalachians as mantle-oceanic crust sequences. *J. geophys. Res.*, 76, 1460-1466.

Clague, D.A. and Straley, P.F., 1977. Petrologic nature of the oceanic Moho. *Geology*, 5, 133-136.

Clague, D.A, Rubin, J. and Brackett, R., 1981. The age of the garnet amphibolite underlying the Thetford Mines ophiolite, Quebec. *Canad. J. Earth Sci.*, 18, 469-486.

Clague, D.A., Frankel, C.S. and Eaby, J.S., 1985. The age and origin of felsic intrusions of the Thetford Mines ophiolite, Quebec. *Canad. J. Earth Sci.*, 22, 1257-1261.

Cloos, M., 1982. Flow mélanges : numerical modeling and geologic constraints on their origin in the Franciscan subduction complex, California. *Geol. Soc. Amer. Bull.*, 93, 330-345.

Cobbold, P., and Quinquis, H., 1980. Development of sheath folds in shear regimes. *J. Struct. Geol.*, 2, 119-126.

Cocker, J.D., Griffin, B.J. and Muehlenbachs, K., 1982. Oxygen and carbon isotope evidence for seawater-hydrothermal alteration of the Macquarie Island ophiolite. *Earth and Planet. Sci. Lett.*, 61, 112-122.

Coish, R.A., 1977. Ocean floor metamorphism in the Betts Cove Ophiolite, Newfoundland. *Contr. Mineral. Petrol.*, 60, 255-270.

Coish, R.A. and Church, W.R., 1979. Igneous geochemistry of mafic rocks in the Betts Cove Ophiolite, Newfoundland. *Contr. Mineral. Petrol.*, 70, 29-39.

Coleman, P.J., 1966. The Solomon Islands as an island arc. *Nature*, 211, 1249-1251.

Coleman, R.G., 1967. Glaucophane schists from California and New Caledonia. *Tectonophysics*, 4, 479-498.

Coleman, P.J., 1970. Geology of the Soloman and New Hebrides Islands, as part of the Melanesian re-entrant, southwest Pacific. *Pacific Sci.*, 24, 289-314.

Coleman, R.G., 1971. Plate tectonics emplacement of upper mantle peridotites along continental edges. *J. geophys. Res.*, 76, 1212-1222.

Coleman, R.G., 1972. The Colebrooke schist of Southwestern Oregon and its relation to the tectonic evolution of the region. *U.S. Geol. Surv. Bull.*, 1339, 1-61.

Coleman, R.G., 1977. *Ophiolites-Ancient oceanic lithosphere ? Minerals and Rocks, Springer-Verlag*, 12, 229 p..

Coleman, R.G.., 1981. Tectonic setting for ophiolite obduction in Oman. *J. geophys. Res.*, 86, 2497-2508.

Coleman, R.G., 1984. The diversity of ophiolites. *Geol. Mijnbouw*, 63, 141-150.

Coleman, R.G., and Peterman, Z.E., 1975. Oceanic plagiogranite. *J. geophys. Res.*, 80, 1099-1108.

Collins, J.A., Brocher, T.M. and Karson, J.A., 1986. Two-dimensional seismic reflection modeling of the inferred fossil oceanic crust/mantle transition in the Bay of Islands ophiolite. *J. geophys. Res.*, 91, 12520-12538.

Collot, J.Y., Malahoff, A., Recy, J., Latham, G. and Missegue, F., 1987. Overthrust emplacement of New Caledonia ophiolite : geophysical evidence. *Tectonics*, 6, 215-232.

Coombs, D.S., Landis, C.A., Norris, R.J., Sinton, J.M., Borns, D.J. and Craw, D., 1976. The Dun Mountain ophiolite belt, New-Zealand, its tectonic setting, constitution, and origin, with special reference to the southern portion. *Amer. J. Sci.,* 276, 561-603.

Cormier, M.H., Detrick, R.S. and Purdy, G.M., 1984. Anomalously thin crust in oceanic fracture zones : new seismic constraints from the Kane fracture zone. *J. geophys. Res.*, 89, 10249-10266.

Cortesogno, L., Galbiati, B., Principi, G., 1981. Descrizione dettagliata di alcuni caratteristici affioramenti di brecce serpentiniche della Liguria orientale ed interpretazione in chiave geodinamica. *Ofioliti*, 6 : 47-76.

Coulon, C., Maluski, H., Bollinger, C. and Wang, S., 1986. Mesozoic and Cenozoic volcanic rocks from Central and Southern Tibet : $^{39}Ar$-$^{40}Ar$ dating, petrological characteristics and geodynamical significance. *Earth Panet. Sci. Lett.*, 79, 281-302.

Courtillot, V., Achache, J., Landre, F., Bonhommet, N., Montigny, R. and Féraud, G., 1984. Episodic spreading and rift propagation : new paleomagnetic and geochronologic data from the Afar nascent passive margin. *J. geophys. Res.*, 89, 3315-3333.

Cowan, D.S., 1978. Origin of blueschist-bearing chaotic rocks in the Franciscan Complex, San Simeon, California. *Geol. Soc. Amer. Bull*, 89, 1415-1423.

Cowan, D.S., 1985. Structural styles in Mesozoic and Cenozoic mélanges in the western Cordillera of North America. *Geol. Soc. Amer. Bull.*, 96, 451-462.

Cowan, D.S. and Silling, R.M., 1978. A dynamic, scaled model of accretion at trenches and its implications for the tectonic evolution of subduction complexes. *J. geophys. Res.*, 83, 5389-5396.

Cowan, D.S., Botros, M. and Johnson, H.P., 1986. Bookshelf tectonics : rotated crustal blocks within the Sovanco fracture zone. *Geophys. Res. Lett.,* 13, 995-998.

Crane, K., 1985. The spacing of rift axis highs : dependence upon diapiric processes in the underlying asthenosphere? *Earth and Planet. Sci. Lett.*, 72, 405-414.

Crane, K., Aikman, F., Embley, R, Hammond, S. and Malahoff, A., 1985. The distribution of geothermal fields on the Juan de Fuca Ridge. *J. geophys. Res.*, 90, 727-744.

Crawford, A.J., Beccaluva, L. and Serri, G., 1981. Tectono-magmatic evolution of the West Philippine-Mariana region and the origin of boninites. *Earth and Planet. Sci. Lett.*, 54, 346-356.

Dahl, R., 1984. Etude géométrique, pétrologique et géochimique de la séquence crustale de l'ophiolite d'Oman, massif de Rustaq (bloc d'Haylayn) ; un modèle tridimensionnel de zone d'accrétion. *Thèse Doc. 3° cycle, Univ. Clermont-Ferrand*, 275 p.

Dallmeyer, R.D. and Williams, H., 1975. $^{40}Ar/^{39}Ar$ ages from the Bag of Islands metamorphic aureole :

their bearing on the timing of Ordovician ophiolite obduction. *Canad J. Earth. Sci.*, 12, 1685-1690.

Darot, M., 1973. Méthodes d'analyse structurale et cinématique. Application à l'étude du massif ultrabasique de la Sierre Bermeja. *Thèse Doc. 3° cycle, Univ. Nantes*, 120 p.

Davies, H.L., 1971. Peridotite-gabbro-basalt complex in Eastern Papua : an overthrust plate of oceanic mantle and crust. *Bur. Mineral. Res. Austr. Bull.*, 128 p.

Davies, H.L., 1980. Crustal structure and emplacement of ophiolite in Southern Papua, New Guinea. *Coll Int. C.N.R.S., Paris*, 272, 17-34.

Davies, H.L., 1980. Folded thrust fault and associated metamorphics in the Suckling-Dayman massif, Papua New Guinea. *Amer. J. Sci.*, 280, 171-191.

Davies, H.L. and Jaques, A.L., 1984. Emplacement of ophiolite in Papua New Guinea. *Geol. Soc. London Sp. Publ.*, 13, 341-350.

Davies, E.E. and Lister, C.R.B., 1977. Heat flow measured over the Juan de Fuca Ridge : evidence for widespread hydrothermal circulation in a highly heat transportive crust. *J. geophys. Res.*, 82, 4845-4860.

Decandia, F.A. and Elter, P., 1972. La zona ofiolitifera del Braco nel settore compresso tra Levanto e la Val Graveglia (Appennino ligure). *Mem. Soc. Geol. it.*, 11, 503-530.

Den Tex, E., 1969. Origin of ultramafic rocks, their tectonic setting and history. *Tectonophysics*, 7, 457-488.

Deng Wanming, Yang Ruiying, Huang Zhongwiang, Jiang Yong, Guo Yiughnan, Luo Shihna, Zhao Zheulan and Feng Xizhang, 1984. Geochimie des éléments en traces du complexe ophiolitique de la région de Xigaze au Xizang (Tibet). In : *Mission Franco-chinoise au Tibet 1980*, J. Mercier ed., CNRS Publ. Paris, 239-253.

Dercourt, J., 1970. L'expansion océanique actuelle et fossile ; ses implications géotectoniques. *Bull. Soc. géol. France*, XII, 261-317.

De Roever, W.P., 1957. Sind die alpinotypen peridotitmassen vielleicht tektonisch verfrachtete bruchstücke der peridotitschale ? *Geol. Rundschau*, 46, 137-146.

Detrick, R.S., Bulh, P., Vera, E., Mutter, J., Orcutt, J., Madsen, J. and Brocher, T., 1987. Multichannel seismic imaging of a crustal magma chamber along the East Pacific Rise. *Nature*, 326, 35-41.

Dewey, J.F., 1976. Ophiolite obduction. *Tectonophysics*, 31, 93-120.

Dewey, J.F. and Bird, J.M., 1970. Mountain belts and the new global tectonics. *J. geophys. Res.*, 75, 2625-2647.

Dewey, J.F. and Bird, J.M., 1971. Origin and emplacement of the ophiolite suite : Appalachian ophiolites in Newfoundland. *J. geophys. Res.*, 76, 3179-3206.

Dewey, J.F., and Kidd, W.S.F., 1977. Geometry of plate accretion. *Geol. Soc. Amer. Bull.*, 88, 960-968.

Dick, H.J.B., 1977. Evidence of partial melting in the Josephine peridotite. *Oregon Dept. Geol. Mineral. Ind.*, 96, 59-62.

Dick, H.J.B. and Fisher, R.L., 1984. Mineralogic studies of the residues of mantle melting : abyssal and alpine-type peridotites. In : *Kimberlites II : The mantle and Crust-mantle relationships*, J. Kornprrobst ed., Elsevier, 295-308.

Dick, H.J.B., Fisher, R.L. and Bryan, W.B., 1984. Mineralogical variability of the uppermost mantle along mid-oceanic ridges. *Earth and Planet. Sci. Lett.*, 69, 88-106.

Dick, H.J.B. and Bullen, T., 1984. Chromian spinel as a petrogenetic indicator in abyssal and alpine-type peridotites and spatially associated lavas. *Contr. Mineral. Petrol.*, 86, 54-76.

Dickey, J.S., 1970. Partial fusion products in alpine type peridotites : Serrania de la Ronda and other examples. *Mineral. Soc. Amer. Spec. Pap.*, 3, 33-49.

Dickinson, W.R., 1970. Relations of andesite, granites and derivative sandstones to arc-trench tectonics. *Rev. Geophys. Space Phys.*, 8, 813-860.

Dickinson, W.R., 1977. Paleozoic plate tectonics and the evolution of the Cordilleran continental margin. *Pac. Sect. Soc. Econ. Paleont. Mineral.*, 1, 137- 156.

Dickinson, W.R., 1979. Mesozoic forearc basin in Central Oregon. *Geology*, 7, 166-170.

Dickinson, W.R., 1981. Plate tectonics and the continental margin of California. In : *The geotectonic development of California*, W.G. Ernst ed., Rubey vol.I., Prentice-Hall Publ., 1-28.

Dietrich, V., Emmermann, R., Oberhansli, R. and Puchelt, H., 1978. Geochemistry of basaltic and gabbroic rocks from West Mariana basin and the Mariana trench. *Earth and Planet. Sci. Lett.*, 39, 127-144.

Donaldson, C.H., 1985. A comment on crystal shapes resulting from dissolution in magma. *Mineral. Mag.*, 49, 129-132.

Dobretsov, N.L., 1978. Glaucophane metamorphism and ophiolites. *Pacific Geol.*, 13, 87-100.

Dobretsov, N.L., Coleman, R.G., Liou, J.G. and Maruyama, S., 1987. Blueschist belts in Asia and possible periodicity of blueschist facies metamorphism. *Ofioliti*, 12, 445-456.

Dubertret, L., 1953. Geologie des roches vertes du Nord-Ouest de la Syrie et du Hatay (Turquie). Mus. Natl. Hist. nat. Paris, *Notes Mém. Moyen-Orient*, 6, 179 p.

Dubuisson, G., Hirn, A., Girardeau, J., Mercier, J.C.C. and Veinante, J.L., 1988. Multiple Variscan nappes in Limousin, western Massif Central, France : geophysical constraints to the geological model and geodynamic implications. *Tectonophysics*, 147, 19-31.

Dungan, M.A. and Avé Lallemant H.G., 1977. Formation of small dunite bodies by metasomatic transformation of harzburgite in the Oregon Mountain ophiolite, Northeast Oregon. *Oregon Dept. Geol. Mineral. Ind. Bull.*, 96, 109-128.

Dunning, G.R. and Pedersen, R.B., 1988. U/Pb ages of ophiolites and arc-related plutons of the Norwegian Caledonides : implications for the development of Iapetus. *Contr. Mineral. Petrol.*, 98, 13-23.

Dupuy, C., Dostal, J. and Leblanc, M., 1981. Geochemistry of an ophiolitic complex from New Caledonia. *Contr. Mineral. Petrol.*, 76, 77-83.

Earle, M.M., 1980. A note on the relationship between inclined isothermal surfaces and subduction-zone metamorphism. *Tectonophysics*, 68, 313-324.

Ecors-Crop Gravity Group., in press. Gravity modelling along the Ecors-Crop vertical seismic reflection profile through the Western Alps. *Tectonophysics*.

Einarsson, P., and Brandsdottir, B., 1980. Seismological evidence for lateral magma intrusion during the July 1978 deflation of the Krafla volcano in NE-Iceland. *J. Geophys.*, 47, 160-165.

Elthon, D., 1979. High magnesia liquid as the parental magma for ocean floor basalts. *Nature*, 278, 514-518.

Elthon, D., Casey, J.F. and Komor, S., 1982. Mineral chemistry of ultramafic cumulates from the North Arm Mountain Massif of the Bay of Islands Ophiolite : implication for high pressure fractionation of oceanic basalts, *J. geophys. Res.*, 87, 8717-8734.

Elthon, D., Casey, J.F. and Komor, S., 1984. Cryptic mineral chemistry variations in a detailed traverse through the cumulate ultramafic rocks of the North Arm Mountain Massif of the Bay of Islands Ophiolite Complex. *Geol. Soc. London Sp. Publ.*, 13, 83-97.

Elthon, D., Lawrence, J.R., Hanson, R.E. and Stern, C., 1984. Modelling of oxygen-isotope data from the Sarmiento ophiolite complex, Chile. *Geol. Soc. London Sp. Publ.*, 13, 185-197.

Elthon, D., Karson, J.A., Casey, J.F., Sullivan, J. and Siroky, F.X., 1986. Geochemistry of diabase dikes from the Lewis Hills Massif, Bay of Islands Ophiolite : evidence for partial melting of oceanic crust in transform faults. *Earth and Planet. Sci. Lett.*, 78, 89-103.

Emerman, S.H. and Turcotte, D.L., 1984. Diapiric penetration with melting. *Phys. Earth Planet. Int.*, 36, 276-284.

Engin, T., Ozkoçak, O. and Artan, U., 1987. General geological setting and character of chromite deposits in Turkey. In : *Evolution of chromium fields*, Stowe C.W. ed., New-York, Van Nostrand Reinhold Co. 195-219.

Ernewein, M., Pflumio, C. and Whitechurch, H., 1988. The death of accretion zone as evidenced by the magmatic history of the Sumail ophiolite (Oman). *Tectonophysics*, 247-274.

Ernst, T.G., 1935. Olivinknollen der basalte als bruchstücke alter olivinfelse. *Nachr. Akad. Wiss. Göttingen, Math. Phys.*, 1, 147-154.

Ernst, W.G., 1975. Schematics of large-scale tectonics and age progressions in Alpine and Circumpacific blueschist belts. *Tectonophysics*, 26, 229-246.

Ernst, W.G., 1981. Petrotectonic settings of glaucophane schist belts and some implications for Taiwan. *Geol. Soc. China Mem.*, 4, 229-267.

Ernst, W.G., 1983. Mountain building and metamorphism : a case history from Taiwan. In : *Mountain Building*, Hsü ed., Academic Press, London, 247-256.

Ernst, W.G. and Liou, J.G., 1984. Summary of oceanic metamorphism and inferred tectonic history of the East Taiwan ophiolite. *Ofioliti*, 9, 223-244.

Espirat, J.J., 1963. Etude géologique de régions de Nouvelle Calédonie septentrionale (extrémité Nord et versant Est). *Thèse Doc. Etat, Univ. Clermont-Ferrand*, 217 p.

Evans, C.A., 1985. Magmatic metasomatism in peridotites from the Zambales ophiolite. *Geology*, 13, 166-169.

Evans, B. and Goetze, C., 1979. The temperature variation of hardness of olivine and its implication for polycrystalline yield stress. *J. geophys. Res.*, 84, 5505-5524.

Evans, C.A. and Hawkins, J.W., 1979. Mariana arc-trench system : petrology of 'seamounts' on the trench-slope break. *EOS*, 60, 968.

Evans, C.A., Hawkins, J.W. and Moore, G.F., 1983. Petrology and geochemistry of ophiolitic and associated volcanic rocks on the Talaud Islands, Molucca sea collision zone, Northeast Indonesia. *Geodynamics series, AGU Washington*, 11, 159-172.

Evans, C.A. and Girardeau, J., in press. Galicia Margin peridotites : undepleted abyssal peridotites from the N. Atlantic. *Proc. ODP, final Reports*, 103-?.

Evarts, R.C. and Schiffman, P., 1983. Submarine hydrothermal metamorphism of the Del Puerto ophiolite, California. *Amer. J. Sci.*, 283, 289-340.

Fedotov, S.A., 1978. Ascent of basic magmas in the crust and the mechanism of basaltic fissure eruptions. *Inter. Geol. Rev.*, 20, 33-48.

Feigenson, M.D., 1986. Constraints on the origin of Hawaiian lavas. *J. geophys. Res.*, 91, 9383-9393.

Feininger, T., 1981. Amphibolite associated with the Thetford Mines ophiolite complex at Belmina Ridge, Quebec. *Canad. J. Earth Sci.*, 18, 1878-1892.

Ferguson, J., and Pulvertaft, T.C.R., 1963. Contrasted styles of igneous layering in the Gardar Province of South Greenland. *Mineral. Soc. Amer. Spec. Pap.*, 1, 10-21.

Ferrière, J., 1982. Paléogéographies et tectoniques superposées dans les Hellénides internes : les massifs de l'Othrys et du Pelion (Grèce continentale). *Soc. Géol. Nord*, 8, 970 p.

Ferrière, J., 1985. Nature et développement des ophiolities hélleniques du secteur Othrys-Pelion. *Ofioliti*, 10, 255-278.

Fisher, R.L. and Engel, C.G., 1969. Ultramafic and basaltic rocks dredged from the nearshore flank of the Tonga trench. *Geol. Soc. Amer. Bull.*, 80, 1373-1378.

Fisher, R.L., Dick, H.J.B., Natland, J.H. and Meyer, P.S., 1984. Mafic/ultramafic suites of the slowly spreading Southwest Indian Ridge : protea exploration of the Antarctic Plate boundary, 24°E-47°E. *Ofioliti*, 9, 147-178.

Fisk, M.R.1986. Basalt magma interaction with harzburgite and the formation of high-magnesium andesites. *Geophys. Res. Lett.*, 13, 467-470.

Fleet, A.J. and Robertson, A.H.F., 1980. Ocean ridge metalliferous and pelagic sediments of the Semail nappe, Oman. *J. Geol. Soc. London*, 137, 403-422.

Fleitout, L. and Froidevaux, C., 1979. Thermal and mechanical evolution of shear zones. *J. Struct. Geol.*, 2, 159-164.

Fleitout, L. and Yuen, D.A., 1984. Steady state, secondary convection beneath lithospheric plates with temperature- and pressure-dependent viscosity. *J. geophys. Res.*, 89, 9227-9244.

Flower, M.F.J. and others, 1977. Cretaceous crust sought. *Geotimes*, 22, 20-22.

Forsyth, D.W., 1977. The evolution of the upper mantle beneath mid-ocean ridges. *Tectonophysics*, 38, 89-118.

Forsyth, D.W. and Wilson, B., 1984. Three dimensional temperature structure of a ridge-transform-ridge system. *Earth and Planet. Sci. Lett.*, 70, 355-362.

Fox, P.J., Schreiber, E., and Peterson, J.J., 1973. The geology of the oceanic crust : compressional wave veolocities of oceanic rocks. *J. Geol. Phys. Res.*, 78, 5155-5172.

Fox, P.J., Detrick, R.S. and Purdy, G.M., 1980. Evidence for crustal thinning near fracture zones : implications for ophiolites. *Proc. Inter. Ophiolite Symp.*, Geol. Survey Dept. Cyprus, 161-168.

Fox, P.J. and Gallo, D.G., 1984. A tectonic model for ridge-transform-ridge plate boundaries : implications for the structure of oceanic lithosphere. *Tectonophysics*, 104, 205-242.

Francheteau, J., Choukroune, P., Hekinian, R., Le Pichon, X. and Needham, H.D., 1976. Oceanic fracture zones do not provide deep sections of the crust. *Canad. J. Earth Sci.*, 13, 1223-1235.

Francheteau, J., Needham, H.D., Choukroune, P., Juteau, T., Seguret, M., Ballard, R.D., Fox, J.P., Normark, W., Carranza, A., Cordoba, D., Guerrero, J., Bougault, H., Cambon, P. and Hékinian, R., 1979. Massive deep-sea sulphide ore deposits discovered on the East Pacific Rise. *Nature*, 277, 523-528.

Francheteau, J. and Ballard, R.D., 1983. The East Pacific Rise near 21°N, 13°N and 20°S : inferences for along-strike variability of axial processes of the Mid-Ocean Ridge. *Earth and Planet. Sci. Lett.*, 64, 93-116.

Franchi, S., 1902. Contribuzione allo studio delle roccie a glaucofane e del metamorfismo onde ebbero origine nelle regione liguro-alpine occidentale. *Boll. R. Comitato Geol. Italia*, 33, 255-318.

Freer, R., 1981. Diffusion in silicate minerals and glasses : a data digest guide to the literature. *Contr. Mineral. Petrol.*, 76, 440-454.

Frey, F.A., Suen, C.J., Stockman, H.W., 1985. The Ronda high temperature peridotite : geochemistry and petrogenesis. *Geochim. Cosmochim. Acta*, 49, 2469-2491.

Fuchs, K. 1983. Recently formed elastic anisotropy and petrological models for the continental subcrustal lithosphere in Southern Germany. *Phys. Earth Planet. Inter.*, 31, 93-118.

Fuis, G.S., Zucca, J.J., Mooney, W.D. and Milkereit, B., 1987. Geologic interpretation of seismic refraction results in northeastern California. *Geol. Soc. Am. Bull.*, 98, 53-65.

Fuis, G.S., Mooney, W.D., Healy, J.H., McMechan G.A. and Lutter, W.J., 1984. A seismic refraction survey of the Imperial Valley Region, California. *J. geophys. Res.*,, 89, 1165-1189.

Fujii, T. and Bougault, H., 1983. Melting relations of a magnesian abyssal tholeiite and the origin of MORBs. *Earth and Planet. Sci. Lett.*, 62, 283-295.

Fujii, T. and Scarfe, C.M., 1985. Composition of liquids coexisting with spinel lherzolite at 10 kbar and the genesis of MORBs. *Contr. Mineral. Petrol.*, 90, 18-28.

Furnes, H., Roberts, D., Sturt, B.A., Thon, A. and Gale, G.H., 1980. Ophiolite fragments in the Scandinavian Caledonides. *Proc. Inter. Ophiolite Symp.*, Geol. Survey Dept. Cyprus, 582-600.

Furnes, H., Pedersen, R.B. and Stillman, C.J., 1988. The Leka Ophiolite Complex, central Norwegian Caledonides : field characteristics and geotectonic significance. *J. Geol. Soc. (London)*, 145, 401-412.

Gagnon, D. and Jamieson, R.A., 1985. Geology of the Mont Albert region, Gaspe Peninsula, Quebec. *Current Research, Part A, Geol. Surv. Canad. Paper*, 85-1A, 783-788.

Gansser, A., 1966. The Indian Ocean and the Himalayas, a geological interpretation. *Eclogae geol. Helv.*, 59, 832-848.

Gansser, A., 1974. The ophiolitic mélange, a world-wide problem on Tethyan examples. *Eclogae geol. Helv.*, 67, 479-507.

Gansser, A., 1977. The great suture zone between Himalaya and Tibet : a preliminary account. *Coll. Inter. CNRS Paris*, 268, 191-

Garcia, M.O., 1982. Petrology of the Rogue River island arc complex, Southwestern Oregon. *Amer. J. Sci.*, 282, 783-807.

Gass, I.G., 1967. The ultramafic volcanic assemblage of the Troodos Massif, Cyprus. In : *Ultramafic and related rocks.* John Wiley, New York, 121-134.

Gass, I.G., 1968. Is the Troodos massif of Cyprus a fragment of the Mesozoic ocean floor ? *Nature*, 220, 39-42.

Gass, I. and Smewing, J.O., 1973. Intrusion, extrusion and metamorphism at constructive plate margins. *Inter. Counc. Sci. Uni.* W.G. 4, Reykjavik Meet., Rep. 5.

Gansser, A., 1977. The great suture zone between Himalaya and Tibet. A prelimary account. *Coll. Inter. CNRS*, Paris, 268, 181-191.

Gealey, W.K., 1980. Ophiolite obduction mechanism. *Proc. Inter. Ophiolite Symp.*, Geol. Surv. Dept. Cyprus, 228-243.

Gente, P., 1987. Etude morphostructurale comparative de dorsales océaniques à taux d'expension variés. *Thèse Doc. Univ. Brest*, 371p.

George, R.P.J., 1978. Structural petrology of the Olympus Ultramafic Complex in the Troodos Ophiolite, Cyprus. *Geol. Soc. Amer. Bull.*, 89, 845-865.

Gerlach, D.C., Avé Lallemant, H.G. and Leeman, W.P., 1981a. An island arc origin for the Canyon Mountain Ophiolite Complex, Eastern Oregon, U.S.A. *Earth and Planet. Sci. Lett.*, 53, 255-265.

Gerlach, D.C., Leeman, W.P. and Avé Lallemant, H.G., 1981b. Petrology and geochemistry of plagiogranite in the Canyon Mountain Ophiolite, Oregon. *Contr. Mineral. Petrol*, 77, 82-92.

Gettrust, J.F., Furukawa, K. and Kempner, W.B., 1982. Variation in young oceanic crust and upper mantle structure. *J. geophys. Res.*, 87, 8435-8445.

Ghent, E.D. and Stout, M.Z., 1981. Metamorphism at the base of the Semail ophiolite, Southeastern Oman mountains. *J. geophys. Res.*, 86, 2557-2573.

Gianelli, G. and Principi, G., 1977. Northern Appennine ophiolite : an ancient transcurrent fault. *Boll. Soc. geol. it.*, 96, 53-58.

Gibert, D. and Courtillot, V., 1988. Geoid roughness and long-wavelength segmentation of the South Atlantic spreading ridge. *Nature*, 333, 255-258.

Gibson, I.L., 1979. Crust of oceanic affinity in Iceland. *Nature*, 281, 347-351.

Gillis, K.M. and Robinson, P.T., 1988. Distribution of alteration zones in the upper oceanic crust. *Geology*, 16, 262-266.

Ginzburg, A., Whitmarsh, R.B., Roberts, D.G., Montadert, L., Camus, A.L., and Avedik, F., 1985. The deep seismic structure of the northern continental margin of the Bay of Biscay. *Ann. Geophys.*, 3, 499-515.

Girardeau, J., 1979. Structure des ophiolites de Terre Neuve et modèle de croûte océanique. *Thèse Doc. Univ. Nantes*, 154 p.

Girardeau, J., 1982. Tectonic structures related to thrusting of ophiolitic complexes : the White Hills Peridotite, Newfoundland. *Canad. J. Earth Sci.*, 19, 709-722.

Girardeau, J. and Nicolas, A., 1981. Structures in two of the Bay of Islands (Newfoundland) ophiolite massifs : a model for oceanic crust and uppermantle. *Tectonophysics*, 77, 1-34.

Girardeau, J. and Mevel, C., 1982. Amphibolitized sheared gabbros from ophiolites as indicators of the evolution of the oceanic crust : Bay of Islands, Newfoundland. *Earth and Planet. Sci. Lett.*, 61, 151-165.

Girardeau, J. Marcoux, J. and Zao Yougong, 1984. Lithologic and tectonic environment of the Xigaze ophiolite (Yarlung Zangbo suture zone, Southern Tibet, China) and kinematics of its emplacement. *Eclogae geol. Helv.*, 77, 153-170.

Girardeau, J. and Mercier, J.C.C, 1985. Structure of the Xigaze ophiolite, Yarlung Zangbo suture zone, Southern Tibet, China : genetic implications. *Tectonics*, 4, 267-288.

Girardeau, J., Mercier, J.C.C. and Wang Xibin, 1985a. Petrology of the mafic rocks of the Xigaze ophiolite, Tibet. Implications for the genesis of the ocenic lithosphere. *Contr. Mineral. Petrol.*, 90, 309-321.

Girardeau, J., Mercier,J.C.C. and Zao Yougong, 1985b. Origin of the Xigaze ophiolite,Yarlung Zangbo suture zone, Southern Tibet. *Tectonophysics*, 119, 407-433.

Girardeau, J., Dubuisson, G. and Mercier, J-C., 1986. Cinématique de mise en place des ophiolites et nappes cristallophylliennes du Limousin, Ouest du Massif central français. *Bull. Soc. géol. France*, 5, 849-860.

Girardeau, J. and Mercier J.C.C., 1988. Petrology and texture of the ultramafic rocks of the Xigaze ophiolite (Tibet), constraints for mantle structure beneath slow spreading ridges. *Tectonophysics*, 147, 33-58.

Girardeau, J., Evans, C.A., and Beslier, M.O., in press. Structural analysis of plagioclase-bearing peridotites emplaced at the end of continental rifting : hole 637D, ODP leg 103 on the Galicia margin. *Proc. ODP, final rept.*

Glasser, E., 1903. Rapport à M. le Ministre des Colonies sur les richesses minérales de la Nouvelle-Calédonie. *Ann. Mines Paris*, 4, 392 p.

Glennie, K.W., Boeuf, M.G.A., Hughes Clarke, M.W., Moody-Stuart, M., Pilaar, W.F.H. and Reinhardt, B.M., 1973. Late Cretaceous nappes in Oman mountains and their geologic evolution. *Amer. Assoc. Petrol. Geol. Bull.*, 57, 5-27.

Glennie, K.W., Boeuf, M.G.A., Hughes Clarke, M.W., Moody-Stuart, M., Pilaar, W.F.H. and Reinhardt, B.M., 1974. Geology of the Oman mountains. *Verhandelingen van het Koninklijk Nederlands Geologisch Mijnbouwkundig Genootschap*, 31, 423p.

Goetze, C., 1975. Sheared lherzolites : from the point of view of rock mechanics. *Geology*, 3, 172-173.

Goffé, B. Michard, A. Kienast, J.R., and Le Mer, O., 1988. A case of obduction-related high-pressure, low-temperature metamosphism in upper crustal nappes, Arabian continental margin, Oman : P-T paths and kinematic interpretation. *Tectonophysics*, 151, 363-386.

Golding, H.G. and Johnson, K.R., 1971. Variation in gross chemical composition and related physical properties of podiform chromite in the Coolac district, Australia. *Econ. Geol.*, 66, 1017-1027.

Goode, A.D.T., 1976. Sedimentary structures and magma current velocities in the Kalka layered intrusion, central Australia. *J. Petrol.*, 17, 546-558.

Göpel, C., Allègre, C.J. and Rong Hua Xu, 1984. Lead isotope study of the Xigaze ophiolite (Tibet) : the problem of the relationship between magmatites (gabbros, dolerites, lavas) and tectonites (harzburgites). *Earth and Planet. Sci. Lett.*, 6, 301-310.

Goullaud, L., 1977. Structure and petrology in the Trinity mafic- ultramafic complex, Klamath Mountains, Northern California. *73rd meet. Geol. Soc. Amer. Cordilleran Sect.*, 112-133.

Graciansky, C., de, 1972. Recherches géologiques dans le Taurus lycien occidental. *Thèse Doc. Etat, Univ. Orsay*, 762 p.

Graham, G.M., 1980 a. Evolution of a passive margin and nappe emplacement in the Oman mountains. *Proc. Inter. Ophiolite Symp.*, Geol. Survey Dept. Cyprus, 414-423.

Grandjacquet, C., and Haccard, D., 1977. Position structurale et rôle paléographique de l'Unité du Bracco au sein du contexte ophiolitique liguro-piémontais (Appenin, Italie). *Bull. Soc. géol. France*, 19, 601-908.

Green, A.G., Clowest, R.M., Yorath, Spencer, C., Kanasewich, E.R., Brandon, M.T., and Sutherland Brown, A., 1986. Seismic reflection imaging of the subducting Juan de Fuca Plate. *Nature*, 319, 210-213.

Green, D.H., 1963. Alumina content of enstatite in a Venezuelan high- temperature peridotite. *Geol. Soc. Amer. Bull.*, 74, 1397-1402.

Green, D.H., 1964. The petrogenesis of the high-temperature peridotite intrusion in the Lizard area, Cornwall. *J. Petrology*, 5, 134-188.

Green, D.H. and Ringwood, A.E. (1967). The genesis of basaltic magmas. *Contr. Mineral. Petrol.*, 15, 103-190.

Green, D.H., Hibberson, W.O. and Jacques, A.L., 1979. Petrogenesis of mid-ocean basalts. In : *The Earth : its origin, structure and evolution*. H.W. Elhinny ed., Academic Press London, 265-299.

Greenbaum, D., 1972. Magmatic processes at ocean ridges : evidence from the Troodos massif, Cyprus. *Nature*, 238, 18-21.

Gregory, R.T., 1980. Oxygen and hydrogen isotope study of the Samail ophiolite, Oman : implications for origin and hydrothermal alteration of the oceanic crust, *Ph.D., Calif. Inst. Tech.*, Pasadena.

Gregory, R.T. and Taylor, H.P., 1981. An oxygen isotope profile in a section of Cretaceous oceanic crust, Samail Ophiolite, Oman : evidence for $\delta^{18}O$ buffering of the oceans by deep (>5 km) seawater-hydrothermal circulation at mid-ocean ridges. *J. geophys. Res.*, 86, 2737-2755.

Griscom, A., 1977. Areomagnetic and gravity interpretation of the Trinity ophiolitic complex, northern California. *Geol. Soc. Amer., Abstr. with Programs* , 9, 426-427.

Gudmunsson, A., 1983. Form and dimensions of dykes in eastern Iceland, *Tectonophysics*, 95, 295-307.
Gudmunsson, A., 1986. Formation of crustal magma chambers in Iceland. *Geology*, 14, 164-166.
Guillon, J.H., and Saos, J.L., 1971. Les règles de distribution des sulfures cupronickélifères dans les massifs péridotiques de Nouvelle-Calédonie. *ORSTOM-SLN, internal report*, 27 p.
Hale, L.D., Morton, C.J. and Sleep, N.H., 1982. Reinterpretation of seismic reflection data over the East Pacific rise. *J. geophys. Res.*, 87, 7707-7717.
Hall, J.M. and Robinson, P.T.. 1979. Deep crustal drilling in the north Atlantic Ocean. *Science*, 204, 573-586.
Hamelin, B., Dupré, B. and Allègre, C.J., 1984. Lead-strontium isotopic variations along the East Pacific Rise and the Mid-Atlantic ridge : a comparative study. *Earth and Planet. Sci. Lett.*, 67, 340-350.
Hamilton, D.L., Burnham, C.W. and Osborn, E.F., 1984. The solubility of water and effects of oxygen fugacity and water content on crystallization of mafic magmas. *J. Petrol.*, 5, 21-39.
Hamilton, W., 1969. Mesozoic California and the underflow of Pacific mantle. *Geol. Soc. Amer. Bull.*, 80, 2409-2430.
Hamlyn, P.R., and Bonatti, E., 1980. Petrology of mantle-derived ultramafics from the Owen fracture zone, Northwest Indian Ocean : implications for the nature of the oceanic upper mantle. *Earth Planet. Sci Lett.*, 48, 65-79.
Hardee, H.C., 1982. Incipient magma chamber formation as a result of repetitive intrusions. *Bull. Volcanol.*, 45-1.
Harper, G.D., 1982. Evidence for large-scale rotations at spreading centers from the Josephine ophiolite. *Tectonophysics*, 82, 25-44.
Harper, G.D., 1983. Comment on 'Formation of uppermost oceanic crust' by Eric Rosencrantz. *Tectonics*, 2, 499-501.
Harper, G.D., submitted. Freezing magma chambers and amagmatic extension in the Josephine ophiolite. *Geology*.
Harper, G.D., Bowman, J.R. and Kuhns, R., 1988. A field, chemical, and stable isotope study of subseafloor metamorphism of the Josephine ophiolite, California-Oregon. *J. geophys. Res.*, 93, 4625-4656.
Hawkins, J.W. 1980. Petrology of back-arc basins and island arcs : their possible role in the origin of ophiolites. *Proc. Inter. Ophiolite Symp.*, Geol. Surv. Dept. Cyprus. 244-254.
Hawkins, J.W. and Evans; C.A., 1983. Geology of the Zambales range, Luzon, Philippine Islands : ophiolite derived from an Island arc-back arc basin pair. *Geophys. Monogr. 27, AGU Washington*, 95-123.
Haymon, R.M., Koski, R.A. and Sinclair, C., 1984. Fossils of hydrothermal vent worms from Cretaceous sulfide ores of the Samail Ophiolite, Oman. *Science*, 223, 1407-1409.
Hébert, R., Bideau, D. and Hekinian, R., 1983. Ultramafic and mafic rocks from the Garret transform fault near 13°30'S on the East Pacific Rise : igneous petrology. *Earth and Planet. Sci. Lett.*, 65, 107-125.
Hékinian, R., 1982. *Petrology of the ocean floor*. Elsevier Oceanography Series, 33, 393 p.
Helgason, J., and Zentilli, M., 1985. Field characteristics of laterally emplaced dikes : anatomy of an exhumed Miocene dike swarm in Reydarfjordur, Eastern Iceland. *Tectonophysics*, 115, 247-274.
Helmstaedt, H., Allen, J.M., 1977. Metagabbro-norite from Deep Sea Drilling Project hole 334 : and example of high-temperature deformation and recrystallisation near the Mid-Atlantic Ridge. *Canad. J. Earth Sci.*, 14(4), 886-898.
Hernandez, M., 1979. Datos preliminaros sobre las caracteristicas petrograficas de las rocas del macizo Sierra Purial. *Minerias en Cuba*, 5, 2-7.
Hernandez-Pacheco, A., 1987. Estudio geografico y geoquimico del macizo ultramafico de Ojen (Malaga). *Estudios Geol.*, 23, 85-143.
Herron, T.J., Stoffa, P.L., and Bulh, P., 1980. Magma chamber and mantle reflection-East Pacific Rise. *Geophys. Res. Lett.*, 7, 989- 992.
Hess, H.H., 1938. A primary peridotite magma. *Amer. J. Sci.*, 35, 321-344.
Hess, H.H., 1955. Serpentines, orogeny and epeirogeny. *Geol. Soc. Amer. Spec. Pap.*, 62, 391-408.

Hess, H.H., 1960. Caribbean Research Project : Progress report. *Geol. Soc. Amer. Bull.*, 71, 235-240.

Hess, H.H., 1962. History of ocean basins. In Petrological Studies : A volume in Honor of A.F. Buddington, (A.E.J. Engel et al. ed.). *Geol. Soc. Amer. Mem. Publ.*, 599-620.

Hess, H.H., 1965. Mid-oceanic ridges and tectonics of the sea floor. In : *Proc. of the 17th Symp. of the Colston Research Society.* Butterworths, London, 317-334.

Hickey, R.L. and Frey, F.A., 1982. Geochemical characteristics of boninite series volcanics ; implications for their source. *Geochim. Cosmochim. Acta*, 46, 2099-2116.

Himmelberg, G.R. and Loney, R.A., 1980. Petrology of ultramafic and gabbroic rocks of the Canyon Mountain Ophiolite, Oregon. *Amer. J. Sci.*, 280, 232-268.

Hock, M. and Friedrich, G., 1985. Structural features of ophiolitic chromitites in the Zambales Range, Luzon, Philippines. *Mineralium Depos.*, 20, 290-301.

Hostetler, P.B., Coleman, R.G., Mumpton, F.A. and Evans, B.W., 1966. Brucite in Alpine serpentinites. *Amer. Mineral.* 51, 75-98.

Honnorez, J., 1981. The aging of the oceanic crust at low temperature. In : *Oceanic lithosphere*, C. Emiliani ed., J. Wiley, London, 525-587.

Hopson, C.A. and Mattinson, J.M., 1973. Ordovician and late Jurassic ophiolitic assemblages in the Pacific NW. *Geol. Soc. Amer. Cordilleran Sect. Abstr*, 5, 57.

Hopson, C.A. and Franno, C.J., 1977. Igneous history of the Point Sal ophiolite, Southern California. *Bull. State Oregon Dep. Geol. Min. Ind.*, 95, 161-183.

Hopson, C.A., Coleman, R.G., Gregory, R.T., Pallister, J.S. and Bailey, E.H., 1981. Geologic section through the Samail ophiolite and associated rocks along a Muscat-Ibra transect. *J. geophys. Res.*, 86, 2527-2544.

Hsü, K.J., 1971. Franciscan mélanges as a model for eugeosynclinal sedimentation and underthrusting tectonics. *J. geophys. Res.*, 76, 1162-1170.

Hussong, D.M., 1972. Detailed structural interpretations of the Pacific oceanic crust using Asper and ocean-bottom seismometer methods. *Ph.D., Univ. Hawaii,* 165 p.

Hynes, A., 1975. Comment on 'The Troodos ophiolitic complex was probably formed in a island arc' by A. Miyashiro. *Earth and Planet. Sci. Lett.*, 25, 213-216.

Irvine, T.N., 1980. Magmatic density currents and cumulus processes. *Amer. J. Sci.*, 280, 1-58.

Irvine, T.N., 1982. Terminology for layered intrusions. *J. Petrol.*, 23, 127-162.

Irvine, T.N. and Findlay, T.C., 1972. Alpine peridotite with particular reference to the Bay of Islands Igneous Complex. *Univ. Ottawa Earth Phys. Branch Publ.*, 42, 97-128.

Irwin, W.P., 1966. Geology of the Klamath Mountains Province. *Bull. Calif. Div. Mines Geol.*, 190, 19-38.

Irwin, W.P., 1981. Tectonic accretion of the Klamath Mountains In : *The Geotectonic development of California*, W.G. Ernst ed., Rubey Vol. 1, 29-49.

Irwin, W.P. and Coleman, R.G., 1972. Preliminary map showing global distribution of alpine-type ultramafic rocks and blueschists. *U.S.G.S. Map 1/40 000 000.*

Irwin, W.P. and Jones, D.L., 1981. Distribution, age and tectonic significance of ophiolites of western North America. In : *IGCP Ophiolite Project*, N. Bogdanov ed., John Wiley, New York.

Ishiwatari, A., 1985. Alpine ophiolites : product of low-degree mantle melting in a Mesozoic transcurrent rift zone. *Earth and Planet. Sci. Lett.*, 76, 93-108.

Ivanov, S.N., Perfiliev, A.S., Efimov, A.A., Smirnov, G.A., Necheukhin, V.M. and Fershtater, G.B., 1975. Fundamental features in the structure and evolution of the Urals. *Amer. J. Sci.*, 275, 107-130.

Ivanov, S.N., Perfiliev, A.S., Puchkov, V.N., Ruzhentsev, S.V. and Samygin, S.G., 1979. The tectonic positions of ophiolites in the Urals. *Mem. Newfoundland Univ. Dept. Geol.*, 8,109-114.

Jackson, E.D., 1961. Primary textures and mineral associations in the ultramafic zone of the Stillwater Complex, Montana. *Geol. Surv. Prof. Pap.*, 358, 106p.

Jackson, E.D., Green, H.W., and Moores, E.M., 1975. The Vourinos ophiolite, Greece : cyclic units of lineated cumulates overlying harzburgite tectonite. *Geol. Soc. Amer. Bull.*, 86, 390-398.

Jackson, H.R., Reid, I. and Falconer, R.K.H., 1982. Crustal structure near the Arctic Mid-Ocean Ridge. *J.*

*geophys. Res.*, 87, 1773-1783.

Jackson, M., 1979. Structures des filons dans les massifs de péridotite : mécanismes d'injection et relations avec la déformation plastique. *Thèse Doc. 3° cycle, Univ. Nantes*, 143 p.

Jacobsen, S.B. and Wasserburg, G.J., 1979. Nd and Sr isotopic study of the Bay of Islands Ophiolite Complex and the evolution of the source of mid-oceanic ridge basalts. *J. geophys. Res.*, 84, 7429-7445.

Jacobsen, S.B., Quick, J.E. and Wasserburg, G.J., 1984. A Nd and Sr isotopic study of the Trinity peridotite ; implications for mantle evolution. *Earth and Planet. Sci. Lett.*, 68, 361-378.

Jahn, B., 1986. Mid-ocean ridge or marginal basin origin of the East Taiwan ophiolite : chemical and isotopic evidence. *Contr. Mineral. Petrol.*, 92, 194-206.

Jaques, A.L. and Chappell, B.W., 1980. Petrology and trace element geochemistry of the Papuan ultramafic belt. *Contr. Mineral. Petrol.*, 75, 55-70.

Jaques, A.L. and Green, D.H., 1980. Anhydrous melting of peridotite at 0-15Kb pressure and the genesis of tholeiitic basalts, 1980. *Contr. Mineral. Petrol.*, 73, 287-310.

Jamieson, R.A., 1980. The formation of metamorphic aureoles beneath ophiolite suites - evidence from the St Anthony Complex, Northwestern Newfoundland. *Geology*, 8, 150-154.

Jamieson, R.A., 1981. Metamorphism during ophiolite emplacement - the petrology of the St. Anthony Complex. *J. Petrol.*, 22, 397-449.

Jamieson, R.A., 1986. P-T paths from high temperature shear zones beneath ophiolites. *J. metam. Geol.*, 4, 3-22.

Javoy, M., Pineau, F. and Delorme, H., 1986. Carbon and nitrogen isotopes in the mantle. *Chem. Geol.*, 57, 41-62.

Javoy, M., Pineau, F. and Agrinier, P., 1988. Volatiles and stable isotopes in recycling. In : *crust-mantle recycling at subduction/collision zones.* L. Gulen and S.R. Härt eds, Reidel, Dordrecht, in press.

Jayko, A.S., Blake, M.C. and Tekla Harms, 1987. Attenuation of the Coast Range ophiolite by extensional faulting, and nature of the Coast Range 'thrust', California. *Tectonics*, 6, 475-488.

Jenkyns, H.C. and Winterer, E.L., 1982. Paleoceanography of Mesozoic ribbon radiolarites. *Earth and Planet. Sci. Lett.*, 60, 351-375.

Ji, S. and Mainprice, D., 1988. Natural deformation fabrics of plagioclase : implications for slip systems and seismic anisotropy. *Tectonophysics*, 147, 154-163.

Johnson, G.L., and Jakobsson, S.P., 1985. Structure and petrology of the Reykjanes Ridge between 62°55'N and 63°48'N. *J. geophys. Res.*, 90, 73-83.

Jurewicz, A.J.G. and Watson, E.B.,1988. Cations in olivine, part 2 : diffusion in olivine xenocrysts, with applications to petrology and mineral physics. *Contr. Mineral. Petrol.*, 99, 186-201.

Juteau, T., 1974. Les ophiolites des nappes d'Antalya (Taurides occidentales, Turquie). Pétrologie d'un fragment de l'ancienne croûte téthysienne. *Thèse Doc. Etat, Univ. Nancy* , 692 p.

Juteau, T., Nicolas, A., Dubessy, J., Fruchard, J.C. and Bouchez, J.L., 1977. Structural relationships in the Antalya ophiolite complex, Turkey : possible model for an oceanic ridge. *Geol. Soc. Amer. Bull.*, 88, 1740-1748.

Juteau, T. and Whitechurch, H., 1980. The magmatic cumulates of Antalya (Turkey) : evidence of multiple intrusions in an ophiolitic magma chamber. *Proc. Inter. Ophiolite Symp.*, Geol Surv. Dept. Cyprus, 377-391.

Juteau, T., Ernewein, M., Reuber, I., Whitechurch, H., and Dahl, R., 1988a. Duality of magmatism in the plutonic sequence of the Sumail nappe, Oman. *Tectonophysics*, 151, 107-135.

Juteau, T., Beurrier, M., Dahl, R., and Nehlig, P., 1988b. Segmentation at a fossil spreading axis. The plutonic sequence of the Wadi Haymiliayah area (Haylayn Block, Sumail Nappe, Oman). *Tectonophysics*, 151, 167-197.

Juteau, T., Cannat, M. and Lagabrielle, Y., 1988. Peridotites in the upper oceanic crust away from transform zones : a comparison of the results of previous DSDP and ODP legs. *Proc. ODP, Init. Repts*, 106-109.

Kappel, E.S. and Ryan, W.B.F., 1986. Volcanic episodicity and a non-steady state rift valley along

Northeast Pacific spreading centers ; evidence from Sea MAR I. *J. geophys. Res.*, 91, 13925-13940.
Karato, S.I., 1984. Grain-size distribution and rheology of the upper mantle. *Tectonophysics*, 104, 155-176.
Karato, S.I., Toriumi, M. and Fujii, T., 198O. Dynamic recrystallization of olivine single crystals during high temperature creep. *Geophys. Res. Lett.*, 7, 649-652.
Karig, D.E., 1980. Ridges and basins of the Tonga-Kermadec island arc systems. *J. Geophys. Res.*, 75, 239-254.
Karpoff, A.M., Walter, A.V. and Pflumio, C., 1988. Metalliferous sediments within lava sequences of the Sumail ophiolite (Oman) : mineralogical and geochemical characterization, origin and evolution. *Tectonophysics*, 151, 223-245.
Karson, J.A., 1977. The geology of the Northern Lewis Hills, Western Newfoundland. *Ph.D. SUNY, Albany*, 474p.
Karson, J.A., 1982. Reconstructed seismic velocity structure of the Lewis Hills massif and implications for oceanic fracture zones. *J. geophys. Res.*, 87, 961-978.
Karson, J.A., 1984. Variations in structure and petrology in the Coastal Complex, Newfoundland : anatomy of an oceanic fracture zone. *Geol. Soc. London Spec. Publ.*, 13, 131-144.
Karson, J.A., 1987. Factors controlling the orientation of dykes in ophiolites and oceanic crust. *Geol. Assoc. Canad. Spec. Pap.*, 34, 229-241.
Karson, J.A. and Dewey, J.F, 1978. Coastal Complex, Western Newfoundland : an early Ordovician oceanic fracture zone. *Bull. Geol. Soc. Amer.*, 89, 1037-1049.
Karson, J.A., Elthon, D.L. and De Long , S.E., 1983. Ultramafic intrusions, in the Lewis Hills massif, Bay of Islands ophiolite Complex, Newfoundland : implications for igneous processes at oceanic fracture zones. *Bull. Geol. Soc. Amer.*, 94, 15-29.
Karson, J.A. and Dick, H.J.B., 1984. Deformed and metamorphosed oceanic crust on the Mid-Atlantic Ridge. *Ofioliti*, 9, 279-302.
Karson, J.A. and Winters, A.T., 1987. Tectonic extension on the Mid-Atlantic Ridge. *EOS*, 68, 1508.
Karson, J.A., Thompson, G., Humphris, S.E., Edmond, J.M., Bryan, W.B., Brown, J.R., Winters, A.T., Pockalny, R.A., Casey, J.F., Campbell, A.C., Klinkhammer, G., Palmer, M.R., Kinzler, R.J. and Sulanowska, M.M., 1987. Along-axis variations in seafloor spreading in the MARK area. *Nature*, 328, 681-685.
Kastens, K., Mascle, J., Auroux, C., Bonatti, E., Broglia, C., Channel, J., Curzi, P., Emeis, K.C., Glaçon, G., Hasegawa, S., Hieke, W., Mascle, G., McCoy, F., McKenzie, J., Mendelson, J., Muller, C., Rehault, J.P., Robertson, A., Sartori, R., Sprovieri, R. and Torii, M., 1986. La campagne 107 du Joides Resolution (ODF) en Mer Tyrrhénienne : premiers résultats. *C.R. Acad. Sci. Paris*, 5, 391-396.
Kay, R.W., and Senechal, R.G., 1976. The rare-earth geochemistry of the Troodos ophiolite complex. *J. geophys. Res.*, 81, 964-970.
Kempner, W.C. and Gettrust, J.F., 1982a. Ophiolites, synthetic seismograms, and ocean crustal structure. 1. Comparison of ocean bottom seismometer data and synthetic seismograms for the Bay of Islands ophiolite. *J. geophys. Res.*, 87, 8447-8462.
Kempner, W.C., and Gettrust, J.F., 1982b. Ophiolites, synthetic seismograms, and oceanic crustal structure. 2. A comparison of synthetic seismograms of the Samail ophiolite, Oman, and the ROSE refraction data from the East Pacific Rise. *J. geophys. Res.*, 87, 8463-8476.
Ketner, K.B., 1984. Recent studies indicate that major structures in northeastern Nevada and the Golconda thrust in north-central Nevada are of Jurassic or Cretaceous age. *Geology*, 12, 483-486.
Kidd, W.S.F., 1977. The Baie Verte lineament, Newfoundland : ophiolite complex floor and mafic volcanic fill of a small Ordovician marginal basin. Maurice Ewing Series 1, *AGU Washington*, 407-418.
Kidd, R.G.W., 1977. A model for the process of formation of the upper oceanic crust. *Geophys. J. r. astron. Soc.*, 50, 149-183.
Kidd, R.G.W. and Cann, J.R., 1974. Chilling statistics indicate an ocean-floor spreading origin for the Troodos complex, Cyprus. *Earth and Planet. Sci. Lett.*, 24, 151-155.

Kienast, J.R., Komor, S.C., Elthon, D. and Casey, J.F., 1985. Mineralogical variation in a layered ultramafic cumulate sequence at the North Arm Mountain massif, Bay of Islands ophiolite, Newfoundland. *J. geophys. Res.*, 90, 7705-7736.

Kimball, K.L., Spear, F.S., and Dick, H.J.B., 1985. High temperature alteration of abyssal ultramafics from the Islas Orcadas Fracture Zone, South Atlantic. *Contr. Mineral. Petrol.*, 91, 307-320.

Klein, E.M. and Langmuir, C.H., 1987. Global correlations of ocean ridge basalt chemistry with axial depth and crustal thickness. *J. geophys. Res.*, 92, 8089- 8115.

Knight, M.D. and Walker, G.P.L.,1988. Magma flow directions in dikes of the Koolau Complex, Oahu, determined from magnetic fabric studies. *J. geophys. Res.*, 93, 4301-4319.

Koepke, J., Kreuser, H. and Seidel, E., 1985. Ophiolites in the southern Aegean arc (Crete, Karpathos, Rhodes) - linking the ophiolite belts of the Hellenides and the Taurides. *Ofioliti*, 10, 343-354.

Koyanagi, R.Y. and Endo, E.T., 1971. Hawaiian seismic events during 1969. *U.S. Geol. Surv. Prof. Pap.*, 750 C, 158-164.

Kundig, E., 1956. The position in time and space of the ophiolites with relation to orogenic metamorphism. *Geologie en Mijnbouw*, 18, 106-114.

Kushiro, I., 1973. Origin of some magmas in oceanic and circum oceanic regions. *Tectonophysics*, 17, 211-222.

Kusznir, N.J., 1980. Thermal evolution of the oceanic crust ; its dependence on spreading rate and effect on crustal structure. *Geophys. J. r. astron. Soc.*, 61, 167-181.

Lachenbruch., A.H., 1976. Dynamics of a passive spreading center. *J. geophys. Res.* 81, 1883-1902.

La Fehr, T.R., 1966. Gravity in the Eastern Klamath Mountains, California. *Geol. Soc. Amer. Bull.*, 77, 1177-1190.

Lagabrielle, Y., Polino, R., Auzende, J.M., Blanchet, R., Caby, R., Fudral, S., Lemoine, M., Mével, C., Ohnenstetter, M., Robert, D., and Tricart, P., 1984. Les témoins d'une tectonique intraocéanique dans le domaine téthysien : analyse des rapports entre les ophiolites et leurs couvertures métasédimentaires dans la zone piémontaise des Alpes franco-italiennes. *Ofioliti*, 9, 67-88.

Lago, B.L., Rabinowicz, M. and Nicolas, A., 1982. Podiform chromite ore bodies : a genetic model. *J. Petrol.*, 23, 103-125.

Lallemant, S., Chamot-Rooke, N., Le Pichon, X. and Rangin, C., 1988. Zenisu Ridge : a deep intraoceanic thrust related to subduction, off South-West Japan. *Tectonophysics*, in press.

Landis, C.A. and Coombs, D.S., 1967. Metamorphic belts and orogenesis in southern New Zealand. *Tectonophysics*, 4, 501-518.

Langmuir, C.H., Bender, J.F., Bence, A.E., Hanson, G.N. and Taylor, S.R., 1977. Petrogenesis of basalts from the FAMOUS area : Mid Atlantic ridge. *Earth and Planet. Sci. Lett.*, 36, 133-156.

Langmuir, C.H., Bender, J.F. and Batiza, R., 1986. Petrological and tectonic segmentation of the East Pacific Rise, 5°30'-14°30' N. *Nature*, 322, 422-429.

Langseth, M.S., Mottl, M., Hobart, M. and Fisher, A., 1986. Hydrothermal circulation in the vicinity of the DSDP 501/504 sites on the south flank of the Costa Rica Rift. *EOS Trans. AGU*, 67, 1222.

Lanphere, M.A., Irwin, W.P. and Hotz, P.E., 1968. Isotopic age of the Nevadan orogen and older plutonic and metamorphic events in the Klamath Mountains, California. *Geol. Soc. Amer. Bull.*, 79, 1027-1058.

Lapierre, H., 1972. Les formations sédimentaires et éruptives des nappes de Mamonia et leurs relations avec le massif de Troodos (Chypre). *Thèse Doc. Etat, Univ. Nancy*, 420 p.

Lapierre, H., Albarède, F., Albers, J., Cabanis, B. and Coulon, C., 1985. Early Devonian volcanism in the eastern Klamath Mountains, California : evidence for an immature island arc. *Canad. J. Earth Sci.*, 22, 214-227.

Laverne, C., 1987. Les altérations des basaltes en domaine océanique, minéralogie, pétrologie et géochimie d'un système hydrothermal : le puits 504B, Pacifique Oriental. *Thèse Doc.Univ. Marseille*, 315 p.

Laurent, R., 1980. Environment of formation, evolution and emplacement of the Appalachian ophiolites from Quebec. *Proc. Inter. Ophiolite Symp.* Geol. Surv. Dept. Cyprus, 628-636.

Laurent, R. and Kacira, N., 1987. Chromite deposits in the Appalachian ophiolites. In : *Evolution of chromium ore fields,* C.W. Stowe ed., Van Nostrand Reinhold Co, New-York, 169-193.

Leblanc, M., 1986. Co-Ni arsenide deposits, with accessory gold, in ultramafic rocks from Morocco. *Canad. J. Earth Sci.*, 23, 1592-1602.

Leblanc, M., 1987. Chromite in oceanic arc environments : New Caledonia. In : *Evolution of chromium ore fields,* C.W. Stowe ed., Van Nostrand Reinhold Co, New-York, 265-296.

Leblanc, M., Dupuy, C., Cassard, D., Moutte, J., Nicolas, A., Prinzhofer, A., Rabinovitch, M. and Routhier, P., 1980. Essai sur la génèse des corps podiformes de chromite dans les peridotites ophiolitiques : étude des chromites de Nouvelle Calédonie et comparaison avec celles de Méditerranée orientale. In : *Proc. Inter. Ophiolite Symp.* Geol. Surv. Dept. Cyprus, 691-701.

Leblanc, M., Cassard, D. and Juteau, T., 1981. Cristallisation et déformation des orbicules de chromite. *Mineralium depos.*, 16, 269-282.

Leblanc, M. and Violette, J.F., 1983. Distribution of aluminium-rich and chromium-rich chromite pods in ophiolite peridotites. *Econ. Geol.*, 78, 293-301.

Lees, G.M., 1928. The geology and tectonics of Oman and of parts of Southeastern Arabia. *Geol. Soc. London Quart. J.*, 84 : 585-67O.

Leg 118 Shipboard Scientific Party, 1988. Plutonic rocks in fracture zones. *Nature,* 333, 115-116.

Lehmamn, J., 1983. Diffusion between olivine and spinel : application to geothermometry. *Earth and Planet. Sci. Lett.*, 64, 123-138.

Le Douaran, S. and Francheteau, J., 1981. Axial depth anomalies from 10° to 50° north the Mid-Atlantic Ridge : correlation with other mantle properties. *Earth and Planet. Sci. Lett.,* 54, 29-47.

Leitche, E.C., 1984. Island arc elements and arc-related ophiolites. *Tectonophysics*, 106, 177-203.

Le Métour J., 1987. Géologie de l'autochtone des montagnes d'Oman dans la fenêtre du Saih Hatat. *Thèse Doc. Etat, Univ. Paris*, 420 p

Le Métour, J., Rabu, D., Tegyey, M., Bechennec, F., Beurrier, M. and Villey, M., 1986. Le métamorphisme régional Crétacé de faciès éclogites-schistes bleus sur la bordure omanaise de la plateforme arabe : conséquence d'une tectogenèse précoce anté- obduction. *C.R. Acad. Sci. Paris,* 302, 905-910.

Lemoine, M., 1980. serpentinites, gabbros and ophicalcites in the Piemont-Ligurian domain of the Western Alps : possible indicators of oceanic fracture zones and of associated serpentinite protusions in the Jurassic-Cretaceous Tethys. *Archives Sci., Geneva,* 33 : 103-115.

Lemoine, M., Tricart, P. and Boillot, G., 1987. Ultramafic and gabbroic ocean floor of the Ligurian Tethys (Alps, Corsica, Apennines) : in search of a genetic model. *Geology*, 15, 622-625.

Lensch, G., 1976. Ariegite und websterite in lherzolite von Balmuccia. *Neues Jb. Mineral. Abh.*, 128, 186-2O8.

Lensch, G., Mihm, A. and Alavi-Tehrani, N., 1977. Petrography and geology of the ophiolite belt North of Sabzevar/Khorassan (Iran). *Neues Jb. Mineral. Abh,* 131, 156-178.

Le Pichon, X., 1969. Models and structure of the oceanic crust. *Tectonophysics,* 7, 385-401.

Le Pichon, X. and Francheteau, J., 1978. A plate tectonic analysis of the Red Sea-Gulf of Aden area. *Tectonophysics,* 46, 369-406.

Le Pichon, X., Angelier, J. and Sibuet, J.C., 1982. Plate boundaries and extensional tectonics. *Tectonophysics*, 81, 239-256.

Le Pichon, X., Iiyama, T., Chamley, H., Charvet, J., Faure, M., Fujimoto, H., Furuta, T., Ida, Y., Kagami, H., Lallemant, S., Leggett, J., Murata, A., Okada, H., Rangin, C., Renard, V., Taira, A. and Tokuyama, H., 1987. The Eastern and Western ends of Nankai Trough : results of Box 5 and Box 7 Kaiko Survey. *Earth and Planet. Sci. Lett.*, 83, 199-213.

Le Pichon, X., Liyama T., Boulègue, J., Charvet, J., Faure, M., Kano K, Lallemant, S., Okada, H., Rangin, C., Taira, A. Urabe, T. and Uyeda S., 1987. Nankai trough and Zenisu Ridge : a deep-sea submersible survey. *Earth and Planet. Sci. Lett.*, 83, 285-299.

Le Sueur, E., Boudier, F., Cannat, M., Ceuleneer, G. and Nicolas, A., 1984. The Trinity mafic-ultramafic complex : first results of the structural study of an untypical ophiolite. *Ofioliti*, 9, 487-498.

Le Sueur, E., and Boudier, F., 1986. Structures du complexe basique et ultrabasique de Trinity, Californie : génèse d'une ophiolite atypique. *Bull. Soc. géol. France*, 6, 1007-1014.

Lewis, B.T.R., 1983. The process of formation of ocean crust. *Science*, 220, 151-157.

Lewis, B.T.R. and Snydsman, W.E., 1979. Fine structure of the lower oceanic crust on the Cocos Plate. *Tectonophysics*, 55, 87-105.

Lindsay-Griffin, N., 1977. The Trinity ophiolite, Klamath Mountains, California. *Oregon Dept. Geol. Mineral. Ind. Bull.*, 95, 107-120.

Liou, J.G. and Ernst, W.G., 1979. Oceanic ridge metamorphism of the East Taiwan ophiolite. *Contr. Mineral. Petrol.*, 68, 335-348.

Lipman, P.W., 1964. Structure and origin of an ultramafic pluton in the Klamath Mountains, California. *Amer. J. Sci.*, 262, 199-222.

Lippard, S.J., 1983. Cretaceous high pressure metamorphism in NE Oman and its relationship to obduction and ophiolite nappe emplacement. *J. geol. Soc. (London)*, 140, 97-104.

Lippard, S.J. and Rex, D.C., 1982. K-Ar ages of alkaline igneous rocks in the northern Oman mountains, NE Arabia, and their relations to rifting, passive margin development and destruction of the Oman Tethys. *Geol. Mag.*, 119 : 497-5O3.

Lippard, S.J., Shelton, A.W. and Gass, I.G., 1986. The ophiolite of Northern Oman. *Geol. Soc. London Mem.*, 11, 178 p.

Lister, C.R.B., 1972. On the thermal balance of a mid-ocean ridge. *Geophys. J. r. astron. Soc.*, 26, 515-535.

Lister, C.R.B., 1980. Heat flow and hydrothermal circulation. *Ann. Rev. Earth Planet. Sci.*, 8, 95-117.

Lister, C.R.B., 1983. On the intermittency and crystallization mechanisms of sub-seafloor magma chambers. *Geophys. J. r. Astron. Soc.*, 73, 351-365.

Lockwood, J.P., 1972. Possible mechanisms for the emplacement of Alpine-type serpentinite. *Mem. Geol. Soc. Amer.*, 132, 273-287.

Lombardo, B. and Pognante, U., 1982. Tectonic implications in the evolution of the western Alps ophiolite metagabbros. *Ofioliti*, 371-394.

Lonsdale, P., 1983. Overlapping rift zones at the 5.5°S offset of the East Pacific Rise. *J. geophys. Res.*, 88, 9393-9406.

Loomis, T.P., 1972. Diapiric emplacement of the Ronda high-temperature ultramafic intrusion, Southern Spain. *Geol. Soc. Amer. Bull.*, 83 : 2475-2496.

Loomis, T.P., 1975. Tertiary mantle diapirism, orogeny, and plate tectonics east of the Strait of Gibraltar, *Amer. J. Sci.*, 275, 1-30.

Loomis, T.P. and Gottschalk R.R., 1981. Hydrothermal origin of mafic layers in Alpine type peridotites : evidence from the Seiad Ultramafic complex, California, U.S.A. *Contr. Mineral. Petrol.*, 76, 1-11.

Lorand, J.P., 1988. Fe-Ni-Cu sulfides in tectonite peridotites from the Maqsad district, Sumail ophiolite, southern Oman : implications for the origin of the sulfide component in the oceanic upper mantle.*Tectonophysics*, 151, 57-73.

Lotti, B., 1886. Paragone fra la roccie ofiolitiche terziare italiane e la roccie basiche pare terziarie della Scozia et dell'Irlande. *Boll. R. Comitato Geol. Italia,* 76.

Loubet, M., Shimizu, N. and Allègre, C.J., 1975. Rare earth elements in alpine peridotites. *Contr. Miner. Petrol.*, 53, 1-12.

Lundeen, M., 1978. Emplacement of the Ronda peridotite, Sierra Bermeja, Spain. *Geol. Soc. Amer. Bull.*, 89, 172-180.

Luyendyk, B.P., and Day, R., 1982. Paleomagnetism of the Samail ophiolite, Oman. 2 : The Wadi Kadir gabbro section. *J. geophys. Res.*, 87, 10903-10917.

Maaloe, S., 1981. Magma accumulation in the ascending mantle. *J. geol. Soc. (London)*, 138, 223-236.

Maaloe, S. and Steel, R., 1980. Mantle composition derived from the composition of lherzolites. *Nature*, 285, 321-322.

MacCulloch, M.T., Wasserburg, G.J., and Taylor, H. P., 1979. A neodymium, strontium and oxyden isotopic study of the cretaceous Samail ophiolite and implications for the petrogenesis and seawater-hydrothermal alteration of oceanic crust. *Earth and Planet. Sci. Lett.*, 46, 201-211.

MacDonald, K.C., 1982. Mid-ocean ridges : fine scale tectonic, volcanic and hydrothermal processes within the plate boundary zone. *Ann. Rev. Earth Planet. Sci.*, 10, 155-190.

MacDonald, K.C. and Luyendyk, B.P., 1977. Deep-tow studies of the structure of the Mid-Atlantic Ridge crest near lat. 37°N. *Geol. Soc. Amer. Bull.* 88, 621-636.

MacDonald, K.C. and Fox, P.J., 1983. Overlapping spreading centers : a new kind of accretion geometry on the East Pacific Rise. *Earth and Planet. Sci. Lett.*, 40, 407-141.

MacDonald K.C., Sempere, J.C. and Fox, P.J., 1984. East pacific rise from Siqueiros to Orozco fracture zones : along-strike continuity of axial neovolcanic zone and structure and evolution of overlapping spreading centers. *J. Geophys. Res.*, 89, 6049-6069.

MacDonald, K.C., Sempere, J.C. and Fox, P.J., 1986. Reply : the debate concerning overlapping spreading centers and mid-ocean ridge processes. *J. geophys. Res.*, 91, 10501-10511.

MacDonald, K.C., Haymon, R.M., Miller, S.P., Sempere, J.C. and Fox, P.J., 1988. Deep-tow and sea beam studies of dueling propagating ridges on the east pacific rise near 20°40'S. *J. geophys. Res.*, 93, 2875-2898.

MacGregor, I.D. and Basu, A.R., 1979. Petrogenesis of the Mont Albert ultramafic massif, Quebec. *Geol. Soc. Amer. Bull.*, 90, 1529-1627.

MacKenzie, D.B., 1960. High-temperature Alpine-type peridotite from Venezuela. *Geol. Soc. Amer. Bull.*, 71, 303-318.

Mackwell, S.J., Kohlstedt, D.L. and Paterson, M.S., 1985. The role of water in the deformation of olivine single crystals. *J. geophys. Res.*, 90, 11313-11333.

Madsen, J., Forsyth, D. and Detrick, R., 1984. A new isostatic model for the East Pacific Rise crest. *J. geophys. Res.*, 89, 9997-10015.

Malpas, J., 1977. Petrology and tectonic significance of Newfoundland ophiolites with examples from the Bay of Islands. *Oregon Dept. Geol. Mineral. Ind.*, 95, 13-23.

Malpas, J., 1978. Magma generation in the upper mantle, field evidence from ophiolite suites and application to the generation of oceanic lithosphere. *Phil. Trans. r. Soc. London.*, 288A, 527-546.

Malpas, J., 1979. The dynamo-metamorphic aureole of the Bay of Islands ophiolite suite. *Canad. J. Earth Sci.*, 16, 2086-2101.

Manghnani, M.H. and Coleman, R.G., 1981. Gravity profiles across the Samail ophiolite, Oman. *J. geophys. Res.*, 86 : 2509-2526.

Mankinen, E.A., Irwin, W.P. and Gramme, C.S., 1982. Tectonic rotation of the Eastern Klamath Mountains terrane, California. *EOS*, 63, 914.

Marcoux, J., De Wever, P., Nicolas, A., Girardeau, J., Xiao Xuchang, Chang Chengfa, Wang Naiwen, Zao Yougong, Bassoulet, J.P, Colchen, M. and Mascle, G., 1982. Preliminary report on depositional sediments on top of volcanic member : the Xigaze ophiolite (Yarlung-Zangbo suture zone, Zizang, China). *Ofioliti*, 6, 31-32.

Marsh, B.D., 1976. Mechanics of Benioff zone magmatism. In : The geophysics of the Pacific Ocean Basin and its margin. *AGU Washington, Geophys. Monogr.*,19, 337-350.

Marsh, N.G. and Wood, D.A., 1980. Ophiolites as ocean crust or marginal basin crust : a geochemical approach. *Proc. Inter. Ophiolite Symp.*, Geol. Surv. Dept. Cyprus, 193-204.

Martin, B., 1987. Etude détaillée de la solidification d'un filon de basalte alcalin du Lodévois. *Thèse Doc. Univ., Paris*, 105 p.

Mattauer, M., 1983. Subduction de lithosphere continentale, décollement croûte-manteau et chevauchements d'échelle crustale dans la chaine de collision himalayenne. *C.R. Acad. Sci. Paris*, 296, 481-486.

Mattauer, M., Proust, F., and Tapponnier, P., 1980. Tectonic mechanism of obduction, in relation with high pressure metamorphism, orogenic mafic and ultramafic association. *Coll. Inter. CNRS, Paris*, 272, 197-201.

Mattinson, J.M., 1976. Ages of zircons from the Bay of Islands ophiolite Complex, Western Newfoundland. *Geology*, 4, 393-394.

Mattinson, J.M. and Hopson, C.A., 1972. Paleozoic ages of rocks from ophiolite complexes in the Washington and Northern California. *EOS*, 53, 543.

Mauffret, A. and Montadert, L., 1987. Rift tectonics on the passive continental margin off Galicia (Spain). *Marine Petrol. Geol.*, 4, 49-70.

Maurel, C. and Maurel, P., 1982. Etude expérimentale de la solubilité du chrome dans les bains silicatés basiques et de sa distribution entre liquide et minéraux coexistants : conditions d'existence du spinelle chromifère. *Bull. Minéral.*, 105, 640-647.

Maynard, G.L., 1970. Crustal layer of seismic velocity 6.9 to 7.6 kilometers per second under the deep oceans. *Science*, 168, 120-121.

McBirney A.R., Baker, B.H. and Nilson, R.H., 1985. Liquid fractionation. Part I : Basic principles and experimental simulations. *J. volcanol. Res.*, 24, 1-24.

McBirney, A.R., and Noyes, R.M., 1979. Crystallization and layering of the Skaergaard intrusion. *J. Petrol.*, 20, 487-554.

McCaig, A., 1983. P-T conditions during emplacement of the Bay of Islands ophiolite Complex. *Earth and Planet. Sci. Lett.*, 65, 459-473.

McCarthy, J., Morton, J.L., Mutter, J.C., Sleep, N.H. and Thompson, G.A., (1988). Relic magma chamber structures preserved within the Mesozoic North Atlantic crust. *Geol. Soc. Amer. Bull.*, 100, 1423-1436.

McClain, K.J., and Lewis, B.T.R., 1982. Geophysical evidence for the absence of a crustal magma chamber under the Northern Juan de Fuca Ridge : a contrast with ROSE results. *J. geophys. Res.*, 87, 8477-8489.

McClain, J.S., Orcutt, J.A., and Burnett, M., 1985. The East Pacific Rise in cross section : a seismic model. *J. geophys. Res.*, 90, 8627-8639.

McCulloch, M.T., Gregory, R.T., Wasserburg, G.J., and Taylor, H.P., 1981. Sm-Nd, RB-Sr, and $^{18}O/^{16}O$ isotopic systematics in an oceanic crustal section : evidence from the Samail ophiolite. *J. geophys. Res.*, 86, 2721-2736.

McCulloch, M.T., and Cameron, W.E., 1983. Nd-Sr isotopic study of primitive lavas from the Troodos ophiolite, Cyprus : evidence for a subduction-related setting. *Geology*, 11, 727-731.

McKenzie, D.P., 1967. Some remarks on heat flow and gravity anomalies. *J. geophys. Res.*, 72, 6261-6273.

McKenzie, D.P., 1984. The generation and compaction of partially molten rock. *J. Petrol.*, 25, 713-765.

McMillan, P. and Holloway J., 1987. Water solubility in aluminosilicate melts. *Contr. Mineral. Petrol.*, 97, 320-332.

Medaris, L.G., 1966. Geology of the Seiad Valley area, Siskiyou County, California, and petrology of the Seiad ultramafic complex. *Ph.D*, *Los Angeles Univ.*, California, 359 p.

Melson, W.G. and Van Andel, T.H., 1968. Volcanism and metamorphism in the Mid-Atlantic Ridge, 22° N. Latitude. *J. geophys. Res*, 73 : 5925-5941.

Menard, H.W. and McNutt, M., 1982. Evidence for and consequences of thermal rejuvenation. *J. geophys. Res.*, 87, 8570-8580.

Menzies, M., 1973. Mineralogy and partial melt textures within an ultramafic-mafic body, Greece. *Contr. Mineral. Petrol.*, 42, 273-285.

Menzies, M., 1976a. Rifting of a Tethyan continent. Rare earth evidence of an accreting plate margin. *Earth and Planet. Sci. Lett.*, 28, 427-438.

Menzies, M. and Allen, C., 1974. Plagioclase lherzolite residual mantle relationships within two eastern Mediterranean ophiolites. *Contr. Mineral. Petrol.*, 45, 197-213.

Mercier, J.C., and Nicolas, A., 1975. Textures and fabrics of upper mantle peridotites as illustrated by basalt xenoliths. *J. Petrol.*, 16, 454-487.

Mercier, J.C., Anderson, D.A. and Carter, N.L., 1977. Stress in the lithosphere : inferences from steady-state flow of rocks. *Pure Appl. Geophys.*, 115, 199-226.

Messiga, B. and Piccardo, G.B., 1974. Rilevamento geopetrografico e strutturale del gruppo di Volvi. *Mém. Soc. Geol. it.*, 13, 301-315.

Mével, C., 1987. Evolution of oceanic gabbros from DSDP Leg 82 : influence of the fluid phase on metamorphic crystallizations. *Earth and Planet. Sci. Lett.*, 83, 67-79.
Mével, C., 1988. Metamorphism in oceanic layer 3, Gorringe Bank, Eastern Atlantic. *Contr. Mineral. Petrol.*, in press.
Mével, C., Caby, R. et Kiénast, J.R., 1978. Lower amphibolite facies conditions in the oceanic crust : example of amphibolitized flaser-gabbro and amphibolites from an alpine ophiolite massif (Chenaillet massif, Hautes Alpes, France). *Earth and Planet. Sci. Lett.*, 39 : 98-108.
Michael, P., 1988. Partition coefficients for rare earth elements in mafic minerals of high silica rhyolites : the importance of accessory mineral inclusions. *Geochim. Cosmochim. Acta,* 52, 275-282.
Michael, P., in press. The concentration, behavior and storage of $H_2O$ in the suboceanic upper mantle : implications for mantle metasomatism. *Geochim. Cosmochim. Acta,* 52.
Michael, P. and Bonatti, E., 1985. Peridotite composition from the North Atlantic : regional and tectonic variations and implications for partial melting. *Earth and Planet. Sci. Lett.*, 73, 91-104.
Michard, A., Bouchez, J.L. and Ouazzani-Touhami, M., 1984. Obduction-related planar and linear fabrics in Oman. *J. struct. Geol.*, 6, 39-49.
Michard, A., Juteau, T. and Whitechurch, H., 1985. L'obduction : revue des modèles et confrontation au cas de l'Oman. *Bull. Soc. géol. France,* 2, 189-198.
Miller, R.B. and Mogk, D.W., 1987. Ultramafic rocks of a fracture-zone ophiolite, North Cascades, Washington. *Tectonophysics,* 142, 261-289.
Misseri, M., 1982. Structures des massifs ophiolitiques de Canyon Mountain (Oregon) et de Wadi Tayin (Oman) : lithosphère d'arc insulaire, lithosphère océanique. *Thèse Doc. 3° cycle, Univ. Nantes,* 165 p.
Misseri, M., 1987. Structure et cinématique des péridotites feldspathiques du Cap Bougaroun (Algérie). *J. Afric. Earth Sci.*, 6, 741-744.
Misseri, M. and Boudier, F., 1985. Structures in the Canyon Mountain ophiolite indicate an island arc intrusion. *Tectonophysics,* 120, 191-209.
Mitchell, A.H.G., 1985. Ophiolite detachment and emplacement related to spreading ridge subduction. *Ofioliti,* 10, 355-362.
Miyashiro, A., 1973. The Troodos Complex was probably formed in an island arc. *Earth and Planet. Sci. Lett.*, 25 : 217-222.
Miyashiro, A., 1975. Origin of Troodos and other ophiolites : a reply to Hynes. *Earth and Planet. Sci. Lett.*, 25, 217-222.
Miyashiro, A., Shido, F., and Ewing, M., 1971. Metamorphism in the Mid-Atlantic Ridge near 24° and 30°N. *Phil. Trans. r. Soc. London,* 268, 589-603.
Molnar, P. and Atwater, T., 1978. Interarc spreading and cordilleran tectonics as alternates related to the age of subducted oceanic lithosphere. *Earth and Planet. Sci. Lett.*, 41, 330-340.
Montigny, R., Le Mer, O., Thuizat, R. and Whitechurch, H., 1988. K-Ar and $^{40}Ar/^{39}Ar$ study of metamorphic rocks associated with the Oman ophiolite : tectonic implications. *Tectonophysics,* 151, 345-362.
Moore, J.G., 1970. Water content of basalt erupted on the ocean floor. *Contr. Mineral. Petrol.,* 28, 272-279.
Moore, D.E., 1984. Metamorphic history of a high-grade blueschist exotic block from the Franciscan Complex, California. *J. Petrol.,* 25, 126-150.
Moore, J.G., Fleming, H.S., and Philipps, J.D., 1974. Preliminary model for extrusion and rifting at the axis of the Mid-Atlantic Ridge, 36°48'N. *Geology,* 2, 437-440.
Moores, E.M., 1969. Petrology and structure of the Vourinos ophiolitic Complex, northern Greece. *Geol. Soc. Amer. Spec. Pap.*, 118, 74 p.
Moores, E.M., 1975. Discussion of 'Origin of Troodos and their ophiolites : a reply to Hynes' by A. Miyashiro. *Earth and Planet. Sci. Lett.*, 25, 223-226.
Moores, E. M., 1982. Origin and emplacement of ophiolites. *Rev. Geophys. (Space Physics*), 20, 735-760.
Moores, E.M. and Vine, F.J., 1971. The Troodos massif, Cyprus, and other ophiolites as oceanic crust :

evaluation and implications. *Phil. Trans. r. Soc. London,* 268, 443-466.
Moores, E.M., and Jackson, E.D., 1974. Ophiolites and oceanic crust. *Nature,* 250, 136-139.
Moores, E.M., Roeder, D.H., Abbas, S.G. and Ahmad, Z., 1980. Geology and emplacement of the Muslimbagh ophiolite, Pakistan. *Proc. Inter. ophiolite Symp.,* Geol. Surv. Dept. Cyprus, 424-429.
Moores, E. M., 1982. Origin and emplacement of ophiolites. *Rev. Geophys. Space Physics,* 20, 735-760.
Moores, E.M., Robinson, P.T., Malpas, J. and Xenophontos, C., 1984. Model for the origin of the Troodos massif, Cyprus and other mideast ophiolites. *Geology,* 12, 500-503.
Morton, D.M., 1959. The geology of Oman. *Proc. 5th World Petrol. Congr. New York,* I, 277-290.
Morton, W.H., and Black, R., 1975. Crustal attenuation in Afar. In : *The Afar region of Ethiopia and related rift problems.* A. Pilger and A. Rosler ed., Schweizerbart, Stuttgart, 55-65 pp.
Morton, J.L., and Sleep, N.H., 1985. A mid-oceanic ridge thermal model : constraints on the volume of axial hydrothermal heat flux. *J. geophys. Res.,* 90, 11345-11353.
Moseley, F. and Abbotts, I.L., 1979. The ophiolite melange of Masirah, Oman : a Cretaceous Indian Ocean transform. *J. geol. Soc. (London),* 136, 713-724.
Moutte, J., 1979. Le massif de Tiebaghi, Nouvelle Calédonie et ses gites de chromite. *Thèse Doc. Ing. Ecole Natl. Sup. Mines,* Paris, 160p.
Moutte, J., 1982. Chromite deposits of the Tiebaghi ultramafic massif, New Caledonia. *Econ. Geol.,* 77, 576-591.
Muller, P.D., 1979. Geology of the Los Amates quadrangle and vicinity, Guatemala, Central America. *PhD thesis,* SUNY Binghamton, 326 p.
Murris, R.J., 198O. Middle East : stratigraphic evolution and oil habitat. *Amer. Ass. Petrol. Geol. Bull.,* 64, 597-618.
Murton, B.J., 1986. Anomalous oceanic lithosphere formed in a leaky transform fault : evidence from the Western Limassol Forest Complex, Cyprus. *J. Geol. Soc. (London),* 143, 845-854.
Mutter, J.C., Detrick, R.S. and North-Atlantic Transect Study Group, 1984. Multichannel seismic evidence for anomalously thin crust at Blake Spur fracture zone. *Geology,* 12, 534-537.
Murton, B.J., and Gass, I.G., 1986. Western Limassol Forest complex, Cyprus : Part of an upper cretaceous leaky transform fault. *Geology,* 14, 255-258.
Mutter, J.C., Buck, W.R. and Zehnder, C.M., 1987. Convective partial melting. I : a model for the development of thick igneous crust during the initiation of spreading. *J. geophys. Res.,* 93, 1031-1048.
Mysen, B.O. and Boettcher, A.L., 1974. Melting of a hydrous mantle. I : Phase relations of natural peridotite at high pressures and temperatures with controlled activities of water, carbon dioxide, and hydrogen. *J. Petrol.,* 16, 520-548.
Mysen, B.O. and Kushiro, I., 1977. Compositional variation of coexisting phases with degrees of melting of peridotite in the upper mantle. *Amer. Mineral.,* 62, 843-865.
Nagle, F., 1974. Blueschists, eclogite, paired metamorphic belts and the early tectonic history of Hispaniola. *Geol. Soc. Amer. Bull.,* 85, 1461-1466.
Naka, J. and Uehara, S., 1983. Igneous rocks dredged at sites KH82-4. In : *Preliminary reports of the Hakuho-Maru Cruise, KH-82-4.* K. Kobayashi ed., Ocean Res. Inst. Univ. Tokyo, 187-94.
Nance, R.D., Worsley, T.R. and Moody, J.B., 1986. Post-Archean biogeochemical cycles and long-term episodicity in tectonic processes. *Geology,* 14, 514-518.
Navon, O., and Stolper, E., 1987. Geochemical consequences of melt percolation : the upper mantle as a chromatographic column. *J. Petrol.,* 95, 285-307.
Nehlig, P., 1989. Etude d'un système hydrothermal océanique fossile : l'ophiolite de Semail (Oman). *Thèse Doc. Univ. Brest,* 308 p.
Nehlig, P. and Juteau, T., 1988. Flow porosities, permeabilities and preliminary data on fluid inclusions and fossil thermal gradients in the crustal sequence of the Sumail ophiolite (Oman). *Tectonophysics,* 151 : 199-221.
Nicholls, I.A., Ferguson, J., Jones, H., Marks, G.P. and Mutter, J.C., 1981. Ultramafic blocks from the ocean floor southwest of Australia. *Earth and Planet. Sci. Lett.,* 56, 362-374.

Nicolas, A., 1969. Serpentinization d'une lherzolite : bilan chimique, implication tectonique. *Bull. Volcanol.*, 32, 499-508.
Nicolas, A., 1978. Stress estimates from structural studies in some mantle peridotites. *Phil. Trans. r. Soc. London*, 288, 49-57.
Nicolas, A., 1984. Lherzolites of the Western Alps : a structural review. In : *Kimberlites II* , J. Kornprobst ed., 333-345.
Nicolas, A., 1985a. Novel type of crust produced during continental rifting. *Nature*, 315, 112-115.
Nicolas, A., 1985b. Discussion de la note 'Les dunites en enclaves dans les basaltes alcalins des iles océaniques : approche pétrologique' par E.T. Berger et M. Vannier. Origine résiduelle d'olivine à contours automorphes. *Bull. Minéral.*, 108, 87-88.
Nicolas, A., 1986a. A melt extraction model based on structural studies in mantle peridotites. *J. Petrol.*, 27,
Nicolas, A., 1986b. Structure and petrology of peridotites : clues to their environment. *Rev. geophys.*, 24, 875-895.
Nicolas, A., 1987. *Principles of rock deformation.* Reidel , Dordrecht, 208p.
Nicolas, A., Bouchez, J.L., Boudier, F. and Mercier, J.C., 1971. Textures, structures and fabrics due to solid state flow in some European lherzolites. *Tectonophysics*, 12, 55-86.
Nicolas, A. and Jackson, E.D., 1972. Répartition en deux provinces des péridotites des chaînes alpines longeant la Méditerranée : implications géotectoniques. *Bull. Suisse Mineral. Petrol..*, 53, 385-401.
Nicolas, A., Boudier, F. and Boullier, A.M., 1973. Mechanisms of flow in naturally and experimentally deformed peridotites. *Amer. J. Sci.*, 273, 853-876.
Nicolas, A. and Boudier, F., 1975. Kinematic interpretation of folds in Alpine-type peridotites. *Tectonophysics*, 25, 233-260.
Nicolas, A. and Poirier, J.P., 1976. *Crystalline plasticity and solid state flow in metamorphic rocks.* Wiley-Interscience, London, 444 p.
Nicolas, A., Boudier, F., and Bouchez, J.L., 1980. Interpretation of peridotite structures from ophiolitic and oceanic environments. *Amer. J. Sci.*, 280, 192-210.
Nicolas, A., Girardeau, J., Marcoux, J., Dupré, B., Wang Xibin, Cao Yougong, Zheng Haixiang and Xiao Xuchang, 1981. The Xigaze ophiolite (Tibet) : a peculiar oceanic lithosphere. *Nature*, 294, 414-417.
Nicolas, A. and Violette, J.F., 1982. Mantle flow at oceanic spreading centers : models derived from ophiolites. *Tectonophysics*, 81, 319-339.
Nicolas, A. and Jackson, M., 1982. High-temperature dikes in peridotites : origin by hydraulic fracturing. *J. Petrol.*, 23, 568-582.
Nicolas, A. and Prinzhofer, A., 1983. Cumulative or residual origin for the transition zone in ophiolites : structural evidence. *J. Petrol.*, 24, 188-206.
Nicolas, A. and Dupuy, C., 1984. Origin of ophiolitic and oceanic lherzolites. *Tectonophysics*, 110, 177-187.
Nicolas, A. and Christensen, N.I., 1987. Formation of anisotropy in upper mantle peridotites - a review. *Geodyn. Ser., 16, AGU Washington*, III-124.
Nicolas, A., Boudier, F. and Montigny, R., 1987. Structure of Zabargad Island and early rifting of the Red Sea. *J. geophys. Res.*, 92, 461-474.
Nicolas, A., Ceuleneer, G., Boudier, F., and Misseri, M., 1988a. A structural mapping in the Oman ophiolites : mantle diapirism along an oceanic ridge. *Tectonophysics*, 151, 27-56.
Nicolas, A., Reuber, I. and Benn, K., 1988b. A new magma chamber model based on structural studies in the Oman ophiolite. *Tectonophysics*, 151, 87-105.
Nisbet, E.G., and Fowler, C.M., 1978. The mid-Atlantic Ridge at 37 and 45°N : some geophysical and petrologic constraints. Geophys. *J. r. astron. Soc.*, 54, 631-660.
Norrel, G.T. and Harper, G.D., 1988. Detachment fauting and amagmatic extension at mid-ocean ridges : the Josephine ophiolite as an example. *Geology*, 16, 827-830.
Obata, M., 1976. The solubility of $Al_2O_3$ in orthopyroxenes in spinel and plagioclase peridotites and spinel pyroxenite. *Amer. Mineral.*, 61, 804-816.

Obata, M., 1980. Ronda peridotite : Garnet-, spinel-, and plagioclase-lherzolite facies and the P-T trajectories of a high-temperature mantle intrusion. *J. Petrol.*, 21 : 533-572.

O'Hara, M.J., 1967. Mineral facies in ultrabasic rocks. In : *Ultramafic and related rocks*. John Wiley, New-York, 7-17.

O'Hara, M.J., 1968. The bearing of phase equilibria studies in synthetic and natural systems on the origin and evolution of basic and ultrabasic rocks. *Earth Sci. Rev.*, 4, 69-133.

O'Hara, M.J., 1977. Geochemical evolution during fractional crystallization of a periodically refilled magma chamber. *Nature*, 266, 503-507.

Orcutt, J.A., 1987. Structure of the Earth : oceanic crust and uppermost mantle. *Rev. Geophys.*, 25, 1177-1196.

Olafsson, M., 1977. The Atftafjordur dike swarm on the Berufjordur coastline, field relations and geochemistry (in Icelandic). Unpublished *BSc-Honors Thesis*, University of Iceland, Reijkjavik.

Oxburgh, E.R., 1972. Flake tectonics and continental collision. *Nature*, 239, 202-204.

Ozawa, K., 1983. Relationships between tectonite and cumulate in ophiolites : the Miyamori ultramafic complex, Kitakami Mountains, Northeast Japan. *Lithos*, 16, 1-16.

Pallister, J.S., 1981. Structure of the sheeted dike complex of the Samail ophiolite near Ibra, Oman. *J. geophys. Res.*, 86, 2661-2672.

Pallister, J.S. and Hopson, C.A., 1981. Samail ophiolite plutonic suite : field relations, phase variation, cryptic variation and layering, and a model of a spreading ridge magma chamber. *J. geophys. Res.*, 86, 2593-2644.

Pallister, J.S. and Knight, R.J., 1981. Rare-earth element geochemistry of the Samail ophiolite near Ibra, Oman. *J. geophys. Res.*, 86, 2673-2698.

Palmason, G., 1980. A continuum model of crustal generation in Iceland ; kinematic aspects. *J. geophys..Res.*, 47, 7-18.

Pamic, J., 1977. Variation in geothermometry and geobarometry of peridotite intrusions in the Dinaride Central Ophiolite zone, Yugoslavia. *Amer. Miner.*, 62, 520-584.

Pamic, J. 1983. Considerations on the boundary between lherzolite and harzburgite subprovinces in the Dinarides and Northern Hellenides. *Ofioliti*, 8, 153-164.

Pamic, J. and Adib, D., 1982. High-grade amphibolites and granulites at the base of the Neyriz peridotites in Southeastern Iran. *N. Jb. Mineral. Abh.*, 143, 113-121.

Pamic, J., Sestini, G. and Abid, D., 1979. Magmatic and metamorphic processes and plate tectonics of the Zagros range. *Geol. Soc. Amer. Bull.*, 90, 569-576.

Paris, J.P., Andreieff, P. and Coudray, J., 1979. Sur l' age Eocène supérieur de la mise en place de la nappe ophiolitique de Nouvelle-Calédonie, unité de charriage océanique périaustralien, déduit d'observations nouvelles sur la série Népoui. *C.R. Acad. Sci. Paris*, 288, 1659-1661.

Parker, R.L. and Oldenburg, D.W., 1973. Thermal model of ocean ridges. *Nature*, 242, 137-139.

Parmentier, E.M. and Forsyth, D.W., 1985. Three-dimensional flow beneath a slow spreading ridge axis : a dynamic contribution to the deepening of the median valley toward fracture zones. *J. geophys. Res.*, 90, 678-684.

Parrot, J.F., 1967. Le cortège ophiolitique du Pinde septentrional (Grèce). *Thèse Doc. 3° cycle, Univ. Paris*, 114 p.

Parrot, J.F. and Guernet, 1972. Le cortège ophiolitique de l'Eubée moyenne (Grèce) : étude pétrographique des formations volcaniques et des roches métamorphiques associées dans les monts Kandilis aux radiolarites. *Cahiers ORSTOM Sér. Géol.*, 2, 153-161.

Parrot, J.F. and Ricou, L.E., 1976. Evolution des assemblages ophiolitiques au cours de l'expansion océanique. *Cahiers ORSTOM Sér. Geol.*, 7, 49-68.

Parrot, J.F. and Whitechurch, H., 1978. Subductions antérieures au charriage nord-sud de la croute téthysienne : facteur de métamorphisme de séries sédimentaires et volcaniques liées aux assemblages ophiolitiques Syro-Turcs, en schistes verts et amphibolites. *Rev. Géog. phys. Géol. dyn.*, 20, 153-170.

Parrot, J.F. and Dugas, F., 1980. The disrupted ophiolitic belt of the southwest Pacific : evidence of an

Eocene subduction zone. *Tectonophysics*, 66, 349-372.

Parson, B. and Sclater, J.G., 1977. An analysis of the variation of ocean floor bathymetry and heat flow with age. *J. geophys. Res.*, 82, 803-827.

Paterson, M.S., 1978. Experimental rock deformation. In : *Mineral and Rocks.* P.J. Wyllie ed., Springer-Verlag Publ., 13, 254 p.

Patriat, P., Segoufin, J., Schlich, R., Goslin, J., Auzende, J.M. Beuzard, P., Bonin, J. and Olivet, J.L., 1982. Les mouvements relatifs de l'Inde, de l'Afrique et de l'Eurasie. *Bull. Soc. géol. France,* XXIV : 363-373.

Patriat, P. and Achache, J., 1984. India-Eurasia collision chronology has implications for crustal shortening and driving mechanisms of plates. *Nature,* 311, 615-621.

Pavlis, T.L., 1986. The role of strain heating in the evolution of megathrusts. *J. geophys. Res.*, 91, 12407-12422.

Pavlis, T.L., and Bruhn, R.L., 1983. Deep-seated flow as a mechanism for the uplift of broad forearc ridges and its role in the exposure of high P/T metamorphic terranes. *Tectonics,* 2, 473-497.

Payne, J.G., and Strong, D.F., 1979. Origin of Twillingate trondhjemite, North-Central Newfoundland : partial melting in the roots of an island arc. In : *Trondhjemites, dacites, and related rocks,* F. Barker ed., Elsevier, Amsterdam, 489-516.

Pearce, J.A., 1975. Basalt geochemistry used to investigate past tectonic environments on Cyprus. *Tectonophysics,* 25, 41-67.

Pearce, J.A., 1979. Geochemical evidence for the genesis and eruptive setting of lavas from Tethyan ophiolites. In : *Proc. Inter. Ophiolite Symp.*, Geol. Surv. Dept. Cyprus, 261-272.

Pearce, J.A. and Cann, J., 1973. Tectonic setting of basic volcanic rocks using trace elements analysis. *Earth and Planet. Sci. Lett.*, 19, 290-300.

Pearce, J.A., Alabaster, T., Shelton, A.W., and Searle, M.P., 1981. The Oman ophiolite as a Cretaceous arc-basin complex : evidences and implications. *Phil. Trans. r. Soc. London,* 300, 299-317.

Pedersen, R.B., 1986. The nature and significance of magma chamber margins in ophiolites : examples from the Norwegian Caledonides. *Earth and Planet. Sci. Lett.*, 77, 100-112.

Pedersen, R.B. and Malpas, J., 1985. The origin of oceanic plagiogranites from the Karmoy ophiolite, Western Norway. *Contr. Mineral. Petrol.*, 88, 36-52.

Peters, Tj., 1969. Rocks of the Alpine ophiolitic suite. *Tectonophysics*, 7, 507-510.

Peselnick, L., Nicolas, A. and Stevenson, P.R., 1974. Velocity anisotropy in a mantle peridotite from the Ivrea zone : application to upper mantle anisotropy. *J. geophys. Res.*, 79, 1175-1182.

Peselnick, L. and Nicolas, A., 1978. Seismic anisotropy in an ophiolitic peridotite : application to oceanic upper mantle. *J. geophys. Res.*, 83, 1227-1235.

Phipps Morgan, J., 1987. Melt segregation beneath mid-ocean spreading centers. *Geophys. Res. Lett.*, 14, 1238-1241.

Phipps Morgan, J., Parmentier, E.M., and Lin, J., 1987. Mechanisms for the origin of Mid-Ocean Ridge axial topography : implications for the thermal and mechanical structure of accreting plate boundaries. *J. geophys. Res.*, 92, 823-836.

Piccardo, G.B., 1977. Le ofioliti dell'areale ligure : petrologia e ambiente geodinamico di formazione. *Rend. Soc.. ital. Mineral. Petrol.*, 33, 221-252.

Pichon, J.F. and Brunn, J.H., 1985. An inverted metamorphism under the Vourinos ophiolitic suite, Greece. *Ofioliti,* 10, 363-374.

Platt, J.P., 1986. Dynamics of orogenic wedges and the uplift of high-pressure metamorphic rocks. *Geol. Soc. Amer. Bull.*, 97, 1037-1053.

Pockalny, R.A., Detrick, R.S. and Fox, P.J., 1988. Morphology and tectonics of the Kane transform from Sea beam bathymetry data. *J. geophys. Res.*, 93, 3179-3193.

Podvin, P., 1983. Remobilisation chimiques successives dans les tectonites ophiolitiques et leurs gisements de chromite. Exemple du Massif du Humboldt, Nouvelle Calédonie. *Thèse Doc. Ing., Ecole Natl. Sup. Mines Paris,* 214p.

Podvin, P., Berger, E.T., et Vannier, M., 1985. Interactions géochimiques liées à la mise en place de dikes pyroxénitiques et gabbroiques dans des harzburgites ophiolitiques. *Bull. Minéral.*, 108, 45-62.

Potts, C.G., White, R.S. and Louden, K.E., 1986. Crustal structure of the Atlantic fracture zones. II The Vema fracture zone and transverse ridge. *Geophys. J. r. astron. Soc.*, 86, 491-513.

Pozzi, J.P., Westphal, M., Girardeau, J., Besse, J., Yao Xin Zhou, Xian Yao Chen and Li Sheng Xing, 1984. Paleomagnetism of the Xigaze ophiolite and flysch (Yarlung Zangbo suture zone, Southern Tibet) : lattitude and direction of spreading. *Earth and Planet. Sci. Lett.*, 70, 383-394.

Presnall, D.C., Dixon, J.R., O'Donnel, T.H. and Dixon, S.A., 1979. Generation of Mid-ocean ridge tholeiites. *J. Petrol.*, 20, 3-35.

Presnall, D.C. and Hoover, J.D., 1984. Composition and depth of origin of primary mid-ocean ridge basalts. *Contr. Mineral. Petrol.*, 87, 170-178.

Prichard, H.H., 1985. The Shetland ophiolite. In : *The Caledonide Orogen ; Scandinavia and related areas*, John Wiley and Sons, 2, 1173-1184.

Prichard, H.M. and Cann, J.R., 1982. Petrology and mineralogy of dredged gabbros from Gettysburg bank, Eastern Atlantic. *Contr. Mineral. Petrol.*, 79, 46-55.

Prinzhofer, A., 1981. Structure et pétrologie d'un cortège ophiolitique : le Massif du Sud (Nouvelle-Calédonie). *Thèse Doc. Ing., Ecole Natl. Sup. Mines, Paris*, 185 p.

Prinzhofer, A., 1987. Processus de fusion dans les zones d'extension océaniques et continentales. *Thèse Doc. Etat, Univ. Paris*, 237p.

Prinzhofer, A., Nicolas, A., Cassard, D., Moutte, J., Leblanc, M., Paris, P. and Rabinovitch, M., 1980. Structures in the New Caledonia peridotites-gabbros : implications for oceanic mantle and crust. *Tectonophysics*, 69, 85-112.

Prinzhofer, A. and Nicolas, A., 1980. The Bogota Peninsula, New Caledonia : a possible oceanic transform fault. *J. Geol.*, 88 : 387-397.

Prinzhofer A., Allègre, C.J., Bao Peisheng and Wang Xibin, 1984. Magmatism in southern Tibet : trace element constraints. In : *Himalayan Geology inter. Symp., Chengdu* (abstract).

Prinzhofer, A. and Allègre, C.J., 1985. Residual peridotites and the mechanism of partial melting. *Earth and Planet. Sci. Lett.*, 74, 251-265.

Proust, F., Burg, J.P., Matte, P. and Tapponnier, P., 1984. Succession des phases de plissement sur une transversale du Tibet Méridionale : implications géodynamiques. In : *Mission Franco-chinoise Himalaya-Tibet 1980*, J.L. Mercier and Li Guancen eds., CNRS Paris, 385-392.

Purdy, G.M. and Detrick, R.S., 1986. Crustal structure of the Mid-Atlantic Ridge at 23°N from seismic refraction studies. *J. geophys. Res.*, 91, 3739-3762.

Quan, W. and Xueya, L., 1986. Paleoplate tectonics between Cathaysia and Angaraland in inner Mongolia of China. *Tectonics*, 5, 1073-1088.

Quick, J.E., 1981a. Petrology and Petrogenesis of the Trinity Peridotite and upper mantle diapir in the Eastern Klamath Mountains, Northern California. *J. geophys. Res.*, 86, 11837- 11863.

Quick, J.E., 1981b. The origin and significance of large, tabular dunite bodies in the Trinity peridotite, Northern California. *Contr. Mineral. Petrol.*, 78, 413-422.

Rabinowicz, M., Nicolas, A. and Vigneresse, J.L., 1984. A rolling mill effect in asthenospheric beneath oceanic spreading centers. *Earth Planet Sci. Lett.*, 67, 97-108.

Rabinowicz, M., Ceuleneer, G. and Nicolas, A., 1987. Melt segregation and flow in mantle diapirs below spreading centers : evidence from the Oman ophiolites. *J. geophys. Res.*, 92 : 3475-3486.

Rabu, D., 1987. Géologie de l'autochtone des Montagnes d'Oman. *Thèse Doc. Etat, Univ. Paris*, 582 p.

Ragan, D.M., 1963. Emplacement of the Twin Sisters dunite, Washington. *Amer. J. Sci.*, 261, 549-565.

Ragan, D.M., 1967. The Twin Sisters dunite, Washington. In : *Ultramafic and related rocks*, P.J. Wyllie ed., John Wiley, New-York, 160-166.

Rahgoshay, M., Juteau, T. and Whitechurch, H., 1981. Kizlyuksek Tépé : un gisement exceptionnel de chromite stratiforme dans un complexe ophiolitique (massif de Pozanti-Karsanti, Taurus, Turquie). C. R. Acad. Sci., 293, 765-770

Raitt, R.W., 1963. Seismic refraction studies of the Mendocino fracture zone. *Mar. Phys. Lab. Scripps Inst. Oceanogr.*, Rept. MPL-U-23/63.

Raleigh, C.B., 1968. Mechanism of plastic deformation of olivine. *J. geophys. Res.*, 73 : 5391-5406.

Raschka, H., Nacario, E., Rammlmair, D., Samonte, C. and Steiner, L., 1985. Geology of the ophiolite of central Palawan Island, Philippines. *Ofioliti*, 10, 375-390.

Rautenschlein, M., Jenner, G.A., Hertogen, J., Hofman, A.W., Kerrich, R., Schmincke, H.U. and White, W.M., 1985. Isotopic and trace element composition of volcanic glasses from the Akaki Canyon, Cyrpus : implications for the origin of the Troodos ophiolite. *Earth and Planet. Sci. Lett.*, 75, 369-383.

Reid, I. and Jackson, H.R., 1981. Oceanic spreading rate and crustal thickness. *Mar. geophys. Res.*, 5, 165-172.

Reid, J.B. and Woods, G.A., 1978. Oceanic mantle beneath the southern Rio Grande Rift. *Earth and Planet. Sci. Lett.*, 41, 303-316.

Reinhardt, B.M., 1969. On the genesis and emplacement of ophiolites in the Oman Mountains geosyncline. *Schweiz Mineral. Petrogr. Mitt.*, 49, 1-3O.

Reuber, I., 1984. Mylonitic ductile shear zones within tectonites and cumulates as evidence for an oceanic transform fault in the Antalya ophiolite, S.W. Turkey. *Geol. Soc. London Sp. Publ.*, 17, 319-334.

Reuber, I., 1986. Geometry of accretion and oceanic thrusting of the Spongtang Ophiolite, Ladakh-Himalaya. *Nature*, 321, 592-596.

Reuber, I., 1988. Complexity of the crustal sequence in northern Oman ophiolite (Fizh and southern Aswad block) : the effect of early slicing ?. *Tectonophysics*, 151, 137-165.

Reuber, I., Michard, A., Chalouan, A., Juteau, T. and Jermoumi B., 1982. Structure and emplacement of the alpine-type peridotites from Beni-Bousera, Rif, Morocco : a polyphase tectonic interpretation. *Tectonophysics*, 82, 231-251.

Ribe, N.M., 1985. The generation and composition of partial melts in the earth's mantle. *Earth and Planet. Sci. Lett.*, 73, 361-376.

Rice, A., 1981. Convective fractionation : a mechanism to provide cryptic zoning (macrosegregation), layering, crescumulates, banded tuffs and explosive volcanism in igneous processes. *J. geophys. Res.*, 86, 405-417.

Richard, P. and Allègre, C.J., 1980. Neodymium and Strontium isotope study of ophiolite and orogenic lherzolite petrogenesis. *Earth and Planet. Sci. Lett.*, 47, 65-74.

Richardson, C.J., Cann, J.R., Richards, H.G. and Cowan, J.G., 1987. Metal depleted root zones of the Troodos ore-forming hydrothermal system, Cyprus. *Earth and Planet. Sci. Lett.*, 84, 243-253.

Richter and McKenzie, D., 1984. Dynamical models for melt segregation from a deformale matrix. *J. Geol.*, 92, 729-740.

Ricou, L.E., 1971. Le croissant ophiolitique peri-arabe, une ceinture de nappes mises en place au Crétacé supérieur. *Rev. Géogr. phys. Géol. dyn.*, 13, 327-35O.

Ricou, L.E., 1976. Evolution structurale des Zagrides. La région clef de Neyriz (Zagros iranien). *Mém. Soc. géol. France*, 125, 140 p.

Ringwood, A.E., 1966. The chemical composition and origin of the Earth. In : *Advance in Earth Science*, P.M. Murley ed., 287-356.

Rise Project Group, 1980. East Pacific Rise : Hot springs and geophysical experiments. *Science*, 207, 1421-1433.

Rittmann, A., 196O (first edition in 1936). Vulkane und ihre Tatigkeit. F. Enke ed., Stuttgart.

Robert, D.G., Backman, J., Morton, A.C., Murray, J.W. and Keene, J.B., 1985., Evolution of volcanic rifted margins : synthesis of Leg 81 results on the west margin of Rockall Plateau. *DSDP Init. Repts..*, 81, 883-911.

Robertson, A.H.F. and Fleet, J., 1986. Geochemistry and palaeo-oceanography of metalliferous and pelagic sediments from the Late Cretaceous Oman ophiolite. *Marine petrol. Geol.*, 13, 315-337.

Rocci, G., Ohnenstetter, D. and Ohnenstetter, M., 1975. La dualité des ophiolites téthysiennes. *Pétrologie*, 1, 172-174.

Rocci, G., Baroz, F., Bebien, J., Desmet, A., Lapierre, H., Ohnenstetter, D., Ohnenstetter, M. and Parrot, J.F., 1980. The Mediterranean ophiolites and their related Mesozoic volcano-sedimentary sequences. In

: *Proc. Inter. Ophiolite Symp.*, Geol. Surv. Dept. Cyprus, 273-286.

Rona, P.A., 1984. Hydrothermal mineralization at seafloor spreading centers. *Earth Sci. Rev.*, 20, 1-104.

Rona, P.A. and Richardson, E.S., 1978. Early Cenozoic global plate reorganization. *Earth and Planet. Sci. Lett.*, 40, 1-11.

Roper, P.J., 1978. Structural fabric of serpentinite and amphibolite along the Motagua fault zone in El Progresso quadrangle, Guatemala. *Trans. Gulf Coast Assoc. geol. Soc.*, 28, 449-458.

Rosencrantz, E., 1982. Formation of uppermost oceanic crust. *Tectonics*, 1, 471-494.

Rosencrantz, E., 1983. The structure of sheeted dikes and associated rocks in North Arm massif, Bay of Islands ophiolite Complex, and the intrusive process at oceanic spreading centers. *Canad. J. Earth Sci.*, 20, 787-801.

Rosencrantz, E., and Nelson, D., 1982. Floating lids on large mid-ocean spreading center magma chambers ? *J. Geol.*, 90, 455-458.

Rosendahl, B.R., Raitt, R.W., Dorman, L.M.,Bibee, L.D., Husson, D.M., and Sutton, G.H. 1976. Evolution of oceanic crust. 1 : a physical model of the East Pacific Rise Crest derived from seismic refraction data. *J. geophys. Res.*, 81, 5294-5304.

Rosenfeld, J.H., 1981. Geology of the Western Sierra de Santa Cruz, Guatemala, Central America. An ophiolite sequence. *PhD Thesis*, SUNY Binghamton, 313 p.

Ross, J.V., Ave Lallemant, H.G., and Carter, N.L., 1980. Stress dependence of recrystallized-grain and subgrain size in olivine. *Tectonophysics*, 70, 39-61.

Ross, D.A., Uchupi, E. and White, R.S., 1986. The geology of the persian Gulf of Oman region : a synthesis. *Rev. Geophysics*, 24, 537-556.

Ross, C. K. and Lister, J.R., 1988. Island arc and mid-ocean ridge volcanism, modelled by diapirism from linear source regions. *Earth and Planet. Sci. Lett.*, 88, 143-152.

Rothery, D.A., 1982. The evolution of the Wuqbah block and the applications of remote sensing in the Oman ophiolite. *Unpubl. PhD Thesis*, Open University, Milton Keynes (U.K.) 414 p.

Rothery, D.A., 1983. The role of Landsat Multispectral scanner (MSS) imagery in mapping the Oman Ophiolite : implications for magma chambers at oceanic spreading axes. *J. Geol. Soc. (London)* 140, 287-296.

Roure, F., 1982. Mise en évidence d'une tectonique majeure du Jurassique supérieur (phase névadienne) dans le Nord-Est de l'Oregon. *C.R. Acad. Sci., Paris*, 294, 921-925.

Roure, F., 1984. Une coupe géologique du Golconde au Pacifique (Oregon, NW du Nevada, Nord de la Californie) : évolution Mésozoique et Cénozoique de la marge Ouest-américaine. *Thèse Doc. Etat, Univ. Paris*, 250 p.

Routhier, P., 1946. Volcano-plutons sous marins du cortège ophiolitique. *C.R. Acad. Sci., Paris*, 22, 192-194.

Routhier, P., 1953. Etude géologique du versant occidental de la Nouvelle-Calédonie. *Mém. Soc. géol. France*, 67, 271 p.

Ryan, M.P., 1987. Neutral buoyancy and the mechanical evolution of magmatic systems. *Geochim. Sp. Publ.*, 1, 259-287.

Ryan, M.P., 1988. The mechanics and three-dimensional internal structure of active magmatic systems : Kilauea Valcano, Hawaii. *J. geophys. Res.*, 93, 4213-4248.

Ryan, M.P., Koyanagi, R.Y. and Fiske, R.S., 1981. Modeling the three-dimensional structure of magma transport systems : applications to Kilauea volcano, Hawaï. *J. geophys. Res.*, 86, 7111-7129.

Sabzehei, M., 1974. Les mélanges ophiolitiques de la région d'Esfandaghen (Iran méridional), étude pétrologique et structurale, interprétation dans le cadre iranien. *Thèse Doc. Etat, Univ. Grenoble*, 306 p.

Saleeby, J.B, 1978. Kings River ophiolite, Southwest Sierra Nevada foothills, California. *Geol. Soc. Amer. Bull.*, 89, 617-636.

Saleeby, J.B., 1982. Polygenetic ophiolite belt of the California Sierra Nevada : geochronological and tectonostratigraphic development. *J. geophys. Res.*, 87, 1803-1824.

Sandwell, T.D., 1984. Thermomechanical evolution of oceanic fracture zones. *J. geophys. Res.*, 89,

401-411.

Sarp, H., 1976. Etude géologique et pétrographique du cortège ophiolitique de la région située au Nord-Ouest de Yesilova (Burdur, Turquie). *Thèse Doc. Univ., Genève*, 408 p.

Saunders, A.D., Marsh, N.G. and Wood, D.A., 1980. Ophiolites as ocean crust or marginal basin crust : a geochemical approach. In : *Proc. Inter. Ophiolite Symp.*, Geol. Surv. Cyprus, 193-204.

Savelyev, A.A. and Savelyeva, G.N., 1979. Ophiolites of the Vogkar massif (Polar Urals). *Mem. Newfoundland Univ. Dept. Geol.*, 8, 127-140.

Savelyev, A.A., Scherbakov, S.A., and Denisova, E.A., 1980. High temperature deformation during origination of dunite bodies in harzburgites (in Russian). *Geotectonika*, 3, 16-26.

Schiffman, P., Williams, A.E. and Evarts, R.C., 1984. Oxygen isotope evidence for submarine hydrothermal alteration of the El Puerto ophiolite, California. *Earth and Planet. Sci. Lett.*, 70, 207-220.

Schiffman, P., Smith, B.M., Varga, R.J. and Moores, E.M., 1987. Geometry, conditions and timing of off-axis hydrothermal metamorphism and ore-deposition in the Solea graben. *Nature*, 235, 423-425.

Schilling, J.G., 1973. Iceland mantle plume : geochemical study of Reykjanes ridge. *Nature*, 242, 565-571.

Sclater, J.G. and Francheteau, J., 1970. The implications of terrestrial heat flow observations and current tectonic and geochemical models of the crust and upper mantle of the earth. *Geophys. J. r. astron. Soc.*, 20, 509-542.

Scott, D.R. and Stevenson, D.J., 1984. Magma solitons. *Geophys. Res. Lett.*, 11, 1161-1164.

Scott, D.R. and Stevenson, D.J., in press. A self-consistent model of melting, magma migration and buoyancy-driven circulation beneath mid-ocean ridges. *J. geophys. Res.*

Schwan, W., 1980. Geodynamic peaks in Alpinotype orogenies and changes in ocean-floor spreading during late Jurassic-late Tertiary time. *Amer. Assoc. petrol. geol. Bull.*, 64, 359-373.

Searle, R.C., 1981. The active part of Charlie-Gibbs fracture zone : a study using sonar and other geophysical techniques. *J. geophys. Res.*, 86, 243-262.

Searle, R.C., 1983. Multiple, closely spaced transform faults in fast-slipping fracture zones. *Geology*, 11, 607-610.

Searle, M.P., 1980. The metamorphic sheet and underlying volcanic rocks beneath the Semail ophiolite in the northern Oman mountains of Arabia. *PhD Thesis, Open Univ., Milton Keynes* (U.K.), 213 p.

Searle, M.P., 1985. Sequence of thrusting and origin of culminations in the northern and central Oman Mountains. *J. struct. Geol.*, 7, 129-143.

Searle, M.P., 1988. Thrust tectonics of the Dibba zone and the structural evolution of the Arabian continental margin along the Musandam mountains (Oman and United Arab Emirates). *J. Geol. Soc. (London)*, 145, 43-53.

Searle, M.P. and Graham, G.M., 1982. Oman exotics : oceanic carbonate build-ups associated with the early stages of continental rifting. *Geology*, 10, 43-49.

Searle, M.P. and Malpas, J., 1982. Petrochemistry and origin of sub-ophiolitic metamorphic and related rocks in the Oman mountains. *J. geol. Soc. (London)*, 139, 235-248.

Searle, M.P. and Stevens, R.K., 1984. Obduction processes in ancient, modern and future ophiolites. *Geol. Soc. London, Spec. Publ.*, 13, 303-320.

Searle, R.C., 1981. The active part of Charlie-Gibbs fracture zone : a study using sonar and other geophysical techniques. *J. geophys. Res.*, 86, 243-262.

Searle, R.C., 1983. Multiple, closely spaced transform faults in fast-slipping fracture zones. *Geology*, 11, 607-610.

Sécher, D., 1981. Les lherzolites ophiolitiques de Nouvelle Calédonie et leurs gisements de chromite-Déformation de la chromite. *Thèse Doc. 3° cycle, Univ. Nantes*, 247 p.

Selig, F., 1965. A theoretical prediction of salt dome pattern. *Geophysics*, 30, 633-643.

Sengör, A.M.C., 1987. Tectonics of the Tethysides. *Ann. Rev. Earth Planet. Sci.*, 15, 213-244.

Shackleton, R.M., 1981. Structure of Southern Tibet : report on a traverse from Lhasa to Khatmandou organised by Academia Sinica. *J. struct. Geol.*, 3,97-105.

Shärer, U., Rong Jua Xu and Allègre, C.J., 1984. U-Pb geochronology of Gandise (Transhimalaya) plutonism in the Lhasa- Xigaze region, Tibet. *Earth and Planet. Sci. Lett.*, 69, 311-320.

Shaw, H.R., 1980. The fracture mechanism of magma transport from the mantle to the surface. In : *Physics of magmatic processes*, R.B. Hargraves ed., Princeton Univ. Press, 201-264.

Shelton, A.W., 1984. Geophysical studies on the northern Oman ophiolite. *Ph. D. Thesis, Open Univ., Milton Keynes (U.K.)*, 353 p.

Shiraki, K., 1971. Metamorphic basement rocks of Yap Islands, Western Pacific : Possible oceanic crust beneath an island arc. *Earth and Planet. Sci. Lett.*, 13, 167-174.

Sheridan, R.E., 1987. Pulsation tectonics as the control of continental breakup. *Tectonophysics*, 143, 59-73.

Silver, E.A., McCaffrey, R., Joyodiwiryo, Y. and Stevens, S., 1983. Ophiolite emplacement by collision between the Sula platform and the Sulawesi island arc, Indonesia. *J. geophys. Res.*, 88, 9419-9435.

Simonian, K.O. and Gass, I.G., 1978. Arakapas fault belt, Cyprus : a fossil transform fault. *Geol. Soc. Amer. Bull.*, 89, 1220-1230.

Sinton, J., 1977. Equilibration history of the basal alpine-type peridotite, Red Mountain, New Zealand. *J. Petrol.*, 18, 216-246.

Siroky, F.X., Elthon, D., Casey, J.F. and Bulter, J.C., 1985. Major element variations in basalts and diabases from the North Arm Mountain Massif, Bay of Islands Ophiolite : implications for magma chamber processes at mid-ocean ridges. *Tectonophysics*, 116, 41-61.

Sleep, N.H., 1974. Segregation of a magma from a mostly crystalline mush. *Geol. Soc. Amer. Bull..*, 85, 1225-1232.

Sleep, N.H., 1975. Formation of oceanic crust ; some thermal constraints. *J. geophys. Res.*, 80, 4037-4022.

Sleep, N.H., 1978. Thermal structure and kinematics of mid-ocean ridge axis, implications to basaltic volcanism. *Geophys. Res. Lett.*, 5, 426-428.

Sleep, N.H., 1984. Tapping of magmas from ubiquitous mantle heterogeneities : an alternative to mantle plumes ? *J. geophys. Res.*, 89, 10029-10041.

Sleep, N.H.,1988. Tapping of melt by veins and dikes. *J. geophys. Res.*, 93, 10255-10272.

Sleep, N.H., and Biehler, S., 1978. Topography and tectonics at the intersections of fracture zones with central rifts. *J. geophys. Res.*, 75, 2748-2752.

Sleep, N.H., and Rosendahl, B.R., 1979. Topography and tectonics of mid-oceanic ridge axis. *J. geophys. Res.*, 84, 3831-3839.

Smewing, J.D., 1980. An upper Cretaceous ridge-transform intersection in the Oman ophiolite. In : *Proc. Inter. Ophiolite Symp.*, Geol. Surv. Dept. Cyprus, 407-413.

Smewing, J.D., 1981. Mixing characteristics and compositional differences in mantle-derived melts beneath spreading axes : evidence from cyclically layered rocks in the ophiolite of North Oman. *J. geophys. Res.*, 86, 2645-2660.

Smewing, J.D. and Potts, P.J., 1976. Rare-earth abundances in basalts and metabasalts from the Troodos massif, Cyprus. *Contr. Mineral. Petrol.*, 57, 245-258.

Smith A.G. and Woodcock, N. H., 1976. Emplacement model for some Tethyan ophiolites. *Geology*, 4, 653-656.

Smith, S.E, and Elthon, D., 1988. Mineral compositions of plutonic rocks from the Lewis Hills massifs, Bay of Islands ophiolites. *J. geophys. Res.*, 93, 3450-3468.

Snoke, A.W., 1977. A thrust plate of ophiolitic rocks in the Preston Peak area, Klamath Mountains, California. *Geol. Soc. Amer. Bull.*, 88, 1641-1659.

Sopaheluwakan, J., Helmers, H., Tjokrosapoetro, S. and Surya Nila, E., in press. Medium pressure metamorphism with inverted thermal gradient associated with ophiolite nappe emplacement in Timor. *Netherland J. Sea Res.*, 23.

Spence, D.A. and Turcotte, D.L., 1985. Magma-driven propagation of cracks. *J. geophys. Res.*, 90, 575-580.

Spera, F.J., 1980. Aspects of magma transport. In : *Physics of Magmatic Processes*, R.B. Hargraves ed.,

Princeton Univ. Press, 263-323.

Spera, F.J., 1984. Carbon dioxide in petrogenesis III : role of volatiles in the ascent of alkaline magma with special reference to xenolith-bearing mafic lavas. *Contr. Mineral. Petrol.*, 88, 217-232.

Spiegelman, M. and Mc Kenzie D., 1987. Simple 2-D models for melt extraction at mid-ocean ridges and island arcs. *Earth and Planet. Sci. Lett.*, 83, 137-152.

Spooner, E.T.C., and Bray, C.J., 1977. Hydrothermal fluids of seawater salinity in ophiolitic sulphide ore deposits in Cyprus. *Nature*, 266, 808-812.

Spray, J.G., 1984. Possible causes and consequences of upper mantle decoupling and ophiolite displacement. *Geol. Soc. London Sp. Publ.* , 13, 255-268.

Spray, J.C. and Roddick, J.C., 1980. Petrology and 40Ar/39Ar geochronology of some Hellenic sub-ophiolite metamorphic rocks. *Contr. Mineral. Petrol.*, 72, 43-55.

Spray, J.C. and Williams, G.D., 1980. The sub-ophiolite metamorphic rocks of the Ballantrae igneous complex, SW Scotland. *J. Geol. Soc. (London)*, 137, 359-368.

Spray, J.C. and Roddick, J.C., 1981. Evidence for Upper Cretaceous transform fault metamorphism in West Cyprus. *Earth and Planet. Sci. Lett.*, 55 : 273-291..

Spudich, P., Salisbury, M.H. and Orcutt, J.A., 1978. Ophiolites found in oceanic crust . *Geophys. Res. Lett.*, 5, 341-344.

Spudich, P. and Orcutt, J., 1980. A new look at the seisimic velocity structure of the oceanic crust. *J. Geophys. Space Phys.*, 18, 627-645.

Stakes, D.S., Shervais, J.W. and Hopson, C.A., 1984. The volcanic-tectonic cycle of the Famous and Amar valleys, Mid-Atlantic Ridge (36°47'N) : evidence from basalt glass and phenocryst compositional variations for a steady state magma chamber beneath the valley midsections, Amar 3. *J. geophys. Res.*, 89, 6995-7028.

Stakes, D.S., Taylor, H.P., and Fisher, R.L., 1984. Oxygen-isotope and geochemical characterization of hydrothermal alteration in ophiolite complexes and modern oceanic crust. *Geol. Soc. London Sp. Publ.*, 13, 199-214.

Steen, D., Vuagnat, M. and Wagner, J.J., 1980. Early deformations in Montgenèvre gabbros. *Coll. Inter. CNRS*, Paris, 272, 97-104.

Stephen, R.A., 1985. Seismic anisotropy in the upper oceanic crust. *J. geophys. Res.*, 90, 11,383-11,396.

Steinmann, G., 1927. Die ophiolithischen Zonen in den Mediterranen Kettengebirgen. *14th Inter., Geol. Congr.*, 2, 637-668.

Stern, C., 1979. Open and closed system igneous fractionation within two Chilean ophiolites and the tectonic implication. *Contr. Mineral. Petrol.*, 68, 243-258.

Stern, C. and Elthon, D., 1979. Vertical variations in the effects of hydrothermal metamorphism in Chilean ophiolites : their implications for ocean floor metamorphism. *Tectonophysics*, 55, 179-213.

Stolper, E., 1980. A phase diagram for mid-ocean ridge basalts : preliminary results and implications for petrogenesis. *Contr. Mineral. Petrol.*, 74, 13-27.

Strong, D.F., 1973. Lushs Bight and Roberts Arm Groups of Central Newfoundland : possible juxtaposed oceanic and island-arc volcanics suites. *Geol. Soc. Amer. Bull.*, 84, 3917-3928.

Strong, D.F., Dickson, W.L., O'Driscoll C.F., Kean, B.F. and Stevens, R.K., 1974. Geochemical evidence for an East-dipping Appalachian subduction zone in Newfoundland. *Nature*, 248, 37-39.

Stroup, J.B.and Fox, P.J., 1981. Geologic investigations in the Cayman trough : evidence for thin crust along the Mid-Cayman Rise. *J. Geol.*, 89, 395-420.

Suen, C.J., Frey, F.A. and Malpas, J., 1979. Bay of Islands ophiolite suite, Newfoundland : petrological and geochemical characteristics with emphasis on rare-earth element geochemistry. *Earth and Planet. Sci. Lett.*, 45, 337-348.

Suppe, J., 1973. Geology of the Leech Lake Mountain-Ball Mountain region, California. *Publ. Geol. Sci. Univ. California*. 107, 82 p.

Sutton, G.H., Maynard, G.L. and Hussong, D.M., 1971. Widespread occurrence of a high-velocity basal layer in the Pacific crust found with repetitive sources and sonobuoys. In : *The structure and Physical*

*Properties of the Earth's Crust.* J.G. Heacock ed., A.G.U., Washington, Geophys. Monogr., 14, 193-209.

Takahashi, E. and Kushiro, I., 1983. Melting of a dry peridotite at high pressures and basalt magma genesis. *Amer. Mineral.*, 68, 859-879.

Talkingston, R.W., Watkinson, D.H., Whittaker, P.J., and Jones, P.C., 1984. Platinum group minerals and other solid inclusions in chromites of ophiolites complexes. Occurrences and petrological significances. *Tschermaks Mineral. Petrogr. Mitt.*, 32 : 285-301.

Tallandier, J. and Buchon, M., 1979. Propagation of high frequency Pn waves at great distances in the Central and South Pacific and its implications for the structure of the lower lithosphere. *J. geophys. Res.*, 84, 5613-5619.

Tapponnier, P. and Francheteau, J., 1978. Necking of the lithosphere and the mechanics of slowly accreting plate boundaries. *J. geophys. Res.*, 83, 3955-3970.

Tapponnier, P., Mattauer, M., Proust, F. and Cassaigneau, C., 1981. Mesozoic ophiolites, sutures and large-scale tectonic movements in Afghanistan. *Earth and Planet. Sci. Lett.*, 52, 355-371.

Tapponnier, P. *et al.*, 1981. The Tibetan side of the India- Eurasia collision. *Nature*, 294, 393-399.

Taylor, R.N., and Nesbit, R.W., 1988. Light rare-earth enrichment of supra subduction-zone mantle : evidence from the Troodos ophiolite, Cyprus. *Geology*, 16, 448-451.

Thayer, T.P., 1956. Preliminary geologic map of the John Day quadrangle, Oregon. *U.S. Geol. Surv. Mineral Ind. Field Map*, MF 51.

Thayer, T.P., 1963. The Canyon Mountain Complex, Oregon, and the Alpine mafic magma stem. *U.S. Geol. Surv. Prof. Pap.*, 475 : 82-85.

Thayer, T.P., 1963. Flow layering in alpine peridotite gabbro complexes. *Mineral. Soc. Amer. Sp.Pap.*, 1, 55-61.

Thayer, T.P., 1964. Geological features of podiform chromite deposits. In : *Methodes de prospection de la chromite*. R. Woodtli ed., OCDE, Paris, 135-146.

Thayer, T.P., 1969. Gravity differentiation and magmatic re-emplacement of podiform chromite deposits. *Econ. Geol. Monogr.*, 4, 132-146.

Thayer,T.P., 1977. Some implications of sheeted dike Swarms in 28 ophiolitic complexes. *Geotectonics*, 11, 419-426.

Thayer, T.P., 1978. The Canyon Mountain Complex, Oregon, and some problems of ophiolites. *Oregon Dept. Geol. Mineral. Ind.*, 95, 93-105.

Thomas, V., Pozzi, J.P., and Nicolas, A., 1988. Paleomagnetic results from Oman ophiolites related to their emplacement.*Tectonophysics*, 151, 297-321.

Thompson, G., Bryan, W.B., Ballard, R., Hamuro, K., and Melson, W.G., 1985. Axial processes along a segment of the East Pacific Rise, 10°-12° N. *Nature*, 318, 429-433.

Tilton, G.R., Hopson, C.A. and Wright, J.E., 1981. Uranium-lead isotopic ages of the Samail ophiolite, Oman, with applications to Tethyan Ocean. *J. geophys. Res., 2763-2775.*

Tippit, P.R., Pessagno, E.A., and Smewing, J.D., 1981 The biostratigraphy of sediments in the volcanic unit of the Samail ophiolite. *J. geophys. Res.*, 86, 2756-2762.

Tobisch, O.T., 1968. Gneissic amphibolite at Las Palmas, Puerto Rico, and its significance in the early history of the Greater Antilles island arc . *Geol. Soc. Amer. Bull.*, 79, 557-574.

Toramaru, A. and Fujii, N., 1986. Connectivity of melt phase in a partially molten peridotite. *J. geophys. Res.*, 91, 9239-9252.

Treloar, P.J., Bluck, B.J., Bowes, D.R. and Dudeck, A., 1980. Hornblende-garnet metapyroxenitc beneath serpentinite in the Ballantrae ophiolite of southwest Scotland and its bearing on the depth provenance of oceanic lithosphere. *Trans. r. Soc. Edinburgh*, 71, 201-212.

Tricart, P., and Lemoine, M., 1986. From faulted blocks to megamullions and megaboudins : Tethyans heritage in the strucutre of the Western Alps. *Tectonics*, 5, 95-118.

Tubia, M., and Cuevas, J., 1987. Structures et cinématique liées à la mise en place des péridotites de Ronda (Cordillères Bétiques), Espagne. *Geodinamica Acta*, 1, 59-69.

Turcotte, D.L. and McAdoo, D.C., 1978. The thickness of the lithosphere as determined form geoid

anomalies. *EOS*, 59, p.388.
Turcotte, D.L. and Ahern, J.L., 1978. A porous flow model for magma migration in the asthenosphere. *J. geophys. Res.*, 83, 767-772.
Turcotte, D.L., Haxby, W.F. and Ockendon, J.R., 1979. Lithospheric instabilities. *Maurice Ewing Series,* 1, AGU Washington, 63-69.
Turner, F.J., 1942. Preferred orientation of olivine crystals in peridotites, with special reference to New Zealand examples. *Trans. r. Soc. New Zealand,* 72, 280-300.
Turner, J.S., and Chen, C.F., 1974. Two-dimensional effects and double-diffusive convection. *J. Fluid Mech.* 63, 577-592.
Turner, J. S., and Campbell, I.H., 1986. Convection and mixing in magma chambers. *Earth Sci. Rev.*, 23, 255-352.
Upadhyay, H.D. and Neale, E.R., 1979. On the tectonic regime of ophiolite genesis. *Earth and Planet. Sci. Lett.*, 43, 93-102.
Usselman, T.M., and Hodge, D.S., 1978. Thermal central of low pressure fractionation processes. *J. Volcanol. geotherm. Res.*, 4, 265-281.
Uyeda, S. and Kanamori, H., 1979. Back-arc opening and the mode of subduction. *J. geophys. Res.*, 84, 1049-1061.
Vail, P.R., Mitchum, R.M. and Thompson, S., 1977. Seismic stratigraphy and global changes in sea level. *Amer. Assoc. petrol. Geol. Mem.*, 26, 83-97.
Vallier, T.L., Brooks, H.C. and Thayer, T.P., 1977. Paleozoic rocks of Eastern Oregon and Western Idaho. *Pac. Sec. Soc. Econ. Paleont. Min.* 1, 455-466.
Van Andel,T.H. and Ballard, R.D., 1979. The Galapagos rift at 86°W. 2 : Volcanism, structure and evolution of the rift valley. *J. geophys. Res.*, 84, 5390-5406.
Van der Molen, I. and Paterson, M.S., 1979. Experimental of partially melted granite. *Contr. Mineral. Petrol.*, 70, 299-318.
Varga, R.J., Moores, E.M., 1985. Spreading structure of the Troodos ophiolite, Cyprus. *Geology,* 13, 846-850.
Vaughan, P.J. and Kohlstet, D.L., 1982. Distribution of the glass phase in hot-pressed olivine basalt aggregates : an electron microscopy study. *Contr. Mineral. Petrol..*, 81, 253-261.
Verosub, K.L., and Moores, E.M., 1981. Tectonic rotations in extensional regimes and their paleomagnetic consequences for oceanic basalts. *J. geophys. Res.*, 86, 6335-6349.
Violette, J.F., 1980. Structure des ophiolites des Philippines (Zambales et Palawan) et de Chypre. Ecoulement asthénosphérique sous les zones d'expansion océaniques. *Thèse Doc. 3° cycle, Univ. Nantes*, 152 p.
Vlaar, N.J. and Wortel, M.J.R., 1976. Lithospheric aging, instability and subduction. *Tectonophysics,* 32, 331-351.
Von Bargen, N., and Waff, H.S., 1988. Wetting of enstatite by basaltic melt at 1350°C and 1.0- to 2.5-GPa pressure. *J. geophys. Res.*, 93, 1153-1158.
Von Huene, R., 1978. Structure of the outer convergent margin off Kodiak Island, Alaska, from multichannel seismic records. *AAPG Mem.* 29, 261-272.
Vuagnat, M., 1963. Remarques sur la trilogie serpentinites-gabbros-diabases dans le bassin de la Méditerranée occidentale. *Geol. Rundschau,* 53, 336-357.
Wadge, G., Jackson, T.A., Isaacs, M.C. and Smith, T.E., 1982. The ophiolitic Bath-Dunrobin Formation, Jamaica : significance for Cretaceous plate margin evolution in the North-Western Caribbean. *J. geol. Soc. (London)*, 139, 321-333.
Wadge, G., Draper, G. and Lewis, J.F., 1984. Ophiolites of the Northern Caribbean : a reappraisal of their roles in the evolution of the Caribbean plate boundary. *Geol. Soc. London Sp. Pub.*, 213, 367-380.
Wadworth, W.J., 1973. Magmatic sediments. *Mineral. Sci. Eng.*, 5, 25-35.
Waff, H.S., 1980. Effects of the gravitational field on liquid distribution in partial melts within the upper mantle. *J. geophys. Res.*, 85, 1815-1825.
Waff, H.S. and Bulau, J.R., 1979. Equilibrium fluid distribution in partial melts within the upper mantle.

*J. geophys. Res.*, 85, 1815-1825.
Wager, L.R. and Brown, G.M., 1968. *Layered igneous rocks*. Oliver and Boyd, Edinburgh, 588p.
Walker, D., Stolper, E.M. and Hays, J.F., 1978. A numerical treatment of melt/solid segregation : size of eucrite parent body and stability of terrestrial low-velocity zone. *J. geophys. Res.*, 83, 6005-6013.
Walcott, R.I., 1969. Geology of the Red Hill Complex, New Zealand. *Trans. r. Soc. New-Zealand, Earth Sci.*, 7, 57-88.
Walker, N.W. and Mattinson, J.M., 1980. The Canyon Mountain Complex, Oregon : U-Pb ages of zircons and possible tectonic correlations. *Geol. Soc. Amer. , Abstr. with Programs* , 12, 544.
Walker, N.W., 1981. U-Pb geochronology of ophiolitic and volcanic-plutonic arc terrains, Northeastern Oregon and Westernmost Central Idaho. *EOS*, 62, 1087.
Wang Naiwen, Wang Sien, Liu Guifang, Bassoulet, J.P., Colchen, M., Mascle, G. and Jaeger, J.J., 1983. The Jurassic-Cretaceous marine terrestrial alternating formations in Lhasa area, Xizang, Tibet. *Acta geol. Sinica*, 57, 83-95 (in chinese).
Wang Xibin, Cao Yougong, Zheng Haixiang, Nicolas, A., Marcoux, J. and Girardeau, J., 1984. La séquence ophiolitique du cours moyen du Yarlung Zangbo et un modèle d'évolution de la croute océanique. *Mission Franco-Chinoise au Tibet 1980*, CNRS Publ. Paris, 199-238.
Watson, E.B., and Brenan, J.M., 1987. Fluids in the lithosphere, 1. Experimentally-determined wetting characteristics of $CO_2$-$H_2O$ fluids and their implications for fluid transport, host-rock physical properties, and fluid inclusion formation. *Earth and Planet. Sci. Lett.*, 85, 497-515.
Watts, A.B., Bodine, J.H., and Steckles, M.S., 1980. Observations of flexure and the state of stress in the oceanic lithosphere. *J. geophys. Res.*, 85, 6369-6376.
Wells, P.R.A., 1977. Pyroxene thermometry in simple and complex systems. *Contr. Mineral. Petrol.*, 62, 129-139.
White, R.S. and Ross, D.A., 1979. Tectonics of the Western Gulf of Oman. *J. geophys. Res.*, 84, 3479-3489.
White, R.S., Detrick, R.S., Sinha, M.C. and Cormier, M.H., 1984. Anomalous seismic crustal structure of oceanic fracture zones. *Geophys. J. r. astron. Soc.*, 79 : 779-798.
White, R.S., Spence, G.D., Fowler, S.R., McKenzie, D.P., Westbrook, G.K., and Bowen, A.N., 1987. Magmatism at rifted continental margins. *Nature*, 330, 439-444.
Whitechurch, H. and Parrot, J.F. ,1974. Les écailles infra-péridotitiques du Baër-Bassit (NW de la Syrie). *Cahiers ORSTOM, Sér. Géol.*, 2, 173-184.
Whitechurch, H. and Parrot, J.F., 1978. Ecailles métamorphiques infra-péridotitiques dans le Pinde septentrional (Grèce) : croûte océanique, métamorphisme et subduction. *C.R. Acad. Sci. Paris*, 286, 1491-1494.
Whitehead, J., Dick, H. and Schouten H., 1984. A mechanism for magmatic accretion under spreading centres. *Nature*, 312, 146-148.
Whitmarsh, R.B. and Calvert, A.J., 1986. Crustal structure of Atlantic fracture zones- I. The Charlie-Gibbs fracture zone., *Geophys. J. r. astron. Soc.*, 85, 107-138.
Wickham; S.M., 1987. The segregation and emplacement of granitic magmas. *J. Geol. Soc. (London)*, 144, 281-297.
Williams, D.L., Von Herzen, R.P., Sclater, J.G., and Anderson, R.N., 1974. The Galapagos spreading center : lithsopheric cooling and hydrothermal circulation. *Geophys. J. r. astron. Soc.*, 38 : 587-608.
Williams, H. 1973. Bay of Islands map-area, Newfoundland. *Geol. Surv. Canad. Pap.*, 72, 1-7.
Williams, H., 1975. Structural succession, nomenclature and interpretation of transported rocks in Western Newfoundland. *Canad. J. Earth Sci.*, 12, 1874-1894.
Williams, H. and Malpas, J., 1972. Sheeted dikes and brecciated dike rocks within transported igneous complexes, Bay of Islands, Western Newfoundland. *Canad. J. Earth Sci.*, 9, 1216-1229.
Williams, H. and Smyth, W.R., 1973. Metamorphic aureoles beneath ophiolite suites and alpine peridotites : tectonic implication with West Newfoundland examples. *Amer. J. Sci.*, 273, 594-621.
Williams, P.R., Pigram, C.J. and Dow, D.B., 1984. Mélange production and the importance of shale diapirism in accretionary terrains. *Nature*, 309, 145-146.

Wilshire, H.G. and Shervais, J., 1975. Al-augite and Cr-diopside ultramafic xenoliths in basaltic rocks from the Western United States. *Phys. Chem. Earth*, 9, 257-272.

Wilson, J.T., 1973. Mantle plumes and plate motions. *Tectonophysics*, 19, 149-164.

Witt (de), M.J. and Stern, C.R., 1976. A model for ocean-floor metamorphism, sesmic layering and magmatism. *Nature*, 264, 615-619.

Witt (de), M.J., Dutch, S., Klingfield, R., Allen, R. and Stern, C., 1977. Deformation, serpentinization and emplacement of a dunite complex, Gibbs Island, South Shetland Islands : possible fracture zone tectonics. *J. Geol.*, 85, 745-762.

Witt (de), G. and Seck, H., 1986. Temperature history of sheared mantle xenoliths from the West Eifel, West Germany : evidence for mantle diapirism beneath the Rhenish Massif. *J. Petrol.*, 28, 475-493.

Wolery, T.J., and Sleep, N.H., 1976. Hydrothermal circulation and geochemical flux at mid-ocean ridges. *J. Geol.*, 84, 249-275.

Wood, B.L., 1972. Metamorphosed ultramafites and associated formations near Milford Sound, New Zealand. *New Zealand J. Geol. Geoph.*, 15, 88-127.

Wright, T.L. and Okamura, 1977. Cooling and crystallization of tholeiitic basalt. *U.S. Geol. Surv. Prof. Pap.*, 1004.

Wright, T.L. and Tilling, R.I., 1980. Chemical variations in Kilauea eruptions 1971-1974. *Amer. J. Sci.*, 280, 777-793.

Wyllie, P.J., 1980. The origin of kimberlite. *J. geophys. Res.*, 85, 6902-6910.

Xiao, X., 1984. Les ophiolites de Xigaze au Tibet méridional et quelques problèmes tectoniques les concernant. *Mission Franco-chinoise Himalaya-Tibet 1980*, CNRS Paris Publ., 167-188.

Xiao, X., Qu, J. Chen G., Shn, Z., and Gu, Q., 1980. Ophiolites of the Tethys-Himalaya of China and their tectonic significiance of the Tethys-Himalayas of China. *26th Int. Geol. Congr.*, Abst. III, p. 1398, Coll. 5 'Geology of alpine chains born of the Thetys'. Bur. Rech. Géol. Mines, Orléans, Mém. 115, 149-151..

Xiao, X., Feng, Y., Wang, G., Ju, B., Yen, B., Zhang, Zh.M. and Coleman, R.G., 1985. Tectonic evolution of the Tangbale ophiolite mélange in West Junggar of Xinjiang Region, China. *EOS, Fall Meeting*, 66, p 1129.

Zeuch, D. and Green, H.W., 1984. Experimental deformation of a synthetic dunite at high temperature and pressure. I. Mechanical behavior, optical microstructure and deformation mechanism. *Tectonophysics*, 110, 233-262.

Zimmermann,J., 1968. Structure and petrology of rocks underlying the Vourinos ophiolitic complex, Northern Greece. *PhD thesis,Princetown Univ.*

**INDEX**


Accretion prism, 289, 308–311
Afar, 157
Age, **43**, 68, 69, 81, 83, 87, 94, 107, 112,129, 145, 161, 161, 302, 308–310, 313–316
Aging (oceanic lithosphere), 8, 32, 199, 256, 284, 313
Alpujata massif, 122–126, 188
Amphibolite, 67, 81, 98, 111, 147, 296
Anatexis
  in isotropic gabbros, 59, 109, 113, 166, 191, 193, 201, 271–272, 284
  in metamorphic aureoles, 52, 68, 80, 87, 126, 148–153, 271, 296, 297
Andaman ophiolite, 308
Andesite, 48, 59, 201, 233
Anisotropy
  seismic -, 33, **207**, 213, 254, 284
  structural -, 185
Anorthosite, 13, 14, 57
Antalya ophiolite, 127, 153, 188, 191, 195, 261
Apennine ophiolite, 115, 127, 188
Arakapas ophiolite (see Cyprus)
Ariegite, 21
Aureole (metamorphic), 54, 67, 80–81, 98, 107, 126, 189, 191, **294–304**, 311
Back-arc (basin, ophiolite), 7, 85–87, 115, 159, 169, 187–188, **197–201**, 302, 313, 320
Baie Verte ophiolite, 201
Ballantrae ophiolite, 298
Basalt
  alkali -, 173, 180, 182, 189, 193, 199, 221
  fluid content, **171**
  fluid saturation, **171** 173
  in ophiolite (see volcanics)
  picritic -, 182–183
  temperature, **175**
  tholeiitic -, 173, 182–183, 189, 193, 197–201, 225, 233, 236, 268
Bay of Islands ophiolite, 10, 12, **142–153**, 188, 199, 200, 231, 238, 251, 256, 277, 292, 296, 298, 306–308
Beni Bousera massif, **120–126**, 237
Bermeja massif, 120, 188
Betts Cove ophiolite, 201
Blow-Me-Down ophiolite (see Bay of Islands)
Blue Mountain ophiolite, 299
Blueschist, 68, 129
Bogota peninsula, **127–135**, 140–142, 157, 187, 195
Bonin ophiolite, 298
Boninite, 200
Boudinage, 15, 16, 20, 224, 243
Boundary layer
  crystallization, 262, 263, 267
  thermal, 271
Breccia (magmatic)
  chromite -, 244
  dolerite -, 57, 271, 279

gabbro -, 55, 110, 116, 161, 191, 271
peridotite -, 64, 65
Brook Range ophiolite, 296, 297
Buoyancy, 17, 184, 213–216, 277
Bushveld Complex, 237
Caledonian belt, 294, 316
California Cordillera ophiolite (see also Trinity), 201, 292, 299, 306, 311
Canyon Mountain ophiolite, 35, 159–167, 187, 201
Chert, 49, 94, 96, 98, 277, 296
Chromite, 64, 71, 113, 139, 142, 159, 163, 166, 181, 182, 189, 192, 193, 223–231, 236, **237–252**
Coastal Complex ophiolite, 12, 127, **142–153**, 157, 195, 267
Collo massif, 120–126, 237
Columbia ophiolite, 299
Compaction (solid) **178**, 180, 183–184, 276
Concordant chromite, 237, **243–247**, 251-252
Convection
magma chamber -, 262-270
mantle - (see also asthenospheric flow), 29–32, 184, 203–220
Collison, 289–294, 298-299, 310, 316
Continental rift (see rift)
Cordilleran ophiolite, **289**, 302, 311
Corsica ophiolite, 188, 191, 210
Crete ophiolite, 188
Crystal settling (see cumulate)
Cumulate, 24, 64, 109, 112, 224, 225, 227, 231–236, 243, 251, 254, 262–270
Cyprus ophiolite (see also Troodos), 3, 20, 127, 157, 225, 279, 299
Dacite, 48, 59, 164, 200
Datation (see age)
Deformation
microstructures, 19, **26–27**, 189 192
plastic -, **13–20**, 64, 65, 163, 247
viscous -, **13-20**, 77–78, 163, 247
Denudation (oceanic), 114, 115, 120, 195, 258, **280–281**
Depth
melting -, **175**, 180, 181, 225, 233–235
water penetration, 256, 258–261, 284, 285, 310
Detachment (oceanic), 80–83, 90, 209, 210, 216, 289, **304–305**, 311
DEVAL, 218
Diabase (see dikes, sill, breccia)
Diabase dike swarm (see sheeted dike unit)
Diapirs (mantle), 33, 71, 77, 163, 166–167, 173–175, 178, 182, 183–185, 195, 197, 201, **203–222**, 223, 238, 247, 251, 265, 268, 274, 277, 281
Dibba line, 40
Diehedral angle, **177**–179
Diffusion
coefficient, 235
double, 262
Dike
andesite-plagiogranite -, 59, 65, 161, 271
chromite (see chromite)
diabase - (see also sheeted dike unit), 65–66, 78, 94, 102, 111, 119, 132, 134, 147, 161, 189, 192, 215,

271
dunite - (see dunite)
gabbro-pyroxenites-, **21–23**, 65–67, 77, 81, 109–111, 132,135, 153, 163, 180, 181, 184, 192, 215, 224, 226, 243, 271, 274
wehrlite -, 50, 59
Dinarides ophiolites, 188
Diorite, 55, 61, 149, 271, 284
Discordant chromite, 237, 243, **244–247**, 251–252
Domingo Belt ophiolite, 297, 299
Dun Mountain ophiolite, 299
Dunite, 21, 24, 51–52, 64, 65, 66, 71, 77, 96, 109–112, 163, 181, 183, 184, 189, 192, 223–224, **225–236**
East Pacific Rise, 7, 38, 203, 218–221, 261, 268, 270, 279, 282
Eclogite, 68
Eifel rift, 205
Emplacement (ophiolite), 46, 90, 129, 200, **289–311**, 320
Episodicity
diapir activity -, 215, **221–222**, 281
melt extraction -(see melt)
ophiolite generation -, 313–321
sea-floor spreading, 253, 258, 274, 279, **281–282**
Exotics, 46
Fabrics
interpretation, (see kinematic analysis)
lattice -, **13–20**, 64, 65, 71, 77, 83, 109, 132, 183, 207, 231
shape -, **13–20**
Flaser gabbros, 187, 189, 274, 282, 285
Flow
asthenospheric -, **27–30**, 70–77, 98, 140, 153, 183–184, ***203–222***, 271, 295
line, **32**
lithospheric -, 27, 70, 140, 196
plane, 12, **27**, 32, 70, 77
viscous (magmatic) -, **13–20**, 70, 77, 78, 224, 243, 263–270, 277
Folds
magmatic -, 13, 14, **20**, 51, 61
plastic -, 16, **21**, 132, 163–164, 243
Foliated gabbro, 49, 55, 61, 109, 161, 274
Foliation
magmatic -, **13**, 14, 55, 70, 77–78, 109, 153, 163, 244, 267–268, 274
plastic -, 16, **21**, 30, 32, 70, 71, 77, 81, 83, 98, 109, 111, 126, 129, 132, 139, 153, 161, 163, 189, 207–213, 224, 243–247, 265, 267
Fore-arc
basin, 104, 200, 300, 302–304, 308, 310
ophiolite, 199
Fracture (tension), 15, 22, 23, 132, 135, 157, 180, 184–185, 210, 221, 235, 260, 274, 275, 284
Fracture zone (see transform fault)
Fugacity, 236, 252
Galicia margin, 260
Geobarometry, 29, 66, 112, 233, 285, 289, 296, 300, 308
Geothermometry, 29, 32, 59, 65, 71, 88, 112, 120, 132, 135, 233, 235, 274, 276, 284, 285, 296, 300, 308–311
Gibraltar Arc ophiolites (see also Beni Bousera, Bermeja, Alpujata), 120–126, 197, 210

Gorringe Bank, 289, 308
Graded bedding, **13**, 14, 57, 262, 263
Gradient
  pressure -, *180*, 183, 184
  thermal -, **173**–175, 177
Granite, 68, 161, 297
Granulite, 86, 126, 147, 296
Gravimetry, 46, 126, 129, 203
Gravity
  settling (see cumulate)
  sliding, 83, 265, 306
Greece ophiolite (see also Vourinos), 299, 316
Greenschist, 67, 83, 111, 296
Greywacke, 201
Guatemala ophiolite, 299
Guleman ophiolite, 238
Harzburgite (see also HOT), 61, 65, 71, 96, 111, 129–157, 161, 183, 223–252, 258–260
Hawaii 175–177, 225
Hawasina Unit, 37, **40–43**, 87
High level gabbro (see isotropic, foliated gabbro, breccia)
Hispaniola ophiolite, 299
Hot spot, 169, 175, 177, 187, 194, 211, 218, 261
HOT, 35, 169, 175, **187–197**, 205, 223, 231, 237, 253, 258, 260, 274, 279, 284, 286, 316
Hydrofracturing, **178**–182, 184, 215, 235, 236
Hydrothermalism, 29, 48, 60, 78, 102–104, 109, 115, 132, 135, 161, 166, 191, 193, 201, 252, 253, 261, 271–273, **283–284**
Iceland, 157, 211, 273–279, 281
Indigenous dikes, **21**, 23, 66, 129, 181, 215, 226, 228, 233, 252
Indus-Yarhung Zangbo sature, 94
Ingalls ophiolite, 142, 155
Iran ophiolite, 299
Island-arc ophiolite, 5, 7, 8, 86–87, 94, 103, 115, 145, 149, 159, 164, 166–167, 187, **197–201**, 260, 292–294, 298, 313
Isotropic gabbro, 49, 55, 61, 96, 106, 145, 161, 265, 267, 271, 274, 282, 284
Japan ophiolite (see also Miyamori), 308, 309
Josephine ophiolite, 279, 299
Karmoy ophiolite, 201, 261
Keratophyre, 109, 161, 164, 166, 201
Kimberlite, 171
Kinematic analysis, **26–27**, 29, 81, 83, 205
Kinzigite, 126
Komatiite, 247
Lamination, **13**
Lanzo massif, 112, 120-126, 163, 188, 191, 192, 197, 210, 233
Las Palmas ophiolite, 299
Laterite, 46, 297
Layered gabbro, **13**, 14, 50, 55, 61, 96, 109, 110, 113, 115, 145, 153, 161, 189, 191, 223–224, 226, 253, 262–271, 274, 277, 279, 285
Layering
  in gabbros, 12, 13, 14, **57**, 77–78, **261,-263**, 265–270
  in peridotites, 16, **20**, 21, 52, 65–67, 77, 109–111, 226, 231, 233, 236

seismic -, 4, **253-256**, 258, 284, 277
Leka ophiolite, 231, 238
Lewis Hill ophiolite, 145–153
Lherzolite (see also LOT), 23–24, 91, 109–126, 139–140, 142, 145, 173–175, 178, 181, 224, 225, 228, 233, 237, 251, 258–261, 296
Liguria ophiolite, 115, 120–126, 187, 188, 191, 192, 197–210
Limassol ophiolite (see Cyprus)
Limousin ophiolite, 298
Lineation
magmatic, **13**, 14, 55, 153, 155, 163, 244, 267, 271
plastic, 16, **21**, 32, 33, 70, 71, 81, 83, 98, 109, 111, 126, 129, 132, 139, 150, 153, 161, 163, 189, 207–212, 224, 228, 243–247, 267
Listric faults, 13, 78, 116, 270, 279
Lizard massif, 112, 298
LOT, 35, 169, 175, 187–199, 205, 223, 237, 253, 258–261, 274, 279–284, 285, 286, 316
Macquarie island, 289, 308
Magma chamber, 4, 61, 153, 155, 166, 176–177, 184, 192, 193–195, 228, 237, 251, 253, **261–275**, 281–282
Magnetic anomaly, 126, 216–221
Major elements, 5, 112, 113, 149, 183, 197, 228, 247–251, 260, 282
Makran subduction zone, 39, 40, 87, 292, 306
Margin
active -, (see subduction)
passive -, 42, 126, 194–195, 260, 277, 289–299, 302–308, 310
Mariana ophiolite, 298
Masirah island, 40, 297
Massif Central rift, 205
Megacryst (see xenocryst)
Mélange, 43, 96, 106, 289–294, **302**, 306, 308–311
Melt and melting
adiabatic -, **173**, 194, 233
conduits (see also dikes), 176, 177, **178–182**, 184, 185, 233, 238, 247, 251–252, 271–277
connectivity, 173, **177–180**
corrosion, 24, 231
episodicity, **181–182**
field evidence, **21**, 69, 111–112, 120, 193
hydrous -, **172**, 175, 178
impregnation, 23–24, 51–52, 64–65, 139–140, 142, 163, 164, 166–167, 183–184, 196, 224, 231, 233, 252, 258
incongruent -, 110, 112, 200, **233–236**
overpressure, 177, **178–181**, 235, 252, 272, 282
percolation, **178, 180**, 183–184, 224, 233, 235–236
segregation, 15, **23**, 112, 140, **171–185**, 192, 235, 274
velocity, **181–182**
Metalliferous sediment (see umber)
Metamorphism
high pressure- (see also obduction), 68, 83, 87, 282, 289–296, 298-301, 306, 308–311, 316
high temperature- (see also aureole), 147, 189, 192, **294–300**, 302,–304, 306, 311
hydrothermal- (see hydrothermalism)
Mid-Atlantic ridge, 211, 216–221, 253, 260, 262, 277, 279–281, 286
Mid-oceanic ophiolite, 7, 87, 169, 187–188, **197–200**

Miyamori ophiolite, 231
Moho, 9, 51, 177, 180, 184, 189, 193, 207, 209, 213, 223, 224, 238, 242, 251, 252, **254–261**, 271, 273, 276, 279
Mont Albert ophiolite, 298
MORB, 58, 60, 100, 101, 145, 173, 181, 199, 200, 222, 213
Muslim Bagh ophiolite, 299
Mylonite, 27, 67, 80–81, 98, 111, 132, 139, 142, 145, 161, 163, 260, 296
Nappes, 83, 209
Neotethys, 43, 86, 316–321
New Caledonia ophiolite, **127–142**, 153, 157, 188, 191, 195, 199, 228, 235, 238–251, 292, 297, 298
New Guinea ophiolite (see Papua)
New Hebrides ophiolite, 229
Neyrig ophiolite, 299
Nodular chromite, 181, 237, **244**–247, 252
Norite, 55, 192
Obduction, 68, 78, 83, 90, 103, 226, 289–298, 302–308, 313
Oceanic rift, 35
Off-axis volcano, 205, 210, 221
Olistostrome, 294, 302, 308
Oman ophiolite, 10, 12, 35, **37–90**, 127, 150–157, 183, 184, 187, 191, 195, 200, **205–210**, 216, 218, 223–226, 236, 238, 244, 256, 263, 267, 270, 271, 272, 274–280, 284, 289, 292, 296, 297, 298, 306, 310, 313
Ophicalcite, 98, 115, 191
Ophiolite, **4**
Orbicular chromite, 237, **244**–247, 252
OSC (see overlapping)
Othris ophiolite, 188, 191, 197, 200, 237
Overlapping spreading center, 203, 210, 216,–221, 277
Overpressure (see melt)
Palawan ophiolite, 298
Paleomagnetism, 10, 90, 98, 103, 157, 277–279, 320
Paleomoho, 10, 57, 70, 79, 98, 109, 113, 153, 192
Paleotethys, 43, 316
Pangean cycle, **316–321**
Papua ophiolite, 37, 187, 188, 289, 292, 297, 298, 306
Particle path, **29**
Pegmatitic gabbro, 109, 161, 191, 283–285
Percolation (threshold) (see also melt), 233, 236, 270
Periodicity
  diapirs (see also segmentation) -, 213,–222, 277
  melt extraction -(see melt)
Phenocryst, 225
Picrite, 49, 55, 59, 61, 87, 149, 200, 225, 236
Pillow lavas, 49, 94, 106, 161, 254, 274
Plagiogranite (see also dike), 55, 57–60, 68, 81, 96, 109, 145, 148, 161, 164, 166, 191, 271, 297
Plateau (oceanic), 306, 310
Plated gabbro, **265**
Podiform chromite (see also chromite), **237**
Pozanti-Karsanti ophiolite, 188, 191, 238, 243
Pressure (see also geobarometry)
  dynamic -, 178–180, 183–184, 223, 235, 274

melt -(see melt)
Preston Peak ophiolite, 229
Pull apart (lineation), 228, 237, 243, 244, 245, 247
Purial ophiolite, 299
Pyroxenite (see also dike, still), 65
Quartzite, 67, 81, 98
Radiogenic isotope data, 6, 59, 100, 112, 200
Radiolarite (see chert)
Ramp tectonics, 83
Rare Earth elements (see trace elements)
Rate (spreading), 8, 32, 40, 86, 103, 115, 157, 169, 175, 182, 187–188, **194–199**, 213, 238, 262, 268, 274, 282, 316, 320
Rayleigh-Taylor instability, 213, 215
Rcd Hill ophiolite, 299
Reflection (seismic), 256, 262, 268–270, 279
Refraction (seismic), 254–256, 262
Reunion, 225
Ridge referential, **9**, 32–33, 71, 98, 113, 145, 203, 207–213
Rift
propagating -, 205, 210, 218
Red Sea -, 115, 188, 194, 197, 203, 260
Rifting
continental -, 174, 188, 196–199, 205, 210
oceanic -, 123, 126, 174, 187, 188, 195, 196, 274
slow spreading - (see also rate), 8, 103, 115, 174, 182, 187–188, 195–199, 203, 217, 218, 238, 253, 260, 272, 274, 277–279
fast spreading - (see also rate), 8, 85, 153, 157, 174, 182, 187–188, 194, 203, 218, 221, 238, 253, 272, 274, 277–279
Rodingite, 29, 96, 115, 192
Ronda massif, 112–126, 237
Rotation (see tilting)
Saint Anthony ophiolite, 298
Sarmiento ophiolite, 200
Segmentation (ridge), 183, 203, **212–218**, 277, 279
Seiad ophiolite, 298
Serpentinite, **29**, 60, 78, 111, 191, 254, 256–258, 279, 285, 300, 308
Serpentinization, ***29***, 60, 96, 102–104, 109, 119, 132, 192, 258, 284–286
Shear sense
in gabbros, 131, 268, 270
in metamorphic aureoles, 81–83
in peridotites, 12, **27**, 70, 71–77, 78, 80, 129, 132, 135, 139, 207, 212
Shear
heating, 181, **304**, 311
zone (see also mylonite), 62, 78, 80–81, 96, 111, 129, 157, 192, 231, 271, 279
Sheeted dike unit, 10–12, 49, 55, 60–61, 102, 106–109, 145, 153, 155–157, 161, 166, 191, 207–212, 218, 253, 254, 267, 270–277, **277–282**, 282–285
Shetland ophiolite, 127, 298
Sierra Nevada ophiolite, 127
Sill
diabase -, 35, 94, 161, 166, 189, 192, 277
gabbro-pyroxenite -, 51, 64, 77, 98, 100, 102, 183, 185, 223, 236, 274–275, 279

wehrlite -, 59
Slip line, **29**
Slump, 20, 262
SNOO, 218
Solidus
peridotite, **171–172**, 175, 215
basalt, 261, 268
Solomon ophiolite, 298
Spilite, 109
Spontang ophiolite, 299
Stable isotope data, 60, 100, 282
Stillwater complex, 228–238, 243
Stokes' Law, 181, 220
Strain
ellipsoid, 21, 27
estimation, **27**, 70, 71–77, 132, 207, 223, 296
Stratiform complex, 8, 227, 237, 238, 243, 247, 251, 262, 263, 268
Stream line, **29**
Stress
intensity, 27, 67, 192, 304, 306
orientation, 23, 135, 157, 185, 220, 275–277
Subconcordant chromite, **243–247**, 251–252
Subduction, 86–87, 94, 104, 129, 145, 281, 289–292, 302–311, 320–321
Subsidence, 265, 279
Sulfur, 236
Sulphide, 60–61, 64, 135, 237, 282
Syria ophiolite, 299
Table Mountain ophiolite (see Bay of Islands)
Tahiti, 177
Taiwan (East) ophiolite, 200, 308
Temperature (see geothermometry)
Tethyan
ophiolite, **289**, 300, 310–311, 320
belt, 40, 113, 292, 316
Thetford ophiolite, 29, 289
Thickness of units, 61, 64, 78, 81, 96, 102, 113, 161, 175, 177, 189, 191, 193, 194, 196, 254–256, 260, 265, 270, 275, 279
Thrusting (oceanic), 78–83, 87–88, 104, 106, 145, 191, 289–308, 310–311, 313, 316, 320
Tibetan ophiolite (see also Xigaze), 289, 308
Tiebaghi-Poum-Belep ophiolite, 127, **135–142**, 157, 195
Tilting, 10, 12, 13, 46, 68, 78, 83, **277–281**, 308
Timor ophiolite, 298
Tinaquillo ophiolite, 112, 299
Tonalite, 126, 149, 271, 297
Tonga ophiolite, 298
Trace elements, 5, 58, 60, 100, 113, 140, 164, 175, 182, 199, 200, 225, 233, 282
Trajectory map, **21**, 46, 70–77, 94, 106, 121–126, 129–132, 139, 153, 163
Transform fault, 29, 33, 35, 40, 115, 127–157, 169, 187, 195–196, 203, 209, 213, 216–221, 238, 258, 261, 267, 277, 280, 306–308
Transition zone
gabbro-diabase -, 55, 78, 263–270, **270–275**

periodotite-gabbro -, 24, 51, 61, 71–77, 78, 96, 102, 109, 161, 183–184, 192, **223**–252, 276
Transposition (tectonic), **20**, 51–52, 66, 67, 153, 157, 226, 231, 236, 237, 252, 263
Trinity ophiolite, 35, **105–115**, 187, 188, 191, 192, 197, 200, 205, 210, 223, 226, 261, 298
Troctolite, 96, 226
Trondjhemite (see plagiogranite), 161
Troodos ophiolite, 5, 127, 157, 188, 200
Turkey ophiolite, 188, 238, 299
Umber, 49, 201, 277
Underplating
basalt -, 197
sediment -, 308, 311
Urals ophiolite, 188, 231, 294, 297, 298
Vancouver island, 306
Velocity
in dike, **181–182**, 252
in diapir, **173–175**, 211–215, 220, 267
seismic -, 254–256, 261, 270
place - (see spreading rate)
Viscosity (asthenospheric), 180, 183, 184, 194, 209, 211–215, 221, 263, 267, 270, 276
Volcanics, 49, 60–61, 87, 94, 274–282
Vourinos ophiolite, 188, 247, 299
Wadi Tayin, 35–90, **150–157**, 187, 195, 209, 218, 238
Websterite, 21, 66, 111, 178
Wehrlite, 13, 57–60, 81, 87, 109, 111, 164, 184, 189, 191, 200, **223–225**, 226–231, 243, 263
Western Alps, 91, 115–126, 188, 258, 281, 300
White Hill ophiolite, 251
Wilson cycle (see Pangean cycle)
Wild flysch, 97–98, 111
Xenocryst (olivine), 52, 182, 184, 200, 225, 236
Xenolith (peridotite), 52, 171, 180, 181, 185, 205, 219, 233, 235
Xigaze ophiolite, 35, 40, **91–105**, 112, 188, 260, 261 294, 299
Yakuno ophiolite, 187, 188
Yap ophiolite, 289, 298, 306
Yougoslavia ophiolite, 188, 299
Zabargad island, ***121–126***, 188, 196–197, 210, 260
Zagros belt, 40
Zambales ophiolite, 188, 191, 251
Zapadnyy Sayan ophiolite, 298
Zendam fault, 40

## PETROLOGY AND STRUCTURAL GEOLOGY

1. J. P. Bard, Microtextures of Igneous and Metamorphic Rocks. 1986.
   ISBN 90–277–2220–X (HB), and ISBN 90–277–2313–3 (PB).
2. A. Nicolas, Principles of Rock Deformation. 1986.
   ISBN 90–277–2368–0 (HB), and ISBN 90–277–2369–9 (PB).
3. J. D. Macdougall (ed.), Continental Flood Basalts. 1988.
   ISBN 90–277–2806–2.
4. A. Nicolas, Structures of Ophiolites and Dynamics of Oceanic Lithosphere.
   1989. ISBN 0–7923–0255–9.